Methods in Molecular Biology™

Series Editor
John M. Walker
School of Life Sciences
University of Hertfordshire
Hatfield, Hertfordshire, AL10 9AB, UK

For other titles published in this series, go to
www.springer.com/series/7651

Yeast Systems Biology

Methods and Protocols

Edited by

Juan I. Castrillo

*Department of Biochemistry and Cambridge Systems Biology Centre,
University of Cambridge, Cambridge, UK*

Stephen G. Oliver

*Department of Biochemistry and Cambridge Systems Biology Centre,
University of Cambridge, Cambridge, UK*

Humana Press

Editors
Juan I. Castrillo
Department of Biochemistry
Cambridge Systems Biology Centre
University of Cambridge
Cambridge, CB2 1GA, UK
jic28@cam.ac.uk

Stephen G. Oliver
Department of Biochemistry
Cambridge Systems Biology Centre
University of Cambridge
Cambridge, CB2 1GA, UK
steve.oliver@bioc.cam.ac.uk

ISSN 1064-3745
e-ISSN 1940-6029
ISBN 978-1-4939-6135-1
ISBN 978-1-61779-173-4 (eBook)
DOI 10.1007/978-1-61779-173-4
Springer New York Dordrecht Heidelberg London

Cover illustration: Scanning electron micrograph of the budding yeast *Saccharomyces cerevisiae* (Photo courtesy of Dr. Alan E. Wheals. Department of Biology and Biochemistry. University of Bath, UK).

Printed on acid-free paper

Humana Press is part of Springer Science+Business Media (www.springer.com)

Preface

The explosion of new technologies in the post-genomic era is allowing the study of biological systems on a genome-wide scale, at the main functional genomic levels: (epi)-genome, transcriptome, proteome, endo- and exometabolome, and their interactions. At the same time that these techniques are being applied and refined, the complexity of biological systems is being rediscovered (1) with thousands of components (e.g., genes, transcripts, proteins, metabolites) participating in finely tuned metabolic and regulatory networks. "Molecular pathways, rather than being linear, independent from or parallel to each other, are instead incredibly intertwined and form very complex biological networks. Phenotypes are probably more directly related to systems properties than they are to DNA and 'genes'" (Marc Vidal in (2)). The idea that multi-scale dynamic complex systems formed by interacting macromolecules and metabolites, cells, organs, and organisms underlie some of the most fundamental aspects of life was proposed by a few visionaries half a century ago (*see* (3) and references therein). We are witnessing a powerful resurgence of this idea made possible by the availability of genome sequences, ever improving gene annotations and interactome network maps, the development of new informatic and imaging tools, and the use of concepts from engineering and physics. Alongside other fundamental "great ideas" as suggested by Paul Nurse (4), systems-level understanding (i.e., systems biology) may materialize as one of the major ideas of post-genomic biology (2–4).

Systems biology aims at deciphering all of the genotype–phenotype relationships at the levels of genes, transcripts (RNAs), peptides, proteins, metabolites, and environmental factors participating in complex cellular networks. This should reveal the mechanisms and principles governing the behavior of complex biological systems. Systems biology is not so much concerned with inventories of working parts but, rather, with how those parts interact to produce working units of biological organization whose properties are much greater than the sum of their parts (3–5).

The purpose of this book is to present (a) an up-to-date view of the optimal characteristics of the yeast *Saccharomyces cerevisiae* as a model eukaryote for systems biology studies (6, 7), (b) a perspective on the latest experimental and computational techniques for systems biology studies, most of which were first designed for and validated in yeast, and (c) selected examples of yeast systems biology studies and their applications in biotechnology and medicine. The main molecular mechanisms, biological networks, and sub-cellular organization are essentially conserved in all eukaryotes, being derived from a complex common ancestor (8). Thus *S. cerevisiae* will continue to make a huge contribution to our understanding of eukaryotic systems biology. New advanced post-genomic techniques are opening the way to the characterization of the core of interactions, modules, architectures, and network dynamics that are essential to all eukaryotes. Yeast systems biology experiments under controlled conditions can uncover the complexity and interplay of biological networks with their dynamics, basic principles of internal organization, and balanced orchestrated functions between organelles in direct interaction with the environment as well as the characterization of short- and long-term effects of perturbations and dysregulation of networks that may illuminate the origin of complex human diseases. These approaches can then be reproduced in systems biology studies of more complex organisms, including higher eukaryotes and, ultimately, humans (2, 3).

This book comprises four sections: In Section I (**Chapter 1**), we present an up-to-date view of the optimal characteristics of the yeast *S. cerevisiae* as a model eukaryote for systems biology studies, with selected examples in biotechnology and medicine.

In Section II (**Chapters 2–18**), we present a perspective on the latest high-throughput and molecular techniques and protocols.

In Section III (**Chapters 19–27**), we present a perspective on advanced computational, in silico, systems biology strategies, and studies for quantitative data analysis and integration, as well as modeling approaches toward a holistic quantitative description of the yeast cell as a model for the essential unit of eukaryotic life.

In Section IV (**Chapter 28**), relevant contributions of *S. cerevisiae* as a tool for mammalian studies are presented.

This book is intended for postgraduate students, post-doctoral researchers, and experts in different fields with an interest in (a) comprehensive systems biology strategies in well-defined model systems with specific objectives; (b) a better knowledge of the latest post-genomic strategies at all "omic levels and computational approaches toward analysis, integration, and modeling of biological systems, from single-celled organisms to higher eukaryotes."

New types of expertise and sustained collaborations between researchers from many different fields both within biology and in the engineering, physical, and computational sciences are essential to advance our knowledge of the exquisite complexity of eukaryotes at the systems level (2–5) (**Chapter 1**). We thank all our authors and advisors; their deep knowledge and expertise in different fields have proven invaluable in this essentially multidisciplinary effort. We are also grateful to our families for their support in enabling us to complete this project.

Juan I. Castrillo
Stephen G. Oliver

References

1. Hayden, E. C. (2010) Human genome at ten: life is complicated. *Nature* **464**, 664–667.
2. Heard, E., Tishkoff, S., Todd, J. A., et al., (2010) Ten years of genetics and genomics: what have we achieved and where are we heading? *Nat. Rev. Genet.* **11**, 723–733.
3. Vidal, M. (2009) A unifying view of 21st century systems biology. *FEBS Lett.* **583**, 3891–3894.
4. Nurse, P. (2003) The great ideas of biology. *Clin. Med.* **3**, 560–568.
5. Kitano, H. (2002) Systems biology: a brief overview. *Science* **295**, 1662–1664.
6. Oliver, S. G. (1997) Yeast as a navigational aid in genome analysis. *Microbiology* **143**, 1483–1487.
7. Castrillo, J. I., and Oliver, S. G. (2006) Metabolomics and systems biology in *Saccharomyces cerevisiae*. In: Esser, K. (ed.), and Brown, A. J. P. (Vol. ed.), *The Mycota XIII. Fungal Genomics* (pp. 3–18). Berlin: Springer.
8. Koonin, E. V. (2010) The incredible expanding ancestor of eukaryotes. *Cell* **140**, 606–608.

Contents

Contributors

TRAUDE H. BEILHARZ • *Department of Biochemistry and Molecular Biology, Monash University, Clayton, VIC, Australia*

ROBERT J. BEYNON • *Protein Function Group, Institute of Integrative Biology, University of Liverpool, Liverpool, UK*

SAKARINDR BHUMIRATANA • *National Science and Technology Development Agency, Ministry of Science and Technology, Pathumthani, Thailand*

ELIZABETH BILSLAND • *Department of Biochemistry, Cambridge Systems Biology Centre, University of Cambridge, Cambridge, UK*

JUAN I. CASTRILLO • *Department of Biochemistry, Cambridge Systems Biology Centre, University of Cambridge, Cambridge, CB2 1GA, UK*

BENOIT CHARLOTEAUX • *Department of Cancer Biology, Center for Cancer Systems Biology (CCSB), Dana-Farber Cancer Institute, Boston, MA, USA; Department of Genetics, Harvard Medical School, Boston, MA, USA*

SANDRA CLAUDER-MÜNSTER • *Genome Biology Unit, EMBL, Heidelberg, Germany*

AMY J. CLAYDON • *Protein Function Group, Institute of Integrative Biology, University of Liverpool, Liverpool, UK*

SURESH CUDDAPAH • *Department of Environmental Medicine, New York University, Tuxedo, NY, USA*

KAIRONG CUI • *Laboratory of Molecular Immunology, NHLBI, NIH, Bethesda, MD, USA*

MICHAEL E. CUSICK • *Department of Cancer Biology, Center for Cancer Systems Biology (CCSB), Dana-Farber Cancer Institute, Boston, MA, USA; Department of Genetics, Harvard Medical School, Boston, MA, USA*

LIOR DAVID • *Department of Animal Sciences, R.H. Smith Faculty of Agriculture, Food, and Environment, The Hebrew University of Jerusalem, Rehovot, Israel*

RONALD W. DAVIS • *Stanford Genome Technology Center, Palo Alto, CA, USA; Department of Genetics, Stanford University, Palo Alto, CA, USA; Department of Biochemistry, Stanford University, Palo Alto, CA, USA*

DANIELA DELNERI • *Faculty of Life Sciences, The University of Manchester, Manchester, UK*

MATIJA DREZE • *Department of Cancer Biology, Center for Cancer Systems Biology (CCSB), Dana-Farber Cancer Institute, Boston, MA, USA; Department of Genetics, Harvard Medical School, Boston, MA, USA*

WARWICK B. DUNN • *School of Chemistry, Manchester Centre for Integrative Systems Biology, Manchester Interdisciplinary Biocentre, University of Manchester, Manchester, UK*

EULA L. FUNG • *Stanford Genome Technology Center, Palo Alto, CA, USA*

CLAIRE GAUGAIN • *University of Bordeaux, Bordeaux, France; Université Paul Sabatier, CNRS, LMGM, Toulouse, France*

GURI GIAEVER • *Donnelly Centre for Cellular and Biomolecular Research, University of Toronto, Toronto, ON, Canada; Department of Molecular Genetics, University of*

Toronto, Toronto, ON, Canada; Department of Pharmaceutical Sciences, University of Toronto, Toronto, ON, Canada

AARON HARLAP • *Newton South High School, Newton, MA, USA*

ANDREW HAYES • *Faculty of Life Sciences, The University of Manchester, Manchester, UK*

DAVID E. HILL • *Department of Cancer Biology, Center for Cancer Systems Biology (CCSB), Dana-Farber Cancer Institute, Boston, MA, USA; Department of Genetics, Harvard Medical School, Boston, MA, USA*

ALAIN JACQUIER • *Unité de Génétique des Interactions Macromoléculaires, Institut Pasteur, CNRS, URA2171, Paris, France*

DANIEL F. JARAMILLO • *Stanford Genome Technology Center, Palo Alto, CA, USA*

ROSS D. KING • *Department of Computer Science, University of Aberystwyth, Ceredigion, UK*

EDDA KLIPP • *Theoretical Biophysics, Humboldt-Universität zu Berlin, Berlin, Germany*

AXEL KOWALD • *Humboldt University Berlin, Institute for Biology, Theoretical Biophysics, 10115 Berlin, Germany*

RONG LI • *Stowers Institute for Medical Research, Kansas City, MO, USA; Department of Molecular and Integrative Physiology, University of Kansas Medical Center, Kansas City, KS, USA*

KATHRYN LILLEY • *Cambridge Centre for Proteomics, Cambridge Systems Biology Centre, University of Cambridge, Cambridge, UK*

ED LOUIS • *Institute of Genetics, Queens Medical Centre, University of Nottingham, Nottingham, UK*

UGRAPPA NAGALAKSHMI • *Department of Molecular, Cellular, and Developmental Biology, Yale University, New Haven, CT, USA*

HELEN NEIL • *Unité de Génétique des Interactions Macromoléculaires, Institut Pasteur, CNRS, URA2171, Paris, France*

JENS NIELSEN • *Department of Chemical and Biological Engineering, Chalmers University of Technology, Gothenburg, Sweden*

COREY NISLOW • *Banting and Best Department of Medical Research, University of Toronto, Toronto, ON, Canada; Donnelly Centre for Cellular and Biomolecular Research, University of Toronto, Toronto, ON, Canada; Department of Molecular Genetics, University of Toronto, Toronto, ON, Canada*

INTAWAT NOOKAEW • *Department of Chemical and Biological Engineering, Chalmers University of Technology, Gothenburg, Sweden*

RICHARD A. NOTEBAART • *Centre for Molecular and Biomolecular Informatics (NCMLS), Radboud University Nijmegen Medical Centre, Nijmegen, The Netherlands; Centre for Systems Biology and Bioenergetics, Radboud University Nijmegen Medical Centre, Nijmegen, The Netherlands*

ROBERTO OLIVARES-HERNÁNDEZ • *Department of Chemical and Biological Engineering, Chalmers University of Technology, Gothenburg, Sweden*

STEPHEN G. OLIVER • *Department of Biochemistry, Cambridge Systems Biology Centre, University of Cambridge, Cambridge, CB2 1GA, UK*

BALÁZS PAPP • *Institute of Biochemistry, Biological Research Center of the Hungarian Academy of Sciences, Szeged, Hungary; Department of Genetics, Cambridge Systems Biology Centre, University of Cambridge, Cambridge, UK*

STEPHEN C.J. PARKER • *Bioinformatics Program, Boston University, Boston, MA, USA*

AVINASH PERSAUD • *Program in Cell Biology, The Hospital for Sick Children, Toronto, ON, Canada; Department of Biochemistry, University of Toronto, Toronto, ON, Canada*

YITZHAK PILPEL • *Department of Molecular Genetics, Weizmann Institute of Science, Rehovot, Israel*

THOMAS PREISS • *Genome Biology Department, The John Curtin School of Medical Research, The Australian National University, Building 131, Garran Rd, Acton (Canberra), ACT 0200, Australia*

MICHAEL PROCTOR • *Stanford Genome Technology Center, Palo Alto, CA, USA; Department of Biochemistry, Stanford University, Palo Alto, CA, USA*

SHUYE PU • *Molecular Structure and Function Program, Hospital for Sick Children, Toronto, ON, Canada*

OLIVER J. RANDO • *Department of Biochemistry and Molecular Pharmacology, University of Massachusetts Medical School, Worcester, MA, USA*

BHARAT M. RASH • *Faculty of Life Sciences, The University of Manchester, Manchester, UK*

OLIVER RAY • *Department of Computer Science, University of Bristol, Bristol, UK*

JOHANNA REES • *Cambridge Centre for Proteomics, Cambridge Systems Biology Centre, University of Cambridge, Cambridge, UK*

DANIELA ROTIN • *Program in Cell Biology, The Hospital for Sick Children, Toronto, ON, Canada; Department of Biochemistry, University of Toronto, Toronto, ON, Canada*

DUSTIN E. SCHONES • *City of Hope, Duarte, CA, USA*

JEAN-MARC SCHWARTZ • *Faculty of Life Sciences, University of Manchester, Manchester, UK*

BRIAN D. SLAUGHTER • *Stowers Institute for Medical Research, Kansas City, MO, USA*

MICHAEL SNYDER • *Department of Molecular, Cellular, and Developmental Biology, Yale University, New Haven, CT, USA; Department of Genetics, Stanford University Medical School, Stanford, CA, USA*

LARS M. STEINMETZ • *Genome Biology Unit, EMBL, Heidelberg, Germany*

BALÁZS SZAPPANOS • *Institute of Biochemistry, Biological Research Center, Szeged, Hungary*

THOMAS D. TULLIUS • *Bioinformatics Program, Department of Chemistry, Boston University, Boston, MA, USA*

JAY R. UNRUH • *Stowers Institute for Medical Research, Kansas City, MO, USA*

MALENE L. URBANUS • *Banting and Best Department of Medical Research, University of Toronto, Toronto, ON, Canada; Donnelly Centre for Cellular and Biomolecular Research, University of Toronto, Toronto, ON, Canada*

MARC VIDAL • *Department of Cancer Biology, Center for Cancer Systems Biology (CCSB), Dana-Farber Cancer Institute, Boston, MA, USA; Department of Genetics, Harvard Medical School, Boston, MA, USA*

JIM VLASBLOM • *Molecular Structure and Function Program, Hospital for Sick Children, Toronto, ON, Canada; Department of Biochemistry, University of Toronto, Toronto, ON, Canada*

KARL WAERN • *Department of Molecular, Cellular, and Developmental Biology, Yale University, New Haven, CT, USA; Department of Genetics, Stanford University Medical School, Stanford, CA, USA*

YUEHUA WEI • *Graduate Program in Cellular and Molecular Pharmacology, Department of Pharmacology, Cancer Institute of New Jersey, UMDNJ-Robert Wood Johnson Medical School, Piscataway, NJ, USA*

KEN WHELAN • *Department of Computer Science, University of Aberystwyth, Ceredigion, UK*

CHRISTOPH WIERLING • *Max Planck Institute for Molecular Genetics, Berlin, Germany*
CATHERINE L. WINDER • *School of Chemistry, Manchester Centre for Integrative Systems Biology, Manchester Interdisciplinary Biocentre, University of Manchester, Manchester, UK*
SHOSHANA J. WODAK • *Molecular Structure and Function Program, Hospital for Sick Children, Toronto, ON, Canada; Department of Biochemistry, University of Toronto, Toronto, ON, Canada; Department of Molecular Genetics, University of Toronto, Toronto, ON, Canada*
LEO A.H. ZEEF • *Faculty of Life Sciences, The University of Manchester, Manchester, UK*
NIANSHU ZHANG • *Department of Biochemistry, Cambridge Systems Biology Centre, University of Cambridge, Cambridge, UK*
X.F. STEVEN ZHENG • *Department of Pharmacology, Cancer Institute of New Jersey, UMDNJ-Robert Wood Johnson Medical School, Piscataway, NJ, USA*
QUAN ZHONG • *Department of Cancer Biology, Center for Cancer Systems Biology (CCSB), Dana-Farber Cancer Institute, Boston, MA, USA; Department of Genetics, Harvard Medical School, Boston, MA, USA*

Section I

Yeast Systems Biology

Chapter 1

Yeast Systems Biology: The Challenge of Eukaryotic Complexity

Juan I. Castrillo and Stephen G. Oliver

Abstract

In this chapter, we present an up-to-date view of the optimal characteristics of the yeast *Saccharomyces cerevisiae* as a model eukaryote for systems biology studies, with main molecular mechanisms, biological networks, and sub-cellular organization essentially conserved in all eukaryotes, derived from a complex common ancestor. The existence of advanced tools for molecular studies together with high-throughput experimental and computational methods, most of them being implemented and validated in yeast, with new ones being developed, is opening the way to the characterization of the core modular architecture and complex networks essential to all eukaryotes. Selected examples of the latest discoveries in eukaryote complexity and systems biology studies using yeast as a reference model and their applications in biotechnology and medicine are presented.

Key words: Yeast systems biology, eukaryotic complexity, interactions, modular architecture, network biology, compartmentalization, multilevel control, yeast systems biology applications.

1. Introduction

Systems biology aims at understanding all of the genotype–phenotype relationships at the levels of genes, transcripts (RNAs), peptides, proteins, metabolites, and environmental factors participating in complex cellular networks. This should reveal the mechanisms and principles governing the behavior of complex biological systems (1, 2). Systems biology is not so much concerned with inventories of working parts (e.g., transcripts, proteins, and metabolites) but, rather, with how those parts interact to produce working units of biological organization whose properties are much greater than the sum of their parts. Here, it is important

J.I. Castrillo, S.G. Oliver (eds.), *Yeast Systems Biology*, Methods in Molecular Biology 759,
DOI 10.1007/978-1-61779-173-4_1, © Springer Science+Business Media, LLC 2011

to note that living systems are the products not of design but of evolution through natural selection (3, 4). "Nothing makes sense in biology except in the light of evolution" (5). Thus, the architecture and complexity of a biological system (e.g., single-celled, free-living prokaryote or eukaryote; parasite–host or multicellular system) will always have to be contemplated in the light of evolution (6, 7) (*see* below).

The main objective of systems biology is the construction of quantitative mathematical/logical models by which to interrogate and iteratively refine our knowledge of the workings of living organisms. Systems biology addresses its task by combining integrative experimental and theoretical approaches. This outlook is illustrated in **Fig. 1.1** in the context of the iterative cycle of knowledge (8). First, well-designed (e.g., hypothesis-driven) high-throughput experimental studies at several "omic" levels (e.g., epigenetic, genome, transcriptome and proteome levels, and metabolome analyses), together with molecular studies, provide system-level sets of data (**Fig. 1.1a**). From here, comprehensive theoretical studies (bioinformatics, "in silico" computational methods), integrative data analysis, network studies, and modeling approaches allow the generation of new knowledge and design principles at a "system" level. This generates predictions, some of which can be selected as new ideas/hypotheses capable of

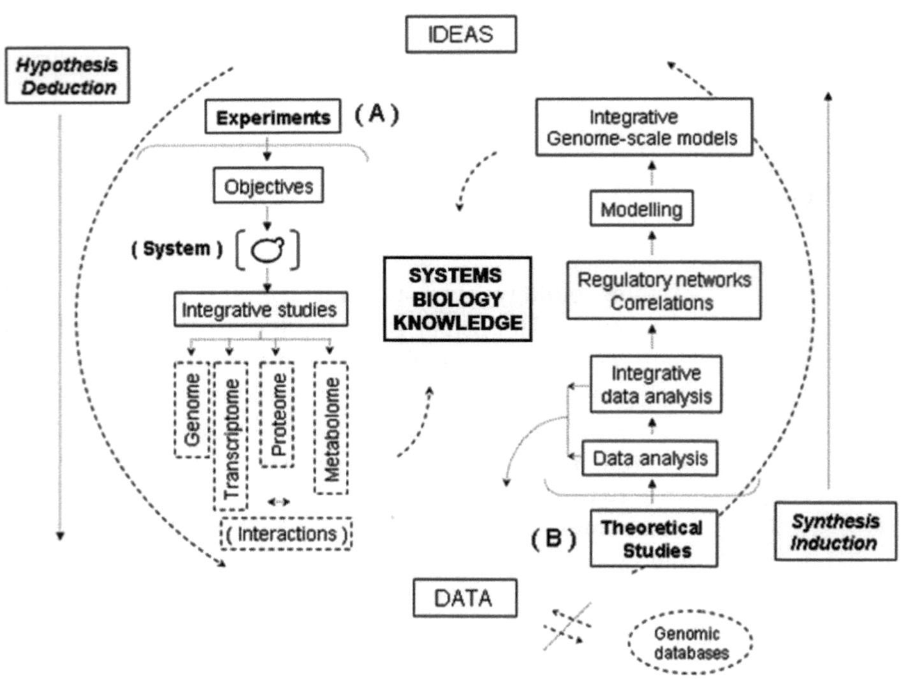

Fig. 1.1. Systems biology. Integration of (**a**) global experimental and (**b**) theoretical in silico studies in the iterative cycle of knowledge (8). Reproduced from (2), with kind permission of Springer Science and Business Media.

being tested in further experimental studies (**Fig. 1.1b**). Hence, systems biology operates in two main ways: (1) by deduction, by perturbing the biological system in specifically designed experiments, monitoring the responses, integrating the data and formulating mathematical models able to reproduce this behavior; and (2) in an inductive way, by perturbing the model, making predictions as to what should happen in response to these perturbations and testing them through new laboratory experiments. The model is refined in the light of the new knowledge derived from new experimental data and new (and more accurate) predictions generated (*see* (2) and references therein).

The ultimate challenge in systems biology would be to integrate all information for a biological system (e.g., a eukaryotic cell) to unveil all biological networks, architecture, and the principles responsible for its behavior, and to construct a global dynamic model able to reliably predict it. The final goal will be to translate this knowledge into direct applications in biotechnology and medicine (e.g., elucidation of bottlenecks in recombinant protein production; diagnostic methods and therapies to treat complex diseases: biological imbalances, cancer, protein quality control disorders and ageing, amongst others). For more complete information on systems biology, its theoretical framework, methodologies, and applications, the reader is referred to the latest reviews and books on the subject (1, 2, 6, 7, 9–22).

Despite the limitations of existing high-throughput techniques, the difficulty of obtaining information at all "omic" levels (and their interactions), the existence of unidentified components and mechanistic uncertainties, and the limitations at the computational level, a number of integrative experimental studies have proved successful in analyzing sub-cellular systems (23), several prokaryotes such as *Escherichia coli* (17) and *Mycoplasma pneumoniae* (24–27), and free-living eukaryotes such as the yeast *Saccharomyces cerevisiae* (28–31). At this point, the optimal characteristics of the yeast *S. cerevisiae* as a touchstone model in the post-genomic era are positioning it at the leading edge of advanced eukaryotic systems biology studies (2, 30, 32) (*see* also next section).

The purpose of this chapter is to present an up-to-date view of the optimal characteristics of the yeast *S. cerevisiae* as a model eukaryote for systems biology studies. Its main molecular mechanisms, biological networks, and sub-cellular organization are essentially conserved in all eukaryotes, being derived from a complex common ancestor (33). The advance of technology is opening the way to the characterization of the core of interactions, modules, architectures, and network dynamics that are essential to all eukaryotes. Examples of latest discoveries in eukaryotic complexity, yeast systems biology studies, and their applications in biotechnology and medicine are presented.

2. Yeast Systems Biology: The Challenge of Eukaryotic Complexity

The explosion of new technologies in the post-genomic era is allowing the study of biological systems on a genome-wide scale, at the main functional genomic levels: (epi)-genome, transcriptome, proteome, endo- and exometabolome, and their interactions. At the same time that these techniques are being applied and refined, the complexity of biological systems is being rediscovered, with thousands of components (genes, transcripts, proteins, and metabolites) participating in finely tuned metabolic and regulatory networks (2, 29). These are beginning to reveal the exquisite complexity and dynamic network architecture of the eukaryotic cell.

2.1. The eukaryotic Cell: Basic Architecture and Levels of Regulation

A schematic representation of the eukaryotic cell, comprising its main levels of regulation at the (epi)-genome, transcriptome, proteome and metabolome, and its interactions with the environment (sensing natural fluctuations in external conditions together with those derived from interaction with other organisms; e.g., competition or cooperation) is presented in **Fig. 1.2**. This system is intrinsically complex and involves the integration of mechanisms and networks in different compartments, at different levels of regulation. These include, among others, environment-sensing signal transduction pathways and regulatory networks; gene expression reprogramming networks at the (epi)-genome, transcriptional, and proteome levels (e.g., DNA–protein; protein–protein, RNA–protein (ribonucleoprotein) networks); and main metabolic networks (e.g., metabolite–protein/enzyme networks), subject to multilevel, distributed control (2, 28, 29). The integration of data from all these levels into models with predictive and explanatory power constitutes one of the most daunting challenges in systems biology (2). Thus, studies on the dynamics of transcriptional regulatory networks of *S. cerevisiae* have revealed large topological changes depending on environmental conditions, with transcription factors altering their subcellular localization and interactions in response to stimuli, some of them serving as permanent hubs but most acting transiently in specific conditions only (34, 35). In this context, the utilization of well-defined model systems, in properly designed experiments under controlled conditions, with well-curated databases and data repositories covering the best updated knowledge of the biology of the system is of central importance, toward the elucidation of essential principles of biological organization (2, 32).

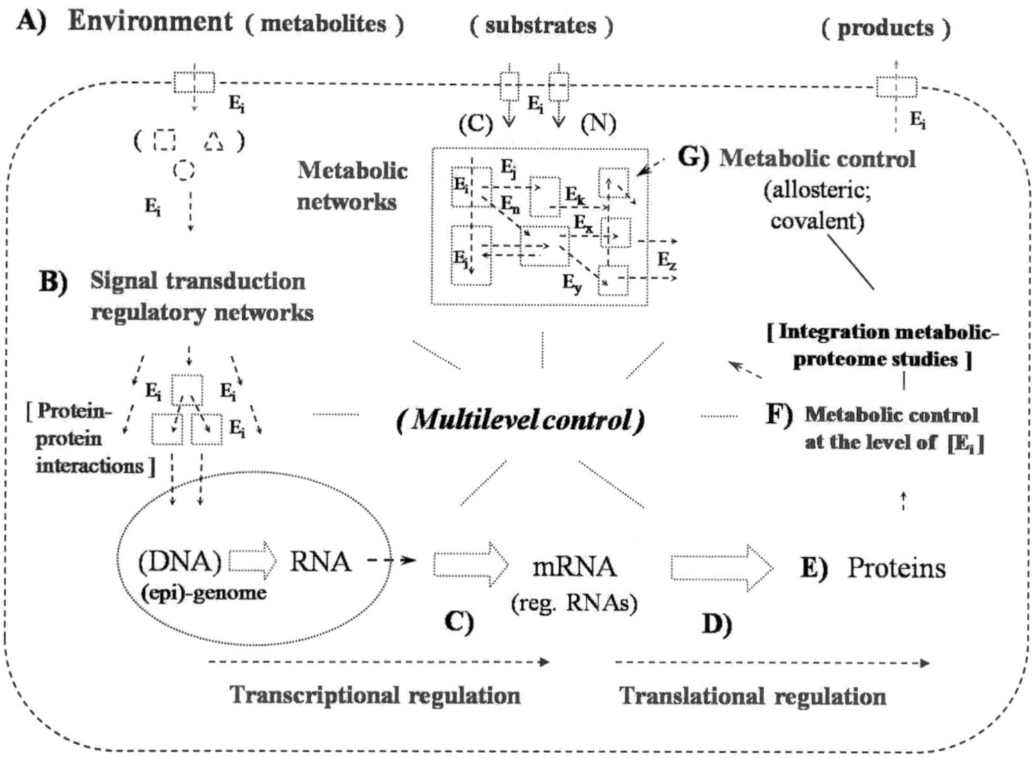

Fig. 1.2. Eukaryotic cell. Levels of regulation. Schematic description of the main regulatory levels and networks subjected to distributed, multilevel control (28, 29), in direct interaction with the environment (sub-cellular organelles, other than the nucleus, are omitted for clarity). (**a**) Environment: the eukaryotic cell is not isolated, sensing fluctuations in external conditions (e.g., concentrations of external compounds, pH, temperature) together with interactions with other organisms (e.g., via external signals; competition or cooperation). (**b**) Signal transduction regulatory networks (e.g., protein–protein networks; DNA–protein, protein–RNA (ribonucleoprotein), and metabolite–protein networks). (**c**) Gene expression at the transcriptional level, resulting in a pool of mRNAs and regulatory RNAs ("transcriptome," balance of synthesis and degradation). (**d**) Gene expression at the translational level, translational regulation. (**e**) Main pool of enzymes and regulatory proteins (i.e. "proteome," balance of synthesis and degradation), responsible for central regulatory and metabolic networks. (**f**) and (**g**) Intracellular enzyme and metabolite concentrations (balance of uptake, intracellular synthesis, and conversion) mainly responsible for in vivo regulation of metabolic fluxes and metabolic networks (e.g., allosteric/covalent modification). Abbreviations: [E_i], pool of enzymes; (C) and (N), fluxes of assimilation of carbon and nitrogen sources, respectively.

2.2. Yeast as a Reference Model for Systems Biology Studies of the Eukaryotic Cell

Saccharomyces cerevisiae is a species of budding yeast, a group of unicellular fungi belonging to the phylum *Ascomycetes.* This yeast is being used as a model eukaryote because the basic mechanisms of DNA and chromosomal replication, cell division, gene expression, translation, signal transduction and regulatory networks, metabolism, and sub-cellular organization are essentially conserved between yeast and higher eukaryotes (32, 36–38). The essential eukaryotic biochemistry, metabolic processes, and the first complete map of central metabolic pathways were first unveiled in *S. cerevisiae* (36, 39–41), and a wide knowledge of the genetics, biochemistry, functional genomics, and physiology of this yeast is presently available (36–38, 42, 43).

Among the properties that make *S. cerevisiae* a particularly suitable organism for biological studies are its generally regarded as safe (GRAS) status, a free-living organism with rapid growth, simple methods of cultivation under defined conditions, facile replica plating, and mutant isolation. It is also a well-defined genetic system with simple techniques of genetic manipulation. The yeast *S. cerevisiae* was the first eukaryotic organism for which the complete genome was sequenced (44) and for which strategies for proper annotation, curation, and standards initiatives for maintenance of high-quality curated databases and data repositories were implemented (e.g., *Saccharomyces* genome database; http://www.yeastgenome.org; *see* also (2) and references therein). Moreover, the majority of high-throughput post-genomic technologies (including new generation sequencing and latest proteomics techniques) were first developed and validated in yeast (32, 45–51). Finally, in a major breakthrough at the molecular level, RNA interference (RNAi), a major gene-silencing regulatory pathway conserved in several eukaryotic species but lost in *S. cerevisiae*, has been restored in this yeast, by reintroducing Dicer and Argonaute proteins from other budding yeast species and bringing the tool of RNAi to the study of budding yeasts and the tools of *S. cerevisiae* to the study of RNAi (52, 53). The yeast *S. cerevisiae* is being used as a model organism to study cell growth (29–31); cell cycle, checkpoints, and their relation to cancer (54, 55); cell polarity (56); control of pre-mRNA splicing (57, 58); eukaryotic translation initiation (59); evolution (60, 61); ageing and extension of lifespan (62); protein folding and chaperone networks (63, 64) and as a model to gain insight into the molecular pathology of neurodegenerative diseases (65, 66). All these advantages are positioning *S. cerevisiae* at the forefront of the post-genomic era, as a touchstone model for eukaryotic systems biology studies (*see* (32) and references therein).

2.3. The Ancestral Eukaryote and the Evolution of Complexity: Yeast as a Reference Model for the Study of Essential Eukaryotic Networks

The early evolution of eukaryotes and the characteristics of the last common ancestor are of major interest for understanding eukaryotic complexity. Comparing the genome sequences of free-living organisms in different eukaryotic supergroups allows predictions to be made about the genome of the last common ancestor. In a relevant study, Fritz-Laylin and coworkers reported the genome sequence of the free-living amoeboflagellate *Naegleria gruberi* which revealed the surprising complexity of this unicellular organism and, by inference, of the last common eukaryotic ancestor (33, 67). The results agree with latest studies inferring an exquisitely complex ancestor (68), with fully formed eukaryote-specific functional systems and with more than 4,000 protein families (~4,133 eukaryotic genes) involved in translation, replication, and other, eukaryote-specific, processes: complex signaling; introns, pre-mRNA splicing, and RNA interference machinery; multifunctional microtubule-based cytoskeleton; anaerobic/aerobic metabolic versatility; and flagellar and amoeboid movement (33, 67, 68).

Specific lineages may then diverge from the complex common ancestor by one of three pathways: (1) genome streamlining, in which numerous genes are lost, the genomes shrink, and functional redundancy decreases; (2) genome stasis, in which limited amounts of genes are lost and gained at roughly the same rate via duplication and other processes; and (3) genome expansion, in which the rate of gene acquisition substantially exceeds the rate of gene loss (33) (**Fig. 1.3**). From the studies of the Fritz-Laylin and Wickstead groups (67, 68), the last common ancestor appears as a complex organism, with no reason to believe that it was simpler than extant free-living unicellular eukaryotes. Here, it is worth noting that the 4,133 genes is a conservative estimate, because it does not account for any loss of ancestral genes, which is likely to be substantial (20–25% in some eukaryotic groups) (33).

The understanding of the processes that led to the emergence of the complex eukaryotic ancestor itself is one of the most exciting challenges in evolutionary biology (33). In this respect, the elucidation of the main core of self-sustained mechanisms, biological network architecture, and sub-cellular organization that have been essentially conserved in single-celled eukaryotes will be of paramount importance. This should preferably be performed with extant, well-annotated, free-living unicellular organisms that may be easily manipulated, with the main essential mechanisms and regulatory networks conserved in all the *eukarya*. At this level, the yeast *S. cerevisiae*, with a genome of ~6,000 genes, its essential complexity and all the favorable characteristics detailed previously, appears as an optimal reference model, helped by comparisons with other organisms (e.g., *Schizosaccharomyces pombe* and other yeast species), for systems biology studies toward the elucidation of this functional core which is shared by all eukaryotes.

2.4. New Eukaryotic Complexity Unveiled in Yeast and Higher Eukaryotes: A New Era of Discoveries

As *S. cerevisiae* is an optimum reference model, the question arises as to whether the present technologies are able to detect and quantify all biological components (with their functions, sub-localization, and interactions) with sufficient accuracy and completeness to allow a comprehensive understanding of biological networks and their dynamics. More importantly, are the latest techniques and integrative strategies revealing new mechanisms or design principles, i.e. is there still room for big discoveries in biology to be made using yeast? The main purpose of this section is to show that the incorporation of advanced molecular and post-genomic techniques is indeed revealing new levels

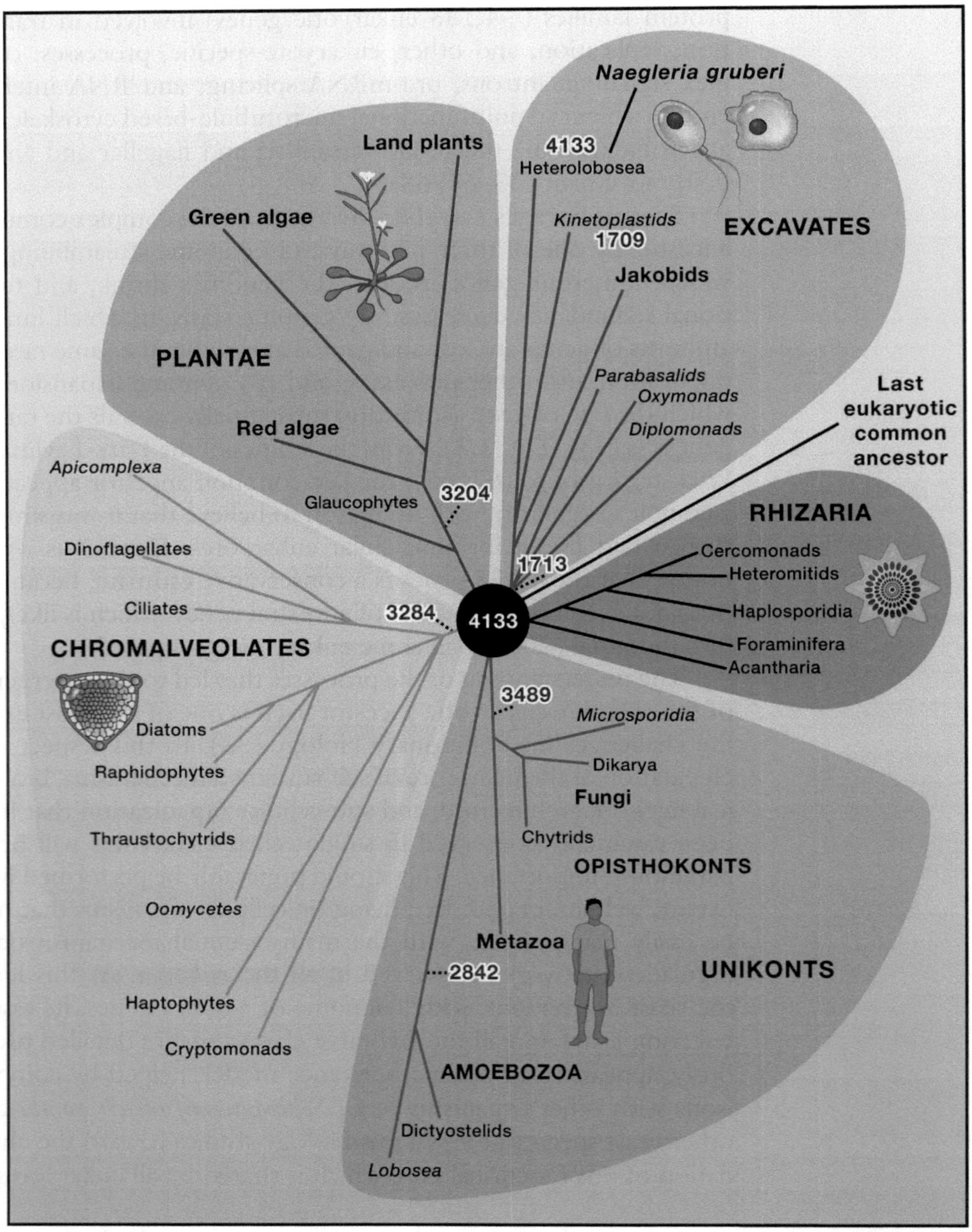

Fig. 1.3. Eukaryotic evolution from the eukaryotic common ancestor. The five eukaryotic supergroups – Excavates, Rhizaria, Unikonts, Chromalveolates and Plantae – are shown to diverge directly from the last common ancestor (*black circle*; 4,133 genes). Branch lengths are arbitrary. Number of putative ancestral genes present in selected major clades (e.g., Unikonts, including Metazoa) indicated separately. Reprinted from *Cell* **140** (5), Koonin, E. V., The incredible expanding ancestor of eukaryotes, pages 606–608 (33). Copyright (2010), with permission from Elsevier.

of hidden complexity in eukaryotes. The more we know, the more the complexity, which presages an exciting era of new discoveries. First and foremost, the incorporation of advanced next-generation sequencing (NGS) technologies, most of them developed and validated in yeast, for genome-wide studies at the (epi)-genome, transcriptome, DNA–protein, and RNA–protein interactions levels, is constituting the main revolution in post-genomic biology, with enormous potential to generate high-throughput data unveiling the dynamics of biological networks at a systems level (69–73). Selected examples of advanced post-genomic technologies and the latest discoveries in eukaryotic biology using *S. cerevisiae* as a main reference model are as follows:

(1) *At the epigenome level*: The Grunstein lab showed in yeast that both histone acetylases (HATs) and deacetylases (HDACs) are associated with active genes at levels which correlated with each gene's transcriptional activity (74), with a similar pattern in human cells demonstrated by the Zhao group (75). Moreover, most genome-wide histone modification studies are being pioneered in yeast (76, 77).

(2) *At the genome level*: Technologies such as ChIP-chip and ChIP-Seq (78, 79) are opening the way to the study of DNA–protein interactions involved in DNA replication and transcriptional gene expression (*see* below). At the replication level, studies in *S. cerevisiae* have revealed novel mechanisms of replication origin recognition (80) and that re-replication can induce the initial step of gene amplification and may be a contributor to gene copy number changes in eukaryotes (81, 82).

(3) *At the high-order, genome organization, level*: A method to globally capture intra- and inter-chromosomal interactions, higher order structures and the first three-dimensional (3D) model of a eukaryotic genome was unveiled in yeast (83).

(4) *At the transcriptional level*: Using *S. cerevisiae*, Steinmetz's group studied the yeast transcriptome landscape and generated a comprehensive genome-wide transcriptional expression atlas of the mitotic cell cycle (84). Moreover, budding yeast has been at the forefront of the development and validation of deep-sequencing technologies such as massively parallel cDNA sequencing (RNA-Seq), providing the first transcriptional landscape of a eukaryotic genome (46). Despite the cost of deep-sequencing technologies, the existence of bias in tiling array studies due to cross-hybridization and noise effects (85, 86) is favoring the use of next-generation sequencing (NGS) techniques. At the transcriptional level, several protocols and data analysis strategies have been reported for RNA-Seq, whose final results may vary, stressing the importance of proper methods comparison. This was addressed by Regev and coworkers in a recent study using *S. cerevisiae* as a benchmark (87). They compared quality metrics of libraries from seven strand-specific RNA-Seq

methods, identifying leading protocols, and provided a computational pipeline for assessment of future protocols in other organisms (87). *Saccharomyces cerevisiae* is also being used as an optimal platform for new advanced techniques such as single-molecule sequencing for the analysis of digital gene expression (smsDGE), resulting in the first comprehensive quantification of the yeast transcriptome by single-molecule sequencing (88), and by direct single-molecule sequencing (direct RNA-Seq; i.e. without prior conversion of RNA to cDNA) (89, 90). Single-molecule sequencing methods are presently being refined in order to apply them to higher eukaryotes (73) and new strategies for analysis of RNA-Seq data are progressively being developed (91, 92).

After all these efforts, the application of latest advanced techniques is finally revealing the complex transcriptional landscape in *S. cerevisiae*, thus shedding light on essential mechanisms present in higher eukaryotes, such as pre-mRNA splicing (57, 58). For instance, the latest studies have provided insights into the origins of protein-dependent eukaryotic spliceosomal introns (93), complex RNA regulatory networks involving non-coding RNAs, and *cis*- and *trans*- RNAi-independent transcriptional gene-silencing mechanisms (94–98). Thus, in a recent study, strand-specific RNA sequencing by Regev's group has revealed extensive regulated long antisense transcripts in *S. cerevisiae*, conserved across yeast species (99). Deep-sequencing techniques have also allowed the first genome-wide measurement of RNA secondary structures in yeast, opening the way to the study of importance of RNA structure in regulation (100).

Application of RNA-Seq and other advanced NGS techniques in yeast and higher eukaryotes has led to major discoveries that reveal the importance of the RNA regulatory world, with non-coding RNAs (RNAs not translated to polypeptide sequences) participating in complex DNA–RNA and RNA–protein regulatory networks, at the signal transduction, (epi)-genome, gene expression, splicing, and other levels. This field is still in its infancy, with exciting challenges ahead (101, 102). The importance of the RNA regulatory networks and their intricate relationships with protein and metabolic networks will remain a big challenge in systems biology. It is important to remark that despite recent studies showing the complex transcriptional landscape of RNA regulatory networks in higher eukaryotes, this may not be the only characteristic underlying high complexity in Metazoa (33, 86). Complex RNA regulatory networks are not exclusive to higher eukaryotes and theoretically "simpler" organisms (e.g., bacteria, single-celled eukaryotes, parasites, and pathogens) have evolved their own complex RNA regulatory networks, in order to readily adapt and survive to fast changes in their specific environments in direct interaction with other organisms (24, 25, 33, 86). The complexity of an organism will make sense only in the light of evolution (3–5). This will be briefly discussed in the next section.

(5) *At the proteome level*: The combination of high-resolution mass spectrometry, stable-isotope labeling by amino acids in cell culture (SILAC) and computational proteomics has allowed the first quantification of a eukaryotic proteome, the yeast proteome, with a low bias against low-abundance proteins. This has revealed a proteome made up of 4,399 individual endogenous proteins in exponentially growing yeast cells (103). More recently, the application of a targeted proteomics approach based on selected reaction monitoring (SRM) has enabled the detection and quantitation of proteins expressed at concentration below 50 copies/cell in total *S. cerevisiae* digests, thus allowing consistent and fast measurement of proteins spanning the entire abundance range (48–50). Together with this, proteome–transcriptome correlation studies in yeast have emphasized that there is no direct correlation between protein and transcript levels, a direct consequence of the widespread role of post-transcriptional regulation (e.g., differential localization of RNAs in nucleus, cytosol and/or p-bodies, polyadenylation states, and level of polysomal occupancy per transcript) during eukaryotic cell growth (*see* (29, 31, 103) and references therein). At the protein network level, methods for the characterization of protein–protein interaction networks, including regulatory networks involved in sensing, signal transduction, gene expression reprogramming, proteostasis, and the control of metabolic fluxes, were mainly discovered first in yeast (36, 39, 40, 41). Recent developments are discussed below.

(6) *At the metabolome level*: *Saccharomyces cerevisiae* has long been the leading model organism for the development of methods and experiments to understand eukaryotic metabolic networks (36, 39, 40). For a review of most relevant techniques and metabolomic approaches toward a systems-level understanding of metabolism, the reader can refer to the literature (2, 32, 104). As a reference, the first reconstruction of a eukaryotic genome-scale metabolic network was performed in *S. cerevisiae* (105), being progressively refined with incorporation of

sub-cellular organelle compartmentalization toward a consensus yeast metabolic network for systems biology studies (106, 107) (http://www.comp-sys-bio.org/yeastnet/). The latest version (v. 4.0; Dobson et al., in revision) included 2,342 reactions and 2,657 chemical species. More importantly, an increasing number of studies are focusing on how changes in the extracellular metabolome (e.g., external environment and nutrient-limiting conditions) result in changes in internal pools of metabolites and patterns of cell growth (29, 31) and how these relationships can be investigated with the help of genome-scale models (108). Apart from this, in the post-genomic era, considerable efforts are focusing on the relevant role of metabolites in regulation, at the signaling, (epi)-genetic, transcriptional, and post-transcriptional levels (*see* (2) and references therein).

(7) *At the DNA–protein networks level*: Immunoprecipitation-based methods, such as ChIP-chip and ChIP-Seq, provide genome-wide information on DNA–protein interactions and promoter occupancy for characterization of DNA–protein transcriptional networks. However, they cannot distinguish whether a transcription factor contacts DNA directly or is tethered by means of another DNA-binding protein and do not measure affinities (78, 79). In an alternative approach, Fordyce and coworkers (109) used a microfluidics-based system for de novo discovery and quantitative biophysical characterization of DNA target sequences and validated the technique by characterizing the binding of 28 yeast transcription factors, six of which had proved intractable to previous approaches. The use of a broad range of experimental conditions under which transcription factors are likely to bind promoters, and combining in vitro methods with advances in whole-genome sequencing, should allow DNA–protein transcriptional networks to be accurately identified and modeled.

(8) *At the RNA–protein (ribonucleoprotein, RNP) networks level*: A rapid affinity purification method for the efficient isolation of the sub-complexes that dynamically organize RNP biogenesis pathways with minimum contamination has been developed in *S. cerevisiae* (110). The latest studies indicate that RNA-binding proteins may interact with several functionally related transcripts, suggesting an extensive regulatory system (111).

(9) *At the protein–protein networks level*: The characterization of all possible interactions and modular network structure of the yeast protein interactome includes methods

for capture and analysis of strong and weak interactions (e.g., pull-down methods and hold-up retention assays for the analysis of weak interactions at equilibrium), direct visualization, information from genetic interaction studies, and inference by application of specific algorithms (112–119). The datasets generated are stored in well-curated databases (e.g., *Saccharomyces* Genome Database (SGD), http://www.yeastgenome.org/ and BioGRID, http://thebiogrid.org/). The main results show a complex, dynamic, modular protein–protein interactome with changes in the sub-cellular localization of regulatory proteins and their interactions (34, 35, 120). More importantly, the present advanced methods are providing comprehensive maps of fully functional regulatory networks, such as the first global protein kinase and phosphatase interaction (KPI) network in yeast (121). Other major developments are the systems analysis of the mechanisms underlying protein kinase specificity (122) and a first atlas of protein–chaperone interactions in *S. cerevisiae* – revealing their role in protein folding and protein homeostasis regulation in eukaryotes (123).

(10) *At the proteome–metabolome interactions level*: Metabolite interactions with proteins are often considered to be confined mainly to interactions with enzymes (e.g., as substrates for their conversion through metabolic networks), or as regulators of enzymatic activity (e.g., allosteric regulation) (41). However, an increasing number of metabolite interactions are being found to be responsible for new regulatory mechanisms in eukaryotes and for covalent modifications of proteins, including the biosynthesis of glycoproteins and lipoproteins (2, 41). At this level, recent studies toward a global analysis of the glycoproteome in *S. cerevisiae* are already revealing new roles for protein glycosylation in eukaryotes (124).

(11) *Finally, at the level of internal compartmentalization and distribution of functions between specialized organelles (e.g., nucleus, cytosol, mitochondria, endoplasmic reticulum, and vacuole)*: This relies on balanced orchestration of functions, finely controlled localization and shuttling of regulators, and the interchange of components under different conditions, an often overlooked level of regulation for which limited information is yet available. In *S. cerevisiae*, the latest studies have identified and characterized a novel mitochondrial carrier for the interchange of citrate and oxoglutarate between mitochondria and cytosol (125). This information and more analysis on the role of compartmentalization such as Klitgord and coworkers' analysis

of *S. cerevisiae* as an "ecosystem of organelles" (126) will need to be progressively incorporated into more refined genome-scale metabolic models.

In all, most advanced studies in *S. cerevisiae* are showing new levels of previously hidden complexity, which anticipate an exciting era of discoveries, toward the elucidation of the main core of biological networks essential to all *eukarya*. From here, it is important to stress a number of facts that should not be overlooked in yeast systems biology studies: (a) the need to define a clear objective, with proper experimental design and protocols, data analysis strategy, and expected results; (b) the reality of existence of unknowns and uncertainties in biology: components and reactions still to be identified, localized, and/or properly annotated (e.g., ATP-producing and -consuming reactions; NAD(P) and NAD(P)H redox balances in different compartments); (c) the possible existence of more than one function per gene, RNA, protein, and/or metabolite. Thus, recent studies have shown that metabolic enzymes play a role in the DNA damage response in the nucleus (127, 128). (d) The correct use of yeast mutants (e.g., auxotrophic and knockout mutants) and interpretation of the results (129–132), and (e) the relevance of compartmentalization and distribution of functions between organelles in the light of evolution (5, 126).

2.5. From Yeast to Human: Toward Principles Underlying Evolution of Eukaryotic Complexity

One of the most important discoveries in the past few years has been the confirmation of the existence of a hidden layer of eukaryotic complexity, the "RNA regulatory world" (101, 102, 133, 134). Although the originally reported "pervasive" transcription of the eukaryotic genome (135, 136) is presently being questioned (85, 86, 137, 138), a new complex universe of non-coding RNAs (e.g., small nucleolar RNAs, snoRNAs; small interfering RNAs, siRNAs; microRNAs, miRNAs; piwi-associated RNAs, piRNAs; and long non-coding RNAs), regulating gene expression at different levels and participating in complex networks, has been reported. These are now the objects of intensive investigation (139–144).

The discovery of particularly complex RNA regulatory networks in higher eukaryotes, at the most recent stages of evolutionary history (**Fig. 1.4**), some of them involved in differentiation, multicellular development, and highly complex processes (e.g., nervous system function) (140, 144, 145), has led to the notion that RNA regulatory networks are one of the most important factors underlying high complexity in eukaryotes (102, 146, 147). Although RNA regulatory networks, together with the essential DNA–protein and protein–protein networks, may indeed constitute a main contributor, the latest studies are unveiling a more complex picture. Thus, complex RNA regulatory networks are not exclusive to higher eukaryotes but have been reported

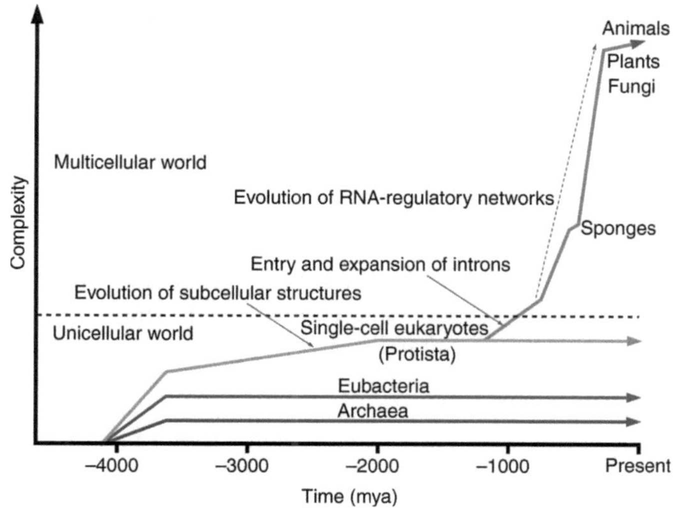

Fig. 1.4. A simplified view of the biological history of the Earth. Reproduced from (102), with permission from Company of Biologist Ltd.

also in simpler organisms (e.g., bacteria, single-celled eukaryotes, parasites) (24, 25, 95, 99, 148, 149), and novel mechanisms and regulatory networks, working in parallel or together with RNA networks, are being reported underlying high complexity in eukaryotes. Examples include the following: (a) short eukaryotic transcripts initially considered to be non-coding RNAs

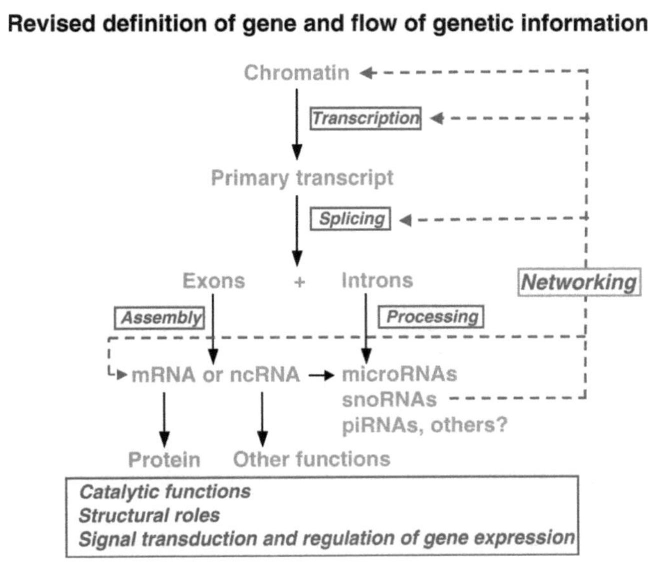

Fig. 1.5. The flow of genetic information in higher eukaryotes. Primary transcripts may be (alternatively) spliced and further processed to produce a range of protein isoforms and/or non-coding RNAs of various types, which are involved in complex networks of structural, functional, and regulatory interactions. Reproduced from (102), with permission from Company of Biologist Ltd.

encode small bioactive peptides controlling differentiation (150); (b) expansion of the eukaryotic proteome by extensive alternative splicing (151) with more than 90% of genes undergoing splicing in human tissue transcriptomes (152) and metabolite-mediated control of alternative splicing (153); (c) the increasing importance of complex RNA–protein (ribonucleoprotein, RNP) networks (154, 155); and (d) proteolytic events controlling functional switches from a nuclear to a cytosolic isoform involved in cell differentiation processes (156).

Independent of all this, most advanced studies are showing that RNA regulatory networks are involved in the regulation of epigenetic reprogramming and alternative splicing events, revealing intricate relationships between RNP and protein–protein regulatory networks (139–141, 157–159) (**Fig. 1.5**). At this point, systems biology studies using *S. cerevisiae* grown under controlled conditions offer the potential to unveil the essential principles

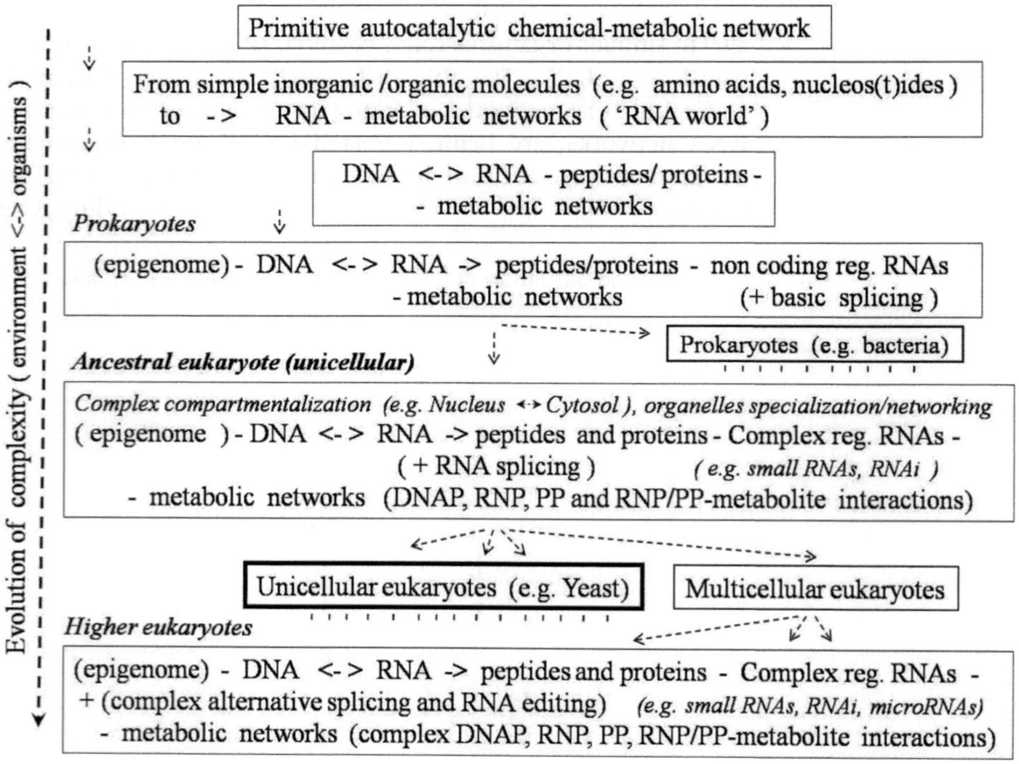

Fig. 1.6. Evolution of complexity. Proposed schematic view of evolution of organisms and biological networks toward eukaryotic complexity. The central position of the complex ancestral eukaryote (33) and its close relationship with present unicellular eukaryotes (e.g., yeast) are shown (frame in *bold*). Evolution toward multicellular and higher eukaryotes with more complex entwined networks is simplified. Abbreviations: DNAP, DNA-protein interactions/networks; RNP, RNA-protein (ribonucleoprotein) interactions/networks; PP, protein-protein interactions/networks; RNP/PP/metabolite interactions, ribonucleoprotein/protein-protein/metabolite interactions/networks. Additional DNA-RNA, RNA-RNA, RNA-metabolite interactions, complex sub-cellular organization and interactions with other organisms are omitted for clarity.

governing the relationships between RNA regulatory networks and the other types of biology networks. The recent incorporation of RNA interference (RNAi) (52, 53) into the armory of yeast researchers should enable the elucidation of the basic mechanisms involved in the interplay of RNAi pathways with essential eukaryotic networks. All this will allow the characterization of the main core of mechanisms and regulatory networks essentially conserved in eukaryotes, from the last common eukaryotic ancestor to extant single-celled organisms and humans (**Fig. 1.6**). This will open the way, with the incorporation of selected model organisms for system-level studies on differentiation and development at multicellular, tissue, organ, and whole-body levels (e.g., *Dictyostelium discoideum*, *Caenorhabditis elegans*, *Drosophila*, mouse, stem cells, and human cell, tissue, and organ culture systems) (20, 160–164), to the elucidation of the main principles underlying the evolution of complexity in higher eukaryotes. This, in turn, will enable new advances in whole-body systems physiology (165) and the development of direct applications (*see* next section).

3. Yeast Systems Biology in Practice

A final goal is to translate the new knowledge generated by systems biology into direct applications. Each systems biology project has the potential to advance progress toward some applied objective (e.g., better characterization of biological networks responsible for a specific disease). Here, the exquisite complexity of eukaryotes should not be a deterrent but rather a stimulus to overcome the challenges and contribute to new applications for benefit of society. Yeast systems biology studies are already making such an impact; selected examples are as follows:

(a) *At the biotechnology and bioprocessing level*: First, in the food and beverage industry, new studies are being applied to the improvement of *Saccharomyces* yeast strains (166) to obtain systems-level information and new insights into essential genome-wide responses for the characterization of standard baking, brewing, and winemaking processes (167–170). In the field of industrial biotechnology toward the production of new compounds (e.g., heterologous proteins), yeast systems biology approaches toward identification of main bottlenecks and networks involved are enabling rational design of strains for the efficient production of drugs (171–173). Thus, relevant examples are the approaches taken toward strain development for the production of artemisinin, an antimalarial drug in *S. cerevisiae* (174), and the latest studies supporting the development of yeast-based therapeutic vaccines where *S. cerevisiae* is engineered to express viral or tumor antigens (175). The incorporation of new yeast systems biology developments is already making an impact (176) and will open the way to the production of new compounds and the optimization of quality and productivity, as well as the minimization of the costs, of biotechnological processes.

(b) *At the human disease/medicine level*: The basic conservation of mechanisms and biological networks from yeast to humans (e.g., the target-of-rapamycin (TOR) pathway; DNA damage response; proteostatic networks) makes yeast a good model for studies of mechanisms underlying several human diseases. Conditions and mechanisms relevant to human health being studied in yeast include cancer, genome stability, apoptosis, cell-cycle control, diseases related to the malfunction of mitochondria, prion diseases, cholesterol metabolism, diabetes, and a number of neurodegenerative diseases (*see* (177) and references therein). However, most complex human diseases are multifactorial in nature, characterized by complex de-regulation of biological networks which, if not counteracted by compensatory mechanisms, lead to a cascade of downstream effects that result in individual-specific phenotypes (diseases) which are difficult to investigate – even with the latest post-genomic techniques. Yeast systems biology offers well-designed experiments toward the characterization of short- and long-term effects of perturbations and dysregulation of networks that may illuminate the origin of complex diseases and open the way to early diagnosis. To reach that goal, advances in the investigation and proper annotation of the main core of regulatory networks (e.g., DNA–protein, RNA–protein, and protein–protein networks) will be of critical importance. Good examples of progress in that direction are the first atlas of chaperone-protein interactions in *S. cerevisiae* with direct implications to protein folding in the cell (123), the latest studies on protein homeostasis networks (178) and a first analysis of the main metabolic imbalances associated with expression of toxic mutant huntingtin in yeast, with their degree of conservation in humans and mice (179). The application of integrative systems biology studies to yeast strains expressing proteins implicated in neurodegenerative diseases (180), together with approaches in other animal models, may constitute a valuable tool to unveil patterns and mechanisms underlying the onset of complex diseases. These approaches may be combined with advanced systems biology studies and genome-wide screens for drug and target discovery (181, 182), and for the dissection of effects of specific drugs at different levels that are conserved between yeast and humans (183–185) (see also Chapter 28 of this volume).

4. Concluding Remarks and Future Perspectives

Eukaryotic life is rich in complexity (186) and yeast can help us to rediscover and appreciate this, and guide us toward new discoveries. New holistic systems biology studies with the incorporation of advanced post-genomic technologies have the potential to unveil the exquisite complexity in the *eukarya* and the development of new applications in biotechnology and medicine.

Yeast systems biology experiments under controlled conditions can uncover the complexity and interplay of biological networks with their dynamics, and basic principles of internal organization and balanced orchestrated functions between organelles in direct interaction with the environment. In order to realize the full potential of yeast as a reference model in systems biology, the high level of international collaboration between yeast research groups seen in the yeast genome sequencing (44) (http://www.yeastgenome.org) and functional genomic projects (187–190) must be continued and even enhanced. To this end, the Yeast Systems Biology Network (YSBN) (http://www.gmm.gu.se/YSBN/), a global consortium of researchers working in Systems Biology of the yeast *S. cerevisiae*, and the International Society for Systems Biology (ISSB) (http://www.issb.org) promote interdisciplinary collaborations, projects, and initiatives between top experts in the different fields. A relevant example is the UNICELLSYS project (http://www.unicellsys.eu/), a systems biology initiative with the overall objective of a quantitative understanding of control of cell growth and proliferation and other fundamental characteristics of single-celled eukaryotes toward a better understanding of the architecture and essential biology of the eukaryotic cell. These joint initiatives and projects can be reproduced in systems biology studies of more complex model organisms, including higher eukaryotes and, ultimately, humans. In this journey of new discoveries, the humble yeast *S. cerevisiae* can accompany us along the path toward the elucidation of the principles underlying the evolution of complexity, from single-celled eukaryotes to humans (191).

Acknowledgments

This work was supported by BBSRC grants BB/C505140/2 and BB/F00446X/1 as well as by a contract from the European Commission under the FP7 Collaborative Programme, UNICELLSYS.

References

1. Kitano, H. (2002) Systems biology: A brief overview. *Science* **295**, 1662–1664.
2. Castrillo, J. I. and Oliver, S. G. (2006) Metabolomics and systems biology in *Saccharomyces cerevisiae*. In: Esser, K. (Ed.), Brown, A. J. P. (Vol. Ed.), *The Mycota XIII. Fungal Genomics*, pp. 3–18. Berlin: Springer.
3. Darwin, C. and Wallace, A. R. (1858) On the tendency of species to form varieties; and on the perpetuation of varieties and species by natural means of selection. *J. Proc. Linnean Soc. London Zool.* **3**, 46–50.
4. Darwin, C. (1859) *On the Origin of Species by Means of Natural Selection, or the Preservation of Favoured Races in the Struggle for Life* (1st edn.). London: John Murray.
5. Dobzhansky, T. (1964) Biology, molecular and organismic. *Am. Zool.* **4**, 443–452.
6. Cain, C. J., Conte, D. A., García-Ojeda M. E., et al. (2008) Integrative biology: What systems biology is (not, yet). *Science* **320**, 1013a.
7. Caetano-Anollés, G. (2010) *Evolutionary Genomics and Systems Biology*. Hoboken, NJ: Wiley-Blackwell.
8. Kell, D. B. and Oliver, S. G. (2004) Here is the evidence, now what is the hypothesis? The complementary roles of inductive and hypothesis-driven science in the post-genomic era. *Bioessays* **26**, 99–105.
9. Klipp, E., Herwig, R., Kowald, A., Wierling, C., and Lehrach, H. (2005) *Systems Biology in Practice: Concepts, Implementation and Application*. Berlin: Wiley-VCH.
10. Alon, U. (2006) *An Introduction to Systems Biology: Design Principles of Biological Circuits*. Boca Raton: Chapman and Hall/CRC.
11. Palsson, B. Ø. (2006) *Systems Biology: Properties of Reconstructed Networks*. Cambridge, UK: Cambridge University Press.
12. Boogerd, F. C., Bruggeman, F. J., Hofmeyr, Jan-Hendrik, S., and Westerhoff, H. V. (2007) *Systems Biology. Philosophical Foundations*. Amsterdam: Elsevier.
13. Choi, S. (2007) *Introduction to Systems Biology*. Totowa, NJ: Humana Press.
14. Ferrell, J. E., Jr. (2009) Q&A: Systems biology. *J. Biol.* **8**, 2.
15. Klipp, E., Liebermeister, W., Wierling, C., Kowald, A., Lehrach, H., and Herwig, R. (2009) *Systems Biology: A Textbook*. Berlin: Wiley-VCH.
16. Maly, I. V. (2009) *Systems Biology. Methods in Molecular Biology*, Vol. 500. New York, NY: Humana Press, Springer.
17. Lee, S. Y. (2009) *Systems Biology and Biotechnology of Escherichia coli*. New York, NY: Springer.
18. Snyder, M. and Gallagher, J. E. (2009) Systems biology from a yeast omics perspective. *FEBS Lett.* **583**, 3895–3899.
19. Coruzzi, G. and Gutierrez, R. (2009) *Plant Systems Biology. Annual Plant Review*, Vol. 35. Chichester, UK: Wiley-Blackwell.
20. Chuang, H. Y., Hofree, M., and Ideker, T. (2010) A decade of systems biology. *Annu. Rev. Cell Dev. Biol.* **26**, 721–744.
21. Grüning, N. M., Lehrach, H., and Ralser, M. (2010) Regulatory crosstalk of the metabolic network. *Trends Biochem. Sci.* **35**, 220–227.
22. Ahn, Y. Y., Bagrow, J. P., and Lehmann, S. (2010) Link communities reveal multiscale complexity in networks. *Nature* **466**, 761–765.
23. Ideker, T., Thorsson, V., Ranish, J. A., et al. (2001). Integrated genomic and proteomic analyses of a systematically perturbed metabolic network. *Science* **292**, 929–934.
24. Ochman, H. and Raghavan, R. (2009) Systems biology. Excavating the functional landscape of bacterial cells. *Science* **326**, 1200–1201.
25. Güell, M., van Noort, V., Yus, E., et al. (2009) Transcriptome complexity in a genome-reduced bacterium. *Science* **326**, 1268–1271.
26. Kühner, S., van Noort, V., Betts. M. J., et al. (2009) Proteome organization in a

genome-reduced bacterium. *Science* **326**, 1235–1240.
27. Yus, E., Maier, T., Michalodimitrakis, K., et al. (2009) Impact of genome reduction on bacterial metabolism and its regulation. *Science* **326**, 1263–1268.
28. Ramanathan, A. and Schreiber, S. L. (2007) Multilevel regulation of growth rate in yeast revealed using systems biology. *J Biol.* **6**, 3.
29. Castrillo, J. I., Zeef, L. A., Hoyle, D. C., et al. (2007) Growth control of the eukaryote cell: A systems biology study in yeast. *J Biol.* **6**, 4.
30. Przytycka, T. M. and Andrews, J. (2010) Systems-biology dissection of eukaryotic cell growth. *BMC Biol.* **8**, 62.
31. Gutteridge, A., Pir, P., Castrillo, J. I., Charles, P. D., Lilley, K. S., and Oliver, S. G. (2010) Nutrient control of eukaryote cell growth: A systems biology study in yeast. *BMC Biol.* **8**, 68.
32. Castrillo, J. I. and Oliver, S. G. (2004) Yeast as a touchstone in post-genomic research: Strategies for integrative analysis in functional genomics. *J. Biochem. Mol. Biol.* **37**, 93–106.
33. Koonin, E. V. (2010) The incredible expanding ancestor of eukaryotes. *Cell* **140**, 606–608.
34. Luscombe, N. M., Babu, M. M., Yu, H., Snyder, M., Teichmann, S. A., and Gerstein M. (2004) Genomic analysis of regulatory network dynamics reveals large topological changes. *Nature* **431**, 308–312.
35. Balaji, S., Babu, M. M., Iyer, L. M., Luscombe, N. M., and Aravind, L. (2006) Comprehensive analysis of combinatorial regulation using the transcriptional regulatory network of yeast. *J. Mol. Biol.* **360**, 213–227.
36. Rose, A. H. and Harrison, J. S. (1987–1995) *The Yeasts*, Vols. 1–6. London, UK: Academic.
37. Sherman, F. (1998) An introduction to the genetics and molecular biology of the yeast *Saccharomyces cerevisiae*. (http://dbb.urmc.rochester.edu/labs/sherman_f/yeast/).
38. Sherman, F. (2002) Getting started with yeast. *Methods Enzymol.* **350**, 3–41 (updated in http://dbb.urmc.rochester.edu/labs/sherman_f/startedyeast.pdf).
39. Lehninger, A. L. (1975) *Biochemistry* (2nd edn.). New York, NY: Worth Publishers.
40. Fell, D. A. (1997) *Understanding the Control of Metabolism*. London: Portland Press.
41. Alberts, B., Johnson, A., Lewis, J., Raff, M., Roberts, K., and Walter, P. (2008) *Molecular Biology of the Cell* (5th edn.). New York, NY: Garland Science, Taylor and Francis Group.
42. Stansfield, I. and Stark, M. J. (2007) *Yeast Gene Analysis* (2nd edn.). London: Academic, Elsevier.
43. Feldmann, H. (2010) *Yeast: Molecular and Cell Biology*. Weinheim: Wiley-VCH.
44. Goffeau, A., Barrell, B. G., Bussey, H., et al. (1996) Life with 6000 genes. *Science* **274**, 546, 563–567.
45. Xiao, W. (2007) *Yeast Protocols* (2nd edn.). *Methods in Molecular Biology*, Vol. 313. New York, NY: Humana Press, Elsevier.
46. Nagalakshmi, U., Wang, Z., Waern, K., et al. (2008) The transcriptional landscape of the yeast genome defined by RNA sequencing. *Science* **320**, 1344–1349.
47. Stagljar, I. (2009) *Yeast Functional Genomics and Proteomics: Methods and Protocols. Methods in Molecular Biology*, Vol. 548. New York, NY: Humana Press, Elsevier.
48. Picotti, P., Bodenmiller, B., Mueller, L. N., Domon, B., and Aebersold, R. (2009) Full dynamic range proteome analysis of *S. cerevisiae* by targeted proteomics. *Cell* **138**, 795–806.
49. Picotti, P., Rinner, O., Stallmach, R., et al. (2010) High-throughput generation of selected reaction-monitoring assays for proteins and proteomes. *Nat. Methods* **7**, 43–46.
50. Kiyonami, R., Schoen, A., Prakash, A., et al. (2011) Increased selectivity, analytical precision, and throughput in targeted proteomics. *Mol. Cell. Proteomics* 10:M110.002931.
51. Mirzaei, H., Rogers, R. S., Grimes, B., Eng, J., Aderem, A., and Aebersold, R. (2010) Characterizing the connectivity of poly-ubiquitin chains by selected reaction monitoring mass spectrometry. *Mol Biosyst.* **6**, 2004–2014.
52. Moazed, D. (2009) Molecular biology. Rejoice – RNAi for yeast. *Science* **326**, 533–534.
53. Drinnenberg, I. A., Weinberg, D. E., Xie, K. T., et al. (2009) RNAi in budding yeast. *Science* **326**, 544–550.
54. Hartwell, L. H. (2004) Yeast and cancer. *Biosci. Rep.* **24**, 523–544.
55. Ferrezuelo, F., Colomina, N., Futcher, B., and Aldea, M. (2010) The transcriptional network activated by Cln3 cyclin at the G1-to-S transition of the yeast cell cycle. *Genome Biol.* **11**, R67.
56. Chang, F. and Peter, M. (2003) Yeasts make their mark. *Nat. Cell Biol.* **5**, 294–299.

57. Meyer, M. and Vilardell, J. (2009) The quest for a message: Budding yeast, a model organism to study the control of pre-mRNA splicing. *Brief Funct. Genom. Proteom.* **8**, 60–67.
58. McGrail, J. C., Krause, A., and O'Keefe, R. T. (2009) The RNA binding protein Cwc2 interacts directly with the U6 snRNA to link the nineteen complex to the spliceosome during pre-mRNA splicing. *Nucleic Acids Res.* **37**, 4205–4217.
59. Altmann, M. and Linder, P. (2010) Power of yeast for the analysis of eukaryotic translation initiation. *J. Biol. Chem.* **285**, 31907–31912.
60. Delneri, D., Colson, I., Grammenoudi, S., Roberts, I. N., Louis, E. J., and Oliver, S. G. (2003) Engineering evolution to study speciation in yeasts. *Nature* **422**, 68–72.
61. Hittinger, C. T. and Carroll, S. B. (2007) Gene duplication and the adaptive evolution of a classic genetic switch. *Nature* **449**, 677–681.
62. Anderson, R. M., Latorre-Esteves, M., Neves, A. R., et al (2003) Yeast life-span extension by calorie restriction is independent of NAD fluctuation. *Science* **302**, 2124–2126.
63. Payne, T., Hanfrey, C., Bishop, A. L., Michael, A. J., Avery, S. V., and Archer, D. B. (2008) Transcript-specific translational regulation in the unfolded protein response of *Saccharomyces cerevisiae*. *FEBS Lett.* **582**, 503–509.
64. Cashikar, A. G., Duennwald, M., and Lindquist, S. L. (2005) A chaperone pathway in protein disaggregation. Hsp26 alters the nature of protein aggregates to facilitate reactivation by Hsp104. *J Biol Chem.* **280**, 23869–23875.
65. Duennwald, M. L., Jagadish, S., Giorgini, F., Muchowski, P. J., and Lindquist S. (2006) A network of protein interactions determines polyglutamine toxicity. *Proc. Natl. Acad. Sci. USA* **103**, 11051–11056.
66. Giorgini, F. and Muchowski, P. J. (2009) Exploiting yeast genetics to inform therapeutic strategies for Huntington's disease. *Methods Mol. Biol.* **548**, 161–174.
67. Fritz-Laylin, L. K., Prochnik, S. E., Ginger, M. L., et al. (2010) The genome of *Naegleria gruberi* illuminates early eukaryotic versatility. *Cell* **140**, 631–642.
68. Wickstead, B., Gull, K., and Richards, T. A. (2010) Patterns of kinesin evolution reveal a complex ancestral eukaryote with a multifunctional cytoskeleton. *BMC Evol. Biol.* **10**, 110.
69. Hayden, E. C. (2009) Genome sequencing: The third generation. *Nature* **457**, 768–769.
70. Nowrousian, M. (2010) Next-generation sequencing techniques for eukaryotic microorganisms: Sequencing-based solutions to biological problems. *Eukaryot. Cell* **9**, 1300–1310.
71. Metzker, M. L. (2010) Sequencing technologies – the next generation. *Nat. Rev. Genet.* **11**, 31–46.
72. Werner, T. (2010) Next generation sequencing in functional genomics. *Brief Bioinform.* **11**, 499–511.
73. Hart, C., Lipson, D., Ozsolak, F., et al. (2010) Single-molecule sequencing: Sequence methods to enable accurate quantitation. *Methods Enzymol.* **472**, 407–430.
74. Kurdistani, S. K., Tavazoie, S., and Grunstein M. (2004) Mapping global histone acetylation patterns to gene expression. *Cell* **117**, 721–733.
75. Wang, Z., Zang, C., Cui, K., et al. (2009) Genome-wide mapping of HATs and HDACs reveals distinct functions in active and inactive genes. *Cell* **138**, 1019–1031.
76. Millar, C. B. and Grunstein, M. (2006) Genome-wide patterns of histone modifications in yeast. *Nat. Rev. Mol. Cell. Biol.* **7**, 657–666.
77. Shahbazian, M. D. and Grunstein, M. (2007) Functions of site-specific histone acetylation and deacetylation. *Annu. Rev. Biochem.* **76**, 75–100.
78. Iyer, V. R., Horak, C. E., Scafe, C. S., Botstein, D., Snyder, M., and Brown, P. O. (2001) Genomic binding sites of the yeast cell-cycle transcription factors SBF and MBF. *Nature* **409**, 533–538.
79. Johnson, D. S., Mortazavi, A., Myers, R. M., and Wold, B. (2007) Genome-wide mapping of in vivo protein–DNA interactions. *Science* **316**, 1497–1502.
80. Eaton, M. L., Galani, K., Kang, S., Bell, S. P., and MacAlpine, D. M. (2010) Conserved nucleosome positioning defines replication origins. *Genes Dev.* **24**, 748–753.
81. Green, B. M., Finn, K. J., and Li, J. J. (2010) Loss of DNA replication control is a potent inducer of gene amplification. *Science* **329**, 943–946.
82. Kaochar, S., Paek, A. L., and Weinert, T. (2010) Genetics. Replication error amplified. *Science* **329**, 911–913.
83. Duan, Z., Andronescu, M., Schutz, K., et al. (2010) A three-dimensional model of the yeast genome. *Nature* **465**, 363–367.

84. Granovskaia, M. V., Jensen, L. J., Ritchie, M. E., et al. (2010) High-resolution transcription atlas of the mitotic cell cycle in budding yeast. *Genome Biol.* **11**, R24.
85. Robinson, R. (2010) Dark matter transcripts: Sound and fury, signifying nothing? *PLoS Biol.* **8**, e1000370.
86. Robertson, M. (2010) The evolution of gene regulation, the RNA universe, and the vexed questions of artefact and noise. *BMC Biol.* **8**, 97.
87. Levin, J. Z., Yassour, M., Adiconis, X., et al. (2010) Comprehensive comparative analysis of strand-specific RNA sequencing methods. *Nat. Methods* **7**, 709–715.
88. Lipson, D., Raz, T., Kieu, A., et al. (2010) Quantification of the yeast transcriptome by single-molecule sequencing. *Nat. Biotechnol.* **27**, 652–658.
89. Ozsolak, F., Platt, A. R., Jones, D. R., et al. (2009) Direct RNA sequencing. *Nature* **461**, 814–818.
90. Rusk, N. (2009) The true RNA-seq. *Nat. Methods* **6**, 790–791.
91. Haas, B. J. and Zody, M. C. (2010) Advancing RNA-Seq analysis. *Nat. Biotechnol.* **28**, 421–423.
92. Auer, P. L. and Doerge, R. W. (2010) Statistical design and analysis of RNA sequencing data. *Genetics* **185**, 405–416.
93. Chalamcharla, V. R., Curcio, M. J., and Belfort, M. (2010) Nuclear expression of a group II intron is consistent with spliceosomal intron ancestry. *Genes Dev.* **24**, 827–836.
94. Camblong, J., Iglesias, N., Fickentscher, C., Dieppois, G., and Stutz, F. (2007) Antisense RNA stabilization induces transcriptional gene silencing via histone deacetylation in *S. cerevisiae*. *Cell* **131**, 706–717.
95. Camblong, J., Beyrouthy, N., Guffanti, E., Schlaepfer, G., Steinmetz, L. M., and Stutz, F. (2009) Trans-acting antisense RNAs mediate transcriptional gene cosuppression in *S. cerevisiae*. *Genes Dev.* **23**, 1534–1545.
96. Harrison, B. R., Yazgan, O., and Krebs, J. E. (2009) Life without RNAi: Noncoding RNAs and their functions in *Saccharomyces cerevisiae*. *Biochem. Cell Biol.* **87**, 767–779.
97. Bumgarner, S. L., Dowell, R. D., Grisafi, P., Gifford, D. K., and Fink, G. R. (2009) Toggle involving *cis*-interfering noncoding RNAs controls variegated gene expression in yeast. *Proc. Natl. Acad. Sci. USA* **106**, 18321–18326.
98. Winston, F. (2009) A transcription switch toggled by noncoding RNAs. *Proc. Natl. Acad. Sci. USA* **106**, 18049–18050.
99. Yassour, M., Pfiffner, J., Levin, J. Z., et al. (2010) Strand-specific RNA sequencing reveals extensive regulated long antisense transcripts that are conserved across yeast species. *Genome Biol.* **11**, R87.
100. Kertesz, M., Wan, Y., Mazor, E. et al. (2010) Genome-wide measurement of RNA secondary structure in yeast. *Nature* **467**, 103–107.
101. Mattick, J. S. (2004) RNA regulation: A new genetics? *Nat. Rev. Genet.* **5**, 316–323.
102. Mattick, J. S. (2007) A new paradigm for developmental biology. *J. Exp. Biol.* **210**, 1526–1547.
103. de Godoy, L. M., Olsen, J. V., Cox, J., et al. (2008) Comprehensive mass-spectrometry-based proteome quantification of haploid versus diploid yeast. *Nature* **455**, 1251–1254.
104. Vaidyanathan, S., Harrigan, G. G., and Goodacre, R. (2005). *Metabolome Analyses: Strategies for Systems Biology*. New York, NY: Springer.
105. Förster, J., Famili, I., Fu, P., Palsson, B. Ø., and Nielsen, J. (2003) Genome-scale reconstruction of the *Saccharomyces cerevisiae* metabolic network. *Genome Res.* **13**, 244–253.
106. Duarte, N. C., Herrgård, M. J., and Palsson, B. Ø. (2004) Reconstruction and validation of *Saccharomyces cerevisiae* iND750, a fully compartmentalized genome-scale metabolic model. *Genome Res.* **14**, 1298–1309.
107. Herrgård, M. J., Swainston, N., Dobson, P., et al. (2008) A consensus yeast metabolic network reconstruction obtained from a community approach to systems biology. *Nat. Biotechnol.* **26**, 1155–1160.
108. Mo, M. L., Palsson, B. Ø., and Herrgård, M. J. (2009) Connecting extracellular metabolomic measurements to intracellular flux states in yeast. *BMC Syst. Biol.* **3**, 37.
109. Fordyce, P. M., Gerber, D., Tran, D., et al. (2010) De novo identification and biophysical characterization of transcription-factor binding sites with microfluidic affinity analysis. *Nat. Biotechnol.* **28**, 970–975.
110. Oeffinger, M., Wei, K. E., Rogers R., et al. (2007) Comprehensive analysis of diverse ribonucleoprotein complexes. *Nat. Methods* **4**, 951–956.
111. Hogan, D. J., Riordan, D. P., Gerber, A. P., Herschlag, D., and Brown, P. O. (2008) Diverse RNA-binding proteins interact with functionally related sets of RNAs, suggesting an extensive regulatory system. *PLoS Biol.* **6**, e255.

112. Cornell, M., Paton, N. W., and Oliver, S. G. (2004) A critical and integrated view of the yeast interactome. *Comp. Funct. Genomics* **5**, 382–402.
113. Charbonnier, S., Zanier, K., Masson, M., and Travé, G. (2006) Capturing protein–protein complexes at equilibrium: The holdup comparative chromatographic retention assay. *Protein Expr. Purif.* **50**, 89–101.
114. Pu, S., Wong, J., Turner, B., Cho, E., and Wodak, S. J. (2009) Up-to-date catalogues of yeast protein complexes. *Nucleic Acids Res.* **37**, 825–831.
115. Benschop, J. J., Brabers, N., van Leenen, D., et al. (2010) A consensus of core protein complex compositions for *Saccharomyces cerevisiae*. *Mol. Cell* **38**, 916–928.
116. Schmitz, A. M., Morrison, M. F., Agunwamba, A. O., Nibert, M. L., and Lesser, C. F. (2009) Protein interaction platforms: Visualization of interacting proteins in yeast. *Nat. Methods* **6**, 500–502.
117. Wang, H., Kakaradov, B., Collins, S. R., et al. (2009) A complex-based reconstruction of the *Saccharomyces cerevisiae* interactome. *Mol. Cell. Proteomics* **8**, 1361–1381.
118. Beltrao, P., Cagney, G., and Krogan, N. J. (2010) Quantitative genetic interactions reveal biological modularity. *Cell* **141**, 739–745.
119. Costanzo, M., Baryshnikova, A., Bellay, J., et al. (2010) The genetic landscape of a cell. *Science* **327**, 425–431.
120. Gavin, A. C., Aloy, P., Grandi, P., et al. (2006) Proteome survey reveals modularity of the yeast cell machinery. *Nature* **440**, 631–636.
121. Breitkreutz, A., Choi, H., Sharom, J. R., et al. (2010) A global protein kinase and phosphatase interaction network in yeast. *Science* **328**, 1043–1046.
122. Mok, J., Kim, P. M., Lam, H. Y., et al. (2010) Deciphering protein kinase specificity through large-scale analysis of yeast phosphorylation site motifs. *Sci. Signal.* **3**, ra12.
123. Gong, Y., Kakihara, Y., Krogan, N., et al. (2009) An atlas of chaperone–protein interactions in *Saccharomyces cerevisiae*: Implications to protein folding pathways in the cell. *Mol. Syst. Biol.* **5**, 275.
124. Kung, L. A., Tao, S. C., Qian, J., Smith, M. G., Snyder, M., and Zhu, H. (2009) Global analysis of the glycoproteome in *Saccharomyces cerevisiae* reveals new roles for protein glycosylation in eukaryotes. *Mol. Syst. Biol.* **5**, 308.
125. Castegna, A., Scarcia, P., Agrimi, G., et al. (2010) Identification and functional characterization of a novel mitochondrial carrier for citrate and oxoglutarate in *Saccharomyces cerevisiae*. *J. Biol. Chem.* **285**, 17359–17370.
126. Klitgord, N. and Segrè D. (2010) The importance of compartmentalization in metabolic flux models: Yeast as an ecosystem of organelles. *Genome Inform.* **22**, 41–55.
127. Scott, E. M. and Pillus, L. (2010) Homocitrate synthase connects amino acid metabolism to chromatin functions through Esa1 and DNA damage. *Genes Dev.* **24**, 1903–1913.
128. Yogev, O., Yogev, O., Singer, E., et al. (2010) Fumarase: A mitochondrial metabolic enzyme and a cytosolic/nuclear component of the DNA damage response. *PLoS Biol.* **8**, e1000328.
129. Pronk, J. T. (2002) Auxotrophic yeast strains in fundamental and applied research. *Appl. Environ. Microbiol.* **68**, 2095–2100.
130. Davies, J. (2009) Regulation, necessity, and the misinterpretation of knockouts. *Bioessays* **31**, 826–830.
131. DeLuna, A., Springer, M., Kirschner, M. W., and Kishony, R. (2010) Need-based up-regulation of protein levels in response to deletion of their duplicate genes. *PLoS Biol.* **8**, e1000347.
132. Gout, J. F., Kahn, D., Duret, L., and Paramecium Post-Genomics Consortium (2010) The relationship among gene expression, the evolution of gene dosage, and the rate of protein evolution. *PLoS Genet.* **6**, e1000944.
133. Baulcombe, D. C. (1996). RNA as a target and an initiator of post-transcriptional gene silencing in transgenic plants. *Plant Mol. Biol.* **32**, 79–88.
134. Fire, A., Xu, S., Montgomery, M. K., Kostas, S. A., Driver, S. E., and Mello, C. C. (1998). Potent and specific genetic interference by double-stranded RNA in *Caenorhabditis elegans*. *Nature* **391**, 806–811.
135. Jacquier, A. (2009) The complex eukaryotic transcriptome: Unexpected pervasive transcription and novel small RNAs. *Nat. Rev. Genet.* **10**, 833–844.
136. Berretta, J. and Morillon A. (2009) Pervasive transcription constitutes a new level of eukaryotic genome regulation. *EMBO Rep.* **10**, 973–982.
137. van Bakel, H., Nislow, C., Blenclowe, B. C., and Hughes, T. R. (2010) Most "dark matter" transcripts are associated with known genes. *PLoS Biol.* **8**, e1000371.

138. Phillips, M. L. (2010) Existence of RNA 'dark matter' in doubt. *Nature*. doi:10.1038/news.2010.248. http://www.nature.com/news/2010/100518/full/news.2010.248.html.
139. Hannon, G. J., Rivas, F. V., Murchison, E. P., and Steitz, J. A. (2006) The expanding universe of noncoding RNAs. *Cold Spring Harb. Symp. Quant. Biol.* **71**, 551–564.
140. Grosshans, H. and Filipowicz, W. (2008) Molecular biology: The expanding world of small RNAs. *Nature* **451**, 414–416.
141. Dieci, G., Preti, M., and Montanini, B. (2009) eukaryotic snoRNAs: A paradigm for gene expression flexibility. *Genomics* **94**, 83–88.
142. MacLean, D., Elina, N., Havecker, E. R., Heimstaedt, S. B., Studholme, D. J., and Baulcombe, D. C. (2010) Evidence for large complex networks of plant short silencing RNAs. *PLoS One* **5**, e9901.
143. Wilusz, J. E., Sunwoo, H., and Spector, D. L. (2009) Long noncoding RNAs: Functional surprises from the RNA world. *Genes Dev.* **23**, 1494–1504.
144. Mercer, T. R., Dinger, M. E., and Mattick, J. S. (2009) Long non-coding RNAs: Insights into functions. *Nat. Rev. Genet.* **10**, 155–159.
145. Qureshi, I. A., Mattick, J. S., and Mehler, M. F. (2010) Long non-coding RNAs in nervous system function and disease. *Brain Res.* **1338**, 20–35.
146. Amaral, P. P., Dinger, M. E., Mercer, T. R., and Mattick, J. S. (2008) The eukaryotic genome as an RNA machine. *Science* **319**, 1787–1789.
147 Mattick, J. (2010) Video Q&A: Non-coding RNAs and eukaryotic evolution – a personal view. *BMC Biol.* **8**, 67.
148. Waters, L. S. and Storz, G. (2009) Regulatory RNAs in bacteria. *Cell* **136**, 615–628.
149. van Vliet, A. H. and Wren, B. W. (2009) New levels of sophistication in the transcriptional landscape of bacteria. *Genome Biol.* **10**, 233.
150. Kondo, T., Plaza, S., Zanet, J., et al. (2010) Small peptides switch the transcriptional activity of Shavenbaby during *Drosophila* embryogenesis. *Science* **329**, 336–339.
151. Nilsen, T. W. and Graveley, B. R. (2010) Expansion of the eukaryotic proteome by alternative splicing. *Nature* **463**, 457–463.
152. Wang, E. T., Sandberg, R., Luo, S., et al. (2008) Alternative isoform regulation in human tissue transcriptomes. *Nature* **456**, 470–476.
153. Apostolatos, H., Apostolatos, A., Vickers, T., et al. (2010) Vitamin a metabolite, all-*trans* retinoic acid, mediates alternative splicing of protein kinase C delta(PKC){delta})VIII isoform via the splicing factor SC35. *J. Biol. Chem.* **285**, 25987–25995.
154. Tsvetanova, N. G., Klass, D. M., Salzman, J., and Brown, P. O. (2010) Proteome-wide search reveals unexpected RNA-binding proteins in *Saccharomyces cerevisiae*. *PLoS One* **5**(9), pii, e12671.
155. Rederstorff, M., Bernhart, S. H., Tanzer, A., et al. (2010) RNPomics: Defining the ncRNA transcriptome by cDNA library generation from ribonucleo-protein particles. *Nucleic Acids Res.* **38**, e113.
156. Conacci-Sorrell, M., Ngouenet, C., and Eisenman, R. N. (2010) Myc-nick: A cytoplasmic cleavage product of Myc that promotes alpha-tubulin acetylation and cell differentiation. *Cell* **142**, 480–493.
157. Mattick, J. S., Amaral, P. P., Dinger, M. E., Mercer, T. R., and Mehler, M. F. (2009) RNA regulation of epigenetic processes. *Bioessays* **31**, 51–59.
158. Alló, M., Buggiano, V., Fededa, J. P., et al. (2009) Control of alternative splicing through siRNA-mediated transcriptional gene silencing. *Nat. Struct. Mol. Biol.* **16**, 717–724.
159. Mattick, J. S., Taft, R. J., and Faulkner, G. J. (2010) A global view of genomic information—moving beyond the gene and the master regulator. *Trends Genet.* **26**, 21–28.
160. Müller, B. and Grossniklaus, U. (2010) Model organisms – A historical perspective. *J. Proteomics* **73**, 2054–2063.
161. Gregor, T., Fujimoto, K., Masaki, N., and Sawai S. (2010) The onset of collective behavior in social amoebae. *Science* **328**, 1021–1025.
162. Srivastava, M., Simakov, O., Chapman, J., et al. (2010) The *Amphimedon queenslandica* genome and the evolution of animal complexity. *Nature* **466**, 720–726.
163. Butler, D. (2010) Human genome at ten: Science after the sequence. *Nature* **465**, 1000–1001.
164. Qin, J., Li, R., Raes, J., et al. (2010) A human gut microbial gene catalogue established by metagenomic sequencing. *Nature* **464**, 59–65.
165. Kuepfer, L. (2010) Toward whole-body systems physiology. *Mol. Syst. Biol.* **6**, 409.
166. Donalies, U. E., Nguyen, H. T., Stahl, U., and Nevoigt, E. (2008) Improvement of *Saccharomyces* yeast strains used in brewing, wine making and baking. *Adv. Biochem. Eng. Biotechnol.* **111**, 67–98.

167. Tanaka, F., Ando, A., Nakamura, T., Takagi, H., and Shima, J. (2006) Functional genomic analysis of commercial Baker's yeast during initial stages of model dough-fermentation. *Food Microbiol.* **23**, 717–728.
168. Pizarro, F., Vargas, F. A., and Agosin E. (2007) A systems biology perspective of wine fermentations. *Yeast* **24**, 977–991.
169. Borneman, A. R., Chambers, P. J., and Pretorius, I. S. (2007) Yeast systems biology: Modeling the winemaker's art. *Trends Biotechnol.* **25**, 349–355.
170. Ando, A., Nakamura, T., Murata, Y., Takagi, H., and Shima, J. (2007) Identification and classification of genes required for tolerance to freeze-thaw stress revealed by genome-wide screening of *Saccharomyces cerevisiae* deletion strains. *FEMS Yeast Res.* 7, 244–253.
171. Lee, S. Y., Kim, H. U., Park, J. H., Park, J. M., and Kim, T. Y. (2009) Metabolic engineering of microorganisms: General strategies and drug production. *Drug Discov. Today* **14**, 78–88.
172. Nielsen, J. and Jewett, M. C. (2008) Impact of systems biology on metabolic engineering of *Saccharomyces cerevisiae. FEMS Yeast Res.* 8, 122–131.
173. Nevoigt, E. (2008) Progress in metabolic engineering of *Saccharomyces cerevisiae. Microbiol. Mol. Biol. Rev.* **72**, 379–412.
174. Ro, D. K., Paradise, E. M, Ouellet, M., et al. (2006) Production of the antimalarial drug precursor artemisinic acid in engineered yeast. *Nature* **440**, 940–943.
175. Ardiani, A., Higgins, J. P., and Hodge, J. W. (2010) Vaccines based on whole recombinant *Saccharomyces cerevisiae* cells. *FEMS Yeast Res.* **10**, 1060–1069.
176. Petranovic, D. and Vemuri, G. N. (2009) Impact of yeast systems biology on industrial biotechnology. *J. Biotechnol.* **144**, 204–211.
177. Petranovic, D. and Nielsen, J. (2008) Can yeast systems biology contribute to the understanding of human disease? *Trends Biotechnol.* **26**, 584–590.
178. Hutt, D. and Balch, W. E. (2010) Cell biology. The proteome in balance. *Science* **329**, 766–767.
179. Joyner, P. M., Matheke, R. M., Smith, L. M., and Cichewicz, R. H. (2010) Probing the metabolic aberrations underlying mutant huntingtin toxicity in yeast and assessing their degree of preservation in humans and mice. *J. Proteome Res.* **9**, 404–412.
180. Khurana, V. and Lindquist, S. (2010) Modeling neurodegeneration in *Saccharomyces cerevisiae*: Why cook with baker's yeast? *Nat. Rev. Neurosci.* **11**, 436–449.
181. Bharucha, N. and Kumar, A. (2007) Yeast genomics and drug target identification. *Comb. Chem. High Throughput Screen.* **10**, 618–634.
182. Smith, A. M., Ammar, R., Nislow, C., and Giaever, G. (2010) A survey of yeast genomic assays for drug and target discovery. *Pharmacol. Ther.* **127**, 156–164.
183. Yu, L., Lopez, A., Anaflous, A., et al. (2008) Chemical–genetic profiling of imidazo [1,2-a]pyridines and -pyrimidines reveals target pathways conserved between yeast and human cells. *PLoS Genet.* **4**, e1000284.
184. Ho, R. L. and Lieu, C. A. (2008) Systems biology: An evolving approach in drug discovery and development. *Drugs R. D.* **9**, 203–216.
185. Stefanini, I., Trabocchi, A., Marchi, E., Guarna, A., and Cavalieri D. (2010) A systems biology approach to dissection of the effects of small bicyclic peptidomimetics on a panel of *Saccharomyces cerevisiae* mutants. *J. Biol. Chem.* **285**, 23477–23485.
186. Hayden, E. C. (2010) Human genome at ten: Life is complicated. *Nature* **464**, 664–667.
187. Oliver, S. G. (1997) Yeast as a navigational aid in genome analysis. *Microbiology* **143**, 1483–1487.
188. Oliver, S. G. (2002) Functional genomics: Lessons from yeast. *Philos. Trans. R. Soc. London B. Biol. Sci.* **357**, 17–23.
189. Winzeler, E. A., Shoemaker, D. D., Astromoff, A., et al. (1999) Functional characterization of the *S. cerevisiae* genome by gene deletion and parallel analysis. *Science* **285**, 901–906.
190. Giaever, G., Chu, A. M., Ni, L., et al. (2002) Functional profiling of the *Saccharomyces cerevisiae* genome. *Nature* **418**, 387–391.
191. Heard, E., Tishkoff, S., Todd, J. A., et al. (2010) Ten years of genetics and genomics: What have we achieved and where are we heading? *Nat. Rev. Genet.* **11**, 723–733.

Section II

Experimental Systems Biology: High-Throughput Genome-Wide and Molecular Studies

Chapter 2

Saccharomyces cerevisiae: Gene Annotation and Genome Variability, State of the Art Through Comparative Genomics

Ed Louis

Abstract

In the early days of the yeast genome sequencing project, gene annotation was in its infancy and suffered the problem of many false positive annotations as well as missed genes. The lack of other sequences for comparison also prevented the annotation of conserved, functional sequences that were not coding. We are now in an era of comparative genomics where many closely related as well as more distantly related genomes are available for direct sequence and synteny comparisons allowing for more probable predictions of genes and other functional sequences due to conservation. We also have a plethora of functional genomics data which helps inform gene annotation for previously uncharacterised open reading frames (ORFs)/genes. For *Saccharomyces cerevisiae* this has resulted in a continuous updating of the gene and functional sequence annotations in the reference genome helping it retain its position as the best characterized eukaryotic organism's genome. A single reference genome for a species does not accurately describe the species and this is quite clear in the case of *S. cerevisiae* where the reference strain is not ideal for brewing or baking due to missing genes. Recent surveys of numerous isolates, from a variety of sources, using a variety of technologies have revealed a great deal of variation amongst isolates with genome sequence surveys providing information on novel genes, undetectable by other means. We now have a better understanding of the extant variation in *S. cerevisiae* as a species as well as some idea of how much we are missing from this understanding. As with gene annotation, comparative genomics enhances the discovery and description of genome variation and is providing us with the tools for understanding genome evolution, adaptation and selection, and underlying genetics of complex traits.

Key words: Gene annotation, comparative genomics, next generation sequencing, novel genes, genetic variation.

1. Introduction

Gene annotation and genome variation are interrelated and each can inform the other (*see* **Fig. 2.1**). To address the question of "When is a gene not a gene?" for dealing with dubious

J.I. Castrillo, S.G. Oliver (eds.), *Yeast Systems Biology*, Methods in Molecular Biology 759,
DOI 10.1007/978-1-61779-173-4_2,

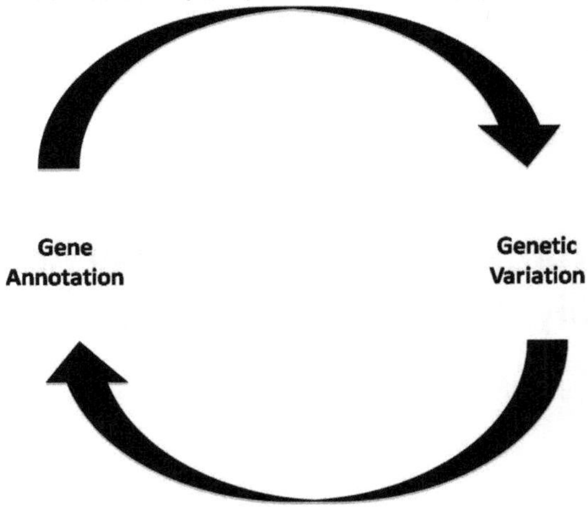

Fig. 2.1. Gene annotation and genetic variation inform each other. It has become clear that the increased number of genome sequences from different strains and species of *Saccharomyces* has been very informative to gene annotation through comparative genomics. This has led to a continuous evolution of the annotation of the reference *S. cerevisiae* genome that has in turn led to the annotations of related genomes. Less well appreciated is the role annotated genomes have played in our understanding of genetic variation, in particular structural variation, due to insertions and deletions as well as rearrangements such as translocations. Annotated genomes allow for quick and efficient alignments, such as through the Yeast Gene Order Browser (*see* text), without the need for multi-sequence alignments. Potential areas of interest are immediately obvious as apparent synteny breakpoints or as duplicated regions.

open reading frames, comparative studies with several strains and species can be a powerful tool, obviating the need for experimental determination of function. Similarly, gene annotations embedded in a gene order or synteny browser can be used to determine areas of genome variation, in terms of insertions, deletions, and rearrangements. Finally, for uncharacterised yet conserved genes, the cross referencing of the vast amount of functional genomic data can inform and improve functional annotation.

2. Gene Annotation

The reference genome, S288C, is constantly changing with updates in both sequences and annotations (1). Many of the updated gene annotations come from functional studies but others come from informatic analysis of genome comparisons with functional follow-ups. These updates can be seen for each annotated gene or sequence element at SGD (*see* http://www.yeastgenome.org/cgi-bin/seqTools) and a global picture of the updates can be found at http://www.yeastgenome.org/cache/genomeSnapshot.html. The rate of change in annotations is not really slowing down as there are still many functions to be determined and many open reading frames still annotated as dubious, as there is no corroborative evidence for them actually being genes. New and improved annotation pipelines are in place to improve functional annotations as more data becomes available (2, 3).

Genome comparisons of related species have given the greatest advance in gene annotation in recent years (4, 5), and numerous ORFs are less likely to be real based on lack of conservation (**Fig. 2.1**). The reduction of the original 6,200 ORFs by 15% is in large part due to comparative genomics. On the flip side, small ORFs, which did not pass the threshold for being annotated as possible ORFs, are now properly annotated due to conservation in related species (1), which is thought to be due to functional constraints. This sequence conservation has also led to the annotation of non-coding sequence elements, either presumed functional elements or discovery of new elements including transcription factor binding sites (4, 6, 7).

2.1. Gene Annotation Informing Genome Variation

Annotated genomes have been useful in comparative genomics, particularly in determining ancestral structural states and the sequence of gross chromosomal rearrangement events that lead to the extant genome of *Saccharomyces cerevisiae* as well as its relatives. The determination of the Whole Genome Duplication by Wolfe and Shields (8) is a case in point, where genome dynamics and variation over evolutionary time have been determined using comparisons of annotated regions which have since developed into an annotated gene order browser (9, 10). Since then the use of a gene order browser has become a standard tool for analysis of genome variation between species.

Sometimes the use of such tools can lead to apparent inconsistencies between studies. When comparing the genomes of the *Saccharomyces sensu stricto* species, it is clear that most of the genome is syntenic and that there are only a few gross chromosomal rearrangements. Mapping the breakpoints by physical means can narrow down the sites of ancestral rearrangements within a few kilo-base pairs. In the case of *Saccharomyces bayanus*, eight breakpoints for four reciprocal translocations were determined (11). When the same isolate was sequenced at low coverage and the reads annotated where ORFs were found, a comparison to the gene order browser for *S. cerevisiae* resulted in the determination of at least 45 breakpoints as ORFs were found in the same clone whose homologues in *S. cerevisiae* were located far apart, mostly on other chromosomes (12). A closer inspection revealed that there was a lot of differential accumulation of divergence in one or the other of a pair of gene duplicates after the whole genome duplication resulting in gene relics that were not annotated (13). What looked like gross chromosomal rearrangements in many cases was loss of open reading frames differentially between duplicated segments in the two species and a sequence comparison revealed that synteny was retained.

Gene annotation has improved but is still an evolving process. Various projects for improving automated gene annotation are underway at many places and these rely in a large part on comparative genomics. It is likely that proper annotation will continue to require human intervention. The state of the art for functional genome annotation in *S. cerevisiae* utilizes high-throughput experimental data as well as computational predictions (2, 3).

2.2. Gene Annotation Challenges

Some genome regions are particularly problematic, as they do not assemble well, making annotations difficult if not impossible. In particular, the subtelomeric regions generally are not included in genome projects due to technical difficulties in cloning, sequencing, and eventual assembly. For *S. cerevisiae* this is of particular importance as many of the genes responsible for important phenotypic variation, such as brewing, baking and general fermentation properties, are located in the subtelomeres. The reference genome for *S. cerevisiae* remains the only finished genome as its subtelomeres were individually marked, cloned, and sequenced during the genome project, an effort that cannot be done efficiently for any other genomes. None of the other yeast strains (14–16) (*see* also http://www.broadinstitute.org/annotation/genome/saccharomyces_cerevisiae.3/Info.html) and species (4, 6, 7) sequenced have assembled and therefore annotated subtelomeres making it difficult to progress with many important studies.

3. Genome Variation

Genome variation has always been underlying studies in *S. cerevisiae* but has not always been taken into account. For example, studies using the reference genome as a template for experiments on another strain may not result in interpretable data for regions of the genomes that are different. In particular the subtelomeric regions as mentioned above vary a great deal between strains and therefore it is not possible to determine which chromosome end or which copy of a gene is responsible for the data. This is illustrated in studies of meiotic double strand breaks, which are generally done in strain SK1, which varies greatly in its subtelomeres from the reference genome (14, 17). The data from array-based analyses therefore cannot be interpreted in these regions (18).

Table 2.1
Techniques for assessing genetic variation

Technique	Uses	Disadvantages
Pulsed field gel analysis	Gross chromosomal rearrangements and chromosome length polymorphisms	Low resolution, not high throughput
AFLPs	Phylo-geographic relationships of closely related strains/species	Not good for distantly related species due to loss of homology with phylogenetic distance
Microsatellites	High-throughput assessment of relatedness of strains using multiple alleles at few loci	Low resolution in terms of genome coverage, identical alleles not necessarily identical by descent
Microarrays – ORFs/long oligos	High throughput, copy number variation (CNV), presence/absence of sequences	Cannot assess unknown sequence, not good for SNP variation
High-density microarrays	High throughput, CNV, presence/absence as well as determination of sequence variants (SNPs)	Cannot assess unknown sequence
Whole genome sequencing	Highest resolution, can assess novel previously unknown sequence	Expensive, throughput depends on ability to multiplex

We now have the capability of assessing variation amongst *S. cerevisiae* strains and related species by a variety of means. Currently in use are physical analysis of structural variation using pulsed field gels and Southern analysis (11, 17, 19–22), amplified fragment length polymorphisms (AFLPs) (23–25), microsatellite variation at several sites across the genome using PCR (26–28), presence/absence as well as SNP detection using microarrays (29–35) and finally sequencing at either several loci (36–39) or the whole genome (14) (*see* **Table 2.1**).

3.1. Genome Variation by Pulsed Field Gel Electrophoresis

The use of pulsed field gel electrophoresis to separate large DNA molecules has proven very useful in looking at structural variation in yeast genomes: assessing variation amongst isolates (17, 19, 21, 22), variation generated by genome instability (20), or variation generated over evolutionary time (11, 17). The resolution of pulsed field gels coupled with Southern analysis is quite low; however, a great deal of effort is required to narrow down breakpoints (11). It is also difficult to scale up to large numbers of samples though 10 s to 100 s are possible (40).

3.2. Genome Variation via AFLPs

A higher throughput method in use is amplified fragment length polymorphisms which is a quick method of assessing relatedness amongst strain isolates (24, 25). This technique is invariably used along with rDNA typing or other genetic characterization to generate a more informed and consistent picture of relationships. One of the problems with the technique is it is difficult to know what is actually being compared as these are randomly amplified fragments, and how much of the genome is being assessed. Another difficulty is that the method is limited to close relatives as increased phylogenetic distances reduce the likelihood that fragments are comparable through homology (23).

3.3. Genome Variation via Microsatellites

One of the most efficient ways of determining general strain variation is by microsatellite analysis. The increased number of alleles available at these loci in part makes up for the lack of number of loci assessed. With only a few markers, large numbers of strains can be genotyped (26–28). Phylogenetic relationships can be inferred and some feel for diversity in the species can be obtained. There are limitations to this approach. One is that only a limited part of the genome is genotyped and in many studies not even every chromosome is marked. This severely limits the use of the genotype data for mapping genetic differences responsible for various phenotypes for example. Another is that the allele state at a microsatellite is not necessarily a good indicator of identity by descent and therefore inferred phylogenetic relationships are compromised. A particular copy number allele could have been arrived at from different "mutational" changes from different alleles. Despite these problems, microsatellite genotyping remains a quick and inexpensive way to assess diversity for large numbers of strains.

3.4. Genome Variation via Microarray Comparative Genome hybridization

A better approach but less high throughput is the use of microarrays and comparative genome hybridization. For large probes on arrays, the resolution can yield information on presence or absence as well as copy number (30). This has generally worked well for some studies but has its limitations. The main limitation is that you cannot look for sequences that are not known. These larger probe arrays are also not very useful for detecting sequence divergence. This approach is particularly suited for genome stability and/or composition studies with hybrids or conditions resulting in aneuploidies where copy number changes are important determinants. Unfortunately the microarrays cannot provide information on location or structure accompanying copy number changes. Complementary analysis with specific probes on pulsed field gels can resolve location and structural differences.

Higher resolution can be obtained with high-density arrays using short oligos that cover the whole known genome (32, 34, 35). With appropriate analysis these can even detect single SNP differences making them almost as good as sequencing (29, 34, 35). Such arrays are more expensive and therefore may be prohibitive for large-scale studies but they provide a genome-wide assessment of variation for SNPs, small- and large-scale deletions as well as copy number to a limited extent. There is still the problem of assaying only previously known sequences.

3.5. Genome Variation via Whole Genome Sequencing (WGS)

By far the most effective assessment of genome variation is whole genome sequencing. Here the issue is balancing the cost with the value of completeness. Only sequencing will reveal novel genes and sequences that were previously unknown. Several individual *S. cerevisiae* strains have been sequenced to near completeness using first generation Sanger sequencing since the reference genome was completed (15, 16) and http://www.broadinstitute.org/annotation/genome/saccharomyces_cerevisiae.3/Info.html. These have each provided insights into genome variation such as novel genes, introgressions from outside the species as well as frequencies and types of variation. Each of these was time consuming and costly and each suffers from incompleteness of the subtelomeres as well as the large tandem arrayed sequences such as the rDNAs. Although a great deal of information can be gleaned from these sequences such as some assessment of population structure, selection and adaptation, and human influence, the limited number of sequences does not represent the species as a whole nor can it really describe population structures which require many individuals within and between to accurately determine.

One way to increase numbers of individuals without the cost of whole genomes is to sequence a few genes from various locations in the genome. This has been used effectively to describe population structures in *Saccharomyces* yeasts (36, 37, 39). This type of analysis resolves some of the issues of population structure but in many cases new questions arise.

Another approach to increase the numbers of individuals without the cost of complete genomes is low level whole genome shotgun sequence coverage that has proven very effective at determining population structure and resolving many issues about selection and human influence (14). With the assumption that there are populations of related individuals, it is possible to determine global genome-wide phylogenetic relationships without having the complete genomes. Although the difficult regions of the genomes are still not resolved, some inferences concerning regions such as subtelomeres and large tandem arrays can be made. By surveying many strains at lower coverage, the discovery of novel genes may be more efficient than complete sequences of fewer strains as populations of related individuals will share these novel genes. The combined sequence coverage within a population reduces the chances of missing novel genes. Using this approach, all six known gene families not in the reference genome were found amongst 35 other *S. cerevisiae* strains as were the novel genes discovered in the individual near complete genomes recently sequenced. In addition 38 new genes/gene families were discovered (14). Despite most of these being in subtelomeric regions that were not assembled, the general composition of novel gene families, i.e. presence and distribution amongst strains/populations could be made. Further complementary analysis with specific probes and pulsed field gels will help with the structural analysis of these novel genes. Annotation of these genes will have the same problems as any novel potential genes and in this case comparative genomics would not be helpful, as the regions are not assembled.

In addition to novel gene discovery, population genomic sequence surveys provide evidence for sequence variation previously unknown. In the case of the rDNA array, it is generally assumed that every copy within an array has the same sequence though different strains can have different variant arrays. The population genomic survey of several strains revealed that in addition to the SNPs that varied between arrays in different strains, there were sequence differences between rDNA copies within arrays with significant frequencies (41). Despite the low coverage overall for the genomes in this survey, the coverage of rDNAs was substantial due to their large copy number. This allowed the determination of sequence variants with high accuracy both between and within arrays. What is not possible from this analysis is the determination of order of variants within an array.

The current state of the art for genome variation determination still includes microsatellite analysis (26–28), as well as microarrays (32, 34, 35), where genotyping by arrays is becoming quite sophisticated (29). The best determination, however, is still whole genome sequencing without which much information on variation is missing. Our current understanding of the population structure of *S. cerevisiae*, with several well-delineated populations and a large number of mosaic strains resulting from interbreeding between these populations (14), could not have been determined without population genomic sequencing.

4. Conclusions

Challenges and future prospects include the assembly and annotation of complex regions of the genome. This includes repetitive regions such as the rDNA arrays, for which we have seen some progress as described above, as well as subtelomeric regions, which we have seen contain many of the genes and gene families responsible for adaptive and phenotypic differences, yet are not well characterized. Second generation sequencing with increased depth of coverage in short timeframes will be a major part of the solution, yet brings with it the additional challenges of quantity and quality of the sequence reads as well as the length of reads available. These challenges in both gene annotation and genome variation will only be met by combined approaches utilizing new sequencing technologies as well as new informatic tools for assembly, comparison, and compilation/cross referencing of diverse data sets.

References

1. Fisk, D. G., Ball, C. A., Dolinski, K., et al. (2006) *Saccharomyces cerevisiae* S288C genome annotation: a working hypothesis. *Yeast* **23**, 857–865.
2. Christie, K. R., Hong, E. L., and Cherry, J. M. (2009) Functional annotations for the *Saccharomyces cerevisiae* genome: the knowns and the known unknowns. *Trends Microbiol.* **17**, 286–294.
3. Hong, E. L., Balakrishnan, R., Dong, Q., et al. (2008) Gene Ontology annotations at SGD: new data sources and annotation methods. *Nucleic Acids Res.* **36**, D577–581.
4. Kellis, M., Patterson, N., Endrizzi, M., Birren, B., and Lander, E. S. (2003) Sequencing and comparison of yeast species to identify genes and regulatory elements. *Nature* **423**, 241–254.
5. Liti, G., and Louis, E. J. (2005) Yeast evolution and comparative genomics. *Annu. Rev. Microbiol.* **59**, 135–153.
6. Cliften, P., Sudarsanam, P., Desikan, A., et al. (2003) Finding functional features in *Saccharomyces* genomes by phylogenetic footprinting. *Science* **301**, 71–76.
7. Cliften, P. F., Hillier, L. W., Fulton, L., et al. (2001) Surveying *Saccharomyces* genomes to identify functional elements by comparative DNA sequence analysis. *Genome Res.* **11**, 1175–1186.
8. Wolfe, K. H., and Shields, D. C. (1997) Molecular evidence for an ancient duplication of the entire yeast genome. *Nature* **387**, 708–713.
9. Byrne, K. P., and Wolfe, K. H. (2005) The yeast gene order browser: combining curated homology and syntenic context reveals gene fate in polyploid species. *Genome Res.* **15**, 1456–1461.
10. Byrne, K. P., and Wolfe, K. H. (2006) Visualizing syntenic relationships among the hemiascomycetes with the yeast gene order browser. *Nucleic Acids Res.* **34**, D452–455.
11. Fischer, G., James, S. A., Roberts, I. N., Oliver, S. G., and Louis, E. J. (2000) Chromosomal evolution in *Saccharomyces*. *Nature* **405**, 451–454.
12. Souciet, J., Aigle, M., Artiguenave, F., et al. (2000) Genomic exploration of the hemiascomycetous yeasts: 1. A set of yeast species for molecular evolution studies. *FEBS Lett.* **487**, 3–12.
13. Fischer, G., Neuveglise, C., Durrens, P., Gaillardin, C., and Dujon, B. (2001) Evolution of gene order in the genomes of two related yeast species. *Genome Res.* **11**, 2009–2019.
14. Liti, G., Carter, D. M., Moses, A. M., et al. (2009) Population genomics of domestic and wild yeasts. *Nature* **458**, 337–341.
15. Novo, M., Bigey, F., Beyne, E., et al. (2009) Eukaryote-to-eukaryote gene transfer events revealed by the genome sequence of the wine yeast *Saccharomyces cerevisiae* EC1118. *Proc. Natl. Acad. Sci. USA* **106**, 16333–16338.
16. Wei, W., McCusker, J. H., Hyman, R. W., et al. (2007) Genome sequencing and comparative analysis of *Saccharomyces cerevisiae* strain YJM789. *Proc. Natl. Acad. Sci. USA* **104**, 12825–12830.
17. Liti, G., Peruffo, A., James, S. A., Roberts, I. N., and Louis, E. J. (2005) Inferences of evolutionary relationships from a population survey of LTR-retrotransposons and telomeric-associated sequences in the *Saccharomyces sensu stricto* complex. *Yeast* **22**, 177–192.
18. Buhler, C., Borde, V., and Lichten, M. (2007) Mapping meiotic single-strand DNA reveals a new landscape of DNA double-strand breaks in *Saccharomyces cerevisiae*. *PLoS Biol.* **5**, e324.
19. Klingberg, T. D., Lesnik, U., Arneborg, N., Raspor, P., and Jespersen, L. (2008) Comparison of *Saccharomyces cerevisiae* strains of clinical and nonclinical origin by molecular typing and determination of putative virulence traits. *FEMS Yeast Res.* **8**, 631–640.
20. Lucena, B. T., Silva-Filho, E. A., Coimbra, M. R., Morais, J. O., Simoes, D. A., and Morais, M. A., Jr. (2007) Chromosome

instability in industrial strains of *Saccharomyces cerevisiae* batch cultivated under laboratory conditions. *Genet. Mol. Res.* **6**, 1072–1084.
21. Schuller, D., Pereira, L., Alves, H., Cambon, B., Dequin, S., and Casal, M. (2007) Genetic characterization of commercial *Saccharomyces cerevisiae* isolates recovered from vineyard environments. *Yeast* **24**, 625–636.
22. Valero, E., Cambon, B., Schuller, D., Casal, M., and Dequin, S. (2007) Biodiversity of *Saccharomyces* yeast strains from grape berries of wine-producing areas using starter commercial yeasts. *FEMS Yeast Res.* **7**, 317–329.
23. Althoff, D. M., Gitzendanner, M. A., and Segraves, K. A. (2007) The utility of amplified fragment length polymorphisms in phylogenetics: a comparison of homology within and between genomes. *Syst. Biol.* **56**, 477–484.
24. Lopandic, K., Gangl, H., Wallner, E., et al. (2007) Genetically different wine yeasts isolated from Austrian vine-growing regions influence wine aroma differently and contain putative hybrids between *Saccharomyces cerevisiae* and *Saccharomyces kudriavzevii*. *FEMS Yeast Res.* **7**, 953–965.
25. MacKenzie, D. A., Defernez, M., Dunn, W. B., et al. (2008) Relatedness of medically important strains of *Saccharomyces cerevisiae* as revealed by phylogenetics and metabolomics. *Yeast* **25**, 501–512.
26. Goddard, M. R., Anfang, N., Tang, R., Gardner, R. C., and Jun, C. (2009) A distinct population of *Saccharomyces cerevisiae* in New Zealand: evidence for local dispersal by insects and human-aided global dispersal in oak barrels. *Environ. Microbiol.* **12**, 63–73.
27. Legras, J. L., Merdinoglu, D., Cornuet, J. M., and Karst, F. (2007) Bread, beer and wine: *Saccharomyces cerevisiae* diversity reflects human history. *Mol. Ecol.* **16**, 2091–2102.
28. Richards, K. D., Goddard, M. R., and Gardner, R. C. (2009) A database of microsatellite genotypes for *Saccharomyces cerevisiae*. *Antonie Van Leeuwenhoek* **96**, 355–359.
29. Bourgon, R., Mancera, E., Brozzi, A., Steinmetz, L. M., and Huber, W. (2009) Array-based genotyping in *S. cerevisiae* using semi-supervised clustering. *Bioinformatics* **25**, 1056–1062.
30. Dunn, B., Levine, R. P., and Sherlock, G. (2005) Microarray karyotyping of commercial wine yeast strains reveals shared, as well as unique, genomic signatures. *BMC Genom.* **6**, 53.
31. Dunn, B., and Sherlock, G. (2008) Reconstruction of the genome origins and evolution of the hybrid lager yeast *Saccharomyces pastorianus*. *Genome Res.* **18**, 1610–1623.
32. Gresham, D., Ruderfer, D. M., Pratt, S. C., et al. (2006) Genome-wide detection of polymorphisms at nucleotide resolution with a single DNA microarray. *Science* **311**, 1932–1936.
33. Primig, M., Williams, R. M., Winzeler, E. A., et al. (2000) The core meiotic transcriptome in budding yeasts. *Nat. Genet.* **26**, 415–423.
34. Schacherer, J., Ruderfer, D. M., Gresham, D., Dolinski, K., Botstein, D., and Kruglyak, L. (2007) Genome-wide analysis of nucleotide-level variation in commonly used *Saccharomyces cerevisiae* strains. *PLoS One* **2**, e322.
35. Schacherer, J., Shapiro, J. A., Ruderfer, D. M., and Kruglyak, L. (2009) Comprehensive polymorphism survey elucidates population structure of *Saccharomyces cerevisiae*. *Nature* **458**, 342–345.
36. Aa, E., Townsend, J. P., Adams, R. I., Nielsen, K. M., and Taylor, J. W. (2006) Population structure and gene evolution in *Saccharomyces cerevisiae*. *FEMS Yeast Res.* **6**, 702–715.
37. Fay, J. C., and Benavides, J. A. (2005) Evidence for domesticated and wild populations of *Saccharomyces cerevisiae*. *PLoS Genet.* **1**, 66–71.
38. Fay, J. C., and Benavides, J. A. (2005) Hypervariable noncoding sequences in *Saccharomyces cerevisiae*. *Genetics* **170**, 1575–1587.
39. Liti, G., Barton, D. B., and Louis, E. J. (2006) Sequence diversity, reproductive isolation and species concepts in *Saccharomyces*. *Genetics* **174**, 839–850.
40. Louis, E. J. (1998) Whole chromosome analysis. In Brown, A. J. P. and Tuite, M. F. (eds.), *Methods in Microbiology 26: Yeast Gene Analysis*, pp. 15–31. San Diego: Academic.
41. James, S. A., O'Kelly, M. J., Carter, D. M., Davey, R. P., van Oudenaarden, A., and Roberts, I. N. (2009) Repetitive sequence variation and dynamics in the ribosomal DNA array of *Saccharomyces cerevisiae* as revealed by whole-genome resequencing. *Genome Res.* **19**, 626–635.

Chapter 3

Genome-Wide Measurement of Histone H3 Replacement Dynamics in Yeast

Oliver J. Rando

Abstract

Chromatin plays critical roles in processes governed in different timescales – responses to environmental changes require rapid plasticity, while long-term stability through multiple cell generations requires epigenetically heritable chromatin. Understanding the dynamic behavior of chromatin is of great interest for fields ranging from transcriptional regulation through meiosis and gametogenesis. Here, we describe a protocol for measuring histone replacement rates genome wide in the budding yeast *Saccharomyces cerevisiae*. With suitable modifications, this protocol could be applied to other organisms, or to replacement dynamics of other DNA-associated proteins.

Key words: Histone H3 replacement, *Saccharomyces cerevisiae*, budding yeast, epigenome, epigenetics, chromatin.

1. Introduction

Chromatin plays critical roles in processes governed in different timescales – responses to environmental changes require rapid plasticity, while long-term stability through multiple cell generations requires epigenetically heritable chromatin. It is of great interest, therefore, to understand the nature of dynamic chromatin states in living cells.

In bulk studies, histones are among the most stably bound DNA-associated proteins. Nonetheless, a number of recent experiments have identified loci where histones are evicted and replaced by different histone molecules in the absence of genomic replication. Furthermore, during DNA replication, histones from the mother chromosome transiently dissociate from the chromosome, raising the question of how chromatin states are stably

J.I. Castrillo, S.G. Oliver (eds.), *Yeast Systems Biology*, Methods in Molecular Biology 759,
DOI 10.1007/978-1-61779-173-4_3, © Springer Science+Business Media, LLC 2011

maintained in the face of the perturbations encountered during genomic replication. The converse question may also be asked – even in the absence of genomic replication, how stable are chromatin states? While most studies paint a picture of stable chromatin states, is this a consequence of a static system in which the same molecules persist, or a consequence of active maintenance by dynamic processes? This question is central to understanding the processes that modify chromatin state, and to understanding the molecular nature of stably heritable chromatin states.

Histones provide a formidable barrier to the elongation of RNA polymerase (1, 2). In yeast, passage of RNA polymerase displaces nucleosomes (assessed by H3 occupancy), which are rapidly reassembled after transcription is shut down (3). In *Drosophila*, RNA polymerase causes exchange of the canonical histone H3 (H3.1) for the "replication-independent" H3 variant H3.3, and H3.3 continues to turn over as long as a coding region is being actively transcribed (4–6). Single-gene and whole-genome studies in yeast have shown that H3 is removed from promoters upon gene activation (7–11, 12). Reassembly of histones at the *PHO5* promoter after shutdown of transcription occurs *in trans*, in the sense that the histones that are reassembled on the promoter are not the same histones that were removed during gene activation (13). Intriguingly, H3 replacement is slow or unmeasurable over silenced coding regions (14–17). H2A/H2B exchange is globally much more widespread than H3 exchange, though it too is diminished over silenced loci.

Detailed understanding of high-resolution chromatin dynamics requires a rapid assay for the determination of chromatin structure over genomic scales. Using genome-wide tiling microarrays coupled with inducible expression of epitope-tagged histones, the dynamics of replication-independent H3 exchange across the yeast genome has been characterized (17, 18). These methods enable the rapid assay of global histone dynamics at single-nucleosome resolution. Below, a protocol for the genome-wide measurement of H3 replacement rates in budding yeast is presented.

2. Materials

2.1. Combined Micrococcal Nuclease and ChIP (MNase-ChIP)

1. YPRaffinose medium (YPRaff): Yeast extract (1%, w/v)/peptone (1%, w/v)/raffinose (2%, w/v) medium.
2. PBS buffer: Phosphate-buffered saline, pH 7.4.
3. Buffer L: 50 mM HEPES–KOH, pH 7.5, 140 mM NaCl, 1 mM EDTA, 1% (w/v) Triton X-100, 0.1% (w/v) sodium deoxycholate.

4. NP buffer: 0.5 mM Spermidine, 1 mM β-mercaptoethanol (β-ME), 0.075% (w/v) Tergitol-type NP-40 detergent (NP-40), 50 mM NaCl, 10 mM Tris–HCl pH 7.4, 5 mM $MgCl_2$, 1 mM $CaCl_2$. Preparation of 5 mL of NP buffer (made fresh on the day of the experiment): 10 μL of 250 mM spermidine, 3.5 μL of 1:10 (v/v) (diluted in water) β-ME, 37.5 μL of 10% (w/v) NP-40. Bring up to 5 mL with NP buffer.
5. Buffer W1: Buffer L with 500 mM NaCl.
6. Buffer W2: 10 mM Tris–HCl, pH 8.0, 250 mM LiCl, 0.5% (w/v) NP-40, 0.5% (w/v) sodium deoxycholate, 1 mM EDTA.
7. Buffer Z: 1 M Sorbitol, 50 mM Tris–HCl, pH 7.4. Preparation of 1 L of buffer Z: 500 mL of 2 M sorbitol, 50 mL of 1 M Tris–HCl, pH 7.4, 450 mL dH_2O.
8. TE buffer: 10 mM Tris–HCl, pH 8.0, 1 mM EDTA.
9. (2×) Proteinase K solution: TE with 0.8 mg/mL glycogen, 2 mg/mL proteinase K.
10. Elution buffer: TE, pH 8.0, with 1% (w/v) SDS, 150 mM NaCl, and 5 mM DTT (note: do not add the DTT until just before use).
11. Zymolyase solution (10 mg/mL in buffer Z; lasts up to 2 weeks at 4°C).
12. Micrococcal Nuclease (Worthington Biochem): Resuspended from lyophilized powder at 20 U/μL in Tris–HCl, pH 7.4. Aliquot into tubes upon first use and store at –80°C.
13. Phenol/chloroform/isoamyl alcohol (PCI) (24:25:1) (v/v).
14. Yellow (heavy) Phase Lock Gel (PLG) tubes (Manual Phase Lock Gel™, PLG, 5Prime GmbH).
15. Protein A beads (Sigma).
16. Isopropanol (100%, v/v).
17. Sigma protease inhibitor cocktail.
18. NEB buffer 3 (New England Biolabs).
19. Calf intestinal phosphatase (CIP) (New England Biolabs, #M0290S).
20. QIAGEN Min Elute Enzyme Cleanup Kit.

2.2. DNA Linear Amplification

1. Calf intestinal phosphatase (CIP) (New England Biolabs).
2. 10× NEB buffer 3 (New England Biolabs).
3. TdT (NEB #M0252S or #M0252L).

4. 5× TdT buffer: 1 M Potassium cacodylate, 125 mM Tris–HCl, pH 6.6, 1.25 mg/mL BSA (Roche).
5. 8% Dideoxynucleotide tailing solution: 92 μM dTTP, 8 μM ddCTP (Invitrogen).
6. Ambion T7 Megascript Kit (Ambion, #1334).
7. DNA polymerase I Klenow fragment (New England Biolabs).
8. NEB buffer 2 (New England Biolabs).
9. QIAGEN RNeasy Mini Kit.
10. β-Mercaptoethanol solution (14.2 M).

2.3. Amplified Antisense RNA (aRNA) Labeling and Cleanup

1. 5 μg/μL Anchored oligodT (22 mer; Integrated DNA Technology, IDT).
2. Cy3 and Cy5 dyes (Amersham): Dye comes dried and should be resuspended in 11 μL of DMSO. This amount of material is sufficient for three labeling runs. If entire amount will not be used, make 3 μL aliquots. Dry down in SpeedVac vacuum concentrator and store at −20°C.
3. (50×) aa-dUTP mixture: Dissolve 1 mg amino-allyl dUTP (Sigma) with the following: 32.0 μL of 100 mM dATP (12.5 mM, final), 32.0 μL of 100 mM dGTP (12.5 mM, final), 32.0 μL of 100 mM dCTP (12.5 mM, final), 12.7 μL of 100 mM dTTP (5 mM, final), 19.3 μL dH_2O.
4. Superscript II (Invitrogen), with SSII buffer and DTT.
5. RNasin (Promega).
6. NaOH solution (1 M).
7. EDTA solution (0.5 M).
8. HEPES solution (1 M, pH 7.5).
9. Sodium bicarbonate solution (50 mM, pH 9.0).
10. QIAGEN Min Elute Enzyme Cleanup Kit (# 28204).

2.4. Microarray Hybridization

1. Cot-1 DNA. Commercially available at 1 mg/mL. Prior to hybridization, concentrate Cot-1 DNA from 1 to 10 μg/μL in a SpeedVac vacuum concentrator.
2. Yeast tRNA (10 μg/μL).
3. PolyA RNA (10 μg/μL).
4. (20×) SSC solution: 0.3 M Trisodium citrate and 3.0 M sodium chloride, pH 7.0.
5. 10% (w/v) Sodium dodecyl sulfate (SDS).
6. 1 M HEPES–KOH (pH 7.0).

7. Microarrays and hybridization chambers (from microarray suppliers such us Agilent, Nimblegen).
8. Axon GenePix scanner (Axon).

2.5. Yeast Strains and Cultivation

The yeast strain required for this experiment carries an inducible copy of an epitope-tagged histone H3. Specifically, we utilize a strain carrying a plasmid with the *GAL1-10* promoter driving expression of both untagged H4 and an N-terminally Flag-tagged H3. In our original published work (17), we used a strain that also carried a constitutively expressed Myc-tagged H3, but in subsequent work we found the Myc tag to be unnecessary (19). We also find it useful to delete the gene encoding the Bar1 protease that degrades the alpha factor from the yeast strain to be used, thus enabling G1 arrest by alpha factor treatment with much lower concentrations of the factor.

A typical experiment involves growth of the yeast strain (carrying a deletion in a chromatin regulator of interest if desired) to mid-exponential phase in YPRaffinose medium. Yeast are arrested with alpha factor (we recommend carrying out a titration in your strain background of interest) for 3 h (more for slow-growing strains). Galactose is added to the medium to a final concentration of 1% (w/v), and at varying times after galactose addition (15, 30, 45, 60, 90, 120, 150 min), mononucleosome-resolution ChIP is carried out as described below. After amplification, Flag ChIP is competitively hybridized against mononucleosomal "input" on whole-genome tiling microarrays.

Below, we provide detailed protocols for four steps:

(1) Nucleosome-resolution ChIP (**Section 3.1**).

(2) Linear amplification (**Section 3.2**).

(3) Labeling for microarrays (**Section 3.3**).

(4) Hybridization to microarrays (**Section 3.4**).

3. Methods

3.1. Combined Micrococcal Nuclease and ChIP Protocol (MNase-ChIP)

Before doing MNase-ChIP, an MNase titration series is needed to determine the proper amount of MNase enzyme to be used. For this purpose, follow the protocol below as follows (starting from day 1) stopping at the end of the MNase digestion and following these steps:

1. Add 150 μL of STOP buffer (0.25 M EDTA, 5% (w/v) SDS).
2. Proteinase K treatment overnight at 65°C. You may use extra-high concentrated proteinase K (20 mg/mL) with no glycogen.

3. Extract with phenol/chloroform/isoamyl alcohol (PCI) reagent once using yellow (heavy) Phase Lock Gel (PLG) tubes.
4. Precipitate with (0.1×) 3 M volume sodium acetate and 1 volume isopropanol.
5. Spin immediately for 10 min at maximum speed (>12,000 × *g*).
6. Wash the pellet in 70% (v/v) ethanol.
7. Resuspend in NEB buffer 3.
8. RNase treatment for 1 h at 37°C.
9. Run gel to determine proper titration point.

3.1.1. Day 1 – Yeast Culture

1. Inoculate starter culture with the cells of interest (full-size colony). Do this 1–2 days in advance.
2. Inoculate 400–550 mL of YPRaffinose medium in a 2-L baffled flask and grow culture shaking at 220 rpm at desired temperature. Target concentration is 8–9 × 10^6 cells/mL. This yields sufficient material for 4–6 immunoprecipitations (IPs).

3.1.2. Day 2 – Cross-linking and MNase Digest

1. Early afternoon: Grow culture to the desired cell density, arrest with alpha factor, and add galactose to a final concentration of 1% (w/v) for the desired amount of time. Add formaldehyde to 1% (w/v) final concentration (*see* volumes below) and shake (~200 rpm) for 15 min.

Cell culture volume (mL)	Formaldehyde (mL)	Glycine (mL)
400	10.7	20.5
450	12.0	23.0
500	13.2	25.3
550	14.8	27.5

2. To quench the formaldehyde, add 2.5 M glycine according to the table above, shake while adding, and then shake or let stand at room temperature for 5 min. Cells can now be left on ice as long as necessary (for example, to collect more samples from a time course experiment).
3. Transfer to 0.5- or 1-L centrifuge bottle and spin down at >3,000 × *g* for 5 min at 4°C.
4. Wash once in 50 mL volume of Milli-Q water (ddH_2O).
5. Resuspend the cell pellet (from the 400–550 mL culture) in 39 mL buffer Z.

6. Add 28 μL of 14.3 M β-ME (final concentration 10 mM) and vortex cells to resuspend.
7. Add 1 mL of zymolyase solution and incubate at 28–30°C, shaking, on tilted tube rack for 35–40 min (time will depend on the assessment of spheroplasting efficiency – *see* below).
8. Spin zymolyase-treated yeast (now spheroplasts) at 5,000 × *g* for 10 min at 4°C:
   (a) During spin, if doing a titration, aliquot the MNase to 4–6 Eppendorf tubes per set. Pipette a dot of MNase on the side of the tube (directly under lid hinge). Suggested concentration range:
      (i) BY4741: 1–10 μL for four aliquots. Alternatively, increase number of aliquots with less amount of enzyme.
   (b) MNase can be subject to three freeze-thaw cycles (–20°C). Store at –20°C after first freeze-thaw.
9. Aspirate the supernatant and resuspend each aliquot in a total of (600 μL × # aliquots + 50 μL) NP buffer by pipetting up and down several times. Cells will be sticky, viscous, and thick. If cells do not stick together in chunks but dissociate easily from each other, spheroplasting may be incomplete.
10. Add 600 μL of cells to each Eppendorf with MNase. Add the cells to the tubes directly on the spot of nuclease, and try to be even and quick between tubes. Start with the most diluted sample and use the same 1,000-μL tip to add to each successive tube to minimize tip changing times.
11. As soon as all the cells have been added, start the timer, close the tubes, invert them once to mix, and incubate at 37°C (in water bath) for 20 min.
12. Stop the reaction after 20 min by transferring the tubes to an ice bath (4°C). This will halt the enzyme activity. Finally, add 12 μL of 0.5 M EDTA (10 mM end concentration) per 600 μL aliquot to inactivate the enzyme.

3.1.3. Day 2 – Immunoprecipitation

1. Prepare protein A beads. Each immunoprecipitation (IP) aliquot takes two protein A bead aliquots, but one of the aliquots will not be used until the following morning. If preparing from a new vial, the beads are 5 mL of slurry in a total volume of 6.3 mL, with the supernatant consisting of 20% (v/v) ethanol. Add 3.7 mL of 20% ethanol to bring up to 10 mL of 50% slurry. This is equivalent to adding 3 mL of water and 750 μL of 100% ethanol:
   a. Pipette 700 μL of slurry per eight IP aliquots into a 1.5-mL tube (enough for the preclear; you will need another

700 μL for eight IP aliquots for the pull-down the following day).

b. Spin down at 3,000 × *g* in microcentrifuge for 1 min.

c. Return the supernatant to stock buffer if slurry is thick, otherwise discard.

d. Wash beads twice in buffer L.

e. Distribute the protein A beads equally to eight 1.5-mL tubes (truncate a pipette tip with a razor blade and use that for easier pipetting, though you may need to dial down the volume, since the pipetman will now draw up slightly more volume).

2. Pool the digested material and add the following to the digestion products to simulate buffer L conditions. Important: Add the salts before the detergents. Amounts below are per 600 μL digestion aliquot; scale accordingly. One IP aliquot will be ~800 μL afterward.

Volume (μL)	**Component**
80	0.5 M HEPES–KOH, pH 7.5
22.4	5 M NaCl
6.4	12.5% (w/v) Sodium deoxycholate
80	10% (w/v) Triton X-100
8	Sigma protease inhibitor cocktail

3. Set aside at least 80 μL (~5%) of pool as non-IP "input"; needed also for gel verification of the MNase digest. This should yield about 0.75–1 μg of DNA in the end.

4. Add adjusted digest product (~800 μL per IP aliquot) to one protein A bead aliquot (pre-equilibrated in buffer L) and rotate at 4°C on tube rotisserie for 1 h to remove proteins that bind nonspecifically to the beads.

5. Spin for 30 s at 3,000 × *g* at 4°C (to avoid MNase reactivation), and transfer the supernatant to another tube containing the appropriate amount of antibody.

6. Incubate with rotation at 4°C for 4 h overnight (up to 16 h).

7. Spin for 30 s at 3,000 × *g* at 4°C and transfer to a tube containing another protein A bead aliquot.

8. Incubate with rotation at 4°C for 1 h (longer is allowable but not necessary).

9. Spin for 30 s at 3,000 × *g* at 4°C and for subsequent pelleting steps in washes.

10. Wash beads successively with 1 mL of the following buffers, for 5 min (on rotisserie at 4°C) each, in the following order: Buffer L (twice), W1 (twice), W2 (twice), and TE, pH 8.0 (twice).

3.1.4. Day 2 – Elution

1. Incubate the beads in 125 μL of elution buffer at 65°C for 10 min with frequent mixing (vortex for a few seconds every 3 min). Be sure to add the DTT in the elution buffer (to 5 mM final concentration) beforehand.
2. Pellet the beads by centrifugation at 10,000 × *g* for 2 min and keep the supernatant.
3. Repeat Steps 1 and 2 above and discard the protein A beads when done.

3.1.5. Day 2 – Reverse Cross-links

1. Add 0.5 volume of (2×) proteinase K solution to both the ChIP input and the eluted samples. (ChIP input samples can receive 10 μL of 20 mg/mL proteinase K instead.)
2. Overlay input and eluted samples with mineral oil (only if volume is <150 μL) and incubate at 65°C overnight.

3.1.6. Day 3 – Protein Degradation and DNA Purification

1. Cool the samples to room temperature and spin down (> 12,000 × *g*).
2. Extract with 1 volume phenol (once), then with 1 volume chloroform/isoamyl alcohol (25:1) (once) using chloroform only (24:25:1 PCI is also acceptable). For the first extraction (all at room temperature), vortex for 10 s, transfer to Phase Lock Gel tube, and then spin for 5 min at maximum speed (>12,000*g*). For subsequent extractions, pipette volume up and down to mix it. The volume will be approximately 400 μL at this stage. Use light (green) Phase Lock Gel tubes for phenol and heavy (yellow) or light gel tubes for chloroform/isoamyl alcohol (extracting twice with 24:25:1 PCI is also acceptable).
3. Add 0.1 volume of 3.0 M sodium acetate, pH 5.3, and 2.5 volumes of ice-cold 100% ethanol.
4. Allow DNA to precipitate overnight at –20°C (to maximize yield).

3.1.7. Day 4 – DNA Purification

1. Pellet the DNA by centrifugation at 14,000 × *g* for 15 min at 4°C.
2. Wash once with cold 70% (v/v) ethanol and spin at 14,000 × *g* for 5 min at 4°C.
3. Aspirate and allow the pellet to dry.
4. Resuspend the pellet in 29 μL of NEB buffer 3 with 0.5 μg of DNase-free RNase A (0.5 mg/mL).

5. Incubate at 37°C for 1 h (can freeze at –20°C at this point).
6. Add 0.75 μL of CIP and leave at 37°C for 1 h. Scale up volume (and enzyme) if more than 50 ng/μL is expected.
7. Clean up the reaction with QIAGEN MinElute Enzyme Cleanup Kit, with elution volume of 20 μL.

3.2. DNA Linear Amplification

While many amplification protocols have been successfully used for ChIP material in genomic studies, it has been shown that for material <500 bp, many extant protocols introduce biased genomic representation. The protocol here is superior to other tested protocols for material of this size, which of course includes mononucleosomal DNA.

3.2.1. Calf Intestinal Phosphatase Treatment of Samples with Terminal 3′ Phosphate Groups

Reagent (for every 10 μL volume)	Volume (μL)	Final concentration
2.5 U CIP enzyme (NEB #M0290S)	0.25	0.25 U/ μL
(10×) NEB buffer 3	1	(1×)
Template DNA (max. 500 ng per 10 μL)	8.75	Max. 50 ng/μL
Total	10	

Each reaction can be scaled up to 100 μL per tube

1. Incubate at 37°C for 1 h.
2. Clean up the reaction with the QIAGEN MinElute Kit (*see* **Section 3.2.6**) Elute in 20 μL.

3.2.2. Tailing Reaction with TdT

Reagent	Volume (μL)	Final concentration
(5×) TdT buffer (*see* **Note 1**)	2	(1×)
8% Dideoxynucleotide tailing solution (*see* **Note 2**)	0.5	5 μM dTTP (8% ddCTP)
5 mM $CoCl_2$	1.5	0.75 mM
Template DNA (max. 75 ng) (*see* **Note 3**)	5	Max. 7.5 ng/μL
20 U TdT enzyme (add this at the end) (*see* **Note 4**)	1	2 U/μL
Total	10	

1. Add 1–2 drops of mineral oil to the top of the mixture, to prevent evaporation loss during incubation.
2. Incubate at 37°C for 20 min.
3. Stop the reaction by adding 2 μL (per 10 μL reaction volume) of 0.5 M EDTA, pH 8.0.

4. Clean up the reaction with the QIAGEN MinElute Kit. If started with 10 μL, it may be necessary to add 10 μL of ddH_2O to bring the volume up to 20 μL. Elute in 20 μL volume.

3.2.3. Second-Strand Synthesis with Klenow Fragment Polymerase

Reagent	Volume (μL)	Final concentration
25 μM T7-A_{18}B primer (*see* **Note 5**)	0.3	300 nM
(10×) NEB buffer 2 (*see* **Note 6**)	2.5	(1×)
5.0 mM dNTP mix (*see* **Note 7**)	1	200 μM
ddH_2O	0.2	
T-tailed DNA	20	
Total	24	

Important: Do not use mineral oil. Trace amounts of mineral oil appear to interfere with cleanup and in vitro transcription (IVT).

Use the following program in a thermal cycler:

Start: 94°C, 2 min to melt; ramp –1°C/s to 35°C, then hold for 2 min to anneal.

Ramp –0.5°C/s to 25°C; hold for 45 s (pause here for as long as needed, up to 6 min).

Add 1 μL of (5 U) Klenow DNA polymerase (NEB# M0210S) during this time and spin down condensation on tube if necessary; 37°C, 90 min to extend. Keep at 4°C to halt enzyme activity until removal of reaction tubes out of the cycler.

Stop the reaction by adding 2.5 μL of 0.5 M EDTA, pH 8.0 (end concentration 50 mM).

Clean up the reaction with the QIAGEN MinElute Kit and elute in 20 μL.

3.2.4. In Vitro Transcription (IVT)

Double-stranded DNA (dsDNA) preparation: The in vitro transcription (IVT) requires that the dsDNA be in 8 μL volume. Dry down the eluate from 20 to 8 μL in a SpeedVac vacuum concentrator for 10–12 min (drying rate approximately 1 μL/min).

Reagent (from Ambion T7 Megascript Kit, #1334)	Volume (μL)
75 mM NTP mix (A, G, C, and UTP) (*see* **Note 8**)	8
Reaction buffer (warm to RT first) (*see* **Note 9**)	2
Enzyme mix (RNase inhibitor and T7 RNA Pol)	2
Template dsDNA	8
Total	20

Incubate at 37°C overnight (5–20 h). Incubate in thermal cycler with heated lid, or in air incubator, and in 0.2-mL RNase-free PCR tubes to minimize vapor volume.

3.2.5. Amplified Antisense RNA (aRNA) Purification Using QIAGEN RNeasy Columns

Use the QIAGEN RNeasy Mini Kit (#74104) for this procedure. You may eventually have to order additional buffer RPE if you use the entire kit for this purpose.

Preparation of master mix:

Reagent	Volume (μL)
β-ME (14.2 M stock solution)	3.5
RNase-free water	80
Buffer RLT (from QIAGEN RNeasy Mini Kit)	350
Total	433.5

1. Pre-aliquot the mix to 1.5-mL RNase-free tubes.
2. Transfer the contents of the IVT mix to the RNase/DNase-free tube and vortex gently and briefly.
3. Add 250 μL of ethanol (95–100%, v/v) and mix well by pipetting. Do not spin here.

   If you have a QIAGEN vacuum manifold, skip to Step 8.

   Without vacuum manifold:
4. Apply the sample (700 μL) to RNeasy mini spin column sitting in a collection tube. Centrifuge for 15 s at ≥8,000 × *g*. Discard the flow-through.
5. Transfer RNeasy column to a new 2-mL collection tube (supplied). Add 500 μL of buffer RPE (which must contain ethanol) and centrifuge for 15 s at ≥8,000 × *g*. Discard the flow-through but reuse the tube.
6. Pipette 500 μL of buffer RPE (included with RNeasy Kit) onto RNeasy column and centrifuge for 2 min at >12,000 × *g*.
7. Remove the flow-through and pipette another 500 μL of buffer RPE onto column. Centrifuge for 2 min at >12,000 × *g* (this is an additional wash that is not in the QIAGEN protocol which we have found necessary because of guanidine isothiocyanate (GITC) contamination in the eluted RNA). Skip to Step 12.

   With vacuum manifold:
8. Apply the sample (700 μL) to RNeasy mini spin column, attached to vacuum manifold. Apply vacuum to the mini spin column.
9. Remove vacuum and pipette 500 μL of buffer RPE onto RNeasy column. Apply vacuum.

10. Repeat Step 9 and transfer columns to 2-mL collection tubes (supplied). Spin for 1 min at >12,000 × *g*.
11. Return columns to vacuum manifold and apply 500 μL of buffer RPE. Apply vacuum.
12. Transfer columns back to the 2-mL tubes. Spin at >12,000 × *g* for 1 min to completely dry column.
13. Transfer RNeasy column into a new 1.5-mL collection tube (supplied) and add 30 μL RNase-free water directly onto membrane. Centrifuge for 1 min at ≥ 8,000 × *g* to elute. Repeat if expected yield is ≥ 30 μg.
14. Check RNA concentration and quality by measuring A_{260} and A_{260}/A_{280}.

3.2.6. QIAGEN MinElute Kit Protocol

The MinElute kit removes free nucleotides below 40 nucleotides (nt). Fragments between 40 and 70 nt might be removed but with a lower efficiency:

1. In a sample of volume 20–100 μL, add 300 μL buffer ERC and mix thoroughly. If the sample is in less than 20 μL, bring up volume with ddH_2O. If the sample is in more than 100 μL, split sample and do in parallel.
2. Add sodium acetate (pH 5.0) if the buffer color is orange or purple (i.e., pH > 7.5). If the buffer is yellow, no additional sodium acetate is necessary.
3. Apply sample to column. Spin for 1 min at >12,000 × *g* in microcentrifuge.
4. Discard the flow-through and add 750 μL buffer PE (which must contain ethanol). Spin for 1 min at >12,000 × *g* in microcentrifuge.
5. Discard the flow-through. Spin for 1 min at >12,000 × *g* in microcentrifuge to dry the column.
6. Transfer the columns to fresh 1.5-mL tubes. Pipette 10–20 μL buffer EB or ddH_2O directly onto column membranes. Let stand for 1 min, then spin for 1 min at >12,000 × *g* in microcentrifuge to elute.
7. *Note on elution volumes*: When working with amounts less than 100 ng of DNA, the 10 μL elution volume in QIAGEN MinElute protocol may recover less than the 80% claimed by QIAGEN. Increase elution volume to 15–20 μL and dry the volume down if necessary. At the post-second-strand synthesis step (**Section 3.2.3**), an elution volume of 20 μL increases yields by 30–40% for a 50 ng sample.
8. *Note on second-strand synthesis with limiting primer amounts*: Limiting amount of primer is highly advisable when amplifying from very small amounts of starting material. Not only will it decrease the amount of primer–dimer product, but

also it may increase the yield of the desired amplification product. The table below shows what single reaction volumes have to be used for a specific amount of starting material.

DNA (ng)	T7 primer (μL)[a]	NEB buffer 2 (μL)	5 mM dNTPs (μL)	Water (μL)	Tailed DNA (μL)	Klenow (μL)	Total volume (μL)
>75	0.60 (25 μM)	5.0	2.0	20.4	20.0	2.0	50
50–75	0.30 (25 μM)	2.5	1.0	0.20	20.0	1.0	25
25	0.15 (25 μM)	2.5	1.0	0.35	20.0	1.0	25
10[a]	1.50 (1 μM)	1.0	0.4	0.20	6.5	0.4	10
5[a]	0.75 (1 μM)	1.0	0.4	0.95	6.5	0.4	10
2.5[a]	0.38 (1 μM)	1.0	0.4	1.32	6.5	0.4	10

[a]If a thermal cycler without a heated lid is used, spin down tubes every 30 min during the 37°C incubation step.

Tailed DNA will have to be dried down in a vacuum centrifuge to the indicated volume for reaction volumes with 10 μL total volume.

3.3. Amplified Antisense RNA (aRNA) Labeling and Cleanup

3.3.1. Reverse Transcription to Prepare aa-dUTP-Labeled cDNAs

(1) Combine 2 μg RNA and dH_2O to a final volume of 14.5 μL.

(2) Add 1 μL of 5 μg/μL anchored oligodT (22 mer; Integrated DNA Technology, IDT). Mix.

(3) Heat at 70°C for 10 min. Snap cool on ice for 10 min.

(4) Quick spin to bring down condensation.

(5) At room temperature, add 14.5 μL of master mix, which has been assembled just before use, adding enzymes at the end.

Master mix:

6.0 μL	(5×) SS II RT buffer
3.0 μL	0.1 M DTT
0.6 μL	(50×) aa-dUTP mix (*see* below)
2.0 μL	dH_2O
1.9 μL	SS II RT
1.0 μL	RNasin

(6) Pipette to mix.

(7) Incubate at 42°C for 2 h.

(8) Incubate at 95°C for 5 min. Snap cool on ice.

(9) Add 13 μL of 1 M NaOH and 1 μL of 0.5 M EDTA to hydrolyze RNA. Mix, quick spin.

(10) Incubate at 67°C for 15 min.

(11) Neutralize the reaction with 50 μL of 1 M HEPES, pH 7.5. Vortex. Quick spin.

(12) Purify the reaction over Zymo column to remove unincorporated nucleotides as follows.

(13) Add 1 mL of binding buffer to reaction. Mix. Load half of this material onto the column. Spin for 10 s at >12,000 × *g*. Discard the flow-through.

(14) Add remainder of the reaction. Spin again for 10 s at >12,000 × *g*. Discard the flow-through.

(15) Add 200 μL of wash buffer to each column. Spin for 30 s at >12,000 × *g*.

(16) Add another 200 μL wash buffer to each column. Spin for 1 min at >12,000 × *g*. Discard the flow-through and spin for 1 min at >12,000 × *g*.

(17) Add 10 μL of 50 mM sodium bicarbonate, pH 9.0, to the filter. Incubate for 5 min at room temperature. Spin for 30 s at >12,000 × *g* to elute cDNA.

(18) Couple aa-dUTP-incorporated cDNAs with dye by adding the eluted material directly to 3 μL Cy5 dye (Amersham) resuspended in DMSO. Incubate at room temp. for 1 h overnight shielded from light.

3.3.2. Cleanup

(1) Keep Cy5 and Cy3 separate for MinElute cleanup (QIAGEN Cat. #28004) to measure CyDye incorporation, if desired. If using thin coverslips with small probe volume, you may need to purify Cy3 and Cy5 together or use a SpeedVac vacuum concentrator to reduce volume after elution.

(2) Add 600 μL of buffer PB (binding buffer) to each sample.

(3) Assemble the MinElute column on the provided 2-mL collection tubes.

(4) Add the entire 720 μL to a MinElute column.

(5) Centrifuge for 1 min at 10,000 × *g*. Discard the flow-through and reuse 2-mL tube.

(6) Add 750 μL of wash buffer PE to the column.

(7) Centrifuge at 10,000 × *g* for 1 min. Discard the flow-through and reuse 2-mL tube.

(8) Centrifuge again at >12,000 × *g* for 1 min to remove residual ethanol.

(9) Place column in a fresh 1.5-mL tube. Add 10 μL of dH_2O to elute.

(10) Allow elution buffer to stand for at least 2 min before spinning.

(11) Centrifuge at >12,000 × *g* for 1 min. Add 10 μL of H_2O to elute.

(12) Allow elution buffer to stand for at least 2 min before spinning.

(13) Centrifuge at >12,000 × *g* for 1 min.

(14) Measure how many microliters eluted for each sample (still keep Cy5 and Cy3 separate) – should be around 18 μL for each column.

(15) Proceed to hybridization (below).

3.4. Microarray Hybridization

3.4.1. Preparation of Hybridization Mix

(1) Prepare Cot, polyA, and yeast tRNA mix:

10 μg/μL Cotl human DNA (Gibco-BRL)	2 μL
10 μg/μL polyA RNA (Sigma, #P9403)	2 μL
10 μg/μL tRNA (Gibco-BRL, #15401-011)	2 μL

(2) Combine Cy5- and Cy3-labeled material from the labeling step.

(3) Add Cot, polyA and yeast tRNA mix, (20×) SSC, and 10% SDS to the combined Cy5 and Cy3 probe, as below*:

Coverslip size	22 × 60 regular thin coverslips	22 × 60 Erie M-series lifter slips
Total Hyb volume (μl)	35	55
Probe volume (μl)	28	36
Cot, polyA, tRNA mix	6	6
(20×) SSC (μl)	5.95	9.35
10% SDS (μl)	1.05	1.65
1 M (pH 7.0) HEPES–KOH	0.84	1.32

For other coverslip sizes, volumes can be adjusted accordingly to maintain the same SSC, SDS, and HEPES concentrations. However, the blocking mix (Cot, polyA, and tRNA) should only be 6 μL in all cases.

(4) Denature probe by heating at 95–100°C for 2 min. Typically, this is done using a heat block set at 100°C with distilled water in the tube holders.

(5) Let probe sit in the dark, at room temperature, for 10 min.

(6) Spin probe at >12,000 × *g* for 5 min at room temperature.

*Avoid introducing bubbles. Do not vortex after adding SDS. HEPES is recommended for all probes.

3.4.2. Apply Hybridization Mixture to Microarray

(1) While probe is sitting at room temperature, set up hybridization chamber. Many different types of chamber are available commercially. Open chamber on clean flat surface and place microarray slide in the chamber with the array side facing up.

(2) Optional: We have found that for some home-printed oligo microarrays (which can often be dim), precipitation of labeled probe can occur at room temperature, which leads to a "pin-prick" morphology with precipitated probe filling in the depressions created during the printing process. To help mitigate this issue, it helps to place the hybridization chamber on a heat block at 55–60°C prior to addition of probe solution.

(3) Once probe is spun down, carefully pipette probe solution onto the microarray surface in one drop toward one end of the array. Make sure not to create any bubbles during pipetting and be careful not to touch microarray surface with pipette tip. Leave ~2 μL of probe in the tube.

(4) Carefully apply a coverslip by placing one edge of the coverslip on the slide near the probe and slowly lowering the other edge, using another coverslip as a lever and wedge to lower it.

(5) Close chamber and immediately submerge in 65°C water bath. Be careful not to tilt the chamber. For safety purposes, metal tongs may be used to place the hybridization chamber in the water bath.

(6) If hybridizing multiple arrays, quickly move on to the next one, as the probes should not sit at room temperature for widely varying times (we try to get all probes onto slides within ~10–15 min of each other).

(7) Incubate chambers at 65°C for 16–20 h.

3.4.3. Array Washing

(1) Prepare dishes with the following solutions from stocks of 20× SSC, 10% SDS, and dH_2O:
Wash 1 – (1×) SSC + 0.03% SDS
Wash 2 – (0.2×) SSC
Wash 3 – (0.05×) SSC

	Wash 1A	Wash 1B	Wash 2	Wash 3
(20×) SSC	20 mL	20 mL	4 mL	1 mL
10% SDS	1.2 mL	1.2 mL	–	–
dH_2O	379 mL	379 mL	396 mL	399 mL
Temperature	RT	RT	37°C	RT

(2) Turn on Axon scanner from the connected black box, located behind the monitor. Let the scanner warm up for at least 15 min to allow the output time of the laser to stabilize.

(3) Launch the GenePix Pro software and let it connect to the scanner and the network hardware key.

(4) Remove one hybridization chamber from water bath. As quickly as possible, wipe the chamber dry with paper towel, open the chamber, remove the slide, and using your gloved hand, tap the slide against the bottom of the wash 1A until coverslip gently slides off the slide. Then transfer the slide to a pre-submerged rack in wash 1B. Maintain slides in wash 1B until all arrays have been transferred.

(5) Quickly move the rack from wash 1B to wash 2 (tilting the rack back and further a couple of times to remove excess wash solution). Plunge up and down several times in wash 2. Incubate for 5 min with occasional plunging up and down.

(6) Quickly move the rack from wash 2 to wash 3 (tilting the rack back and further a couple of times to remove excess wash solution). Plunge up and down several times in wash 3. Incubate for 5 min with occasional plunging up and down.

(7) As quickly as possible, move the rack of slides from wash 3 to place in Beckman tabletop centrifuge, with paper towels under the rack, and spin at 500–600 rpm for 5 min. Place microarrays in an opaque box and begin scanning immediately. Typically, 4–5 arrays are washed at a time, and if more than 5 were hybridized, then the next set of 4–5 should be washed as the last array is being scanned from the first batch.

4. Notes

1. Do not use the NEB buffer 4 supplied with the NEB enzyme: Use the cacodylate buffer (1 M potassium cacodylate, 125 mM Tris–HCl, and 1.25 mg/mL BSA, pH 6.6) supplied with the Roche enzyme.

2. Ensure that the dNTP mixes do not go through more than three freeze-thaw cycles. Additional freeze-thaw cycles will further degrade the dNTPs and will reduce the efficiency of the reaction.

3. Aim for ~1 pmol of template molecules. Tested range is 2.5–75 ng DNA per 10 mL reaction volumes. Scale up the reaction volume accordingly for higher starting amounts.

4. The use of the NEB enzyme, as indicated, is strongly recommended. TdT enzyme from other sources may not perform optimally. If using the Roche recombinant TdT, use double volume of enzyme.

5. T7-A_{18}B primer: (5′- GCATTAGCGGCCGCGAAATTAAT ACGACTCACTATAGGGAG$(A)_{18}$[B], where B refers to C, G, or T). This should be obtained with HPLC, PAGE, or equivalent purification.

   If production of template-independent product is a significant problem, scale down the reaction volume, while keeping the reagent concentrations (except for the T-tailed DNA) constant.

6. New England Biolabs (NEB) in early 2004 switched the supplied buffer for Klenow enzyme from EcoPol buffer to NEB buffer 2. This buffer should provide at least comparable yields to the old buffer and may actually increase yields up to ~14%.

7. Ensure that the dNTP mixes do not go through more than three freeze-thaw cycles. Additional freeze-thaw cycles will degrade the dNTPs and will reduce the efficiency of the reaction.

8. If new kit, combine NTPs into one tube, then aliquot back out into the four tubes. In the first three freeze-thaw cycles, yields drop approximately 10–15% after each cycle. If the NTPs go through more than three freeze-thaw cycles, each subsequent freeze-thaw cycle may drop the yield by as much as 50%.

9. If you add cold buffer and dsDNA, you may risk precipitation of your DNA. Also, if there is precipitate present, warm buffer to 37°C until the precipitate dissolves.

Acknowledgments

These protocols were largely developed by C.L. Liu and have been modified to meet the formatting requirements of this volume. This work was supported by grant GM079205 from National Institute of General Medical Sciences (NIGMS).

References

1. Gottesfeld, J. M., Belitsky, J. M., Melander, C., Dervan, P. B., and Luger, K. (2002) Blocking transcription through a nucleosome with synthetic DNA ligands. *J. Mol. Biol.* **321**, 249–263.
2. Izban, M. G., and Luse, D. S. (1992) Factor-stimulated RNA polymerase II transcribes at physiological elongation rates on naked DNA but very poorly on chromatin templates. *J. Biol. Chem.* **267**, 13647–13655.
3. Schwabish, M. A., and Struhl, K. (2004) Evidence for eviction and rapid deposition of histones upon transcriptional elongation by RNA polymerase II. *Mol. Cell. Biol.* **24**, 10111–10117.
4. Ahmad, K., and Henikoff, S. (2002) Histone H3 variants specify modes of chromatin assembly. *Proc. Natl. Acad. Sci. USA* **99**(Suppl 4), 16477–16484.
5. Ahmad, K., and Henikoff, S. (2002) The histone variant H3.3 marks active chromatin by replication-independent nucleosome assembly. *Mol. Cell* **9**, 1191–1200.
6. Schwartz, B. E., and Ahmad, K. (2005) Transcriptional activation triggers deposition and removal of the histone variant H3.3. *Genes Dev.* **19**, 804–814.
7. Adkins, M. W., Howar, S. R., and Tyler, J. K. (2004) Chromatin disassembly mediated by the histone chaperone Asf1 is essential for transcriptional activation of the yeast PHO5 and PHO8 genes. *Mol. Cell* **14**, 657–666.
8. Bernstein, B. E., Liu, C. L., Humphrey, E. L., Perlstein, E. O., and Schreiber, S. L. (2004) Global nucleosome occupancy in yeast. *Genome Biol.* **5**, R62.
9. Boeger, H., Griesenbeck, J., Strattan, J. S., and Kornberg, R. D. (2003) Nucleosomes unfold completely at a transcriptionally active promoter. *Mol. Cell* **11**, 1587–1598.
10. Lee, C. K., Shibata, Y., Rao, B., Strahl, B. D., and Lieb, J. D. (2004) Evidence for nucleosome depletion at active regulatory regions genome-wide. *Nat. Genet.* **36**, 900–905.
11. Reinke, H., and Horz, W. (2003) Histones are first hyperacetylated and then lose contact with the activated PHO5 promoter. *Mol. Cell* **11**, 1599–1607.
12. Hogan, G. J., Lee, C. K., and Lieb, J. D. (2006) Cell cycle-specified fluctuation of nucleosome occupancy at gene promoters. *PLoS Genet.* **2**, e158.
13. Schermer, U. J., Korber, P., and Horz, W. (2005) Histones are incorporated in trans during reassembly of the yeast PHO5 promoter. *Mol. Cell* **19**, 279–285.
14. Choi, E. S., Shin, J. A., Kim, H. S., and Jang, Y. K. (2005) Dynamic regulation of replication independent deposition of histone H3 in fission yeast. *Nucleic Acids Res.* **33**, 7102–7110.
15. Jamai, A., Imoberdorf, R. M., and Strubin, M. (2007) Continuous histone H2B and transcription-dependent histone H3 exchange in yeast cells outside of replication. *Mol. Cell* **25**, 345–355.
16. Thiriet, C., and Hayes, J. J. (2005) Replication-independent core histone dynamics at transcriptionally active loci in vivo. *Genes Dev.* **19**, 677–682.
17. Dion, M. F., Kaplan, T., Kim, M., Buratowski, S., Friedman, N., and Rando, O. J. (2007) Dynamics of replication-independent histone turnover in budding yeast. *Science* **315**, 1405–1408.
18. Rufiange, A., Jacques, P. E., Bhat, W., Robert, F., and Nourani, A. (2007) Genome-wide replication-independent histone H3 exchange occurs predominantly at promoters and implicates H3 K56 acetylation and Asf1. *Mol. Cell* **27**, 393–405.
19. Kaplan, T., Liu, C. L., Erkmann, J. A., et al. (2008) Cell cycle- and chaperone-mediated regulation of H3K56ac incorporation in yeast. *PLoS Genet.* **4**, e1000270.

Chapter 4

Genome-Wide Approaches to Studying Yeast Chromatin Modifications

Dustin E. Schones, Kairong Cui, and Suresh Cuddapah

Abstract

The genomes of eukaryotic organisms are packaged into nuclei by wrapping DNA around proteins in a structure known as chromatin. The most basic unit of chromatin, the nucleosome, consists of approximately 146 bp of DNA wrapped around an octamer of histone proteins. The placement of nucleosomes relative to a gene can influence the regulation of the transcription of this gene. Furthermore, the N-terminal tails of histone proteins are subjected to numerous post-translational modifications that are also known to influence gene regulation. In recent years, a number of genome-scale approaches to identify modifications to chromatin have been developed. Techniques combining chromatin immunoprecipitation (ChIP) with microarrays (ChIP-chip) and second-generation sequencing (ChIP-Seq) have led to great advances in our understanding of how chromatin modifications contribute to gene regulation. Many excellent protocols related to ChIP-chip have been published recently (Lieb, J. D. (2003) Genome-wide mapping of protein-DNA interactions by chromatin immunoprecipitation and DNA microarray hybridization. *Methods Mol. Biol.* **224**, 99–109.). For this reason, we will focus our attention here on the application of second-generation sequencing platforms to the study of chromatin modifications in yeast. As these genome-scale experiments require both wet-lab and bioinformatic components to reach their full potential, we will detail both the wet-lab protocols and bioinformatic steps necessary to fully conduct genome-scale studies of chromatin modifications.

Key words: Chromatin, histone modifications, ChIP-Seq, nucleosomes, genomics, epigenomics.

1. Introduction

The last several years have witnessed the development of multiple platforms for high-throughput, short read sequencing (referred to here as second-generation sequencing).

These platforms allow researchers to easily probe molecular events on a genome scale. Some of the earliest uses of this

J.I. Castrillo, S.G. Oliver (eds.), *Yeast Systems Biology*, Methods in Molecular Biology 759,
DOI 10.1007/978-1-61779-173-4_4, © Springer Science+Business Media, LLC 2011

technology involved examining chromatin modifications on a genomic scale (1, 2). These studies combined standard molecular protocols such as ChIP assays with second-generation sequencing to map histone modifications across genomes at nucleosome resolution. We describe here the steps necessary to perform such experiments, beginning with the isolation of yeast nuclei, continuing to the techniques such as MNase digestion and ChIP as well as preparing samples for sequencing with the Illumina/Solexa platform. Lastly, we describe the basic bioinformatic tasks necessary to complete such experiments. This protocol is specifically for sequencing with the Illumina/Solexa platform, although a similar protocol – prior to sequencing library preparation – could be used for other platforms.

2. Materials

2.1. General

1. Phenol/chloroform (1:1) (v/v).
2. Chloroform solution (100%, v/v).
3. Sodium acetate (NaOAc) solution (3 M, pH 5.3).
4. 10% (w/v) Sodium dodecyl sulfate (SDS) solution.
5. Ribonuclease A (RNase A).
6. Proteinase K.
7. E-Gels (2% agarose, w/v) (Invitrogen).
8. QIAGEN Gel Extraction Kit (QIAGEN).
9. Ethanol (100%, v/v).
10. Glycerol (100%, v/v).
11. Sodium chloride (NaCl) solution (3 M).
12. Phosphate buffered saline (PBS) solution (1×): 137 mM NaCl; 2.7 mM KCl; 4.3 mM Na_2HPO_4; 1.47 mM KH_2PO_4. Adjust to a final pH of 7.4.
13. Tris–EDTA buffer, TE (1×), pH 7.4 (10 mM Tris–HCl, 1 mM EDTA).
14. Triton X-100 solution.

2.2. Spheroplast Preparation

1. Sorbitol solution (1 M).
2. Spheroplast lysis solution: 1 M Sorbitol, 5 mM 2-mercaptoethanol (add to sorbitol immediately before using).
3. Zymolyase 100T solution (20 mg/mL).

2.3. Isolation of Nuclei

1. Dounce hand homogenizer.
2. Ficoll solution: 18% (w/v) Ficoll, 20 mM phosphate buffer, pH 6.8, 1 mM $MgCl_2$, 0.25 mM EGTA, 0.25 mM EDTA.

2.4. Native Chromatin Preparation by MNase Digestion for Nucleosome Mapping

1. MNase digestion buffer: 15 mM Tris–HCl, pH 7.5, 75 mM NaCl, 3 mM $MgCl_2$, 1.5 mM $CaCl_2$, 1 mM 2-mercaptoethanol.
2. MNase stop buffer: 10 mM Tris–HCl, pH 7.6, 0.5 M EDTA.

2.5. Native Chromatin Preparation by MNase Digestion for ChIP-Seq

1. RIPA buffer: 10 mM Tris–HCl, pH 7.6, 1 mM EDTA, 0.1% (w/v) SDS, 0.1% (w/v) sodium deoxycholate, 1% (w/v) Triton X-100.
2. DNA END-Repair Kit (Epicentre Biotechnologies).

2.6. Chromatin Immunoprecipitation

1. Dynabeads Protein A (Invitrogen).
2. RIPA buffer (*see* above).
3. LiCl wash buffer: 0.25 M LiCl, 1% (w/v) Nonidet P-40, 1% (w/v) sodium deoxycholate, 1 mM EDTA, 10 mM Tris–HCl, pH 8.1.

2.7. DNA END-Repair and Solexa Library Preparation

1. DNA END-Repair Kit (Epicentre Biotechnologies).
2. MinElute PCR Purification Kit (QIAGEN), including elution buffer (ER buffer): 10 mM Tris–HCl, pH 8.5.
3. Taq DNA polymerase and 10× Taq polymerase buffer.
4. T4 DNA ligase (400 U/μL) and 10× ligase buffer.
5. Illumina adapters, PCR primers, and enzyme mix.
6. Qubit spectrometer.

2.8. Computational Infrastructure Requirements

Raw data produced by second-generation sequencing machines is on the scale of terabytes. Labs conducting such experiments will need to have high-end analysis servers with large amounts of RAM (at least 8 Gb) attached to large hard drive arrays (at least 5 Tb) with a fast connection.

3. Methods

3.1. Cell Culture

Sections 3.1, 3.2, and 3.3 are adapted from (3):

1. Inoculate 5 mL of starter culture of an appropriate medium overnight at 30°C on a roller wheel or a shaker.
2. Dilute culture into 500 mL (*see* **Note 1**) of appropriate medium and grow culture to $OD_{600} = 0.8–1.0$.
3. When the correct density is reached, collect cells by centrifugation (e.g., 850×*g* for 5 min). Decant the medium and place cells on ice. Wash cells once with cold water and once with cold 1 M sorbitol.

3.2. Spheroplast Preparation

1. Resuspend the cells in 5 mL spheroplast lysis solution.
2. Add 2 mg of Zymolyase 100T per gram of cells (*see* **Note 2**). Incubate at 30°C with gentle shaking for 20 min. Check completion of the digestion by examining 5 μL of cells on a glass slide with a light microscope (*see* **Note 3**).
3. Collect spheroplasts by centrifugation (e.g., 850×*g* for 5 min) and decant lysis solution.
4. Wash with cold 1 M sorbitol, collect by centrifugation, and decant (*see* **Note 4**).

3.3. Isolation of Nuclei

1. Resuspend spheroplasts in 7 mL Ficoll solution. Transfer cells to a dounce hand homogenizer and lyse spheroplasts with one stroke.
2. Aliquot 3 mL of spheroplasts in Ficoll solution to centrifuge tubes and centrifuge for 30 min at 30,000×*g* at 4°C.
3. Aspirate the cellular debris and Ficoll solution from nuclear pellet. Pool the nuclear pellets together with 5 mL of MNase digestion buffer. Centrifuge and decant.
4. Continue on to **Section 3.4** for nucleosome mapping or **Section 3.5** for chromatin immunoprecipitation (ChIP).

3.4. Native Chromatin Preparation by MNase Digestion for Nucleosome Mapping

To begin, isolate approximately 50 million nuclei as described in Sections 3.1, 3.2, and 3.3:

1. Wash nuclei with 1 mL of MNase digestion buffer at room temperature, split into eight separate 1-mL microcentrifuge tubes, adding 100 μL to each tube.
2. Digest the eight separate tubes with the following concentrations of MNase: 0.005, 0.01, 0.05, 0.1, 0.2, 0.5, 1.0 U, and no MNase as a control (*see* **Note 5**). After adding MNase to the cells, incubate the tubes at 37°C for 8 min.
3. Stop each reaction by adding MNase stop buffer to final concentration of 10 mM of EDTA.
4. Add 10 μg of RNase A and incubate for 5 min at room temperature. Add SDS to a final concentration of 1% (w/v) and 5 μL of 20 mg/mL proteinase K. Incubate at 65°C overnight.
5. Purify DNA by 2× phenol–chloroform extraction followed by 1× chloroform extraction to remove the residual phenol.
6. Precipitate DNA by adding 0.1 vol. of 3 M sodium acetate, pH 5.3, and 2.5 vol. of 100% ethanol.
7. Incubate at –20°C for 30 min and spin down at max. speed for 10 min.

8. Wash once with 70% ethanol, dry the pellet, and resuspend in 100 μL of 1× TE, pH 7.4.
9. Load the samples on a 2% agarose gel (*see* **Note 6**) and excise the mononucleosome-sized DNA bands from each MNase digestion. Combine the samples.
10. Purify the DNA and prepare sequencing library as described in **Section 3.7**.

3.5. Native Chromatin Preparation by MNase Digestion for ChIP-Seq

To begin, isolate approximately 50 million nuclei as described in Sections 3.1, 3.2 , and 3.3:

1. Wash the cells with 1 mL MNase digestion buffer at room temperature.
2. Add 0.2 U MNase and incubate the tubes at 37°C for 8 min.
3. Stop the reaction by adding MNase stop buffer at a final concentration of 10 mM EDTA.
4. Sonicate in ice water three times for 20 s (*see* **Note 7**).
5. Dialyze against 400 mL of RIPA buffer for 2 h at 4°C.
6. Centrifuge for 10 min at 4°C and transfer the supernatant to a new tube, aliquoting 20 μL to check the size. Total supernatant can be stored at –80°C after adding glycerol to a final concentration of 5% (w/v).
7. To check the chromatin size, add 10 μg of RNase A to 20 μL of supernatant and incubate for 5 min at room temperature. Add SDS to a final concentration of 1% (w/v) and 5 μL of 20 mg/mL proteinase K. Incubate at 65°C overnight. Purify the DNA by phenol–chloroform extraction and ethanol precipitation. Run on a 2% (w/v) agarose gel (*see* **Note 8**).

3.6. Chromatin Immunoprecipitation (ChIP)

1. Add 40 μL of Dynabeads Protein A into each of two microcentrifuge tubes. Wash with 600 μL of 1× PBS. Place tubes on magnetic stand and aspirate off PBS. Then add 100 μL of 1× PBS and 4 μg of antibody or preimmune serum to the beads. Rotate for 4–6 h at 4°C.
2. Place the tubes on magnetic stand and remove the supernatant. Wash the beads two times for 5 min with 200 μL of 1× PBS to remove free immunoglobulin G molecules (IgGs).
3. Place the tubes on magnetic stand and remove the supernatant. Add 500 μL of chromatin extracts to the beads and rotate at 4°C overnight.
4. Perform the following washes, 10 min each wash:

   Two times with 1 mL of RIPA buffer,

   Two times with 1 mL of RIPA buffer + 0.3 M NaCl,

   Two times with 1 mL of LiCl buffer,

One time with 1 mL of 1× TE + 0.2% Triton X-100,

One time with 1 mL of 1× TE.

5. Resuspend the beads in 100 μL of 1× TE. Add 3 μL of 10% SDS and 5 μL of 20 mg/mL proteinase K. Incubate overnight at 65°C.
6. Vortex briefly, place the tube on magnetic stand, and transfer the eluate to new tube. Resuspend the beads with 100 μL of 1× TE + 0.5 M NaCl. Combine this new eluate with the last. Perform extraction with 200 μL phenol–chloroform. Add 2 μL of 20 mg/mL glycogen, 20 μL of 3 M NaOAc, pH 5.3, and 500 μL of ethanol to the supernatant for precipitation of DNA.
7. Wash the pellet once with 70% (v/v) ethanol.
8. Resuspend the pellet in 50 μL of 1× TE. Use 2 μL of diluted (1:5) DNA for real-time PCR confirmation. Before continuing with sequencing library preparation, perform real-time PCR to confirm that the ChIP has worked.

3.7. DNA END-Repair and Solexa Library Preparation

1. Repair DNA ends to generate blunt-ended DNA using Epicentre DNA END-Repair Kit. Starting with 1–34 μL DNA (0.3 mg), add the following: 5 μL of 10× END-Repair buffer, 5 μL of 2.5 mM each dNTP, 5 μL of 10 mM ATP.
2. Adjust the final volume to 49 μL by adding H_2O.
3. Add 1 μL END-Repair enzyme mix.
4. Incubate at room temperature for 45 min, purify using MinElute PCR Purification Kit or phenol–chloroform extraction to precipitate DNA. Elute with or resuspend DNA in 30 μL of 1× TE, pH 7.4.
5. Add "A" to 3′-ends: Starting with 30 μL DNA from above, add 2 μL H_2O, 5 μL of 10× Taq buffer, 10 μL of 1 mM dATP, 3 μL of 5 U/μL Taq DNA polymerase.
6. Incubate at 70°C for 30 min and purify using QIAGEN MinElute PCR Purification Kit or phenol–chloroform extraction. Elute with or resuspend in 10 μL of 1× TE, pH 7.4.
7. Linker ligation: Starting with 10 μL DNA (300 ng), add the following: 9.9 μL H_2O, 2.5 μL of 10× T4 DNA ligase buffer, 0.1 μL Adaptor oligo mix (*see* **Note 9**), 2.5 μL T4 DNA ligase (400 U/μL).
8. Incubate at 20–23°C for 30 min and then at 16°C overnight. Purify using MinElute PCR Purification Kit and protocol. Elute in 20–25 μL of elution buffer (EB).
9. Size selection with 2% E-Gel (Invitrogen): Load 20–30 μL linker-ligated DNA onto a 2% E-Gel (*see* **Note 10**). Excise

the gel at the 200–400-bp region (DNA will not be visible). Extract the DNA using MinElute Gel Extraction Kit. Elute in ~12 μL EB.

10. Amplify DNA using Illumina PCR primers and enzyme mix: 10.5 μL of DNA, 12.5 μL of master mix, 1 μL of PCR primer 1.0 (two times diluted with 1× TE, pH 7.4), 1 μL of PCR primer 2.0 (two times diluted with 1× TE, pH 7.4).
11. Denature at 98°C for 30 s, followed by 18 amplification cycles (98°C, 10 s; 65°C, 30 s; 72°C, 30 s).
12. After amplifying for 18 cycles, check 2.5 μL of product on a 2% agarose gel. If the band is not clearly visible, perform three more cycles and check again.
13. Purify amplified products: Load amplified DNA on 2% agarose gel. Excise the band at the 200–400-bp region and purify the DNA using QIAGEN Gel Extraction Kit. Measure the DNA concentration using Qubit spectrometer.

This DNA can now be used for cluster generation and sequencing with Illumina's platform.

3.8. Post-sequencing Analysis

High-throughput sequencing experiments produce large amounts of raw data and have significant bioinformatic needs at every step. For the standard ChIP-Seq type experiment, the usual steps after processing the raw data will consist of the following: (1) aligning sequenced reads back to a reference genome, (2) visualizing the results on a genome browser, (3) evaluating the sequencing results, and (4) identifying enriched regions from the background signal.

3.8.1. Aligning Reads to a Reference Genome

Second-generation sequencing machines currently produce tens of millions of reads per experiment, and the throughput is increasing. Standard alignment algorithms were not designed for such throughput. At the time of this writing, there are several alignment algorithms that have been developed specifically for second-generation sequencing data. These aligners are specifically designed to map many short reads to a reference genome. Issues to consider when aligning short reads to the genome include the following:

1 – Incorporation of base-call quality – some aligners (such as MAQ (4) and RMAP (5)) consider the quality of each individual base when aligning to a reference genome.

2 – Ability to do gapped alignments – certain aligners (e.g. SOAP (6)) are capable of doing gapped alignments; this is useful when sequencing mRNA.

3 – Paired-end read capability – certain alignment algorithms, such as MAQ, SOAP, and ELAND (7), can align mate-pair

reads (Illumina author Anthony Cox, personal communication).

4 – Treatment of non-unique mapping reads – popular strategies for the treatment of non-unique mapping reads include the following: (a) ignoring all non-unique aligning reads, (b) picking an alignment location at random when a read maps to more than one locations, or (c) assigning "hits" to all possible locations that a read maps to, and normalizing by the total number of hits. The optimal choice depends on the experiment.

3.8.2. Visualizing Results on a Genome Browser

The most convenient way to visualize results from ChIP-Seq experiments is to view the genomic coordinates of the aligned reads in a genome browser. Popular genome browsers include the UCSC Genome Browser (8), the Affymetrix Integrated Genome Browser (9), Ensembl (10), and GeneTrack (11) (*see* **Note 11**). An example of hypothetical ChIP-Seq data from *Saccharomyces cerevisiae* as displayed in the UCSC Genome Browser is shown in **Fig. 4.1**. While the individual reads can be viewed by strand as shown in the **Fig. 4.1a**, a more useful representation when viewing large domains is to create "summary" tracks that count the number of reads in a window as shown in **Fig. 4.1b**. Ideally, this window size should be equal to the average chromatin fragment size. The coordinates of read positions can be adjusted to represent the center of a given chromatin fragment prior to creating summary tracks.

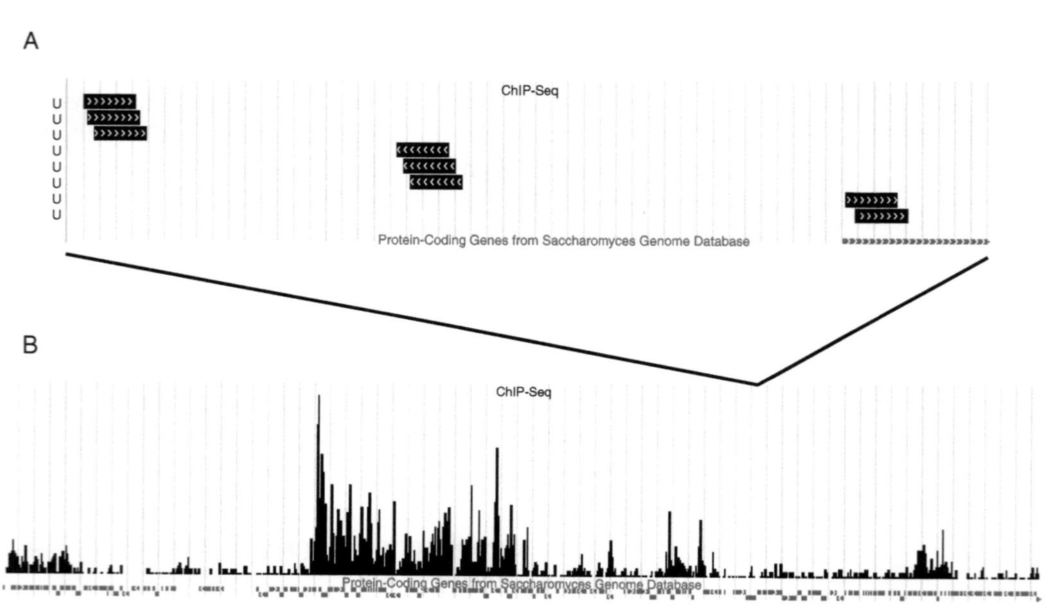

Fig. 4.1. Example of ChIP-Seq data displayed as custom tracks on the UCSC Genome Browser (aligned reads vs. "*Saccharomyces* Genome Database" DNA sequences). (**a**) Individual aligned reads can be viewed by strand. (**b**) A more useful representation for viewing data in large domains is by creating summary tracks. *See* text for details.

3.8.3. Estimating Quality of ChIP-Seq Experiments

The quality of the chromatin immunoprecipitation experiment should be checked prior to sequencing as detailed in **Section 3.6**. In order to ensure that sequencing results are meaningful, it has to be ensured that the amount of sequencing that has been performed is sufficient. The profiling of certain histone modifications that spread to large domains (e.g., H3K27me3) will require more sequencing compared to histone modifications with more localized patterns (e.g., H3K4me3). A saturation analysis can be performed to determine if sufficient sequencing has been achieved by sampling from the sequenced read library at successive levels and finding enriched regions at each level. The point at which the number of enriched regions plateaus as more reads are added to the sequencing library is the saturation point.

3.8.4. Identifying Enriched Regions

One of the most crucial steps of any ChIP-Seq experiment is the identification of enriched regions, separating the positive signal from the background. One of the most common and simple methods to do this is to first adjust all the read locations in the genome to "center" them on their representative chromatin fragment. If the chromatin fragments produced were mononucleosomes, this would involve shifting every read 75 bp in the direction that it is aligned to. One can then scan through the genome with a window size equivalent to the chromatin fragment size (e.g., 150 bp) and count the number of adjusted reads that map to each window. By assuming a given background model of read distribution throughout the genome (e.g., Poisson), one can calculate the number of reads necessary in each window for a desired level of statistical significance.

3.8.5. Nucleosome Positioning Determination from Sequenced Reads

For profiling nucleosome positions with second-generation sequencing, chromatin is digested with MNase, which preferentially cuts in linker regions. The sequenced fragments will represent the borders of nucleosomes. To create nucleosome-scoring profiles, slide a window along the genome and count all reads aligning to the sense strand within ~75 bp upstream and all reads aligning to the antisense strand within ~75 bp downstream of the window. These scoring profiles can also be constructed for histone modification ChIP-Seq data to find the positions of all nucleosomes that contain a certain type of modification.

4. Notes

1. 500 mL is a good amount for nucleosome-mapping experiments; 50 mL is enough for chromatin immunoprecipitation.

2. After Zymolyase digestion, keep cells on ice between each step.
3. Add 0.5 μL of SDS to the slide and look for the appearance of "ghost" cells that have lysed.
4. Spheroplasts are fragile, so be gentle when resuspending cells (i.e., pipet up and down instead of vortexing to resuspend).
5. MNase concentrations should be standardized. This will depend on the concentration/activity of the MNase.
6. When working with small amounts of DNA, it is difficult to avoid contamination; we use Invitrogen E-Gels for this step to minimize the risk of contamination.
7. Keep the tubes on ice for ~1 min after each sonication to prevent the lysate from heating up.
8. This procedure should produce primarily mononucleosomes. The MNase concentration and digestion time can be adjusted to produce different fragment sizes.
9. For a more accurate measure, dilute Adaptor oligo mix 1:10 with H_2O and then add 1 μL of dilution.
10. To avoid contamination, load only one sample per gel.
11. The choice of which genome browser to use is largely dependent on the problem at hand. Lightweight browsers such as the Affymetrix IGB or GeneTrack can be downloaded locally and run on standard lab desktops. The UCSC Genome Browser is useful for uploading data as "custom tracks," but this becomes infeasible if a large number of data sets need to be analyzed together. Alternatively, a local mirror of the UCSC Genome Browser can be set up.

Acknowledgments

The authors would like to thank Keji Zhao and other members of the Zhao lab for assistance in all steps of this work. The authors would also like to thank Amy Leung for critical reading of the manuscript and assistance with protocol development. The authors are supported by the Intramural Research Program of the NIH, NHLBI.

References

1. Barski, A., Cuddapah, S., Cui, K., et al. (2007) High-resolution profiling of histone methylations in the human genome. *Cell* **129**, 823–837.
2. Mikkelsen, T. S., Ku, M., Jaffe, D. B., et al. (2007) Genome-wide maps of chromatin state in pluripotent and lineage-committed cells. *Nature* **448**, 553–560.

3. Moreira, J. M., and Holmberg, S. (1998) Nucleosome structure of the yeast CHA1 promoter: analysis of activation-dependent chromatin remodeling of an RNA-polymerase-II-transcribed gene in TBP and RNA pol II mutants defective in vivo in response to acidic activators. *EMBO J.* **17**, 6028–6038.
4. Li, H., Ruan, J., and Durbin, R. (2008) Mapping short DNA sequencing reads and calling variants using mapping quality scores. *Genome Res.* **18**, 1851–1858.
5. Smith, A. D., Xuan, Z., and Zhang, M. Q. (2008) Using quality scores and longer reads improves accuracy of Solexa read mapping. *BMC Bioinformatics* **9**, 128.
6. Li, R., Li, Y., Kristiansen, K., and Wang, J. (2008) SOAP: short oligonucleotide alignment program. *Bioinformatics* **24**, 713–714.
7. Efficient Large-Scale Alignment of Nucleotide Databases (ELAND). Whole genome alignments to a reference genome. Written by Illumina author Anthony J. Cox for the Solexa 1G machine. http://bioinfo.cgrb.oregonstate.edu/docs/solexa/
8. Kent, W. J., Sugnet, C. W., Furey, T. S., et al. (2002) The human genome browser at UCSC. *Genome Res.* **12**, 996–1006.
9. http://www.affymetrix.com/
10. http://www.ensembl.org
11. Albert, I., Wachi, S., Jiang, C., and Pugh, B. F. (2008) GeneTrack – a genomic data processing and visualization framework. *Bioinformatics* **24**, 1305–1306.

Chapter 5

Absolute and Relative Quantification of mRNA Expression (Transcript Analysis)

Andrew Hayes, Bharat M. Rash, and Leo A.H. Zeef

Abstract

In this protocol, we describe a pipeline for transcript analysis in yeast via the quantification of mRNA expression levels. In the first section, we consider the well-established, proprietary Affymetrix GeneChip® approach to generating transcriptomics data. In the next section, we concentrate on providing a detailed protocol for the validation of these data using quantitative reverse transcriptase-polymerase chain reaction (qRT-PCR). The protocol provides suggested examples of hardware, software, and consumables/reagents required to perform these experiments. There are of course many other options available using alternative approaches (or indeed suppliers), but this protocol is intended to provide an approach that is flexible, inexpensive, sensitive, and easy to use.

Key words: Transcriptome, transcript analysis, validation, qRT-PCR, microarray, mRNA profiling, *Saccharomyces cerevisiae*.

1. Introduction

Quantification of mRNA expression levels has, historically, been performed using a broad range of techniques and these have been widely reviewed. Indeed the *Saccharomyces* Genome Database (SGD) (1) currently curates some 837 records of publications under the literature topic of "Genomic expression study." Most of these are microarray-based papers. Whilst emerging techniques, most notably using next-generation sequencing (2–5), look set to supersede these in the future – *see* **Chapter 8** – the microarray remains a powerful, amenable tool for transcript analysis. One such technology platform (Affymetrix GeneChip) has been used extensively, for yeast transcript analysis, since its inception in the

J.I. Castrillo, S.G. Oliver (eds.), *Yeast Systems Biology*, Methods in Molecular Biology 759,
DOI 10.1007/978-1-61779-173-4_5, © Springer Science+Business Media, LLC 2011

1990s (6) and remains a useful tool in more recent, up-to-date studies (7, 8). At the time of writing, the Affymetrix publication repository (9) contains 616 articles searchable under the organism "yeast."

A crucial aspect of this microarray approach however, and one whose methodology is not so widely reported, is the need to validate the results of these transcript analyses with an independent technique.

In this chapter we consider the Affymetrix GeneChip microarray protocol and then concentrate on providing a protocol to validate the results from such experiments using the technique of quantitative reverse transcriptase-polymerase chain reaction (qRT-PCR).

Many different approaches to qRT-PCR are available, and again these have been widely reviewed (10–12). Briefly though, three main detection chemistries are routinely used and each has its own merits and drawbacks (*see* **Note 1** for more details). Herein we describe a protocol using the SYBR Green approach (13). SYBR® Green I is a fluorogenic, minor-groove binding dye whose fluorescence emission is enhanced upon binding to double-stranded DNA. Thus, as the PCR product accumulates, fluorescence increases. Whilst this approach lacks the specificity of probe-based approaches, it has the advantages of being sensitive, flexible, inexpensive, and easy to use.

2. Materials

2.1. Microarray Analyses

1. Total RNA (minimum 100 ng per sample) from "experimental" and "control" contexts (*see* **Note 2**).
2. GeneChip® Yeast Genome 2.0 Array (Affymetrix, Inc.).
3. GeneChip® 3′ In Vitro Transcription (IVT) Express Kit (Affymetrix, Inc.).
4. GeneChip® Hybridization, Wash, and Stain Kit (Affymetrix, Inc.).
5. Veriti 96-Well Thermal Cycler (Applied Biosystems Ltd).
6. Fluidics Protocol Mini_euk2v3 (Affymetrix, Inc.).
7. GeneChip® Scanner 3000 enabled for high-resolution scanning (Affymetrix, Inc.).
8. Affymetrix GeneChip® Command Console® (AGCC) Software.
9. GeneChip® Fluidics Station 450 (Affymetrix, Inc.).
10. Hybridization Oven 645 (Affymetrix, Inc.).

11. dChip software package (14).
12. Genomics Suite software (Partek, Inc.).

2.2. qRT-PCR

1. Total RNA (minimum 1 μg), ideally aliquots from the same samples used for the microarray analyses.
2. Oligonucleotide PCR primers (nominally 20-mers).
3. MJResearch Chromo4 Real-Time PCR Machine (Bio-Rad Laboratories, Inc.).
4. MJResearch Opticon Monitor analysis software v.3.1 (Bio-Rad Laboratories, Inc.).
5. iScript One-Step RT-PCR Kit with SYBR Green, 200 μL × 50 μL reactions (Bio-Rad Laboratories, Inc.).
6. Plates: Multiplate low-profile, 96-well, unskirted, white PCR plates (Bio-Rad Laboratories, Inc.).
7. Sealers: Microseal "B" adhesive seals (Bio-Rad Laboratories, Inc.) or optical, flat, ultraclear, eight-cap strips (Bio-Rad Laboratories, Inc.).

3. Methods

3.1. Labeling and Fragmentation

The Affymetrix GeneChip® protocol is followed exactly according to the manufacturer's instruction using the 3′ In Vitro Transcription (IVT) Express Kit. Briefly:

1. First-strand cDNA is synthesized by reverse transcription (RT) via priming with an oligo(dT)-T7 primer to yield complementary DNA (cDNA) containing a T7 promoter sequence. Use 100 ng total RNA in <5 μL nuclease-free water.
2. Second-strand cDNA synthesis converts this single-stranded cDNA into double-stranded DNA (dsDNA) template for transcription. RNase H and DNA polymerase simultaneously degrade the RNA and synthesize the second-strand cDNA.
3. In vitro transcription (IVT). In this, the amplification step, the labeling master mix is used to synthesize multiple copies of biotinylated amplified RNA (aRNA) from the double-stranded cDNA.
4. Unincorporated nucleotide triphosphates (NTPs), salts, enzymes, and inorganic phosphate are removed from the aRNA in the purification step. This enhances the stability of the biotin-modified aRNA.

5. The purified, labeled aRNA is then fragmented prior to hybridization to the Yeast Genome 2.0 array. Use 5 μg fragmented and labeled aRNA to make the hybridization cocktail (total volume 100 μL).

3.2. Hybridization

Hybridize for 16 h at 45°C, rotate at 60 rpm in an Affymetrix hybridization oven 645.

3.3. Washing and Staining

Wash and stain the arrays using Fluidics Protocol Mini_euk2v3 on a GeneChip® Fluidics Station 450 for Affymetrix GeneChip® Command Console® Software (AGCC).

3.4. Scanning and Analyses

1. Using a GeneChip® Scanner 3000 enabled for high-resolution scanning, the scanned image (.DAT) file is analyzed for probe intensities (*see* **Note 3**). The software then generates a probe intensity level file (.CEL) upon which the further analyses are continued. At this stage, it is advisable to perform an optional, simple, quality control (QC) step on these files using the dChip software package (14). Following this, image processing and normalization then produce a spreadsheet of expression values for each gene. The data are commonly transformed into logarithmic scale (typically log base 2) for further analyses.

   It is easy to be overwhelmed by the wide variety of microarray analysis tools currently available. However, if one looks beyond the specific software packages and algorithms, commonalities are apparent in the steps that need to be taken during microarray analysis (15). For this task, we favor the commercial package, Genomics Suite from Partek, Inc., although many other open source software tools are also available (most notably the Bioconductor package) (*see* **Note 4**).

2. *Quality assessment of the experiment*: Perform a principal component analysis (PCA) using Genomics Suite (or the chosen software tool) to assess the performance of the experiment (*see* **Note 5**).

3. *Inference*: A statistical test is performed on a gene-by-gene basis to assess if differential expression has occurred. Using fold change alone is not acceptable. A parametric test such as the Student's *t*-test or ANOVA can be used on the data in logarithmic scale. Due to economic considerations, a typical microarray experiment contains a smaller number of replicates than is optimal. Variance shrinkage methods are useful in making use of the parallel-testing nature of microarrays to bolster statistics (16). The parallel-testing nature of microarrays leads to a different statistical problem when it comes to choosing a set of genes that show differential expression

(referred to as a genelist). Selecting a genelist "with p values less than 0.05" leads to a multiple-testing error and so a false-discovery correction should be applied (17).

4. *Genelist analysis*: Once a set of statistically significant differentially expressed genes have been identified, the next challenge is to organize the list of genes. Analyzing the genes into classes, often based on gene ontology categories, is helpful in this regard. An intermediate step of comparative genomics may be required to make use of the information available for the yeast model species. Prior to this, however, it is at this stage that these results should be validated. Candidate genes (both up- and downregulated) should be selected and their expression profiles validated using the following protocol.

3.5. Primer Design for qRT-PCR and Optimization of PCR

1. *Primer design*: Oligonucleotide primers (nominally 20-mers) should be designed to yield products between 70 and 200 base pairs (bp) from the selected "target" (T) genes with all sets of primers having a similar t_m. Many commercial options are available for primer design but for *Saccharomyces*, we favor the online OligoPerfect^{TM} Designer service provided by Invitrogen Corporation (18).

   Also, an appropriate "reference" gene (R) should always be incorporated into the experimental design. A commonly used example is the *ACT1* gene (systematic name, YFL039C) which encodes the single essential gene for actin (19). Actin is a ubiquitous, conserved cytoskeletal element critical for many cellular processes and the gene is often used as a reference (*see* **Note 6** for more details).

2. *Optimize the PCR conditions*:
   Using genomic DNA as a template, the concentration of each of the oligonucleotide primers should be optimized such that all the primers are exhausted in the course of the reactions.
   A useful guideline would be to dilute the primers to give 100 pmol/μL as a stock solution. Take a 1:10 (v/v) dilution of this to give 10 pmol/μL as working solution and use 2 μL per reaction.

3. *DNase treatment*: For one-step reactions, take 1 μg of total RNA, add 10 μL of 10× reaction buffer (RB), 2 units of DNase I, and adjust the volume to 100 μL.

   Incubate: 37°C for 15–30 min.

   Add 1 μL of 0.5 M EDTA.

   Inactivate DNase I by heating at 75°C for 10 min (*see* **Note 7**).

   Make up to 1.0 mL with nuclease-free or DEPC-treated water.

4. Set up reactions in the 96-well plates (*see* **Note 8**) as below and use optimized cycling parameters:
   - 2 μL Forward primer.
   - 2 μL Reverse primer.
   - 5 μL DNase-treated RNA (from previous step).
   - 2 μL iScript RT (Bio-Rad).
   - 25 μL SYBR mix (Bio-Rad).
   - 14 μL Water.

   Thermal cycling protocol:
   - 45°C for 1 h (RT step).
   - 95°C for 15 min (activation of *Taq* polymerase).
   - 95°C for 15 s.
   - "*X*"°C for 1 min (annealing temperature optimized from 3.5.2).
   - 72°C for 30 s.
   - Plate read.
   - Go to Step 3 for 40 times.
   - Melting curve from 55°C to 95°C, every 0.2°C hold for 1 s (*see* **Note 9**).

3.6. Data Analysis

Quantification strategies: Two strategies are routinely used for quantification: absolute and relative. In **Note 10**, we expand on this concept, but for the purposes of this protocol, we concentrate on the latter:

1. Establish the position of the cycle threshold (C_t) line. Using the Opticon Monitor software, the position of the C_t line is set *by the operator*. That is to say that position of the line is manually adjusted until it sits in the linear portion of the curve. It is useful to use the logarithmic view of the curves to facilitate this. Also, the calibration curve can be useful to inform the decision where to position the line. Outliers from the standard curve may be selected/deselected to improve the fit, and the slope of the standard curve should be as close to –0.3 as possible.
2. Once the threshold has been set, then certain QC metrics should be checked:
   - The melt curve graph should show a single discrete peak (**Fig. 5.1**).
   - A large difference (several C_ts) should be evident between the +/– RT controls.
3. Finally, using the software, the C_t values for the target and reference genes from the control and experimental samples are calculated. A graphical representation of this is shown

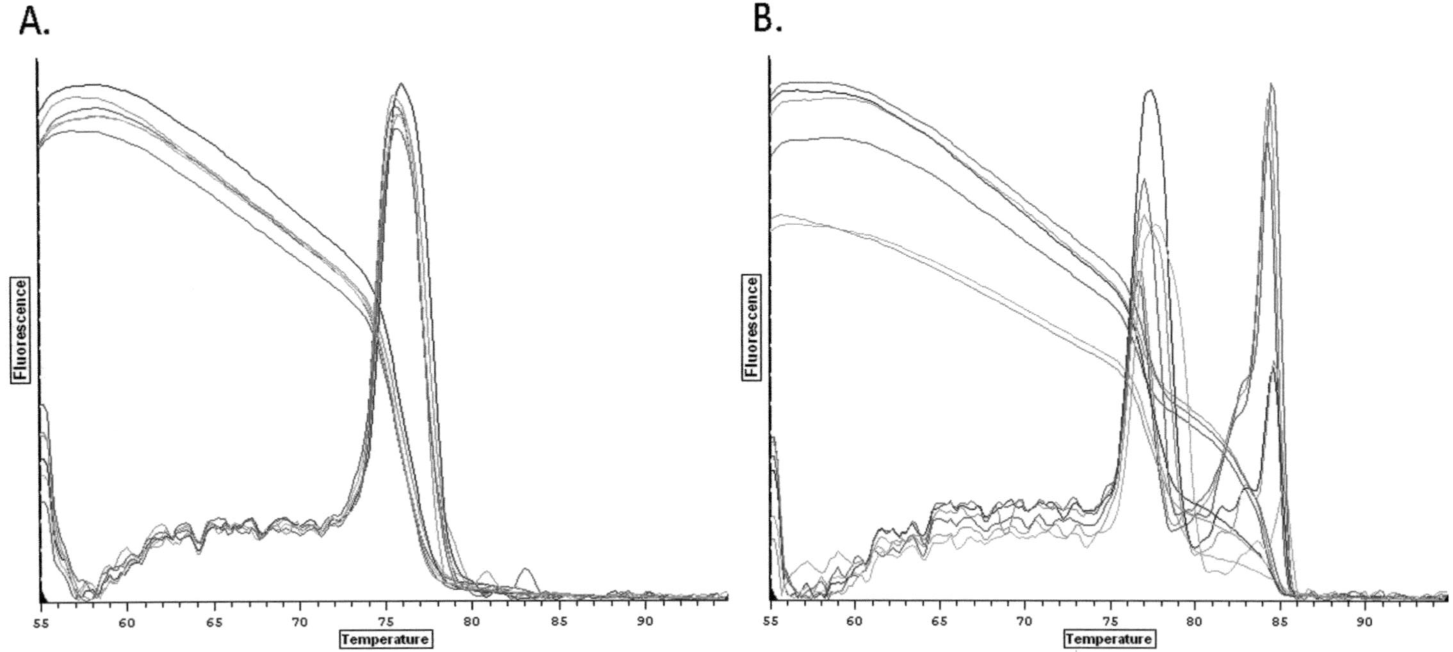

Fig. 5.1. Melting curve analyses of two of the transcripts quantified by qRT-PCR. Plotting the negative, first derivative of the melting curves may make it easier to visualize the peak formed at the temperature odissociation. (**a**) (First transcript) shows a single discrete peak at the predicted t_m (in this case about 76°C). In contrast, however, the analysis of the second transcript (**b**) showed another peak which will affect the fluorescence signal (and consequently the C_t value) for that transcript.

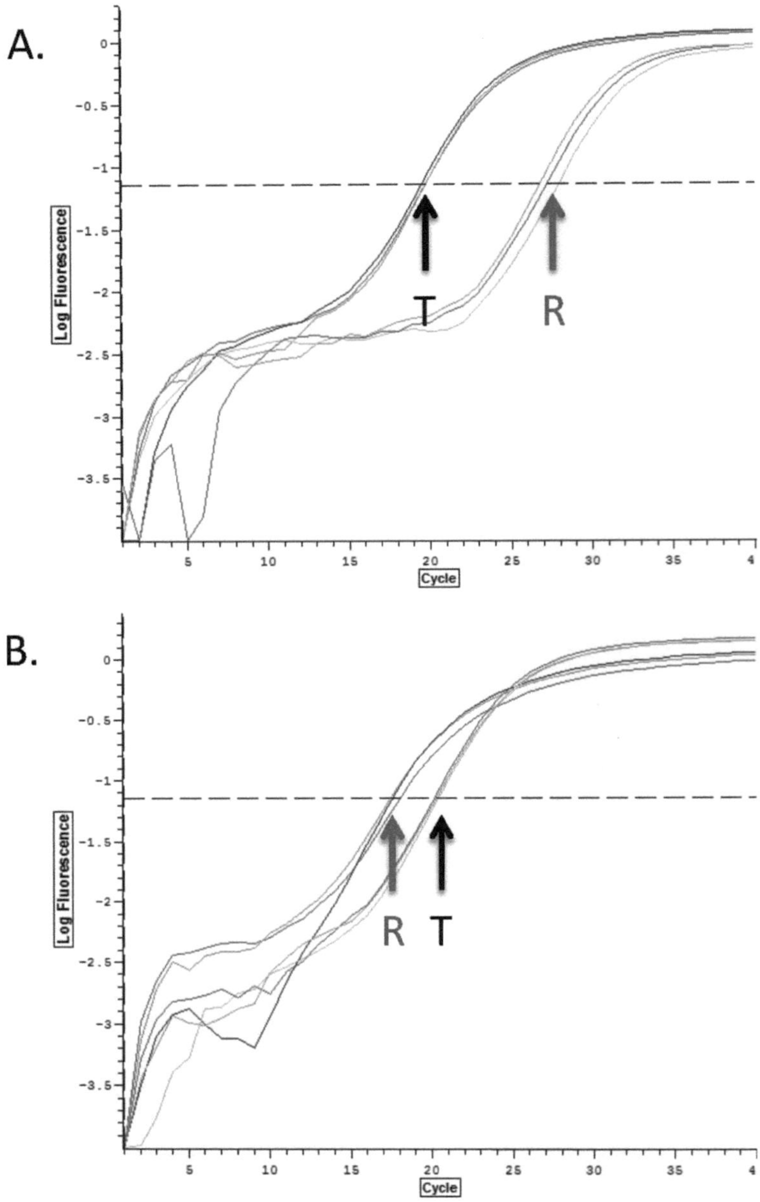

Fig. 5.2. Graphs of a representative qRT-PCR experiment. Expression profiles of two genes, termed target (T) and reference (R), were compared under (**a**) *control* and (**b**) *experimental* conditions. The C_t values are highlighted by the *arrows* and were used in the subsequent worked example (**Table 5.1**).

in **Fig. 5.2**. The C_t values can then be exported as tab-delimited ASCII files for the next phase of the analysis.

4. Differential expression calculation:
   Using the C_t values obtained from the qRT-PCR experiment, the change in relative expression levels is calculated as follows:

- Calculate mean of replicate (at least triplicate) C_t values.
- Calculate ΔC_t:
  - $\Delta C_{t[\text{Control}]} = C_t T - C_t R$ [i]
  - $\Delta C_{t[\text{Experimental}]} = C_t T - C_t R$ [ii]
- Calculate $\Delta\Delta C_t$:
  - $\Delta\Delta C_t =$ [i] – [ii] [iii]
- Fold change $= 2^{(\Delta\Delta C_t)} = 2^{([\text{iii}])}$ (*see* **Note 11**).

A worked example of this procedure is illustrated in **Table 5.1**.

Using an independent assay technique in this manner allows the investigator to validate the results obtained from the microarray experiments.

It is also worth noting that standards are currently emerging for the reporting of qPCR experiments. The MIQE standards (20), analogous to the MIAME standards developed in the microarray community, provide a framework to formalize the reporting of the metadata pertaining to a qPCR experiment. Compliance with these standards is likely to become a prerequisite for publication and must be strongly recommended.

Table 5.1
C_t values obtained from the qRT-PCR experiment (illustrated in Fig. 5.2) and a worked example to calculate the fold change in relative expression levels

Well no.	Condition	Gene	C_t	Mean C_t	ΔC_t [T–R]	$\Delta\Delta C_t$ [7.64–(–2.58)]	Fold change $[2^{(10.23)}]^a$
A7	Control	Target (T)	27.94				
B7	Control	Target (T)	27.23	27.32			
C7	Control	Target (T)	26.79				
					7.64		
A4	Control	Reference (R)	19.75				
B4	Control	Reference (R)	19.76	19.68			
C4	Control	Reference (R)	19.52				
						10.23	1198
A5	Experimental	Target (T)	18.05				
B5	Experimental	Target (T)	17.45	17.70			
C5	Experimental	Target (T)	17.59				
					–2.58		
A2	Experimental	Reference (R)	20.40				
B2	Experimental	Reference (R)	20.17	20.28			
C2	Experimental	Reference (R)	20.27				

[a] *See* **Note 11**

4. Notes

1. *Detection chemistries*: Essentially, three categories of detection chemistry are routinely used in qPCR:
   - Hybridization probe-based methods which rely on the sequence-specific detection of a desired PCR product (e.g., TaqMan hydrolysis probes; Molecular Beacons).
     - *Advantages*: Specific, as products are detected only if the probes hybridize to the appropriate amplification products; readily amenable to multiplexing.
     - *Disadvantages*: Cost, complexity, potential fragility of probe synthesis, primer–dimers can adversely affect amplification efficiency.
   - Intercalating dyes which fluoresce upon light excitation when bound to (any) double-stranded DNA (e.g., SYBR Green I).
     - *Advantages*: Very economical, sensitive, readily amenable to verification by melt curve analyses, easily adaptable, and flexible.
     - *Disadvantages*: Non-specific, detection level is a function of amplicon length.
   - Fluorophores attached to primers (e.g., Lux, from Invitrogen; Plexor, from Promega):
     - *Advantages*: Moderately inexpensive, readily amenable to verification by melt curve analyses.
     - *Disadvantages*: Only moderately specific (dependant on the primers). Less flexible than using intercalating dyes. Often tied to commercial design software.
2. *Experimental design*: Knowing the possible sources of error is important. Since microarrays do not report absolute quantification but only comparative hybridization, this places great importance on the "control" to which the "experimental" (or treated) sample must be compared. Potential sources of difference between the treated sample and control, extraneous factors such as culture volumes (when that is not the factor under investigation), or the manner and time of harvesting, processing RNA samples, or running arrays should be eliminated or at least brought to a minimum. To have any confidence in differential expression of genes being measured by microarray, replication should be performed and this should be biological replication rather than technical (where the same RNA is placed on two or more arrays). Pooling is an acceptable way of improving statistical power without hugely increasing

cost; however, replicates should still be performed. Pooling destroys information about the variability of RNA expression between individuals, so verification of results must be performed on unpooled samples (using qRT-PCR, for example).

3. *Quality control of microarrays*: Following on from the need to control extraneous factors emphasized in the previous paragraph, when it comes to quality analysis, one seeks consistency between the arrays. There may be particular target metrics that the microarray manufacturer recommends, but in an experiment it is important to check that metrics like median intensity, distribution of intensities, and background intensities are consistent. Small differences can be compensated for by normalization methods as we shall see in the next paragraph, but there is a limit to how effective these corrections can be. Checks should be done for technical artifacts and spatial defects on the array such as blobs and spot irregularities.
4. *Image processing and normalization*: For a given array type, several methods may be available for estimation of the amount of RNA from fluorescent array images, while trying to minimize the extraneous variation that occurs owing to technical artifacts. The Bioconductor package, based on the R programming language, is an excellent open source and open development software project, making these methods freely available (21). For example, the RMA method of image processing and normalization which is commonly used for analysis of Affymetrix arrays (22). Image processing and normalization will produce a spreadsheet of expression values for each gene. The data are commonly transformed into logarithmic scale (typically log base 2) for further analysis.
5. Performing a principal component analysis (PCA) is a valuable way of assessing the performance of the experiment. If replicates have been performed, then grouping of replicates indicates that reliable differential expression has occurred between the conditions studied. If on the other hand a random distribution of the samples in PCA space is found, then great caution needs to be exercised in the statistical analysis described in the next paragraph (when thousands of genes are analyzed at once, there will always be some genes that look differentially expressed in a statistically significant way due to random chance events). When more than one factor is being studied, for example, mutation and nutrient limitation, a PCA can indicate the relative strengths of these effects on gene expression.

6. It is crucially important to incorporate the relevant controls in the experimental design. The terminology can become confusing here. For example, different authors use the terms "control" and "reference" interchangeably. To clarify this, in this article we use the terms "control" and "experimental" to refer to the conditions (or contexts) being compared in the experiment. For example, this may refer to a wild-type strain (control) vs. a mutant strain (experimental). Within each 96-well plate layout, we also need to incorporate the relevant controls. Here, we refer to "target" and "reference" genes. The "target" gene (T) is the gene whose expression level is being measured relative to the "reference" gene (R). The reference gene is an mRNA that is naturally present in each experimental and control sample and whose expression level must be invariant throughout all conditions under investigation. Thus, using such an invariant, *endogenous* reference allows quantification of an mRNA target to be normalized for differences in the amount of RNA in the reaction and correct for inter-sample variations in assay efficiency.

   Also, an *exogenous* active reference may be incorporated. This may be a characterized RNA (or DNA) – often an in vitro construct – spiked into each sample at a known concentration and serves to distinguish true target negatives from PCR inhibition. Furthermore, it can serve to normalize for variations in sample preparation efficiency or the RT step in cDNA synthesis by reverse transcriptase. A *passive* reference dye (usually ROX) may also be incorporated to normalize for non-PCR-related fluctuations in fluorescence signal.

   For the reference gene, it is advisable to include a "no-enzyme" negative control, by omitting the reverse transcriptase. This gives an indication if any genomic DNA contamination remained after the DNase I treatment.

   Finally, a standard (or calibration) curve should be incorporated into the plate layout. We routinely use the *ACT1* gene as a template, and a serial dilution spanning at least five orders of magnitude.

7. If EDTA is not added, the RNA will undergo chemical scission when heated.

8. White 96-well plates (or strips) should be used for carrying out the PCR. This is to prevent light scattering and consequently cross talk between the wells during the plate reads. Also, it is important to ensure that the seal of the wells remains intact throughout the thermal cycling process. Leakage is often a problem particularly with the Chromo4 system and the adhesive films described herein – if

leakage does occur, when using the adhesive sealing film, then optical, flat, ultraclear, eight-cap strips should be used to alleviate the problem.

9. *Melting curve (dissociation) analysis*: SYBR Green I does not bind to single-stranded DNA but binds to *any* double-stranded DNA product. Consequently, its use necessitates incorporating a melting curve analysis to assure specificity of the results. The melting point (T_m) is defined as the temperature at which 50% of the DNA is single stranded. Several factors impact on this figure: length of the DNA; nucleotide sequence; G:C content; and Watson–Crick pairing. When DNA-binding dyes such as SYBR Green I are used, as the fragment is heated, the DNA strands dissociate and the dye is released. This leads to a sudden decrease in fluorescence when T_m is attained. This point is determined from the inflection point of the melting curve. Visualization of this point is facilitated by plotting the negative first derivative of the melting curve (**Fig. 5.1**). Melting curve analysis can also be useful in mutation analysis as a new mutation will impact on the peak area/profile or create additional peaks. See (23) for details of melting curve analysis.

10. *The quantification strategy: Absolute vs. relative quantification.*

    One of the two quantification strategies is generally used in qRT-PCR:

    - Absolute quantification relates the fluorescence intensity to that of an input copy number using a calibration curve. The reliability of an absolute assay depends explicitly on *equivalence* of the amplification efficiencies (and the RT step) for both the target and the calibration curve. A range of templates may be used for the calibration curve, for example, recombinant RNA/DNA; synthetic oligonucleotide; purified PCR product (10, 24).

    - Relative quantification measures the relative change in mRNA expression levels. It is based on the expression levels of a target gene *vs.* a reference (often referred to as "housekeeping") gene and in theory is usually adequate to validate genome-wide expression profiling experiments. Also, theoretically, these relative metrics may be compared across multiple qRT-PCR experiments. Strictly speaking, relative quantification experiments do not require a calibration curve – although the inclusion of this step does afford an additional aid in the manual positioning of the threshold line. Mathematical models are well established to calculate the expression of a target gene in relation to an adequate reference gene (25).

11. The "fold-change" calculation assumes 100% efficiency of the PCR. If this is not the case, then the actual efficiency must be factored in to the final calculation (25).

References

1. Saccharomyces Genome Database (2009) (http://www.yeastgenome.org/).
2. Nagalakshmi, U., Wang, Z., Waern, K., et al. (2008) The transcriptional landscape of the yeast genome defined by RNA sequencing. *Science* **320**, 1344–1349.
3. Wilhelm, B. T., Marguerat, S., Watt, S., et al. (2008) Dynamic repertoire of a eukaryotic transcriptome surveyed at single-nucleotide resolution. *Nature* **453**, 1239–1243.
4. Shendure, J. (2008) The beginning of the end for microarrays? *Nat. Methods* **5**, 585–587.
5. Shendure, J., and Ji, H. L. (2008) Next-generation DNA sequencing. *Nat. Biotechnol.* **26**, 1135–1145.
6. Wodicka, L., Dong, H. L., Mittmann, M., Ho, M. H., and Lockhart, D. J. (1997) Genome-wide expression monitoring in *Saccharomyces cerevisiae. Nat. Biotechnol.* **15**, 1359–1367.
7. Taddei, A., Van Houwe, G., Nagai, S., Erb, I., van Nimwegen, E., and Gasser, S. M. (2009) The functional importance of telomere clustering: global changes in gene expression result from SIR factor dispersion. *Genome Res.* **19**, 611–625.
8. Castrillo, J. I., Zeef, L. A., Hoyle, D. C., et al. (2007) Growth control of the eukaryote cell: a systems biology study in yeast. *J. Biol.* **6**, 4.
9. Affymetrix Scientific Publications (2009). (Accessed at http://www.affymetrix.com/publications/full_list.affx).
10. Bustin, S. A. (2000) Absolute quantification of mRNA using real-time reverse transcription polymerase chain reaction assays. *J. Mol. Endocrinol.* **25**, 169–193.
11. Logan, J. M. J., Edwards, K. J., and Saunders, N. A. (2009) *Real-time PCR: Current Technology and Applications.* Wymondham, UK: Caister Academic Press.
12. Bustin, S. A., Benes, V., Nolan, T., and Pfaffl, M. W. (2005) Quantitative real-time RT-PCR – a perspective. *J. Mol. Endocrinol.* **34**, 597–601.
13. Morrison, T. B., Weis, J. J., and Wittwer, C. T. (1998) Quantification of low-copy transcripts by continuous SYBR (R) green I monitoring during amplification. *Biotechniques* **24**, 954–958, 960, 962.
14. dChip software (2009) (Accessed at http://biosun1.harvard.edu/complab/dchip/).
15. Allison, D. B., Cui, X., Page, G. P., and Sabripour, M. (2006) Microarray data analysis: from disarray to consolidation and consensus. *Nat. Rev. Genet.* **7**, 55–65.
16. Cui, X., Hwang, J. T., Qiu, J., Blades, N. J., and Churchill, G. A. (2005) Improved statistical tests for differential gene expression by shrinking variance components estimates. *Biostatistics* **6**, 59–75.
17. Storey, J. D. (2003) The positive false discovery rate: a Bayesian interpretation and the *q*-value. *Ann. Statistics* **31**, 2013–2035.
18. OligoPerfect™ Designer (2009) (Accessed at http://tools.invitrogen.com/content.cfm?pageid=9716).
19. Ng, R., and Abelson, J. (1980) Isolation and sequence of the gene for actin in *Saccharomyces cerevisiae. Proc. Natl. Acad. Sci. USA* **77**, 3912–3916.
20. Bustin, S. A., Benes, V., Garson, J. A., et al. (2009) The MIQE guidelines: minimum information for publication of quantitative real-time PCR experiments. *Clin. Chem.* **55**, 611–622.
21. Gentleman, R. C., Carey, V. J., Bates, D. M., et al. (2004) Bioconductor: open software development for computational biology and bioinformatics. *Genome Biol.* **5**, R80.
22. Bolstad, B. M., Irizarry, R. A., Astrand, M., and Speed, T. P. (2003) A comparison of normalization methods for high density oligonucleotide array data based on variance and bias. *Bioinformatics* **19**, 185–193.
23. Ririe, K. M., Rasmussen, R. P., and Wittwer, C. T. (1997) Product differentiation by analysis of DNA melting curves during the polymerase chain reaction. *Anal. Biochem.* **245**, 154–160.
24. Pfaffl, M. W., and Hageleit, M. (2001) Validities of mRNA quantification using recombinant RNA and recombinant DNA external calibration curves in real-time RT-PCR. *Biotechnol. Lett.* **23**, 275–282.
25. Pfaffl, M. W. (2001) A new mathematical model for relative quantification in real-time RT-PCR. *Nucleic Acids Res.* **29**, e45.

Chapter 6

Enrichment of Unstable Non-coding RNAs and Their Genome-Wide Identification

Helen Neil and Alain Jacquier

Abstract

Cryptic unstable transcripts (CUTs) have been recently described as a major class of non-coding RNAs. These transcripts are, however, extremely unstable in normal cells and their analyzes pose specific technical problems. In this chapter, after a brief introduction discussing general aspects associated with the analysis of non-coding RNAs, we provide details of methods to enrich, map, and quantify this unconventional class of transcripts.

Key words: Non-coding RNAs, CUTs, RNPs, affinity purification, 3′ Long-SAGE, deep sequencing, RNA-Seq.

1. Introduction

The term "transcriptome" most often refers to the whole set of messenger RNA (mRNA) molecules in a cell. However, this is quite different from the whole set of transcripts that are actually transcribed (whole transcriptome). Thus, for example, there are many transcripts that are not translated into proteins, non-coding RNAs (ncRNAs), with functional information, and involved in relevant cellular functions and complexes (e.g., ribonucleoproteins). Together with this, it is important to note that the majority of transcriptome studies usually measure the actual amount of transcripts present in a cell at a specific steady state, a balance between transcription and degradation rates. In the yeast *Saccharomyces cerevisiae*, the rate of mRNA degradation can vary by up to two orders of magnitude (1) with some short-lived transcripts difficult to detect in some cases. This shows that steady-state

J.I. Castrillo, S.G. Oliver (eds.), *Yeast Systems Biology*, Methods in Molecular Biology 759,
DOI 10.1007/978-1-61779-173-4_6, © Springer Science+Business Media, LLC 2011

transcriptional studies may reflect only partially the rich complexity of the whole transcriptome and the actual rates of transcript synthesis.

The distinction between actual rates of transcription and RNA levels becomes more relevant when it comes to the analysis of ncRNAs. Indeed, while many ncRNAs such as small nucleolar RNAs (snoRNAs) or transfer RNAs (tRNAs) are very stable, others, such as the recently discovered "cryptic unstable transcripts" (CUTs), have very short half-lives and are hardly detectable if not stabilized by mutations impairing RNA degradation machineries (2). In all, the latest studies are beginning to show a new picture in which the complete set of sequences actually transcribed is significantly wider than the originally annotated genomic features. Global analyzes of ncRNAs pose some specific problems which will be briefly discussed. After that, we will describe the technical approaches used to analyze a particular class of unstable ncRNAs, the cryptic unstable transcripts (CUTs).

1.1. Problems Specific to Non-coding RNA Global analyzes

1.1.1. Non-coding RNAs are Poorly Annotated

Messenger RNA (mRNA) coding genes are well annotated in yeast thanks to a relatively limited number of introns and the availability of a number of sequences of different species, allowing comparative studies of conserved protein coding sequences (3). The annotation of ncRNAs is more complex, even for ncRNAs belonging to well-described classes. With independence of this, thanks to the use of a combination of bioinformatics (4, 5) and genome-wide experimental studies (6), the majority of stable ncRNAs have been annotated in yeast (*see* "*Saccharomyces* Genome Database": http://www.yeastgenome.org/).

Stable ncRNAs do not, however, represent all reported ncRNAs. For example, several concurrent (7) or antisense (8–11) ncRNAs have been found implicated in the regulation of gene transcription. These unstable ncRNAs are poorly annotated in databases. In fact, in addition to those transcripts discovered through the analysis of their regulatory functions, a multitude of ncRNAs of yet unknown function have been discovered from genome-wide transcriptome analyzes in yeast (12–15). Cryptic unstable transcripts (CUTs), very short-lived PolII transcripts, are likely to represent the major fraction of these ncRNAs (2, 16–18). Because they are difficult to detect and highly heterogeneous, they have not yet been annotated in databases and are not taken into account in common DNA microarrays. Their global analysis requires the use of tiling arrays or RNA-Seq approaches (see below).

1.1.2. Microarrays and RNA-Seq

Many transcriptome analyzes focus on mRNAs and largely ignore ncRNAs. Thus, many of the yeast microarrays that have been used include a limited number of ncRNA probes. Specialized arrays have been developed to study ncRNAs and in particular their maturation (19). Tiling arrays, though still significantly more expensive, now favorably compete with these specialized arrays in term of versatility and amount of information they can gather. However, even the most sophisticated tiling arrays bear some limitations: (i) they do not reach one nucleotide resolution; (ii) they remain sensitive to variations of GC contents in the sequences even after normalization with DNA hybridization (12); (iii) segmentation algorithms used to define transcripts boundaries are difficult to handle; (iv) they are unable to efficiently discriminate overlapping transcripts.

In theory, exhaustive sequencing of complementary DNAs (cDNAs) should provide the ultimate description of a transcriptome. Such an effort has been performed with traditional sequencing technologies for the *S. cerevisiae* mRNA transcriptome (13), but the amount of work it requires made it unsuitable to repeated analyzes. The emergence of new "deep-sequencing" technologies is making these approaches accessible to more routine assays. These technologies are evolving so fast that the specific techniques that we will describe here are likely to be outdated very rapidly. Nevertheless, the overall principles should remain similar. These new sequencing technologies are presently characterized by the fact that they produce large numbers of sequence reads but of relatively small to even very small lengths. At this moment, there are mainly three main technologies available, the Genome Sequencer FLX from 454 Life Sciences/Roche, the Illumina Genome Analyzer, and the Applied Biosystems SOLiD. While the first one produces, per run, about 400,000 sequence reads of a few hundred nucleotides (200–400 nt), the last two generate very large amounts (several tens of millions) of sequence reads, but of very short lengths (a few tens of nucleotides).

The ability to read billions of base pairs per run opens the possibility to characterize the entire yeast transcriptome by deep cDNA sequencing with reasonable effort, an approach now called RNA-Seq. In the most straightforward approach (14), yeast poly(A) RNAs were converted to double-stranded cDNAs (by oligo-dT and hexamer priming) that were essentially treated as genomic DNA, giving rise to ~30 million reads of 35 bp, out of which 15.8 million mapped to unique genomic sequences. About 91.5% of the yeast annotated ORFs were found to generate tags above background, a number similar to the 90% percentage obtained from tiling arrays (12), indicating that both techniques exhibited similar sensitivity. Although RNA-Seq was significantly more accurate in precisely defining both ends of transcripts (attaining single nucleotide resolution), a major drawback of this approach is that it did not allow direct RNA strandedness determination (transcript strandednesses were inferred indirectly by looking for sequence tags associated to short non-coded adenosine stretches that were interpreted as a signature of 3′ ends).

In order to circumvent this limitation, the 3′ and 5′ Illumina-PCR/sequencing primers can be directly and sequentially ligated to fragmented RNAs and used to generate short double-stranded cDNAs that are directly sequenced (20). The sequential ligation of the 3′ and 5′ primers allows RNA strand determination and the discrimination of transcripts overlapping on opposite strands. Finally, previously described techniques, called 5′ and 3′ long serial analysis of gene expression (5′ and 3′ Long-SAGE) (21), were efficient approaches to provide a detailed description of transcriptomes by defining the ends of transcripts genome wide and at the nucleotide level. Their adaptation to deep sequencing technologies should boost their utility (see below).

1.1.3. Non-coding RNA Enrichment

When performing global transcriptome analyzes, it is often advisable to enrich the analyzed RNA fraction for the type of transcripts one is interested in. For example, tiling arrays analyzes of mRNAs have been shown to provide much cleaner results (in term of signal-over-noise ratio) when starting from a poly(A) fraction than when starting from total RNA preparations (12). Likewise, genome-wide cDNA sequencing requires poly(A) enrichment to avoid sequencing mostly ribosomal RNAs (rRNAs). Unfortunately, a number of ncRNAs are not polyadenylated. It is thus important to find an alternative way to enrich the RNAs of interest, in particular to lower the amount of rRNA as much as possible. Subtractive hybridization of the rRNA, as often used for prokaryotic transcriptome analyzes, could in principle be used, but, to our knowledge, its usage is infrequent in yeast. Because many ncRNAs are short, size fractionation could be used, but will unavoidably be associated with strong bias. Affinity purification of ribonucleoprotein complexes (RNPs) is an alternative route. While it will restrict the analysis of RNAs associated with only a specific protein, it has been used with success to determine the complete set of RNAs of a given class (6, 22, 23).

1.2. Analysis of an RNA Fraction Enriched for the CUTs

CUTs are widespread transcripts first described in yeast as very unstable ncRNAs (2). They are relatively short (a few hundred nucleotides) and highly heterogeneous at their 3′ ends, as a result of their particular mode of transcription termination that involves the Nrd1/Nab3/Sen1 complex (24, 25). In order to study the potential function of this pervasive transcription, or its "raison d'être," we devised an approach to generate their complete genomic map.

CUTs are polyadenylated by the TRAMP complex that targets these transcripts for rapid degradation by the nuclear exosome (2). In a strain deleted for Rrp6, a factor specific to the nuclear exosome, and upon depletion of Trf4, the poly(A) polymerase of TRAMP, CUTs accumulate in the nucleus as an unpolyadenylated form. Because they are capped, they can be highly enriched from such a strain by co-purification with the nuclear cap-binding complex. An RNA fraction highly enriched in CUTs was thus obtained by extracting the RNAs associated with Cpb20 (Cpb20/Cbc2 is one of the two subunits of the nuclear cap-binding complex) purified by tandem affinity purification (TAP) (26) from yeast cells deleted for *RRP6*, depleted for Trf4, and harboring a TAP tagged version of the Cbp20 factor (17).

Because CUTs were anticipated to overlap mRNAs, in both sense and anti-sense directions, we wanted to use an approach that would unambiguously discriminate overlapping transcripts. Thus, in addition to hybridization on tiling arrays according to an adapted protocol (12), we developed a robust 3′ Long-SAGE approach adapted to the Genome Sequencer FLX from 454 Life Sciences/Roche to precisely describe the heterogeneous 3′ ends of CUTs. The use of the 3′ Long-SAGE approach turned out to be indeed determinant to distinguish overlapping CUTs and mRNA 5′-UTRs.

In this chapter, we will provide details of the procedure we used to enrich the RNAs associated with Cpb20 adapted from (26); a similar procedure that incorporates some potentially useful modifications can be found in (23). Then, we will provide the experimental details allowing the robust synthesis of 3′ Long-SAGE tags in a format adapted to sequencing with the FLX/454 technology. Note that we will not describe in this chapter the relevant bioinformatics steps necessary for the appropriate analyzes of the sequencing output. These procedures, that can turn out to be rate limiting in this type of project, are described in detail in the supplementary material in (17).

2. Materials

2.1. Cell Culture

1. YPD broth (BD Difco).
2. Doxycycline (Sigma) is dissolved in H_2O at 10 mg/mL, stored at −20°C and then added to cell cultures at a final concentration of 10 μg/mL.

2.2. Tandem Affinity Purification Adapted to Preserve RNA Integrity

1. Breaking buffer: 0.1 M Tris–HCl pH 8.0, 0.1 M NaCl, 2X protease inhibitor cocktail (Complete, Roche), 20 mM ribonucleoside vanadyl complex (New England Biolabs) and 300 U/mL rRNasin (Promega).
2. Cryogenic impact grinder supplied (Freezer Mill 6770, Spex) with grinding beads (e.g., 425–600 μm acid-washed glass beads, Sigma).

3. Igepal® CA-630 (Sigma).
4. Binding buffer: 20 mM Tris–HCl pH 7.4, 0.1 M NaCl, and 0.1% (v/v) Igepal.
5. IgG Sepharose™ 6 Fast Flow (GE Healthcare).
6. DTT-binding buffer: binding buffer supplemented with 1 mM DTT.
7. AcTEV™ Protease (Invitrogen).
8. $CaCl_2$-binding buffer: binding buffer supplemented with 2 mM $CaCl_2$.
9. Calmodulin Sepharose™ 4B (GE Healthcare).
10. Elution buffer: 20 mM Tris–HCl pH 8, 50 mM NaCl and 5 mM EGTA.
11. Phenol–chloroform pH 5.2 premixed with isoamyl alcohol (25:24:1) (v/v) (Amresco).
12. Glycogen RNA grade (20 μg/μL).
13. Ammonium acetate solution (7.5 M) mixed with absolute ethanol at ratio 1:6 (v/v).

2.3. Polyadenylation of Purified RNA

1. RNeasy columns (Qiagen).
2. Poly(A) Tailing Kit (Ambion), supplied with reagents.
3. Cordycepin 5′-triphosphate (3′-deoxyadenosine 5′-triphosphate, Sigma).

2.4. 3′ Long-SAGE

1. SuperScript Double-Stranded cDNA Synthesis Kit (Invitrogen).
2. 5-Methyl 2′-deoxycytidine 5′-triphosphate (methyl-dCTP, Roche).
3. [α-^{32}P] dATP (3,000 Ci/mmol, Perkin Elmer).
4. 1X "First Strand" buffer: (50 mM Tris–HCl pH 8.3, 75 mM KCl, and 3 mM $MgCl_2$), 10 μM DTT, 0.5 mM dNTP mix (with methyl-dCTP replacing dCTP; *see* **Section 3.4**).
5. 1X "Second Strand" buffer (20 mM Tris–HCl pH 6.9, 90 mM KCl, 4.6 mM $MgCl_2$, 0.15 mM β-NAD^+, and 10 mM $(NH_4)_2SO_4$), 0.2 mM dNTP mix (with dCTP instead of methyl-dCTP), 0.6 mM dCTP (*see* **Section 3.4**).
6. *Escherichia coli* DNA ligase (New England Biolabs).
7. *E. coli* DNA polymerase (New England Biolabs).
8. *E. coli* RNase H (New England Biolabs).
9. QIAquick PCR Purification Kit (Qiagen).
10. Dynabeads® MyOne™ Streptavidin T1 magnetic beads (Dynal), 1X binding and washing buffer (1X B&W: 10 mM

Tris–HCl pH 7.5, 1 mM EDTA, and 2 M NaCl), and 2X B&W buffer.

11. *Gsu*I (5 U/μL, Fermentas) supplied with 10X B buffer.
12. $T_{10}N_{50}E_{0.1}$ solution: 10 mM Tris–HCl pH 8, 50 mM NaCl, and 0.1 mM EDTA pH 8.
13. T4 DNA ligase (2,000 U/μL, New England Biolabs) supplied with 10X T4 DNA ligase buffer.
14. *Mme*I (2 U/μL, New England Biolabs) supplied with 10X NEB4 buffer and *S*-adenosylmethionine (SAM).
15. Non-denaturing polyacrylamide gel: 8% (w/v) of 4X acrylamide–2X bisacryl (mix 19:1, Eurobio), 1X Tris borate EDTA (TBE), 5% (w/v) glycerol, 0.04% (w/v) ammonium persulfate (APS), and 0.15% (w/v) *N*,*N*,*N*′,*N*′-tetramethylethylenediamine (TEMED).
16. Gel extraction buffer: 0.3 M NaCl, 0.2 M Tris–HCl pH 7.5, 25 mM EDTA, and 2% (w/v) sodium dodecyl sulfate (SDS).
17. *Bsa*I (10 U/μL, New England Biolabs) supplied with 10X NEB3 buffer.
18. NucleoSpin Extract II Kit (Macherey-Nagel).

3. Methods

The 3′ Long-SAGE technique allows, like the first described SAGE technique (27), to obtain strand-specific tags that identify transcripts while their number reflects the relative abundance of these transcripts. The 3′ Long-SAGE tags are characterized by the fact that (i) they are longer (18 nt compared to 14 nt), facilitating the precise genomic location with fewer ambiguities and (ii) they are located upstream of the poly(A) tail, thus defining the termination sites at a single nucleotide resolution.

Different steps of the initially described protocol (21) were modified to improve efficiency and the concatenation of the tags was adapted to 454 pyro-sequencing with final products of about 150 bp containing four tags framed by defined linkers (tetratags). The overview of the technique is described in **Fig. 6.1**. Note that when this protocol was designed, the FLX/454 was the only deep sequencing technology available. Presently, the use of the Solexa/Illumina or SOLiD technologies might be more appropriate for tag sequencing. The oligonucleotide sequences used here can be easily modified to adapt to these new technologies. Because these technologies generate shorter reads, the concatenation steps could then be advantageously skipped.

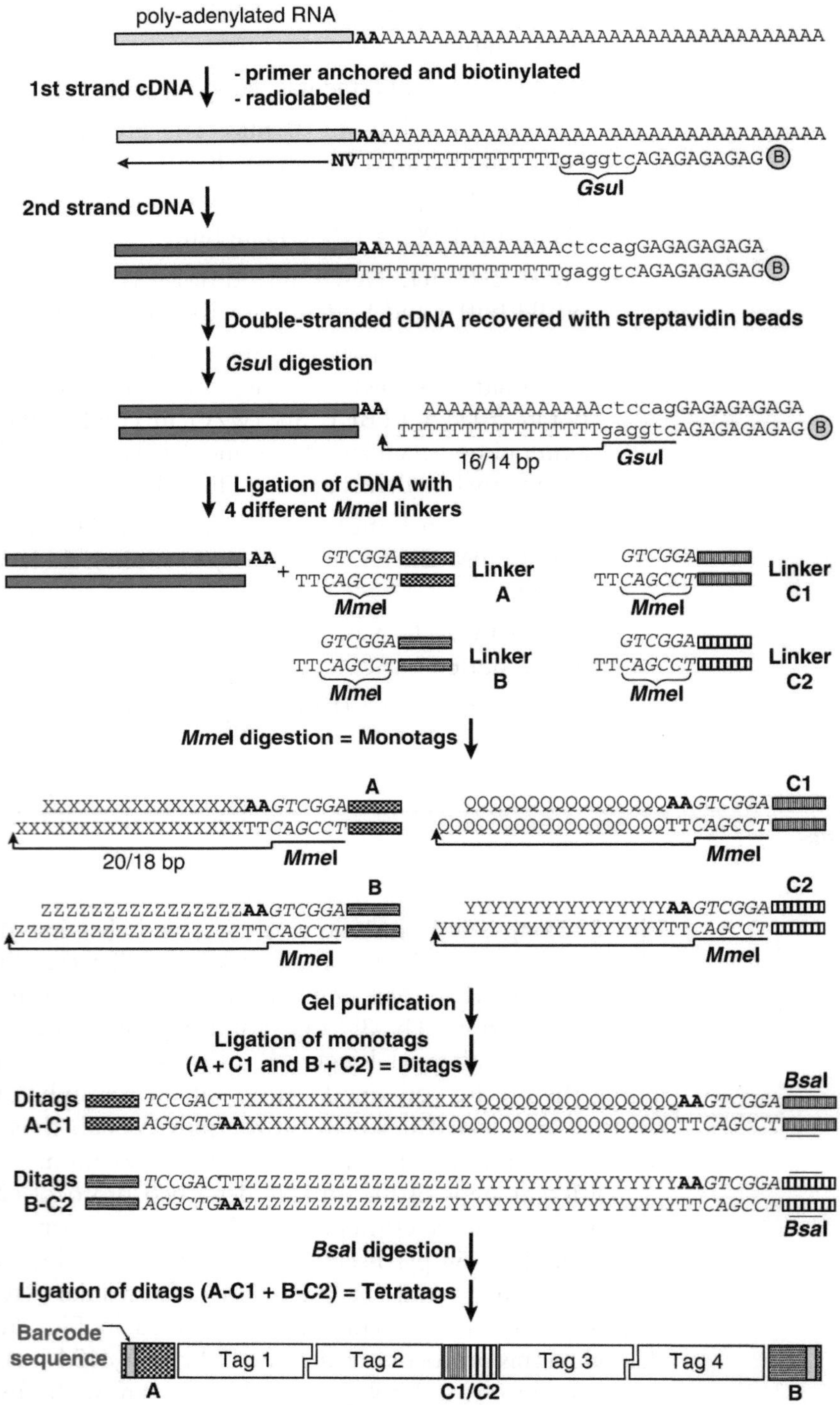

Fig. 6.1. Overview of the technique to synthesize 3′ Long-SAGE tags. The oligo-d(T) primer is 5′-biotinylated (*circled* B) and contains a *GsuI* site (*in lower case*). After *GsuI* digestion, the overhung "AA" (*in bold face*) is released and indicates the transcript orientation. The linkers A, B, C1, and C2 contain a *MmeI* site (*in italics*).

3.1. Preparation of Cell Extracts by Cryogenic Grinding

In our case study, 6 L of yeast cells is grown in YPD at 25°C (temperature depends on the strain, usually 30°C for a wild-type strain) up to an OD_{600} of 0.07, at which point doxycycline is added in order to repress the *TRF4* gene that has been placed under the control of tetracycline-sensitive transcription activator. The culture is incubated for further 16 h at 25°C to reach the mid-log phase with OD_{600} about 0.6. The cells are collected by centrifugation at 4°C at 3,000×*g* in a high-capacity centrifuge, washed in cold (4°C) water, re-suspended in 12 mL of cold breaking buffer, and frozen in liquid nitrogen as drops that can be stored at –80°C until use.

The frozen cells are broken with a cryogenic impact grinder (*see* **Note 1**). The equivalent of 3 mL of grinding beads undergoes two cycles of 30 s of grinding, at a rate of 10 cycles per second with a pause of 30 s between the two cycles. The powder is recovered and it can be kept at –80°C until use.

3.2. Tandem Affinity Purification of Ribonucleoprotein Complexes (RNPs)

The frozen powder is thawed in a cold-water bath and centrifuged in a Nalgene® tube at 22,000×*g* for 45 min. The supernatant is recovered in a 50 mL Falcon™ conical tube and Igepal is added to a final concentration of 0.1% (w/v).

The tandem affinity purification entails (1) incubation of the extracts with IgG sepharose beads to allow binding to the IgG domain. (2) After washing, the elution of the bound complexes by incubation with TEV protease, which cleaves the protein at the TEV recognition site. (3) The binding of the eluted complexes to a calmodulin sepharose column via the calmodulin-binding domain epitope. (4) After washing, since this binding is dependent on the presence of calcium, the purified complex can be eluted by addition of a calcium chelator (e.g., EGTA) (26).

In our protocol, 700 μL of IgG Sepharose™ suspension is equilibrated in a 15 mL Falcon™ conical tube by washing twice with 10 mL of cold binding buffer. Centrifugation of the beads must be performed at 100×*g* for 2 min at 4°C. The beads are then added to the total extract and binding is allowed during 2 h at 4°C on a rotating wheel at low speed. After centrifugation, the beads are washed twice by incubation in 10 mL of binding buffer for 10 min at 4°C on a rotating wheel at low speed and once in 10 mL of DTT-binding buffer.

The beads are re-suspended in 600 μL of DTT-binding buffer and 200 U of AcTEV protease is added to the suspension. The reaction is incubated for 2 h at 16°C on a rotating wheel at low speed. The supernatant is recovered and the beads are washed twice with 200 μL of cold binding buffer. Each supernatant is recovered, pooled with the first one in a 2 mL Eppendorf tube, and $CaCl_2$ is added to a final concentration of 2 mM.

450 μL of Calmodulin Sepharose™ suspension is equilibrated in a 15 mL Falcon™ conical tube by washing twice with 10 mL of cold $CaCl_2$-binding buffer. Centrifugation of the beads must be performed at 100×*g* for 2 min at 4°C. The beads are then added to the pooled supernatants and binding is allowed for 50 min at 4°C on a rotating wheel at low speed. The beads are recovered by centrifugation and washed three times with 1 mL of cold $CaCl_2$-binding buffer by inverting the tube several times. Elution is performed by incubating the beads in 500 μL of elution buffer for 5 min at room temperature, twice.

Finally, the supernatants are pooled in a 2 mL Eppendorf tube and treated with 1 mL of acidic phenol/chloroform: vortexed for 30 s and centrifuged for 5 min at 16,000×*g* at room temperature and this treatment is performed twice. The aqueous phase is precipitated (*see* **Note 2**) and the pellet, containing the TAP-purified RNA, is re-suspended in 50 μL of H_2O.

3.3. In Vitro Polyadenylation of TAP-Purified RNA

The 3′ Long-SAGE procedure uses polyadenylated RNA. But in our case, the CUTs were enriched from a strain depleted for the poly(A) polymerase Trf4, component of the TRAMP complex that polyadenylates RNAs for targeting to exosome degradation (2), and are thus not polyadenylated. Thus, the RNAs from the CUT fraction must first be submitted to an in vitro polyadenylation step prior to be used in the 3′ Long-SAGE protocol. This step is obviously not required when using naturally polyadenylated RNAs such as mRNAs.

Purified RNA is first cleaned up using RNeasy columns (*see* **Note 3**) as described in the manufacturer's protocol and elution is performed twice consecutively with 50 μL of H_2O, pre-heated to 65°C to increase elution efficiency.

Polyadenylation is performed by incubating 1 μg of RNA from the previous step in H_2O at 65°C for 10 min. If necessary, the volume of eluted RNA can be reduced by centrifugation in a speed vacuum. 1X E-PAP buffer, 2.5 mM $MnCl_2$, and 1 mM ATP are added on ice. Cordycepin is also added to the reaction at 50 μM (*see* **Note 4**). Finally, 4 U of *E. coli* PAP (poly(A)-polymerase) is added and the reaction is incubated for 1 h at 37°C. The reaction is precipitated (*see* **Note 2**).

3.4. Synthesis of 3′ ong-SAGE Tags

The first cDNA strand is synthesized by incubating the RNA from previous step, re-suspended in H_2O to a final volume of 10 μL, with 1 μg of oligo-d(T) primer (with a 5′-biotin and a *Gsu*I site, **Table 6.1**; *see* **Note 5**) for 10 min at 65°C and then put on ice. 1X "First Strand" buffer (50 mM Tris–HCl pH 8.3, 75 mM KCl, and 3 mM $MgCl_2$), 10 mM DTT, 0.5 mM dNTP mix (with methyl-dCTP replacing dCTP; *see* **Note 6**), 20 μCi [α-^{32}P] dATP (*see* **Note 7**), and 200 U of SuperScript™ II RT are added to the tube to a final volume of 20 μL and the reaction is incubated

Table 6.1
Oligonucleotide sequences. *GsuI* site is in lower case, *MmeI* sites are italicized, labels are in bold face and *BsaI* sites are underlined. The oligo-d(T) primer and the linkers must be PAGE purified

Name	Sequence
Oligo-d(T) primer	5′-(biotin)GAGAGAGAGActggagTTTTTTTTTTTTTTTTTTVN-3′
Linker A–a	5′-CCCGCCTCCCTCGCGCCATCAG**CGTCGAC***TCCGAC*TT-3′
Linker A–b	5′-(phosphate)*GTCGGA***GTCGACG**CTGATGGCGCGAGGGAGGC-3′
Linker B–a	5′-CCCGCCTTGCCAGCCCGCTCAG**AGATCTC***TCCGAC*TT-3′
Linker B–b	5′-(phosphate)*GTCGGA***GAGATCT**CTGAGCGGGCTGGCAAGGC-3′
Linker C1–a	5′-CCCTCACTCAGGCTCTAGCTCATCTATCACACAT<u>GGTCTC</u>GAG *TCCGAC*TT-3′
Linker C1–b	5′-(phosphate)*GTCGGA*CTC<u>GAGACC</u>ATGTGTGATAGATGAGCTAGAG CCTGAGTGA-3′
Linker C2–a	5′-CCCTCACTCAGGCTCTAGCTCATCTATCACACAT<u>GGTCTC</u>GGAC *TCCGAC*TT-3′
Linker C2–b	5′-(phosphate)*GTCGGA*GTCC<u>GAGACC</u>ATGTGTGATAGATGAGCTAGA GCCTGAGTGA-3′
Primer A	5′-GCCTCCCTCGCGCCATCAG-3′
Primer B	5′-GCCTTGCCAGCCCGCTCAG-3′
Primer C	5′-TCACTCAGGCTCTAGCTCATCTA-3′

for 1 h at 43°C and put on ice. The second cDNA strand is synthesized by adding to the previous reaction 1X "Second Strand" buffer (20 mM Tris–HCl pH 6.9, 90 mM KCl, 4.6 mM $MgCl_2$, 0.15 mM β-NAD^+ and 10 mM $(NH_4)_2SO_4$), 0.2 mM dNTP mix (with dCTP instead of methyl-dCTP), 0.6 mM dCTP (*see* **Note 6**), 10 U of *E. coli* DNA ligase, 40 U of *E. coli* DNA polymerase, 2 U of *E. coli* RNase H, and H_2O to a final volume of 150 μL and incubating for 2 h at 16°C. 10 U of T4 DNA polymerase is added to the reaction and the incubation is continued for 5 min at 16°C. 10 μL of EDTA 0.5 M is finally added to stop the reaction. A volume/volume phenol/chloroform extraction is performed and the aqueous phase is purified on a QIAquick column as described in the manufacturer's protocol, except that elution is performed twice consecutively with 50 μL of elution buffer pre-heated to 65°C to enhance the efficiency of elution.

200 μL of Dynal (Dynabeads MyOne) magnetic beads is prepared by washing three times in 200 μL of 1X B&W buffer with a magnet, as described in the manufacturer's protocol. 100 μL of 2X B&W buffer is added to the 100 μL of elution from the previous step, this 200 μL is added to the washed beads and incubated on a rotating wheel for 30 min at room temperature. The

beads are washed twice with 200 μL of 1X B&W buffer and twice with 200 μL of 1X B buffer.

The *Gsu*I digestion is performed by adding 30 U of enzyme to the beads from the previous step re-suspended in 200 μL of 1X B buffer. The reaction is incubated for 2 h at 30°C and the beads are maintained in suspension by occasionally flicking the tube. The supernatant is recovered and a phenol/chloroform extraction is performed, followed by a chloroform extraction. The aqueous phase is precipitated (*see* **Note 2**) and the pellet is re-suspended in 40 μL of H_2O.

The four different linkers A, B, C1, and C2 (*see* **Note 8**) are prepared independently by annealing 5 μg of primer "a" with 5 μg of its complementary primer "b" (**Table 6.1**). This mix of primers is dried completely in a speed vacuum, re-suspended in 10 μL of $T_{10}N_{50}E_{0.1}$ solution and incubated in a thermal cycler with the following annealing program: 2 min at 95°C, 2 min at 75°C, 2 min at 60°C, 2 min at 50°C, 2 min at 40°C, and 2 min at 20°C. The cDNA from the previous step, cut by *Gsu*I, is then shared to perform four independent ligations by incubating 10 μL of cDNA with 10 μL of annealed linker, 1X T4 DNA ligase buffer, 4,000 U of T4 DNA ligase, and H_2O to a final volume of 30 μL, overnight at 16°C. The four reactions are purified on Qiaquick columns as described in the manufacturer's protocol, except that elution is performed twice consecutively with 50 μL of elution buffer pre-heated to 65°C.

Elutions from the previous step are digested with *Mme*I by adding 50 μM SAM, 1X NEB4 buffer, 4 U of *Mme*I enzyme, H_2O to a final volume of 100 μL and incubating for 1 h 30 min at 37°C. 100 μL of H_2O is added to the reactions and a phenol/chloroform extraction is performed. The aqueous phases are then precipitated (*see* **Note 2**) and the pellets are re-suspended in 5 μL of H_2O and 5 μL of 2X DNA loading buffer. The samples are loaded on a 8% (w/v) non-denaturing polyacrylamide gel and the migration is performed in 1X TBE until the bromophenol blue is about 10 cm from the top. The expected bands are 50 bp + 5 nt long for the monotags A and B, and 64 bp + 5 nt long for the monotags C1 and C2 (**Fig. 6.2**). These bands are recovered from the gel (**Fig. 6.3**; *see* **Note 9**) and then incubated independently in 200 μL of gel extraction buffer overnight at room temperature. The supernatants are recovered and a phenol/chloroform extraction is performed. The aqueous phases are precipitated (*see* **Note 2**) and the pellets are re-suspended in 8 μL of H_2O.

The pool of monotags A and B is ligated with the pool of monotags C1 and C2, respectively, to produce A–C1 and B–C2 ditags (*see* **Note 10**). Two independent reactions are then performed by incubating 8 μL of gel-purified monotags A with 8 μL of gel-purified monotags C1 (and the same for B and C2), 1X T4

Fig. 6.2. Design of the monotags and ditags. The locations of primers A, B, C1 and C2 in linkers A, B, C1 and C2, respectively, are indicated and the *Mme*I sites are italicized. The labels in linkers A and B are in *bold face*. X, Z, Q, and Y letters correspond to the 18/16 nt-long tags released after *Mme*I cleavage. The *Bsa*I sites in linkers C1 and C2 are underlined with the sequences distinguishing C1 and C2 delimited by a *dotted line*. The *slash* represents the cleavage after *Bsa*I digestion of ditags.

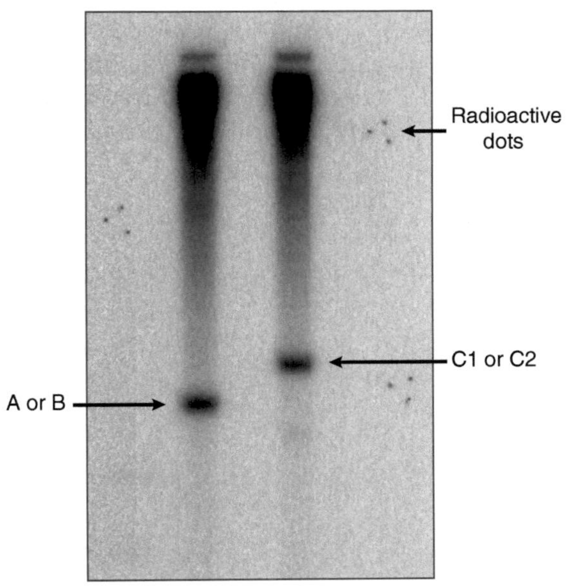

Fig. 6.3. Gel purification of monotags. After *Mme*I digestion, the cDNAs are loaded on a 8% (w/v) non-denaturing polyacrylamide gel to purify the monotags thus released. Note that the monotags A and B are shorter than C1 and C2. The radioactive marks, allowing the recover of these products, are indicated.

DNA ligase buffer, and 4,000 U of DNA ligase in a final volume of 20 μL, overnight at 16°C. A phenol/chloroform extraction is performed, the aqueous phases are precipitated (*see* **Note 2**) and the pellets are re-suspended in 5 μL of H_2O and 5 μL of 2X DNA loading buffer. The samples are loaded on a 8% (w/v) non-denaturing polyacrylamide gel and the migration is performed in 1X TBE until the bromophenol blue is about 10 cm from the top. The expected bands are 116 bp + 6 nt and 117 bp + 6 nt long for the heterogeneous A–C1 and B–C2 ditags, respectively (*see* **Note 11** and **Fig. 6.2**) and are recovered from the gel (**Fig. 6.4**; *see* **Notes 9–11**). After incubation in gel extraction buffer, phenol/chloroform extraction, and precipitation (*see* **Note 2**), the ditags are re-suspended in 20 μL of H_2O.

The ditags A–C1 and B–C2 are PCR-amplified using primers A and C and primers B and C, respectively (**Table 6.1**), and 1 μL of a 1:10 (v/v) dilution of DNA matrix in a 50 μL final volume reaction, twice for each. The following program is applied: 2 min at 95°C [30 s at 95°C, 30 s at 49°C, and 30 s at 72°C] 20 times, and 5 min at 72°C. At the end of this step, Taq polymerase, dNTP mix, and each primer are added again to the tube and a second program is performed: [2 min at 95°C, 1 min 30 s at 49°C, and 5 min at 72°C] once (*see* **Note 12**). The PCR products are then purified on QIAquick columns as described in the manufacturer's protocol and elution is performed once with 50 μL of elution buffer pre-heated to 65°C.

The PCR products are digested with *Bsa*I by incubating the elution from previous step with 1X NEB3 buffer, 10 U of *Bsa*I enzyme, and H_2O to a final volume of 20 μL for 1 h at 50°C. The reactions are precipitated (*see* **Note 2**) and the pellets are

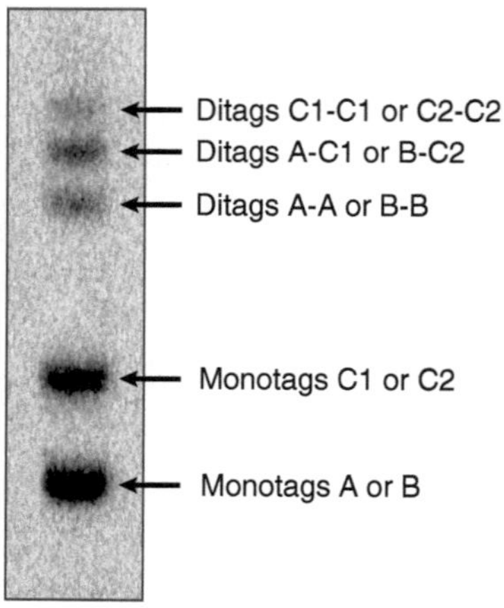

Fig. 6.4. Gel purification of ditags. After ligation of the monotags, the reactions are loaded on a 8% (w/v) non-denaturing polyacrylamide gel to separate the different ditags that are obtained, homogeneous (A–A, B–B, C1–C1 or C2–C2) and heterogeneous (A–C1 or B–C2). Note that a majority of tags remain as monotags.

re-suspended in 20 μL of H_2O. 20 μL of 2X DNA loading buffer is added and the samples are loaded on a 3% (w/v) agarose gel. The sizes of the expected bands are 116 and 117 bp for the full-length amplified A–C1 and B–C2 ditags, 74 bp + 7 nt and 75 bp + 4 nt, respectively, and 38 bp + 4 nt for the digested ditags (**Fig. 6.2**). The 74 and 75 bp long bands are excised from the gel, purified on a NucleoSpin column, and eluted with 50 μL of elution buffer pre-heated to 65°C.

The purified ditags from previous step are pooled in the same tube and the volume is reduced in a speed vacuum. The ligation of these ditags into tetratags A-C1-C2-B is performed by adding 1X T4 DNA buffer, 4,000 U of T4 DNA ligase, and H_2O to a final volume of 50 μL, and incubating overnight at 25°C. The reactions are precipitated (*see* **Note 2**) and the pellets are re-suspended in 10 μL of H_2O. This 10 μL of tetratag matrix is then PCR amplified using primers A and B (**Table 6.1**) with the following program: 2 min at 95°C [30 s at 95°C, 30 s at 50°C, and 30 s at 72°C] five times and 5 min at 72°C. The reaction is mixed with 2X DNA loading buffer and loaded on a 3% (w/v) agarose gel. The band, with an expected size of 154 bp, is excised from the gel, purified on a QIAquick column, and eluted with 50 μL of elution buffer pre-heated to 65°C. These final products can be directly used for 454 sequencing (specific adapters will be ligated).

3.5. Assessment of Transcript Integrity

When this technique was performed using an oligo-d(T)-enriched RNA fraction (mRNA fraction), the distribution of the vast majority of tags peaks around 150 nt downstream of the stop codon of open reading frames (17), corresponding to the average size of the 3′ UTRs (14). This constitutes a good control for the fact that the 3′ specificity of the SAGEs obtained by our procedure was good. This 3′ end specificity relies on the first cDNA synthesizing steps, which uses oligo-d(T) priming. In the case of mRNAs, the poly(A) tail is added naturally, in vivo, prior to RNA purification. If some limited degradation of the RNA occurs during the purification procedure, this should thus not affect the 3′ end specificity of the SAGE tags that are obtained by the described procedure.

The situation is radically different for ncRNAs such as the CUTs. Indeed, the purification is performed after incubation in whole cell extracts, which maximizes the risks of experiencing RNA degradation. Moreover, and most importantly, polyadenylation is performed subsequently in vitro. The poly(A) tail thus does not mark the natural 3′ end of these RNAs but rather the 3′ end of all forms after purification, including the partially degraded ones. It is thus essential to have some means to evaluate to what extent this might adversely affect the final description of these transcripts. One way to assess the specificity and the quality of the determination of termination sites is to analyze the results obtained for some well-reported ncRNAs. For example, the U1 small nuclear ribonucleoprotein (snRNP), which is known to bind the nuclear cap binding complex, is indeed highly enriched in our TAP experiment. More than 90% of the 3′ Long-SAGE tags issued from this RNA correspond exactly to the natural 3′ end of the U1 small nuclear RNA. However, U1 is partially protected from degradation because it is embedded in ribonucleoprotein particles. This control is thus not very demanding and we looked for a more stringent control, such as the 3′ ends of RNAs not expected to be protected from degradation. For example, such RNA 3′ ends could be expected from the products of the Rnt1 RNA cleavages that represent intermediates of nuclear degradation or maturation pathways. Some of these were mapped in vitro and in vivo for snoRNA precursors such as *snR50* (28) or from pre-mRNAs that are down-regulated by Rnt1 such as *RPS22b* (29). As shown in **Fig. 6.5**, the expected cleavage pattern of Rnt1 at these sites is detected in our experiment. This indicates that, if

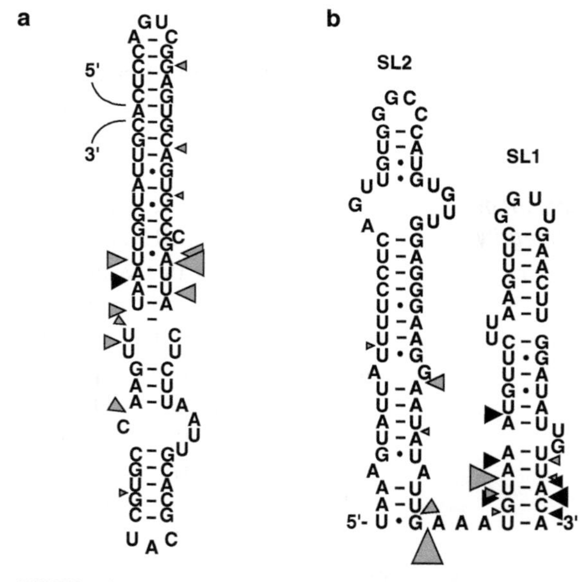

Fig. 6.5. Rnt1 cleavage sites of the *snR50* precursor (**a**) and the *RPS22b* pre-messenger (**b**). *Black triangles* correspond to Rnt1 cleavage sites previously identified in vitro and in vivo (28, 29) in comparison with the *gray triangles* corresponding to the 3′ ends we identified by 3′ Long-SAGE (17). Their surface is proportional to the number of tags mapping the corresponding sites. The 3′ ends expected for the cleavage sites were identified in our experiment. Moreover, additional 3′ ends are observed within a dozen nucleotides upstream of the expected cleavage sites. We could not distinguish if they result from a partial in vivo degradation involving the compromised exosome or from an in vitro degradation happening during the experiment. In any events, these results show that intact RNAs are preserved and that, if some partial degradation occurs during the enrichment steps, it must remain very limited.

some RNA degradation occurs during the procedure, it must be limited and does not preclude the identification of the authentic ncRNA 3′ ends.

4. Notes

1. It is crucial to perform the grinding of cells in liquid nitrogen because it considerably prevents RNA degradation, compared to French press lysis, even in the presence of RNase inhibitors.
2. For precipitation, 1 μL of glycogen (20 μg/μL) is first added to the RNA extract and mixed. Then, 3.5 volumes of ammonium acetate/ethanol mix are added and mixed. The tube is incubated overnight at –20°C, or 30 min at –80°C, and centrifuged for 20 min at 16,000×*g* at 4°C. The pellet is washed with 1 mL of ethanol 70% (v/v) and dried at room temperature.
3. Even after phenol extraction, precipitation, and ethanol washing of the TAP-purified RNA, the polyadenylation reaction may be inhibited by remaining contaminants (this can be analyzed by Northern blot by checking the length of a transcript). This can be caused, for example, by the presence of residual EGTA from the elution buffer which may inhibit enzymatic reactions. A step in which RNA is first cleaned up on specific RNA columns may be necessary.
4. Cordycepin (3′-deoxyadenosine) is added to the polyadenylation reaction, with a ratio of 1:20 (v/v) relative to ATP, in order to limit the final length of the poly(A) tail.
5. The oligo-d(T) primer finishes by the nucleotides "VN," as an anchor at the 3′ end, allowing the primer to exactly anneal at the beginning of the poly(A) tail. It also contains a 5′ biotin in order to recover on beads, at an early stage, only the cDNAs that are specifically produced from the annealing of this primer. Finally, a *Gsu*I site is located upstream from 16 "T" meaning that after the cleavage by *Gsu*I, that occurs 18 and 16 nt downstream of the recognition site, the poly(A) tail is completely removed from the double-stranded cDNA, leaving the overhung "AA" as a mark of transcript orientation and a platform to anneal the linkers.
6. For the synthesis of the first-strand DNA, methyl-dCTP is used instead of dCTP in order to protect the double-strand cDNA from an ectopic *Gsu*I digestion (*Gsu*I is a

methylation-sensitive enzyme). However, during the second-strand synthesis, an excess of dCTP is added to ensure that the complementary sequence of the linkers will not be methylated to maintain a recognizable *Gsu*I site.

7. The radioactive labeling of the cDNA is very convenient to monitor and check each step of the protocol, particularly the gel purifications.

8. The linkers A, B, C1, and C2 correspond to two complementary primers, "a" and "b," that are annealed together. The primer "b" is phosphorylated at the 5′ end to allow ligation of the overhang "TT" from the primer "a" with the overhang "AA" from the double-stranded cDNA released after *Gsu*I digestion. A *Mme*I site is also present for the release of the 18 bp long tags. Finally, the linkers A and B bear a label upstream of the *Mme*I site, corresponding to a barcode sequence that can eventually allow to distinguish the libraries after a batch sequencing.

9. The gel is left on one of the glass plates, covered with a Saran wrap and radioactive marks are placed on it. These marks correspond to white correction fluid in which a very small amount of radioactive element is incorporated and we make three dots arranged in triangle on a piece of colored tape. After 20 min of contact with a phosphor screen, the gel is scanned. The picture is printed on a transparency film and this film is incised at the locations of the bands. The film is then placed on the gel according to the dots and the gel is cut with a scalpel according to the pattern to recover the bands.

10. As described in **Fig. 6.1**, the final product of this experiment results in four linker-associated tags concatenated together with the unique configuration A-C1-C2-B. While all the linkers have an equivalent structure, the linkers C1 and C2 also bear a *Bsa*I site upstream of the *Mme*I site. Because the *Bsa*I recognition site is partially degenerated, we could design two different *Bsa*I sites for C1 and C2, so that only C1 and C2 could ligate together, and not C1 together nor C2. Thus, the monotags A and C1 and the monotags B and C2 are first ligated together depending on the probability to find a complementary partner at the two overhung nucleotides left after *Mme*I digestion. Then, after *Bsa*I digestion, these ditags are pooled for ligation and the design of the *Bsa*I sites allows obtaining only A-C1-C2-B tetratags.

11. As explained in **Note 10**, the ditags result from the compatibility of two overhung nucleotides. Therefore, homogeneous ditags can be obtained (for instance, A–A

and C1–C1) as well as heterogeneous ones (A–C1). It is thus necessary to purify the desired products on gel and this step is possible because the linkers are designed with different sizes (*see* **Fig. 6.4**).

12. During the first PCR program, the primers get exhausted quite rapidly, allowing the 117 nt single-stranded PCR products to randomly anneal together at their homologous linker sequences, located at the extremities, but not in the central region. The heterogeneous duplexes thus produced present a disturbed migration on agarose gel, preventing the correct recovery after *Bsa*I digestion. Therefore, a second program with only one cycle is crucial to homogenize the products.

Acknowledgments

We thank C. Saveanu, F. Feuerbach and M. Fromont-Racine for stimulating discussions and the staff of the Genoscope (Evry) for sequencing. This work was supported by the Institut Pasteur, the CNRS, the ANR (CUT program) and the European Science Foundation (RNA quality program). H.N. is supported by the ANR (CUT program).

References

1. Wang, Y., Liu, C. L., Storey, J. D, Tibshirani, R. J., Herschlag, D., and Brown, P. O. (2002) Precision and functional specificity in mRNA decay. *Proc. Natl. Acad. Sci. USA* **99**, 5860–5865.
2. Wyers, F., Rougemaille, M., Badis, G., et al. (2005) Cryptic pol II transcripts are degraded by a nuclear quality control pathway involving a new poly(A) polymerase. *Cell* **121**, 725–737.
3. Dujon, B. (2005) Hemiascomycetous yeasts at the forefront of comparative genomics. *Curr. Opin. Genet. Dev.* **15**, 614–620.
4. Lowe, T. M., and Eddy, S. R. (1999) A computational screen for methylation guide snoRNAs in yeast. *Science* **283**, 1168–1171.
5. Schattner, P., Decatur, W. A., Davis, C. A., Ares, M., Jr., Fournier, M. J., and Lowe, T. M. (2004) Genome-wide searching for pseudouridylation guide snoRNAs: analysis of the Saccharomyces cerevisiae genome. *Nucleic Acids Res.* **32**, 4281–4296.
6. Torchet, C., Badis, G., Devaux, F., Costanzo, G., Werner, M., and Jacquier, A. (2005) The complete set of H/ACA snoRNAs that guide rRNA pseudouridylations in *Saccharomyces cerevisiae*. *RNA* **11**, 928–938.
7. Martens, J. A., Wu, P. Y., and Winston, F. (2005) Regulation of an intergenic transcript controls adjacent gene transcription in *Saccharomyces cerevisiae*. *Genes Dev.* **19**, 2695–2704.
8. Berretta, J., Pinskaya, M., and Morillon, A. (2008) A cryptic unstable transcript mediates transcriptional trans-silencing of the Ty1 retrotransposon in *S. cerevisiae*. *Genes Dev.* **22**, 615–626.
9. Camblong, J., Iglesias, N., Fickentscher, C., Dieppois, G., and Stutz, F. (2007) Antisense RNA stabilization induces transcriptional gene silencing via histone deacetylation in *S. cerevisiae*. *Cell* **131**, 706–717.
10. Hongay, C. F., Grisafi, P. L., Galitski, T., and Fink G. R. (2006) Antisense transcription

controls cell fate in *Saccharomyces cerevisiae*. *Cell* **127**, 735–745.
11. Uhler, J. P., Hertel, C., and Svejstrup, J. Q. (2007) A role for noncoding transcription in activation of the yeast PHO5 gene. *Proc. Natl. Acad. Sci. USA* **104**, 8011–8016.
12. David, L., Huber, W., Granovskaia, M., et al. (2006) A high-resolution map of transcription in the yeast genome. *Proc. Natl. Acad. Sci. USA* **103**, 5320–5325.
13. Miura, F., Kawaguchi, N., Sese, J., et al. (2006) A large-scale full-length cDNA analysis to explore the budding yeast transcriptome. *Proc. Natl. Acad. Sci. USA* **103**, 17846–17851.
14. Nagalakshmi, U., Wang, Z., Waern, K., et al. (2008) The transcriptional landscape of the yeast genome defined by RNA sequencing. *Science* **320**, 1344–1349.
15. Samanta, M. P., Tongprasit, W., Sethi, H., Chin, C. S., and Stolc, V. (2006) Global identification of noncoding RNAs in *Saccharomyces cerevisiae* by modulating an essential RNA processing pathway. *Proc. Natl. Acad. Sci. USA* **103**, 4192–4197.
16. Davis, C. A., and Ares, M., Jr. (2006) Accumulation of unstable promoter-associated transcripts upon loss of the nuclear exosome subunit Rrp6p in *Saccharomyces cerevisiae*. *Proc. Natl. Acad. Sci. USA* **103**, 3262–3267.
17. Neil. H., Malabat, C., d'Aubenton-Carafa, Y., Xu, Z., Steinmetz, L. M., and Jacquier, A. (2009) Widespread bidirectional promoters are the major source of cryptic transcripts in yeast. *Nature* **457**, 1038–1042.
18. Xu, Z., Wei, W., Gagneur, J., et al. (2009) Bidirectional promoters generate pervasive transcription in yeast. *Nature* **457**, 1033–1037.
19. Peng, W. T., Robinson, M. D., Mnaimneh, S., et al. (2003) A panoramic view of yeast noncoding RNA processing. *Cell* **113**, 919–933.
20. Lister, R., O'Malley, R. C., Tonti-Filippini, J., et al. (2008) Highly integrated single-base resolution maps of the epigenome in Arabidopsis. *Cell* **133**, 523–536.
21. Wei, C. L., Ng, P., Chiu, K. P., et al. (2004) 5' Long serial analysis of gene expression (LongSAGE) and 3′ LongSAGE for transcriptome characterization and genome annotation. *Proc. Natl. Acad. Sci. USA* **101**, 11701–11706.
22. Inada, M., and Guthrie, C. (2004) Identification of Lhp1p-associated RNAs by microarray analysis in *Saccharomyces cerevisiae* reveals association with coding and noncoding RNAs. *Proc. Natl. Acad. Sci. USA* 101, 434–439.
23. Oeffinger, M., Wei, K. E., Rogers, R., et al. (2007) Comprehensive analysis of diverse ribonucleoprotein complexes. *Nat. Methods* **4**, 951–956.
24. Arigo, J. T., Eyler, D. E., Carroll, K. L., and Corden, J. L. (2006) Termination of cryptic unstable transcripts is directed by yeast RNA-binding proteins Nrd1 and Nab3. *Mol. Cell* **23**, 841–851.
25. Thiebaut, M., Kisseleva-Romanova, E., Rougemaille, M., Boulay, J., and Libri, D. (2006) Transcription termination and nuclear degradation of cryptic unstable transcripts: a role for the nrd1-nab3 pathway in genome surveillance. *Mol. Cell* **23**, 853–864.
26. Rigaut, G., Shevchenko, A., Rutz, B., Wilm, M., Mann, M., and Seraphin, B. (1999) A generic protein purification method for protein complex characterization and proteome exploration. *Nat. Biotechnol.* **17**, 1030–1032.
27. Velculescu, V. E., Zhang, L., Zhou, W., et al. (1997) Characterization of the yeast transcriptome. *Cell* **88**, 243–251.
28. Lee, C. Y., Lee, A., and Chanfreau, G. (2003) The roles of endonucleolytic cleavage and exonucleolytic digestion in the 5′-end processing of *S. cerevisiae* box C/D snoRNAs. *RNA* **9**, 362–1370.
29. Danin-Kreiselman, M., Lee, C. Y., and Chanfreau, G. (2003) RNAse III-mediated degradation of unspliced pre-mRNAs and lariat introns. *Mol. Cell* **11**, 1279–1289.

Chapter 7

Genome-Wide Transcriptome Analysis in Yeast Using High-Density Tiling Arrays

Lior David, Sandra Clauder-Münster, and Lars M. Steinmetz

Abstract

In the last decade, it became clear that transcription goes far beyond that of protein-coding genes. Most RNA molecules are transcribed from intergenic regions or introns and exhibit much variability in size, expression level, secondary structure, and evolutionary conservation. While for several types of non-coding RNAs some cellular functions have been reported, like for micro-RNAs and small nucleolar RNAs, for most others no indications of function or regulation have so far been found. Therefore, the RNA population inside a cell is diverse and cryptic and, thus, demands powerful methods to study its composition, abundance, and structure. DNA oligonucleotide microarrays have proven to be of great utility to study transcription of genes in various organisms. Recently, due to advancement in microarray technology, tiling microarrays that extend transcription measurement to genomic regions beyond protein-coding genes were designed for several species. The *Saccharomyces cerevisiae* yeast tiling array contains overlapping probes across the full genomic sequence, with consecutive probes starting every 8 bp on average on each strand, enabling strand-specific measurement of transcription from a full eukaryotic genome. Here, we describe the methods used to extract yeast RNA, convert it into first-strand cDNA, fragment, and label it for hybridization to the tiling array. This protocol will enable researchers not only to study which genes are expressed and to what levels, but also to identify non-coding RNAs and to study the structure of transcripts including their untranslated regions, alternative start, stop, and processing sites. This information will allow understanding their roles inside cells.

Key words: Tiling microarray, transcription, gene expression, gene structure, non-coding RNA, whole-genome microarray, strand-specific transcription, cDNA, yeast.

1. Introduction

Genomes are blueprints that contain the necessary information to develop and operate life forms in different environments. But there is still much to learn about how this information is read and

J.I. Castrillo, S.G. Oliver (eds.), *Yeast Systems Biology*, Methods in Molecular Biology 759,
DOI 10.1007/978-1-61779-173-4_7, © Springer Science+Business Media, LLC 2011

utilized to facilitate the concerted functions required for life. One initial step in reading the genomic information is transcription, the process that generates RNA molecules. Two general classes of RNA molecules exist: messenger RNAs (mRNAs), which are translated into proteins; and non-coding RNAs (ncRNAs), some of which have regulatory and structural roles (1). Only since recently have scientists realized that a much larger than expected portion of eukaryotic genomes is transcribed (2–4). Thus, characterizing the transcribed portion of the genome, namely the transcriptome, aids significantly in identifying and annotating protein-coding genes, as well as in enumerating ncRNA molecules, most of which have (thus far) no known function in the biology of the organism.

DNA microarrays are powerful tools for genome-wide analyses. In their earlier versions, DNA microarrays contained only limited sets of probes to interrogate the expression of known genes (5–9). With the emergence of fully sequenced genomes, microarray designs included probes to interrogate also other regions such as predicted genes, exons and promoters. Since the potential of this technology to revolutionize the field of gene expression and genomics became clear early on, there was an intensive effort to improve the efficiency and utilization of DNA microarrays. One desirable outcome of these developments was a reduction in feature size, namely of the area used for synthesis of a DNA probe on the array. The reduction in feature size and the availability of genome sequences eventually led to the development of tiling arrays (2, 10–13).

Under the term tiling arrays, various densities of genomic sequence coverage by the array probes exist. Particularly informative are tiling array designs in which an unbiased representation of the complete genome is achieved by overlapping sets of probes. Fluorescently labeled complementary DNA (cDNA) or cRNA target is then allowed to bind the array probes to determine which and to what level different genomic regions are transcribed. Although they have many advantages, tiling arrays also have several shortcomings. First, they can be designed only for organisms with a known genome sequence. Second, the choice in probe sequences becomes limited as the genomic coverage of the array increases, as probes will be determined by the genomic sequence. Third, the microarray sensitivity, specificity, and dynamic range (the ratio of the smallest to the largest fluorescent signal) make it difficult to reliably measure the expression level of very low-abundance RNAs and to distinguish between highly similar RNA sequences such as those of duplicated genes. Finally, the number of DNA probes that fit on a microarray is limited, putting constraints on the coverage of probes as a function of the genome size, and thus on the resolution at which a given genome can be analyzed by one microarray.

However, tiling arrays present a significant advancement over other array-based technologies as we have demonstrated in our own applications of this method in yeast (13). The budding yeast, *Saccharomyces cerevisiae,* has roughly a 12 Mbp genome size and, thus, the current feature size technology allowed placing overlapping probes across the whole genome rather than probes spaced every fixed size interval with gaps in between. Moreover, since transcription is strand specific, overlapping probes were tiled separately for each strand allowing strand-specific measurement of transcription simultaneously from both strands at 8 bp resolution. Moreover, the tile of the second strand is offset by 4 bp compared to the first strand allowing to measure double-stranded target samples (like genomic DNA or double-stranded cDNA) at 4 bp resolution. Having probes densely tiled on both strands increases the likelihood that targets (whether single or double stranded) will hybridize well, since each target can hybridize to probes with different nucleotide composition. Thus, the benefit of having many diverse probes measuring the same transcriptional unit makes this design a highly sensitive and accurate platform for transcriptome studies.

Transcripts in a genome could be long or short, constitutively expressed or condition specific, isolated or overlapping other transcripts, and expressed at high or low levels. The yeast tiling array provides a sensitive platform to measure all of these transcript classes, provided that the sample preparation is of appropriate quality. To that end, we provide here a protocol that produces first-strand cDNA for hybridization to measure transcription in a strand-specific manner and minimize the processing steps that might introduce bias in representation of transcripts. It is important to note that standard protocols that had been applied previously to low-resolution microarrays did not suffice for application to tiling arrays. The high resolution of the tiling array revealed artifacts introduced during sample preparation that had gone unnoticed before, but that did confound the previous microarray data. Reverse transcription generates spurious second-strand cDNA, even during a first-strand cDNA synthesis, and therefore distorts the conversion step of RNA into cDNA (14). Therefore we incorporated the use of actinomycin D, which specifically inhibits DNA-dependent but not RNA-dependent DNA synthesis, during cDNA synthesis to avoid artifacts of second-strand synthesis. Altogether, the high-density coverage of the genome with the careful sample preparation method allows to measure not only transcripts levels but also structural features of transcription, such as untranslated regions (UTRs), splicing, sense and antisense transcription and more, providing a reliable and detailed map of the full transcriptional landscape of a eukaryotic genome.

2. Materials

2.1. Extraction of Total RNA from Yeast Cells

1. Oak Ridge centrifugation tubes, PPCO (Nalgene).
2. Acid-washed glass beads (Sigma).
3. Phase Lock Gel 50 mL tubes, Heavy (Eppendorf).
4. SYBR Green II RNA gels stain (Molecular Probes).
5. Bead buffer (75 mM NH_4OAc, 10 mM EDTA).
6. Phenol, acidic unbuffered water saturated (pH 4.3, Applichem).
7. Phenol:chloroform:isoamylalcohol; 25:24:1 (v/v/v) (Applichem).
8. Isopropanol solution (100%, v/v).
9. Ethanol solution (100%, v/v).
10. NH_4OAc solution (7.5 M).
11. SDS (10%, w/v).
12. DEPC-treated (RNase-free) water.

2.2. Purification of Poly(A) RNA

1. Oligotex mRNA maxi kit (Qiagen) (*see* **Note 1**).
2. DEPC water.

2.3. Residual Genomic DNA Digestion

1. TURBO DNA-free kit, (Ambion).

2.4. Synthesis of First-Strand cDNA

1. Random Primers 3 μg/μL (Invitrogen).
2. Oligo(dT) 12–18 Primer, 0.5 μg/μL stock (Invitrogen).
3. dNTP + dUTP mix: dCTP, dATP, and dGTP each 10 mM; dTTP 8 mM; dUTP 2 mM.
4. SuperScript II Reverse Transcriptase (Invitrogen).
5. RNase H (Epicentre).
6. RNase cocktail (Ambion).
7. Actinomycin D stock solution, 1.25 mg/mL (Sigma).

2.5. Purification of First-Strand cDNA

1. Affymetrix GeneChip Sample Cleanup module (Affymetrix, 900371). A kit for 10 reactions is also available. Alternatively, MinElute PCR Purification kit can be used (Qiagen).

2.6. Fragmentation of the cDNA and Labeling with Biotin

1. Affymetrix GeneChip WT Terminal labeling kit (Affymetrix, 900671). A kit for 10 reactions is also available.

2.7. Target Hybridization on the Tiling Microarray

1. *S. cerevisiae* yeast tiling array (Affymetrix, PN 520055) (*see* **Note 2**).
2. GeneChip hybridization, wash, and stain kit (Affymetrix, 900720).
3. GeneChip control oligoB2 (Affymetrix, 900301).

3. Methods

Genome-wide analysis of the yeast transcriptome in a given condition: The following protocol includes extraction of RNA, synthesis of first-strand cDNA, and processing it for hybridization on the high-density tiling array. In our experience, the signal-to-noise ratio obtained from hybridizing cDNA derived from poly(A)-enriched RNA captures transcriptional information for most RNAs under standard (e.g., exponential) growth conditions (including mRNAs, ncRNAs, tRNAs, and rRNAs). The protocol described below is mainly structured around preparation of such samples. For some applications, researchers might want to look at total RNA directly, thus we also describe the preparation of first-strand cDNA from total RNA samples. The preparation of cDNA directly from total RNA essentially skips the enrichment of the poly(A) RNA step.

3.1. Extraction of Total RNA from Yeast Cells

1. Before starting: Warm a water bath up to 65°C. Prepare a high-speed centrifuge (e.g., Sorvall), assemble the appropriate rotor (e.g., SS-34) to use with the Oak Ridge tubes, and set its temperature to 10°C. Also prepare a refrigerated centrifuge with a suitable rotor for 50 mL tubes to precipitate the cells and spin the Phase lock Gel tubes. The extraction procedure is better carried out continuously from start to end and in a timely manner.
2. Per sample prepare:
   a. Oak Ridge tube with 5 g of glass beads, 1 mL of 10% SDS and 12 mL of phenol. Place in the 65°C water bath.
   b. A 15 mL tube with 10 mL of bead buffer. Place in the 65°C water bath as well.
   c. A 50 mL gel phase lock tube with 10 mL of phenol:chloroform:isoamylalcohol.
   d. Oak Ridge tube for precipitation with 24 mL of isopropanol and 800 μL of 7.5 M NH_4OAc.
3. Harvest 100 mL culture with an OD_{600} of ~1 by spinning down the cells at room temperature (RT) and discarding

the medium. This step can be easily carried out in two 50 mL plastic tubes per sample. Proceed directly to extract total RNA or snap freeze the cell pellets in liquid N_2 and store at −80°C for later use (*see* **Note 3**). The number of cells to harvest from different conditions should be calculated to ensure a similar RNA yield (*see* **Note 1**). Lower the temperature of the centrifuge to 4°C.

4. Re-suspend cell pellets in 65°C bead buffer (by vortex or pipette) and transfer cell suspension into the pre-warmed Oak Ridge tube that contains the glass beads, SDS, and hot phenol. If the cells of one sample were collected in two separate 50 mL tubes, use the same 10 mL bead buffer to re-suspend both pellets and join them together into one hot phenol Oak Ridge tube.
5. Vortex vigorously the sample for 1 min at maximum speed to break the cells' wall and incubate back at 65°C for 1 min. Repeat vortex/incubation cycles three more times. Incubate for additional 5 min after which vortex for 1 min. Place tubes for 5 min on ice.
6. Centrifuge the Oak Ridge tubes in an appropriate high-speed rotor (e.g., SS34 for Sorvall) for 15 min, 12,000–16,000×*g*, 10°C. Transfer the top aqueous phase into the phase lock 50 mL tube by pipetting gently. Avoid the white interphase as much as possible. Lower the centrifuge temperature to 4°C.
7. Shake the phase lock gel tube vigorously to get a suspension of the aqueous sample with the organic phenol:chloroform:isoamylalcohol solution. Do not vortex. Centrifuge the tubes for 10 min, 1,000–2,000×*g*, 4°C.
8. Pour the top, gel separated, aqueous phase into the Oak Ridge tube that contains isopropanol and NH_4OAc. Shake vigorously to mix. Centrifuge the tubes for 10 min at maximum speed (12,000–16,000×*g*) at 4°C to precipitate the RNA.
9. Decant the liquid leaving the pellet adhered to the bottom/side of the tube. Add 10 mL of 70% ethanol, shake heavily to detach the pellet, and wash it. Centrifuge again for 5 min, 12,000–16,000×*g*, 4°C to precipitate the RNA.
10. Decant the 70% ethanol, leaving as few drops as possible. Air-dry the pellet on ice. Add 500 μL DEPC water to dissolve the RNA (leave on ice until pellet is dissolved). Transfer into a 1.5 mL Eppendorf tube.
11. Dilute a small aliquot 1:100–1:500 to determine the concentration and quality of the RNA sample (*see* **Notes 4** and **5**). You should get above 2 mg total RNA per sample. The absorbance ratio of 260/280 nm should be around

2.0 and that of 260/230 nm around 2.0 (the lower ratio could indicate organic contaminations like phenol).

12. In addition, analyze 1 μL per sample on a 2% (w/v) agarose gel (stained with SYBR Green II RNA gel stain or ethidium bromide). The gel should show a light smear with two bands for the high molecular weight ribosomal subunits and three bands for the low molecular weight RNAs (*see* **Fig. 7.1a** and **Note 6**). The quality of the extracted RNA can also be determined on a dedicated instrument such as the Bioanalyzer (Agilent technologies).
13. Store the total RNA samples at –80°C or aliquot and continue to the next step.

3.2. Purification of Poly(A) RNA

In the case of preparing total RNA for hybridization move directly to **Section 3.3**. This step is carried out using the Oligotex mRNA maxi kit in accordance with the manufacturer's protocol (*see* **Note 1**).

1. Before starting with the purification of the poly(A) fraction, prepare the following: Set the heating block temperature to 70°C. Place a 1.5 mL Eppendorf tube with DEPC water in the 70°C heating block for the elution step.

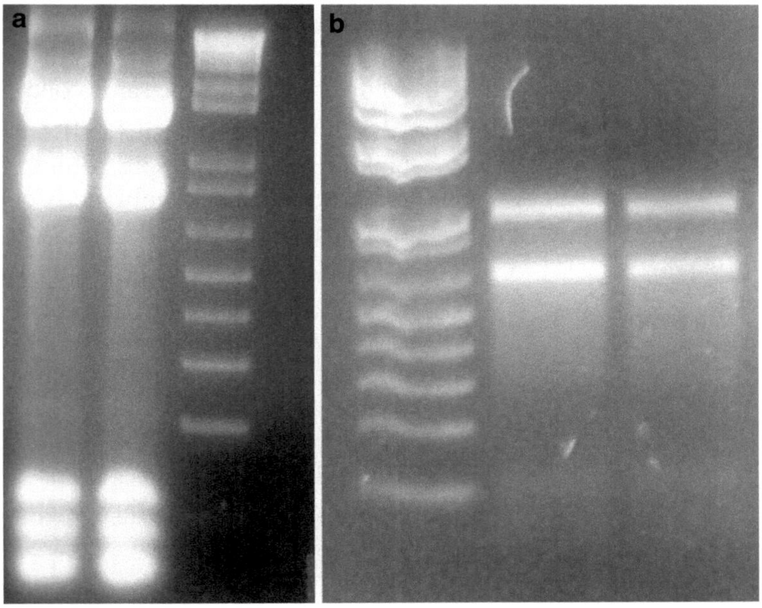

Fig. 7.1. Total RNA samples and poly(A) enrichment. (**a**) Yeast total RNA samples separated on 2% (w/v) agarose gel. Note the bright distinct bands of the high molecular weight ribosomal RNA (rRNA) and of the low weight RNAs that indicate the integrity of the RNA sample. The poly(A) RNA molecules comprise the faint smear observed in the background of each sample. (**b**) Poly(A) RNA after the enrichment step. Note the integrity of the sample and the reduction in the rRNA quantity relative to the background smear that contains the rest of the transcripts. Both gels use a 1 kb DNA ladder as a size marker.

2. Try starting with 2.5 mg of total RNA to ensure appropriate yield of poly(A)-enriched RNA. Make up the volume of the total RNA sample to 650 μL with DEPC water. It is recommended not to use total RNA older than 2 weeks.
3. Pipette up and down to re-suspend the Oligotex beads in their tube. Mix the OBB buffer well before use. Precipitated salts can be dissolved at 37°C. Add 650 μL of OBB buffer to the RNA sample and 135 μL of re-suspended Oligotex beads. If less RNA is used, less Oligotex bead suspension can be used (see manufacture's manual).
4. Flick the tube to mix the sample and incubate at 70°C for 7–10 min to denature secondary structures of the RNA.
5. Flick the tube again and leave at room temperature for 10 min to anneal the RNA to the poly(T)-coated beads.
6. Mark and place a spin column into a collection tube.
7. Spin the sample for 2 min at full speed in a tabletop microcentrifuge (12,000–16,000×*g*) to precipitate the beads with the RNA. Discard the liquid and re-suspend the beads in 600 μL of buffer OW2 by pipetting. Load the sample onto the spin column.
8. Centrifuge for 1 min at the full speed (12,000–16,000×*g*). Discard the flow-through from the tube and place the spin column back into the same collection tube.
9. Wash the sample again with 600 μL of buffer OW2 by pipetting inside the spin column. Be careful not to puncture the column membrane. Centrifuge for 1 min at full speed (12,000–16,000×*g*).
10. Transfer the column into a new and labeled RNase-free tube that is placed in the 70°C heating block. To elute the poly(A) RNA, add 50 μL of 70°C DEPC water into the column and re-suspend the resin by pipetting while the tube is in the heating block. Centrifuge for 1 min at full speed and repeat with another 50 μL of 70°C DEPC water. If multiple samples are processed in parallel, leave the re-suspended ones in the 70°C heating block until all are ready for centrifugation.
11. Usually the yield of poly(A)-enriched RNA is 1–1.5% of the total RNA (*see* **Note 1**). Determine the concentration and purity of the sample by absorbance measurements.

3.3. Digestion of Residual Genomic DNA from the RNA Sample

To avoid detection of residual genomic DNA in the hybridization, a DNase treatment is applied prior to the cDNA synthesis step (*see* **Note 7**). This protocol employs the reliable and easy to use Turbo DNase kit that is quick and does not require heat inactivation or purification of the sample from the protein at the end:

1. Start the protocol with 10–30 μg of poly(A) RNA or 25 μg of total RNA. The exact amount to begin with depends on the level of residual DNA and the procedure should leave at least 9 μg of pure poly(A) RNA or 20 μg of total RNA to start the next step with. Before starting, set a heating block to 37°C.
2. Perform the DNase reaction in up to 110 μL total. To the RNA sample, add 11 μL of (10X) buffer, 3 μL DNase, and fill the reaction up to 110 μL with water. Incubate 30 min at 37°C. The volume and the reagent amounts can be scaled down according to the amount of residual DNA and the required volume for the next step.
3. Mix the inactivating suspension prior to use. Add 0.15 volumes of the inactivation reagent. Inactivate the DNase for 2 min at RT. Flick the tube occasionally to keep the inactivating reagent from precipitating.
4. Centrifuge for 2 min at full speed (12,000–16,000×*g*) in a tabletop microcentrifuge and collect the sample by carefully aspirating the liquid avoiding the pellet of the inactivating reagent.
5. Determine the concentration of the sample. It is recommended to load 0.1–0.2 μg on a 2% (w/v) agarose gel to ensure the sample was not degraded prior to cDNA synthesis (*see* **Fig. 7.1b** and **Note 8**).

3.4. Synthesis of First-Strand cDNA

This step is performed slightly differently for poly(A) and total RNA. Ideally start with 9 μg of poly(A) RNA or 20 μg of total RNA. Sometimes it may be necessary to perform two reactions per sample to ensure sufficient yield and this is generally recommended for total RNA (*see* **Note 1**). If the volume of the DNA-free sample is too big for the cDNA synthesis reaction, mix the RNA with an equal volume of 100% ethanol and precipitate it by centrifugation (>12,000×*g*). Dry and re-suspend the pellet in the required volume of DEPC–water.

3.4.1. Synthesis of First-Strand cDNA from Poly(A) RNA

a. Set three heating blocks to 37, 42, and 70°C. Alternatively, the reaction can be carried out in a thermal cycler.
b. Dilute the oligo(dT) primer to 0.05 μg/μL.
c. Keep the RNA:random primer μg ratio at 1:0.5 and RNA:oligo(dT) primer ratio at 1:0.01.
d. Take a 9 μg aliquot of DNA-free poly(A) RNA. Add 1.5 μL (4.5 μg) random primers and 1.8 μL (0.09 μg) diluted oligo(dT) primer. Bring to 124 μL with water. Mix and incubate at 70°C for 10 min for denaturation. Return to ice.
e. Add 40 μL of (5X) first-strand buffer, 20 μL of 0.1 M DTT, 5 μL of dNTP + dUTP mix, 1 μL of Actinomycin D (final

conc. of 6.25 μg/mL), and 10 μL of superscript II RT to a final volume of 200 μL. Incubate at 42°C for 60 min.

f. Inactivate the enzyme for 15 min at 70°C.

g. To digest the RNA template molecules, add 3 μL of RNase H and 3 μL of RNase cocktail. Incubate at 37°C for 20 min.

3.4.2. Synthesis of First-Strand cDNA from Total RNA

a. The reaction is carried out in a thermal cycler in 0.2 mL PCR tubes.

b. Dilute the oligo(dT) primer to 0.05 μg/μL.

c. Keep the RNA:random primer microgram ratio at 1:0.086 and RNA:oligo(dT) primer ratio at 1:0.0017.

d. Take a 20 μg aliquot of DNA-free total RNA. Add 0.6 μL (1.8 μg) random primers and 0.68 μL (0.034 μg) diluted oligo(dT) primer. Bring to 58.32 μL with water. Mix and incubate at 70°C for 10 min followed by 10 min at 25 and 4°C on hold for denaturation and annealing, respectively. Alternatively, return to ice instead of the 4°C holding step.

e. Add 20 μL of (5X) first-strand buffer, 10 μL of 0.1 M DTT, 5 μL of dNTP + dUTP mix, 1.68 μL of actinomycin D (final concentration of 20 μg/mL), and 10 μL of Superscript II RT to a final volume of 105 μL.

f. Place the tube in a thermal cycler and run the following program: 25°C for 10 min; 37°C for 30 min; 42°C for 30 min, and 70°C for 10 min to inactivate the enzyme.

g. To digest the RNA template molecules, add 3 μL of RNase H and 3 μL of RNase cocktail. Incubate at 37°C for 20 min.

3.5. Purification of First-Strand cDNA

This step uses the cleanup kit from Affymetrix.

1. Add 500 μL of binding buffer to the reaction and mix by vortexing. The color should stay yellow similarly to the original binding buffer color. If the color changed to pink add 2 μL of 3 M sodium acetate to adjust the pH for optimal yield. Place a cleanup column in a 2 mL tube, load the sample, and spin for 1 min at >10,000×*g* in a tabletop microcentrifuge. Discard the flow-through.
2. If the cDNA synthesis was carried out in two separate reactions per sample, combine each with 500 μL of binding buffer, load the first part on the column, spin as above, and discard the flow-through. Then load the other reaction on the same column and spin again.
3. To wash the sample, add 750 μL of washing buffer and spin for 1 min at >10,000×*g*. Discard the flow-through. To dry the membrane after the wash, open the column cap and place it on the rotor in a direction that will not allow the cap to close during spinning. Spin with open caps for 5 min.

4. Transfer the dry column into a clean labeled 1.5 mL tube. To elute the clean cDNA apply 15 μL of elution buffer onto the membrane. Allow 1 min to release the sample and spin for 1 min at maximum speed (12,000–16,000×*g*). Repeat the elution into the same tube with additional 15 μL of elution buffer.
5. Measure absorbance to estimate the quantity and concentration.

3.6. Fragmentation of the cDNA and Labeling with Biotin

This step uses the Affymetrix GeneChip WT Terminal labeling kit (*see* **Note 9**). Reactions are performed in 0.2 mL PCR tubes in a thermal cycler.

1. Start with an aliquot of 4.5 μg of clean cDNA obtained from poly(A) RNA or 5.5 μg cDNA obtained from total RNA. Add water to a total of 31.2 μL. In addition, aliquot 300 ng from the original cDNA sample and save it on ice for later use as a control for the fragmentation gel analysis.
2. Prepare a mix according to the number of reactions. Mix per reaction contains: 10 μL of water, 4.8 μL of cDNA fragmentation buffer, 1.0 μL of UDG (10 U/μL), 1.0 μL APE (1,000 U/μL). The total volume per reaction is 16.8 μL.
3. Combine 16.8 μL of the fragmentation mix with the 31.2 μL of the cDNA. Spin down. Place the reaction in a thermal cycler and run the following program: 37°C for 60 min, 93°C for 2 min, and 4°C for at least 2 min. Store on ice if labeling will soon follow or in –20°C if the labeling will be carried out later.
4. Check the quality of the fragmentation by loading 3 μL of the fragmented sample on a 2% (w/v) agarose gel alongside with 300 ng of the saved untreated cDNA (*see* **Fig. 7.2**). Do not continue the protocol if the intact cDNA was degraded or if the sample was over- or under fragmented.
5. For terminal labeling of the fragmented cDNA use the 45 μL sample. Add 12 μL of (5X) TdT buffer, 2 μL of TdT, and 1 μL of DNA labeling reagent (5 mM) to a total volume of 60 μL. Perform the following reaction on a thermal cycler: 37°C for 60 min, 70°C for 10 min, and 4°C for at least 2 min.

3.7. Target Hybridization on the Tiling Microarray

These steps use the yeast tiling microarray (*see* **Note 2**), the GeneChip Hybridization, Wash and Stain kit, and the GeneChip control oligoB2 from Affymetrix. The hybridization, washing, staining, and scanning are carried out using Affymetrix equipment.

1. Set a heating block to 99°C or boil water. Turn on the hybridization oven and set the temperature to 45°C.

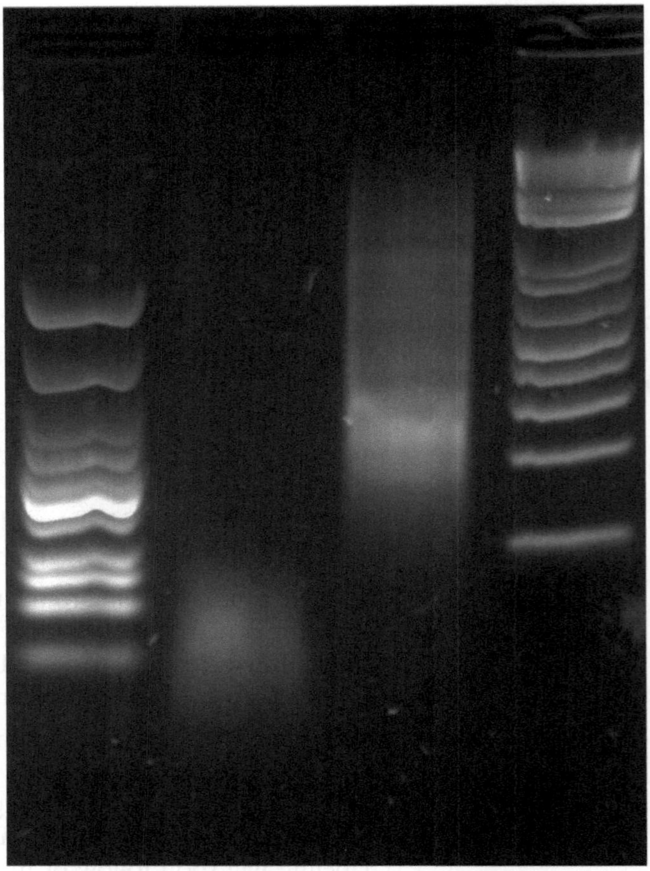

Fig. 7.2. Complementary DNA (cDNA) sample and fragmentation quality analysis. Separation of cDNA samples before (*third lane from the left*) and after fragmentation (*second lane*) on a 2% (w/v) agarose gel. Equal amounts of samples from before- and after fragmentation were loaded side by side. As a reference, the lowest band of the left DNA ladder is 25 bp and that of the right one is 100 bp in size. The cDNA sample was not degraded as can be seen from the size of the smear that is similar to that of the poly(A) RNA sample. Note that the cDNA was fragmented enough since all of it is smaller than 100 bp and not too much since the intensity remained similar to that of the sample before the fragmentation.

2. Take out the packed microarrays from the 4°C storage and allow them to equilibrate to RT. Take the array out and mark it by the sample name and date on the front label. Place the microarray face down on a Kimwipe or a paper towel to avoid scratching the glass.
3. For hybridization, to each sample of 60 μL add 5 μL of 3 nM oligoB2 to a final concentration of 0.05 nM, 150 μL of (2X) hybridization mix, and 85 μL of water to a final volume of 300 μL per sample. If more than one sample is processed for hybridization, the hybridization mix should

be prepared together for all and added individually to each sample.

4. Place the sample in the 99°C heating block (or in boiling water) for 10 min to denature the cDNA. Place denatured sample on ice until loading it on the microarray.
5. While denaturing, wet the microarray with 250 μL (1X) pre-hybridization mix and place it in the 45°C hybridization oven at 60 rpm for pre-hybridization until the sample is ready to load. The injection of liquid into the microarray is carried out from the bottom hole on the back side by puncturing the rubber septa with a small 10 μL tip. Before injecting liquid into the chamber, the upper septa must be punctured by another tip to allow a way out for the air. To inject the liquid place a larger tip inside a small tip or use a large tip with a narrow edge to avoid large punctures in the septa. Injection of liquid into the microarray should be done gently and slowly to avoid damage. Injecting 250 μL will leave a visible air bubble inside the chamber. Make sure if this air bubble is free to rotate by turning and tapping gently on the side of the microarray.
6. While the microarray is pre-hybridizing, take the sample out from the ice and spin it at maximum speed in a tabletop microcentrifuge for 5 min (12,000–16,000×*g*) to precipitate any particles.
7. While the sample is spinning, get back the microarray, place it on a Kimwipe and insert a small tip into the top septa. Take the pre-hybridization buffer out from the microarray. Aspirate 220 μL for poly(A) or 250 μL for total RNA-derived cDNA slowly from the top part of the target sample and inject it slowly into the microarray. Avoid the bottom part of the sample and the possible pellet. Ensure that the air bubble inside the microarray is free to rotate. Cover the two septa with Microtube Tough Spots adhesive labels to avoid leaks during hybridization. Place the microarray in the 45°C oven for 16 h at 60 rpm. Hybridizing 220 μL out of the 300 μL sample corresponds to 3.3 μg cDNA from poly(A) RNA. Hybridizing 250 μL out of the 300 μL sample corresponds to 4.58 μg cDNA from total RNA per microarray. Freeze the rest of the target sample in –20°C.
8. Before the end of the hybridization step (16 h), turn the Affymetrix fluidics station on and prime the required number of positions according to the number of arrays used. Use wash buffers A and B of the kit. Make sure to place enough buffers and water according to the number of arrays processed.

9. While running the priming protocol on the fluidics station, get the microarray from the hybridization oven. Peel off the Tough Spots, take the target sample out slowly from the microarray and return the sample into the tube with the frozen remaining sample. The sample can be hybridized again so keep it at –80°C at least until you see the scanning results are of sufficient quality or keep it longer if needed.
10. Immediately after thoroughly emptying the microarray from the target sample, refill the chamber slowly with wash buffer A and perform the following manual washing steps: Inject 220 μL and rotate the microarray 180° by hand for three times allowing the bubble to rotate up and down. Draw slowly the wash buffer A out and discard it. Repeat this wash two more times. Removing the sample and performing this manual wash step allow keeping the labeled sample for possible use in the future instead of washing it into the lines of the fluidics station. *Note*: The microarray after the hybridization should not be kept without wash buffer A, so fill up the microarray chamber with wash A buffer before attaching it to the fluidics station especially when processing several microarrays in parallel. To avoid trapping small air bubbles inside the chamber take an excess of buffer (~300 μL) and inject it slowly holding the microarray in an upright position allowing all the air to exit until the buffer flows out from the upper tip. Check carefully that no tiny bubbles remained trapped. If there is a bubble, draw half of the volume back out and fill the chamber again in the same way.
11. Prepare three tubes per microarray with 600 μL of each stain cocktail 1, stain cocktail 2, and array holding buffer. Place each tube in its corresponding position on the fluidics station. Upload and run the fluidics protocol FS450_0001 (if using fluidics station 450 or its matching protocol on fluidics station 400). The last step of the protocol leaves a filled microarray (with wash buffer A) ready to scan. Take the microarray off the station and perform the shutdown protocol if required.
12. Turn on the scanner ahead of time to warm it up. Before scanning, clean the glass of the microarray using water and Kimwipe. Check that no tiny air bubbles were trapped behind the glass as those will obstruct the scanning. If you find a bubble, take it out manually by replacing the wash buffer A as in Step 10 above.
13. For scanning, the Tough Spots can be placed again on the back of the microarray to protect from leaks inside the scanner. Use the scanner according to the manufacturer

instructions. For the yeast tiling array at least an Affymetrix 3000 7G scanner is required to scan at the right resolution. The specific files that are supplied with the microarrays have to be installed prior to the scan for the program to recognize the array type (*see* **Note 2**). For a rough evaluation of the results *see* **Note 10**.

4. Notes

1. Amounts of RNA – Lower amounts of RNA transcripts are produced under some growth conditions or by certain mutant strains. Thus, we recommend using the Oligotex maxi kit that can process more than 1 mg of total RNA in order to obtain sufficient amounts of poly(A) RNA. Under more standard conditions the midi kit can be used to obtain 9 μg of poly(A) RNA. Similarly, the efficiency of cDNA synthesis can vary by conditions and in some cases two synthesis reactions per sample have to be carried out to obtain 4.5 μg of clean first-strand cDNA from poly(A) RNA or 5.5 μg cDNA from total RNA.
2. Yeast tiling microarray design – Affymetrix manufactures two designs of *S. cerevisiae* tiling arrays, both of which are commercially available. The authors use and recommend the design (PN 520055) that tile a probe every 8 bp on average separately for each strand on the same microarray. The other design includes a tile of 5 bp but of one strand only.
3. Although RNA samples can be stored at –80°C after the extraction, for quality reasons it is recommended to continue the poly(A) RNA enrichment protocol or DNase treatment soon after the extraction and within a couple of weeks at most.
4. RNA and cDNA quality – The key for accurate transcriptome measurement is obtaining high-quality RNA. The RNA extraction and poly(A) RNA enrichment steps (**Sections 3.1** and **3.2**) should be carried out promptly and in an RNase-free environment including the reagents and equipment (e.g., gel buffer and box). Extracting RNA from freshly harvested cells is recommended over using frozen cell pellets.
5. It is recommended that the time from harvesting of the cells to the first vortexing in the hot phenol tube should not exceed 8 min to capture reliably the RNA repertoire.

Therefore, because of the timely protocol, avoid processing too many samples in parallel.

6. Gel analysis of the total RNA – This analysis should indicate the quality of the sample. A heavy smear down the lane or very bright bands at low molecular weight may be indicative of RNA degradation or RNase contamination. Ideally the high molecular weight ribosomal RNA bands should be strong and distinct as they comprise most of the RNA in the sample. Faint bands and brighter smear at a size lower than the ribosomal RNA bands indicate degradation and require repeating the extraction of the RNA sample.
7. Genomic DNA contamination – Some genomic DNA could be extracted along with the RNA sample. To avoid reading signals from the genomic DNA while measuring the transcriptome, a DNase treatment is applied. The DNase treatment could be carried out either after the poly(A) RNA enrichment step (as described above) or after the total RNA extraction. The later is necessary in the case of measuring the transcriptome from total RNA samples or recommended if multiple poly(A) RNA samples will be prepared from the same total RNA sample. In any case, the above protocol works well also if the poly(A) enrichment step follows the DNase treatment rather than preceding it.
8. Gel analysis before cDNA synthesis – Regardless of whether poly(A) enriched or total RNA samples are used for cDNA synthesis, the quality of the sample should be confirmed on a 2% (w/v) agarose gel right before the cDNA synthesis step. In the case of DNase treatment after the poly(A) RNA enrichment (as described above), it is recommended to run an analysis gel only once after the DNase treatment. But if the order of poly(A) enrichment and DNase treatment is swapped the gel analysis should be carried out after the poly(A) enrichment step before the cDNA synthesis. As indicated in **Fig. 7.1b**, cDNA synthesis should be carried out only if the RNA sample is not degraded.
9. Target labeling – If problems with the target cDNA labeling are suspected, the labeling efficiency can be tested by a gel shift assay according to the GeneChip® Whole Transcript (WT) Sense Target Labeling Assay Manual from Affymetrix.
10. A rough sanity check of the scan image – A quick quality control check of the scan image can indicate the quality of the data. Check for the checkerboard pattern of the borders that was created by hybridizations of the synthetic oligoB2. The intensity of the off (dark) features will indicate the background signal. High contrast and sharp borders of the

checkerboard pattern with high intensity of the oligoB2 probes indicate that the labeling, hybridization, washing, staining, and scanning were good. If very low intensity signal is found for the genomic probes compared to the oligoB2 synthetic checkerboard probes, the sample preparation was poor suggesting over- or under-fragmentation or too little material to begin with. Check also the overall picture for dim areas that indicate trapped air bubbles, dust particles, or manufacturing defects. Some of the analysis algorithms will discard these dim regions automatically. Check for the decrease between the perfect match and mismatch probes that reflects the specificity of the hybridization signal.

References

1. Amaral, P. P., Dinger, M. E., Mercer, T. R., and Mattick, J. S. (2008) The eukaryotic genome as an RNA machine. *Science* **319**, 1787–1789.
2. Kapranov, P., Cawley, S. E., Drenkow, J., et al. (2002) Large-scale transcriptional activity in chromosomes 21 and 22. *Science* **296**, 916–919.
3. Carninci, P., Kasukawa, T., Katayama, S., et al. (2005) The transcriptional landscape of the mammalian genome. *Science* **309**, 1559–1563.
4. Birney, E., Stamatoyannopoulos, J. A., Dutta, A., et al. (2007) Identification and analysis of functional elements in 1% of the human genome by the ENCODE pilot project. *Nature* **447**, 799–816.
5. Wodicka, L., Dong, H., Mittmann, M., Ho, M. H., and Lockhart, D. J. (1997) Genome-wide expression monitoring in *Saccharomyces cerevisiae*. *Nat. Biotechnol.* **15**, 1359–1367.
6. Schena, M., Shalon, D., Davis, R. W., and Brown, P. O. (1995) Quantitative monitoring of gene expression patterns with a complementary DNA microarray. *Science* **270**, 467–470.
7. Lockhart, D. J., Dong, H., Byrne, M. C., et al. (1996) Expression monitoring by hybridization to high-density oligonucleotide arrays. *Nat. Biotechnol.* **14**, 1675–1680.
8. Lashkari, D. A., DeRisi, J. L., McCusker, J. H., et al. (1997) Yeast microarrays for genome wide parallel genetic and gene expression analysis. *Proc. Natl. Acad. Sci. USA* **94**, 13057–13062.
9. Fodor, S. P., Read, J. L., Pirrung, M. C., Stryer, L., Lu, A. T., and Solas, D. (1991) Light-directed, spatially addressable parallel chemical synthesis. *Science* **251**, 767–773.
10. Yamada, K., Lim, J., Dale, J. M., et al. (2003) Empirical analysis of transcriptional activity in the *Arabidopsis* genome. *Science* **302**, 842–846.
11. Tjaden, B., Saxena, R. M., Stolyar, S., Haynor, D. R., Kolker, E., and Rosenow, C. (2002) Transcriptome analysis of *Escherichia coli* using high-density oligonucleotide probe arrays. *Nucleic Acids Res.* **30**, 3732–3738.
12. Selinger, D. W., Cheung, K. J., Mei, R., et al. (2000) RNA expression analysis using a 30 base pair resolution *Escherichia coli* genome array. *Nat. Biotechnol.* **18**, 1262–1268.
13. David, L., Huber, W., Granovskaia, M., et al. (2006) A high-resolution map of transcription in the yeast genome. *Proc. Natl. Acad. Sci. USA* **103**, 5320–5325.
14. Perocchi, F., Xu, Z., Clauder-Munster, S., and Steinmetz, L. M. (2007) Antisense artifacts in transcriptome microarray experiments are resolved by actinomycin D. *Nucleic Acids Res.* **35**, e128.

Chapter 8

RNA Sequencing

Karl Waern, Ugrappa Nagalakshmi, and Michael Snyder

Abstract

This chapter describes the RNA sequencing (RNA-Seq) protocol, whereby RNA from yeast cells is prepared for sequencing on an Illumina Genome Analyzer. The protocol can easily be altered to use RNA from a different organism. This chapter covers RNA extraction, cDNA synthesis, cDNA fragmentation, and Illumina cDNA library generation and contains some brief remarks on bioinformatic analysis.

Key words: RNA-Seq, RNA, transcriptome, transcription profiling, high-throughput sequencing.

1. Introduction

Defining the RNA content of a population of cells – their 'transcriptome' – has long been of interest to biologists. Currently, microarrays are the dominant high-throughput method of doing so, but, as costs continue to come down, high-throughput sequencing of the transcriptome will become the norm (1–3). Already, RNA-Seq data can be substituted for microarray data, while yielding additional information (4).

A limitation of RNA-Seq today is that not all protocols provide strand-specific data, i.e., it is not known from which DNA strand a particular transcript originated. An increasing number of protocols do not have this limitation, however (2, 5), but care must be taken to validate each new protocol.

RNA-Seq has several advantages over microarrays. It is possible to annotate exons and introns precisely and identify transcript ends. In addition, RNA-Seq is more sensitive than microarrays, as evidenced by its higher dynamic range (1). Unlike arrays, further sequencing can also be performed on a prepared sample to

J.I. Castrillo, S.G. Oliver (eds.), *Yeast Systems Biology*, Methods in Molecular Biology 759,
DOI 10.1007/978-1-61779-173-4_8, © Springer Science+Business Media, LLC 2011

generate additional data if they are needed. Several reviews (6–9) are available.

The following protocol describes RNA-Seq with RNA from yeast cells being prepared for sequencing on an Illumina Genome Analyzer following the protocol employed in (1). RNA extraction can easily be altered to use RNA from a different organism. Preparing the cDNA for sequencing on platforms other than the Illumina Genome Analyzer is also possible, but is a non-trivial modification to the protocol.

This protocol describes how to perform high-throughput sequencing on a cDNA library made from doubly poly-A-enriched RNA. The steps in the protocol can be viewed in outline in **Fig. 8.1** and follow this general procedure: RNA extraction, cDNA synthesis, cDNA fragmentation, Illumina library generation, Illumina sequencing, and bioinformatic analysis.

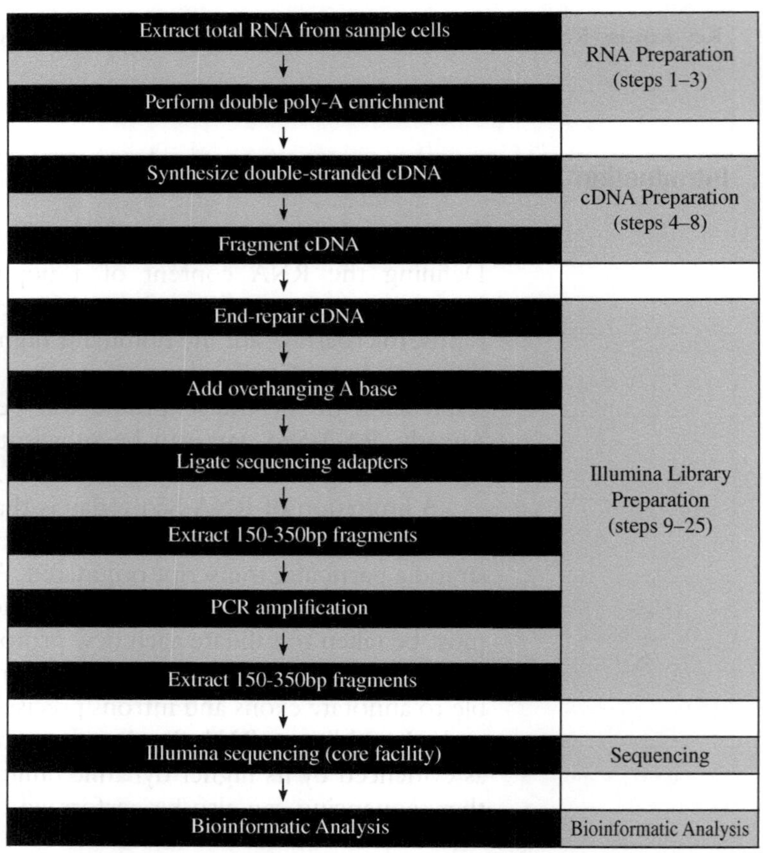

Fig. 8.1. Flowchart of the RNA-Seq protocol.

2. Materials

1. RiboPure™-Yeast Kit (Applied Biosystems).
2. MicroPoly(A)Purist™ Kit (Applied Biosystems).
3. SuperScript® Double-Stranded cDNA Synthesis Kit (Invitrogen).
4. Random Primers (Invitrogen).
5. DNase I (New England Biolabs).
6. QIAquick PCR Purification Kit (Qiagen).
7. QIAquick Gel Extraction Kit (Qiagen).
8. MinElute PCR Purification Kit (Qiagen).
9. End-It™ DNA End-Repair Kit (Epicentre).
10. Klenow Fragment (3′–5′ exo-) (New England Biolabs).
11. dATP (New England Biolabs).
12. LigaFast Rapid DNA Ligation System (Promega Corporation).
13. Genomic Adapter Oligo Mix (Illumina).
14. Phusion™ High-Fidelity PCR Master Mix (New England Biolabs).
15. PCR Primer 1.1 and PCR Primer 2.1 (Illumina).

3. Methods

1. Follow the manufacturer recommended protocol to extract RNA from yeast grown under your specified conditions using the RiboPure™-Yeast Kit. For best results, start with close to the recommended maximum of 3×10^8 yeast cells (*see* **Note 1**).
2. To remove the ribosomal RNA, follow the manufacturer recommended protocol for the MicroPoly(A)Purist™ Kit. Start with Step B.2b, i.e., there is no need to ethanol precipitate the total RNA obtained at Step 1. The protocol can be terminated after Step E.1., i.e., there is no need to ethanol precipitate the RNA at this stage.
3. Repeat the above step to perform one further round of poly-A enrichment. As above, start at Step B.2b and terminate at Step E.1.

4. Follow the manufacturer recommended protocol to synthesize double-stranded cDNA from 10 μL or 200 ng of the poly-A-enriched RNA obtained in Step 3. The Random Primers can be used as the 100 pmol/μL primer called for in the cDNA synthesis protocol. The procedure can be terminated after Step 3 on page 3, i.e., there is no need to ethanol precipitate the cDNA at this step (*see* **Note 2**).
5. Perform a QIAquick PCR Purification Kit cleanup on the cDNA from Step 4. Follow the manufacturer recommended protocol, but elute in a final volume of 35 μL of Buffer EB.
6. Mix the following: 8 μL of water, 1 μL of DNase I buffer, and 1 μL of DNase I. Add 2 μL of that mixture to the 35 μL of cDNA obtained in Step 5.
7. Incubate for 10 min at 37°C. Terminate the reaction by incubating for 10 min at 95°C (*see* **Note 3**).
8. Perform a QIAquick PCR Purification Kit cleanup. Elute in a final volume of 35 μL of Buffer EB.
9. Add the following, from the End-It™ DNA End-Repair Kit, to the fragmented cDNA from Step 8: 5 μL End-Repair Buffer, 5 μL dNTP mix, 5 μL ATP, 1 μL End-Repair Enzyme Mix.
10. Incubate for 45 min at room temperature.
11. Perform a QIAquick PCR Purification Kit cleanup. Elute in a final volume of 35 μL of Buffer EB.
12. If not already made, prepare a 1 mM stock of dATP. If the New England Biolabs' 100 mM dATP was purchased, this will require a 1:100 dilution in water or Buffer EB (*see* **Note 4**).
13. Add the following to the products of Step 11: 1 μL of Klenow Fragment, 5 μL NEBuffer 2 (supplied with Klenow Fragment), and 10 μL of 1 mM dATP.
14. Incubate for 30 min at 37°C.
15. Perform a MinElute PCR Purification Kit cleanup. Elute in a final volume of 12 μL of Buffer EB.
16. Prepare a fresh 1:30 dilution of the Illumina Genomic Adapter Oligo Mix in water. Do not reuse this dilution in subsequent RNA-Seq experiments (*see* **Note 5**).
17. Add the following, from the LigaFast Rapid DNA Ligation System kit, to the products of Step 15: 15 μL of DNA ligase buffer, 2 μL of DNA ligase, and 1 μL of diluted Adapter Oligo Mix.
18. Incubate for 15 min at room temperature.

19. Perform agarose gel electrophoresis with a 2% (w/v) agarose gel on the products of Step 18. Cut out a slice with DNA of approximate length 150–350 base pairs (*see* **Note 6**).
20. Perform a QIAquick Gel Extraction Kit cleanup. Elute in a final volume of 25 μL of Buffer EB.
21. Prepare fresh 1:1 dilutions of the Illumina PCR Primers 1.1 and 2.1 in water. Do not reuse these dilutions in subsequent RNA-Seq experiments.
22. Mix the following: 23 μL of product from Step 20, 25 μL Phusion™ High-Fidelity PCR Master Mix, 1 μL of diluted PCR Primer 1.1, and 1 μL of diluted PCR Primer 2.1.
23. Amplify on a thermal cycler, using the following protocol (*see* **Note 7**):
    1. 30 s at 98°C
    2. 10 s at 98°C
    3. 30 s at 65°C
    4. 30 s at 72°C
    5. Go to Step 2 fourteen times.
    6. 5 min at 72°C
24. Perform gel electrophoresis with a 2% (w/v) agarose gel on the products of Step 23. Cut out a slice with DNA of approximate length 150–350 base pairs (*see* **Note 6**).
25. Perform a QIAquick Gel Extraction Kit cleanup. Elute in a final volume of 30 μL of Buffer EB. The DNA is now ready for sequencing on an Illumina Genome Analyzer (*see* **Note 8**).
26. Bioinformatic analysis (*see* **Note 9**).

4. Notes

1. RNA purification is highly straightforward. An RNA yield of greater than 20 ng/μL at the end of the second round of poly-A purification is desirable, as this permits starting the cDNA synthesis with 200 ng or more of RNA, which yields optimal results. However, good results have been obtained with as little as 80 ng of starting material. The use of the RiboPure™-Yeast Kit makes this a yeast-specific protocol. However, extracting total RNA before poly-A enriching with the MicroPoly(A)Purist™ Kit is a general method, and

RNA from any organism can be used as input at this stage. Care should, of course, be taken when working with RNA, but no extraordinary precautions are needed.

2. cDNA synthesis is, again, a straightforward matter of following a manufacturer recommended protocol. Oligo-dT primers can be substituted for random hexamers, with the tradeoff that there will be fewer reads at the 5′ ends of transcripts, making those harder to map, but more polyadenylated reads, which help map 3′ ends (*see* **Note 9**).
3. cDNA fragmentation is a key step. As each sequenced read is short, and originates at the end of a strand of cDNA, the cDNA strands must be fragmented to ensure that the entire transcriptome is sampled during sequencing. This requires fragmenting the cDNAs to an appropriate length, approximately 100–300 bp. The protocol in this chapter is optimized for yeast, as in (1); if RNA from a different organism is used, where average transcript sizes differ, this step may require further optimization. Also note that the DNase I mixture used is very dilute. It is possible to use a more concentrated mixture and adjust the incubation time downward. A 10 min incubation allows enough margin of error that multiple samples can be processed in parallel, however. It bears pointing out that too long an incubation will destroy the cDNA sample. All due care should be taken to ensure that this step is done with no deviation from the protocol.
4. dATP is very sensitive to freeze–thaw cycles, so ensure that the 1 mM dATP stocks have been subjected to as few freeze–thaw cycles as possible. One suggestion is to prepare many small (e.g., 50 μL) aliquots from the stock solution.
5. The optimal Genomic Adapter Oligo Mix (v/v) dilution to make is somewhere between 1:10 and 1:50. 1:30 is recommended as a good starting point if upward of 200 ng of RNA was used to start with; a more diluted or more concentrated cDNA sample may require a weaker or stronger Genomic Adapter Oligo Mix. Upon the subsequent gel purification, an adapter–adapter band should be visible at about 120 base pairs; if this band looks excessively bright, too much adapter is being used, and this will interfere with an optimal library preparation.
6. At Step 19 it is normal to be unable to see any DNA. At Step 24, however, a normally distributed smear from 150 to 350 base pairs should be visible if the library preparation has been successful. If starting quantities of RNA are low, yields can be improved by using E-Gels® (Invitrogen) instead of agarose gels cast in the lab. The E-Gels® are thinner and result in substantially better recovery during the gel

purification process. Two caveats exist: first, E-Gels® take a maximum of 20 μL of product, so MinElute PCR Purification steps should be added between Steps 18 and 19, as well as between Steps 23 and 24. While loading dye is not, strictly speaking, necessary, eluting in 18 μL of Buffer EB and adding 2 μL of loading dye helps visualize the DNA's progress in the gel. Second, E-Gels® are difficult to open, even with Invitrogen's E-Gel® Opener. A clamp, hammer, and chisel provide an easier solution.

7. If library yields are low, the number of PCR cycles can be increased. Increases of more than two cycles are not recommended, however, due to the risk of bias in amplified sequences and subsequent potential loss of gene quantitation accuracy.
8. In most labs, sequencing will at this point be done by a core facility. Best results are to be expected if the final yield of library after gel purification is greater than 15 ng/μL.
9. Currently, most labs use an in-house software suite to analyze RNA-Seq data, and, in lieu of using commercial software such as Illumina's Genome Studio, collaborating with a bioinformatics group is recommended. Briefly, however, the main bioinformatic analyses that can be performed are as follows and were performed in this exact fashion in (1) and (6).

   To ascertain transcript abundance in the sample, the number of reads present mapping to 10–30 base pairs at the 3′ end of any given transcript can be counted. These values can then be used to compare gene expression levels. If multiple samples have been run, it is also necessary to normalize by the total number of sequencing reads generated for each sample; see e.g., (7).

   In addition, further annotation can be obtained by looking for gapped reads and poly-adenylated reads. Gapped reads are candidates for intron-spanning reads, and, if multiple reads have a gap across the same base pairs, that is strong evidence for the presence of an intron. Poly-adenylated reads are reads that have a non-genomic run of adenines, and these mark the 3′ end of a transcript, i.e., where that RNA was poly-adenylated. In addition, these poly-adenylated reads are strand-specific and indicate which DNA strand an RNA transcript was produced from. This is because a transcript produced from the + strand will be sequenced so as to either appear to come from the + strand with a poly-A tail or appear to come from the – strand with a poly-T tail. A transcript produced from the – strand, on the other hand, will appear to come from the – strand with a poly-A tail or from the + strand with a poly-T tail.

Lastly, searching for sudden drops in expression allows the pinpointing of 5′ ends or of 3′ ends where poly-adenylated reads were not detected. Searching for these sudden drops in un-annotated regions can turn up potential novel transcripts, as well.

References

1. Nagalakshmi, U., Wang, Z., Waern, K., et al. (2008) The transcriptional landscape of the yeast whole genome defined by RNA sequencing. *Science* **320**, 1344–1349.
2. Lister, R., O'Malley, R. C., Tonti-Filippini, J., Gregory, B. D., Millar, A. H., and Ecker, J. R. (2008) Highly integrated single-base resolution maps of the epigenome in Arabidopsis. *Cell* **133**, 523–536.
3. Wilhelm, B. T., Marguerat, S., Watt, S., et al. (2008) Dynamic repertoire of a eukaryotic transcriptome surveyed at single-nucleotide resolution. *Nature* **453**, 1239–1243.
4. Marioni, J. C., Mason, C. E., Mane, S. M., Stephens, M., and Gilad, Y. (2008) RNA-seq: an assessment of technical reproducibility and comparison with gene expression arrays. *Genome Res.* **18**, 1509–1517.
5. Parkhomchuk, D., Borodina, T., Amstislavskiy, V., et al. (2009) Transcriptome analysis by strand-specific sequencing of complementary DNA. *Nucleic Acids Res.* **37**, e123.
6. Wang, Z., Gerstein, M., and Snyder, M. (2009) RNA-Seq: a revolutionary tool for transcriptomics. *Nat. Rev. Genet.* **10**, 57–63.
7. Mortazavi, A., Williams, B. A., McCue, K., Schaeffer, L., and Wold, B. (2008) Mapping and quantifying mammalian transcriptomes by RNA-Seq. *Nat. Methods* **5**, 621–628.
8. Shendure, J. (2008) The beginning of the end for microarrays? *Nat. Methods* **5**, 585–587.
9. Morozova, O., and Marra, M. A. (2008) Applications of next-generation sequencing technologies in functional genomics. *Genomics* **92**, 255–264.

Chapter 9

Polyadenylation State Microarray (PASTA) Analysis

Traude H. Beilharz and Thomas Preiss

Abstract

Nearly all eukaryotic mRNAs terminate in a poly(A) tail that serves important roles in mRNA utilization. In the cytoplasm, the poly(A) tail promotes both mRNA stability and translation, and these functions are frequently regulated through changes in tail length. To identify the scope of poly(A) tail length control in a transcriptome, we developed the *p*oly*a*denylation *st*ate micro*a*rray (PASTA) method. It involves the purification of mRNA based on poly(A) tail length using thermal elution from poly(U) sepharose, followed by microarray analysis of the resulting fractions. In this chapter we detail our PASTA approach and describe some methods for bulk and mRNA-specific poly(A) tail length measurements of use to monitor the procedure and independently verify the microarray data.

Key words: Poly(A) tail, polyadenylation, deadenylation, post-transcriptional regulation, translational control, mRNA stability, polysome, microarray, transcriptome, proteome.

1. Introduction

Translation of mRNA is a highly controlled step in eukaryotic gene expression with control being exerted *en masse*, for instance in response to growth promoting or stress signals, or *en détail* for transcripts carrying cis-acting regulatory elements that enable a selective response (1). Often, these elements act through recruiting specific factors that organize mRNAs into functionally related subsets and may not only regulate their translation but also stability and localization (2, 3). During translation, the initiation phase is most commonly the target of regulatory intervention

J.I. Castrillo, S.G. Oliver (eds.), *Yeast Systems Biology*, Methods in Molecular Biology 759,
DOI 10.1007/978-1-61779-173-4_9, © Springer Science+Business Media, LLC 2011

(1, 4). The 3′ poly(A) tail of eukaryotic mRNA is implicated in translation and its regulation, displaying synergy with the 5′ cap structure to promote efficient initiation (5–9). This suggests that mRNAs adopt a closed-loop configuration during translation (10, 11), and the two mRNA ends can be brought together through bridging interactions of the factor eIF4G with cap-bound eIF4E and the poly(A)-binding protein (12, 13). The closed-loop concept has interesting implications for models of polysome function. For instance, it would ensure that only intact mRNAs are translated and allow the redirection of terminating ribosomes to the 5′ end of the same mRNA template.

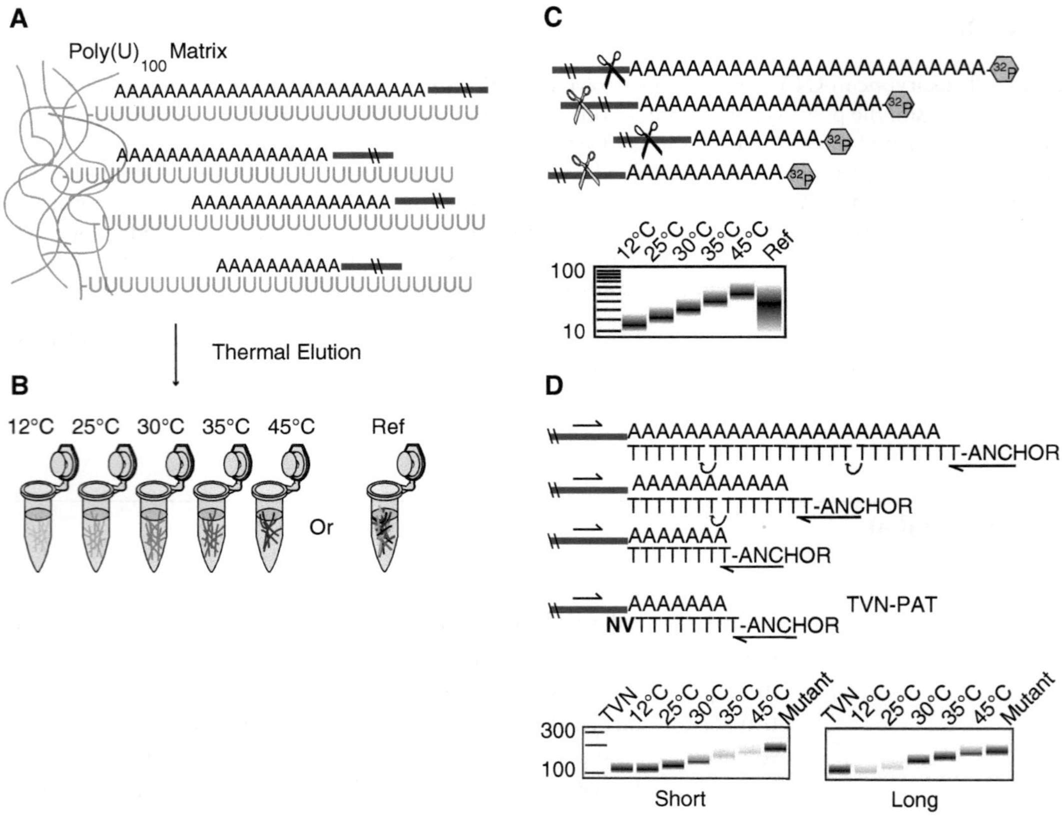

Fig. 9.1. Schematic of poly(U) chromatography and methods to control the procedure. (**a**) mRNA is bound to poly(U)$_{100}$ sepharose beads at 12°C and then eluted by stepwise increases in temperature as detailed in the main text. (**b**) Each elution fraction contains mRNA species of a defined poly(A) tail length range. (**c**) Bulk poly(A) tail length analysis is done by 3′ end-labeling mRNAs with α-[^{32}P]-pCp, digestion with RNases A and T1, and analysis of labeled poly(A) tracts by denaturing PAGE (shown here in schematic). A ^{32}P-5′ end-labeled 10(nt) DNA ladder is run along side for approximate sizing. (**d**) The distribution of poly(A) tail lengths on individual mRNA species can be measured by the LM-PAT assay. To approximate the size of a short-tailed mRNA, RT-PCR amplification from an aliquot of the same RNA is also done using an anchor–(dT)$_{12}$ primer (*left lane*) to generate a TVN-PAT product (on left). As a marker for maximal mRNA tail length an LM-PAT product from cells lacking deadenylase enzymes is also run (labelled 'Mutant'; see main text for details). Mock gels with expected results for a short and a long-tailed mRNA are shown. PCR products are sized against a 100 bp DNA ladder to determine the range of poly(A) lengths for each mRNA.

Many known translational control paradigms target some aspects of the molecular bridge between the two mRNA ends (14–16); typically this involves a block to cap function and/or the shortening of the mRNA poly(A) tail. Well-studied examples are not only the maternal mRNAs during oocyte maturation but also mRNAs targeted by microRNAs (17–20). The poly(A) tail length can thus be seen as an mRNA feature that reports on the functional state of mRNAs. To further test this concept, we developed the PASTA method (21), applied it to the transcriptomes of budding and fission yeast, and revealed a widespread interdependence between 3′-untranslated region-mediated poly(A) tail length control, Pab1 binding, and mRNA translation (22, 23). The PASTA approach is thus capable of determining the scope of mRNA-specific polyadenylation state control in a transcriptome under investigation and PASTA data are a rich source for further study of gene-regulatory networks and mechanisms. Given the correlation between polyadenylation and translation states of mRNA, PASTA analysis could furthermore serve as a surrogate for the more established *t*ranslation *s*tate micro*a*rray *a*nalysis (TSAA), a combination of sucrose density gradient centrifugation to separate polysomal complexes with microarray analysis of gradient fractions (24, 25). This may be of particular interest in situations where an overlap in sedimentation of polysomal complexes and aggregates that sequester translationally repressed mRNA, such as processing bodies, is suspected (e.g., (26)).

The general PASTA procedure begins with the isolation of yeast total RNA by the hot acid phenol method and binding of the adenylated species to poly$(U)_{100}$ sepharose beads in suspension. This is done by heat denaturing the total RNA in the presence of poly$(U)_{100}$ sepharose, followed by cooling, washing of the beads, and batch elution by stepwise increases in temperature (**Fig. 9.1a, b**). To test fractions for the degree of tail length separation, aliquots of the eluted RNA are labeled at the 3′ end with α-[^{32}P] pCp, followed by digestion of the mRNA body with RNases T1 and A, leaving only stretches of poly(A) intact. These poly(A) tracts can be resolved by denaturing PAGE (**Fig. 9.1c**). Gene-specific poly(A) tail length distribution can be monitored by the RT-PCR-based *l*igation-*m*ediated *p*oly(A)-*t*est (LM-PAT), a sensitive assay yielding PCR product sizes that reflect the poly(A) tail lengths present on a given mRNA (**Fig. 9.1d**). It is used to monitor the efficacy of the poly(U) chromatography step as well as to verify tail lengths on candidate mRNAs. To generate the PASTA data, we perform dual-color microarray analyses of the poly(U) sepharose eluates (**Fig. 9.2**).

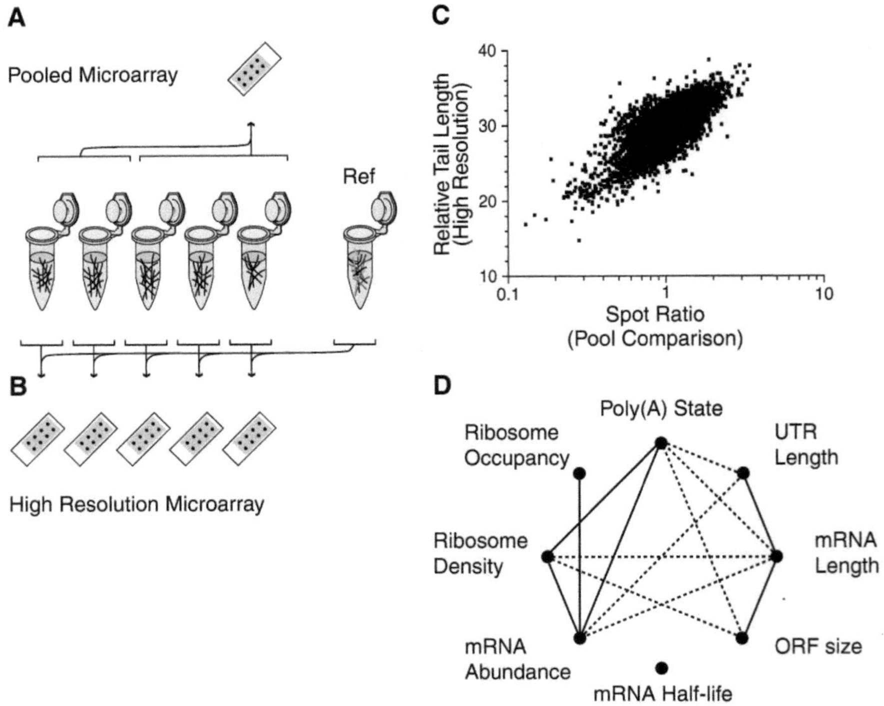

Fig. 9.2. Microarray analysis of poly(U) chromatography fractions. (**a**) For low-resolution PASTA analyses, pools of elution fractions are compared to each other by microarray (30, 35, and 45°C [e.g., Cy5-labeled cDNA] versus 12 and 25°C [e.g., Cy3-labeled cDNA]). (**b**) For high-resolution PASTA analyses, each temperature eluate (e.g., reverse-transcribed into Cy5-labeled cDNA) is compared against reference mRNA eluted in a single step at 45°C (e.g., as Cy3-labeled cDNA) on separate dual-color microarrays. (**c**) To generate a relative value of poly(A) tail length from high-resolution PASTA data, ratios of each mRNA from the five elution fractions can be transformed into percentages of mRNA for each fraction. The percentage of each fraction was multiplied by arbitrary weights of 0.1, 0.2, 0.3, 0.4, and 0.5 for the respective fractions 1 to 5 and summated so that all mRNAs receive a relative poly(A) tail length measure between 10 and 50 [arbitrary units; the approach is detailed in (23)]. The graph shows a plot of this measure ($N = 5{,}698$) against the spot ratio for the corresponding mRNA in a pool comparison experiment. The Spearman rank correlation coefficient for this data was ($r = 0.6921$, $p < 0.0001$). (**d**) Global correlation between mRNA characteristics in *S. cerevisiae*. *Solid lines* represent positive and *dashed lines* inverse correlations. Data in C and D taken from (22) and references therein.

2. Materials

2.1. Equipment

1. Clinical bench-top centrifuge capable of spinning 15 mL tubes (Eppendorf model 5702).
2. Rotating wheel (Ratek RSM6).
3. Bench-top microcentrifuges, operating either at ambient temperature (23°C) or cooled (4°C) (Eppendorf models 5415 D and 5415).
4. NanoDrop 2000 spectrophotometer (Thermo Scientific).
5. Standard equipment for running horizontal slab gels.

6. Eppendorf Thermomixer Compact.
7. Disposable insulin syringes (Becton Dickinson, # 326719).
8. Bio-Rad Micro Bio-Spin 6 columns (# 732-6221).
9. Large format sequencing gel apparatus (≥20 cm in length).
10. Electrophoresis power supply with temperature probe (Bio-Rad Power PAC 3000).
11. FLA-5100 fluorescent and radioisotopic imaging system and MultiGauge software (Fujifilm, Tokyo, Japan).

2.2. Reagents

1. Poly(U)$_{100}$ sepharose 4B (GE Healthcare, # 17-0610-01).
2. Molecular biology-grade water (Sigma-Aldrich).
3. Hydration buffer (HB): 0.1 M NaCl; 10 mM Tris–HCl at pH 7.4.
4. Elution buffer (EB): 0.1 M NaCl, 0.01 M EDTA, 0.5 M Tris–HCl at pH 7.4, 0.2 (w/v) SDS, 25% (v/v) molecular biology-grade formamide (Sigma-Aldrich).
5. HSBB (high-salt binding buffer): 0.7 M NaCl, 0.01 M EDTA, 0.5 M Tris–HCl at pH 7.4, 0.2% (w/v) sodium lauryl sarcosine, 12.5% (v/v) formamide.
6. NaCl solution (5 M).
7. Pellet Paint co-precipitant (Novagen, # 69049-4).
8. Polynucleotide kinase (PNK, 10 U/μL) and corresponding 10x buffer (New England Biolabs, # M020L).
9. 100 and 80% (v/v) molecular biology-grade ethanol.
10. Cytidine 3′-monophosphate (Cp, Sigma-Aldrich, # C1133) made up to 3 mM in water.
11. [γ-^{32}P]-ATP (3000 μCi/mmol, 10 mCi/mL, Perkin Elmer, # NEG502A500UC).
12. T4 RNA ligase and supplied 10x buffer (Promega, # M105A).
13. Dimethyl sulfoxide (DMSO, Sigma-Aldrich, # D2650).
14. Acetylated bovine serum albumin (BSA). The stock 100 μg/mL acetylated BSA supplied with restriction enzymes from New England Biolabs is a convenient source.
15. RNase T1 100,000 U/mL (Roche, # 10109495001).
16. RNase A (10 mg/mL made up in water) (Roche, # 109169).
17. 10x RNase Digest Buffer: 100 mM Tris–HCl, pH 7.5 and 3 M NaCl.
18. tRNA 10.5 mg/mL (Sigma-Aldrich, # 10109495001).
19. RNase stop solution: 2 mg/mL proteinase K (Roche, # 745723), 130 mM EDTA, and 2.5% (w/v) SDS.

20. Acid phenol, UltraPure™ Phenol:Water (3.75:1, v/v) (Invitrogen, # 15594-047).
21. Chloroform: isoamylalcohol (24:1 v/v Fluka, # 2466).
22. Phase Lock Gel Heavy 1.5 mL tubes (5 Prime, # 2302810).
23. SequaGel sequencing system kit (National Diagnostics, # EC-840).
24. 2x formamide RNA loading dye made freshly prior to use as follows: 250 μL formamide, 1 μL loading dyes (250x solution is 10 mg/mL bromophenol blue, 10 mg/mL xylene cyanol, dissolved in methanol), 25 μL 10 × TBE, 5 μL 0.5 M EDTA.
25. 10 bp ladder (Promega, # G4471).
26. 100 μM stock solution of pd(T)12-18. This can be prepared in house by mixing custom synthesized 5' phosphorylated oligo nucleotides. We use a 1:1:1:1 mix of pd(T)12, pd(T)14, pd(T)16 and pd(T)18 each disolved to 100 μM prior to mixing.
27. Primers (made up to 100 μM in water and stored in aliquots at –80°C): anchor–$(dT)_{12}$ primer (5′ GCGAGC TCCGCGGCCGCGTTTTTTTTTTTT); anchor–$(dT)_{12}$ VN primer (5′ GCGAGCTCCGCGGCCGCGTTTTTT TTTTTTVN) where V stands for G, A, or C and N is any nucleotide.
28. Superscript III (Invitrogen, # 18080044).
29. dNTPs, 10 mM each (Roche, # 1969064).
30. 10 mM ATP (Promega, # P1132).
31. T4 DNA ligase (New England Biolabs, # M0202L).
32. RNAseOUT (Invitrogen, # 10777-019).
33. Fast-start Taq (Roche, # 12032937001).
34. Agarose 1000 (Invitrogen, # 10975-035).
35. 100 bp DNA ladder (New England Biolabs, # N3231S).
36. Ethidium bromide 10 mg/mL (Sigma-Aldrich).

3. Methods

3.1. Fractionation of mRNA by Poly(U) Chromatography

Binding to poly(U) sepharose at low temperature, followed by stepwise thermal elution can be used to fractionate cellular mRNAs by the length of their poly(A) tail (**Fig. 9.1a, b**) (27, 28). In our work (22, 23), we have employed a commercial sepharose matrix that is coupled to $poly(U)_{100}$ ligands, a length well suited to the fractionation of yeast mRNAs, as their poly(A) tails do not exceed ~70–90 adenosines in wild-type genetic backgrounds. It should be possible to change the type of beads, length of poly(U) ligand, or the source RNA material used, but this will require re-tuning of the experimental procedure detailed below. Total RNA of high quality and purity is required so that it retains integrity throughout the lengthy manipulations at elevated temperature. RNA can be rapidly prepared from yeast cells according to the hot acid phenol method. Total RNA prepared by other methods or from other sources should also be suitable, provided the resulting RNA has $OD_{260\ nm}/OD_{280\ nm}$ and $OD_{260\ nm}/OD_{230\ nm}$ ratios of >2. All solutions are prepared from RNase-free stocks in new or RNase-free plastic or glassware. *See* **Note 1** regarding the preparation of buffers.

1. Suspend ~500 mg of dry poly(U)$_{100}$ sepharose in 10 mL of HB at room temperature in disposable 15 mL tubes. Hydration should be performed for at least 30 min at room temperature or overnight at 4°C using a rotating wheel to keep the matrix in suspension. Prior to use wash a further three times in 10 mL HB, each time collecting the beads by spinning at 500×*g* for 5 min in a bench-top clinical centrifuge.
2. Transfer the equivalent of 150 μl settled wet gel volume per binding/elution reaction to a 1.5 mL microfuge tube and wash once with 1 mL of EB followed by 1 mL HSBB for 5 min at 70°C in a thermomixer set at 1,100 rpm to keep matrix in suspension. For all wash and elution steps, sediment matrix by spinning at 5,000×*g* for 30 s. Use disposable insulin syringes to remove as much liquid as possible from the matrix and to avoid bead loss (*see* **Note 2**).
3. Add 600 μl of HSBB and up to 100 μg of total yeast RNA to the prepared poly(U)$_{100}$ matrix. Denature the RNA in the presence of beads for 5 min at 70°C in a thermomixer at 1,100 rpm.
4. Subsequent binding and wash steps are performed in the thermomixer (set to 1,100 rpm) in a cold room. Transfer the microfuge tube to the thermomixer set at 55°C, then change setting to 12°C and leave for 90 min (~25 min of ramp-down time and ~65 min incubation at 12°C).
5. After binding, the matrix is washed four times by addition of 1 mL ice-cold HSBB and incubation at 12°C for 5 min in the thermomixer. Between wash steps the matrix is collected by centrifugation as in Step 2.
6. Each thermal elution step is performed by resuspending the matrix in 600 μl EB at the specified temperature for 5 min in the thermomixer (set to 1,100 rpm). *See* **Note 3** for suitable elution temperatures.

7. Each elution fraction is transferred to a new 1.5 mL microfuge tube and re-spun. 550 μL of the fraction is then transferred (avoiding any carryover of beads) to a 2 mL microfuge tube containing 1/10 vol of 5 M NaCl and co-precipitant. The microfuge tubes are then filled with ice-cold ethanol (~1.4 mL), inverted to mix, and stored at –80°C for at least 1 h prior to collection of the RNA by centrifugation for 10 min at 16,000×*g* in a cooled microcentrifuge. The pellet is washed once with 80% (v/v) ethanol and air-dried.
8. The mRNA is resuspended in 22 μL water and desalted using Micro Bio-Spin 6 columns according to the manufacturer's instructions prior to any further enzymatic analyses.

3.2. Measurement of mRNA Tail Length in Bulk or for Individual mRNAs

We routinely test whether the fractionation procedure efficiently separates mRNA into pools with distinct poly(A) tail lengths, prior to genome-wide composition analyses. To this end, we monitor the adenylation state of bulk mRNAs as well as that of individual mRNAs in each fraction.

3.2.1. Bulk Poly(A) Tail Length Analysis

For bulk poly(A) tail length analysis in a sample, the RNA is 3′ end-labeled with α-[^{32}P] pCp and digested with RNAses T1 and A (which cleave after G, or after C and U, respectively), as previously described (29). This results in radio-labeled poly(A) tracts which can be separated by 16% urea-PAGE and visualized by autoradiography (**Fig. 9.1c**). α-[^{32}P] pCp can be obtained from a commercial stock or prepared in-house as described in **Section 3.2.1.1**.

3.2.1.1. Preparation of α-[^{32}P] pCp

1. Assemble the following for a total reaction volume of 20 μL: 5 μL dH_2O, 2 μL 10x PNK buffer, 2 μL 3 mM Cp, 10 μL γ-[^{32}P]-ATP, and 1 μL T4 PNK.
2. Incubate for 30 min at 37°C, then heat inactivate enzyme at 70°C for 5 min.
3. Store at –80°C if not using immediately.

3.2.1.2. RNA 3′ End-Labeling and RNase Digestion for Poly(A) Length Analysis

1. A 30 μL reaction volume is prepared containing the following: 1 μg of total RNA or 90% of an elution fraction (up to a volume of 17.5 μL), 3 μL 10x T4 ligase buffer, 3 μL DMSO, 3 μL BSA, 2.5 μL α-[^{32}P] pCp prepared as detailed in **Section 3.2.1.1** (or the equivalent amount from a commercial source), and 1 μL T4 RNA ligase. Incubate for 2 h at 37°C.
2. Inactivate enzyme at 70°C for 5 min.
3. Remove unincorporated nucleotides using Micro Bio-Spin 6 columns.
4. The RNA is then subjected to simultaneous degradation by RNase A and RNase T_1 for 2 h at 37°C. Combine 30 μL

of labeled RNA from Step 3; 40 μg yeast tRNA (as "ballast"); 80 U RNase T_1 and 4 μg RNase A; 8 μL 10x RNase digest buffer and water to a total reaction volume of 80 μL. This is most conveniently added as 50 μL from a master mix.

5. Stop the reaction by addition of 20 μL stop solution and incubation for 30 min at 37°C.
6. Phenol–chloroform extract once by addition of 100 μL of a solution of 50% (v/v) acid phenol and 50% chloroform:isoamylacohol (24:1, v/v). Vortex thoroughly and spin in a microcentrifuge at 16,000×*g* for 5 min. Phase Lock Gel tubes can be used to simplify this step but are not essential.
7. Transfer the aqueous phase to a new 1.5 mL microfuge tube containing 20 μg of yeast tRNA as carrier.
8. Precipitate RNA by addition of 250 μL 100% ethanol and incubation at –80°C for at least 1 h or overnight. Wash the pellet once with 80% ethanol and resuspend in 20 μL of dH_2O.
9. Pre-prepare a 16% (w/v) polyacrylamide sequencing gel (*see* **Note 4**). The gel is pre-run at 45°C (controlled using a temperature probe) and 100 W for 30 min prior to loading of the RNA.
10. In new microfuge tubes combine 5 μL RNA and 5 μL 2x formamide RNA loading dye. Heat samples to 80°C for 5 min immediately prior to loading.
11. Load 4–6 μL of denatured sample per lane using disposable sequencing tips. Size markers should be run in parallel (*see* **Note 5**).
12. The gel is run at the same parameters as in Step 9 until the bromophenol blue dye has migrated ~two-thirds down the gel.
13. Transfer the gel onto Whatman blotting paper, dry, and visualize by autoradiography or phosphorImager analysis using the FLA-5100 imager (*see* **Note 6**).

3.2.2. Ligation-Mediated Poly(A) Test (LM-PAT)

The combination of oligonucleotide-mediated RNAse H cleavage of the mRNA and high-resolution northern blotting for the 3′ cleavage fragment is the benchmark for measuring poly(A) tail lengths of individual mRNAs (30). However, the LM-PAT assay (**Fig. 9.1d**) (22, 31, 32) can generate equivalent results and has advantages in many situations due to its higher sensitivity and ease of use. Briefly, RNA is first incubated with oligo(dT)$_{12-18}$ primers in the presence of T4-DNA ligase at 42°C, creating a poly(dT) copy of each mRNA's poly(A) tails within the sample. Addition of excess anchor–$(dT)_{12}$ primer and further incubation at 12°C favors annealing of the anchor primer to unpaired poly(A) ends and ligation to the poly(dT) stretch. This assembly is then used to prime synthesis of first-strand cDNA by reverse transcription. PCR reactions with primers specific to the mRNA of choice and the anchor region generate LM-PAT products, which are visualized by agarose gel electrophoresis.

Due to the limited range of primer lengths in the oligo$(dT)_{12–18}$ mixture some LM-PAT product "laddering" is often visible. The use of a anchor–$(dT)_{12}$ primer further determines that the first "rung" in the PCR product ladder represents all short tails that can accommodate a single anchor–$(dT)_{12}$ but not an additional oligo$(dT)_{12–18}$ primer. To control for any bias due to incomplete saturation of poly(A) tails with oligo$(dT)_{12–18}$ prior to cDNA synthesis or a drift toward shorter amplicons during PCR, we routinely include control reactions with total RNA from a deadenylase double mutant *Δpan2/ccr4-1* strain (33), in which all mRNAs possess long poly(A) tails (34, 35). We also recommend inclusion of a separate reaction, where cDNA synthesis is primed with an anchor–$(dT)_{12}$VN primer which we term TVN-PAT. PCR from this cDNA will generate a size marker for the shortest possible LM-PAT product for the mRNA being analyzed (**Fig. 9.1d**). Finally, LM-PAT products can be cloned and sequenced to guard against miss-priming and other PCR artifacts.

3.2.2.1. Preparation of LM-PAT cDNA

1. Dilute 1 μg total RNA or 2–10% of a poly$(U)_{100}$ chromatography fraction to a total of 6 μL with dH_2O and add 1 μL of a 1/20 dilution of the stock oligo$(dT)_{12–18}$ primers (*see* **Note 7**).
2. Denature the RNA and primers at 65°C for 5 min.
3. Meanwhile, prepare and pre-warm to 42°C, a master mix containing following components for each reaction : 4 μL dH_2O, 4 μL 5x Superscript III reaction buffer, 2 μL 0.1 M DTT, 1 μL dNTP mix, 1 μL 10 mM ATP, 1 μL T4 DNA ligase, and 0.5 μL RNaseout.
4. Flash spin down the denatured RNA, transfer to a 42°C heat block, and add 13 μL pre-warmed master mix. Incubate for 30 min.
5. Add 1 μL anchor–$(dT)_{12}$ primer while tubes remain at 42°C, then mix, flash-spin, and transfer to 12°C for 2 h.
6. Return to 42°C for 2 min before addition of 1 μL Superscript III. Incubate at 42°C for 1 h and then increase temperature to 52°C for a further 1 h.
7. Inactivate the reverse transcriptase by incubation at 70°C for 10 min.

3.2.2.2. Preparation of TVN-PAT cDNA (*See* **Note 8**)

1. Assemble 1 μg of total RNA, 1 μL of the anchor–$(dT)_{12}VN$ primer and water in a total volume of 11.5 μL.
2. Denature at 65°C for 5 min.
3. Meanwhile, assemble in a separate tube and pre-warm to 42°C the following: 3.5 μL water, 4 μL 5x Superscript III reaction buffer, 2 μL 0.1 M DTT, 1 μL dNTP mix, 0.5 μL RNAseout, and 1 μL Superscript III.
4. Flash spin down the denatured RNA, transfer to a 42°C heat block adding the 12 μL of pre-warmed components from Step 3 to the denatured RNA/anchor–$(dT)_{12}VN$ primer mix.
5. Mix flash-spin and incubate at 42°C for 1 h, after which the temperature is increased to 52°C and incubation continued for a further hour.

3.2.2.3. PCR Amplification of LM-PAT and TVN-PAT cDNAs

1. Dilute PAT and TVN-PAT reactions 5- and 20-fold, respectively in water.
2. Prepare 25 μL PCR reactions using 5 μL diluted cDNA as template and standard PCR components and buffers supplied with the Fast-Start Taq. Assemble master mixes containing 1.25 U Fast-Start Taq per reaction and both the gene-specific forward primer and the PAT assay primer at a concentration of 1 μM (*see* **Note 9**).
3. Cycling parameters are as follows: 2 min at 95°C; then 30 s each at 95, 60, and 72°C for 25–35 cycles (template dependent); and a final 2 min at 72°C (*see* **Note 10**).
4. Run one-third of the PCR product on 2% (w/v) high-resolution agarose gels using thin (0.75 mm) gel combs for finer resolution. Gels are cast containing 0.5 μg/mL ethidium bromide for visualization. A 100 bp ladder is used to size the products and to estimate the poly(A) tail length. The migration of the TVN sample represents a short tail of ~12 adenosines.
5. High-resolution images are best obtained by laser scanning of the gels on the FLA-5100 imager.

3.3. Microarray Analysis of Poly(U) Chromatography Fractions

We use dual-color spotted oligonucleotide microarrays for this purpose, employing generic microarray and data analysis methods as detailed elsewhere (22, 23). We implemented two experimental designs for analysis of poly(U) sepharose eluates. In the first, we compare pooled low temperature fractions against a pool of high-temperature eluates on a single microarray (**Fig. 9.2a**). One chromatography run at the scale detailed above should generate suitable amounts of labeled cDNAs for this. The poly(A) tail status of a given mRNA is then represented by the corresponding array spot ratio, which after normalization will be ~1 for an mRNA with an average elution pattern. For this pool comparison we then empirically choose spot ratio cut-offs to generate mRNA candidate lists or simply rank the transcriptome by tail length ratio. The second experimental design uses five separate arrays to compare each eluate against reference mRNA (**Fig. 9.2b**). To obtain this reference, bound mRNA is eluted from beads in one batch at 45°C. It may be necessary to combine corresponding elution fractions from several parallel poly(U) chromatography runs at the scale detailed above to achieve suitable amounts of labeled cDNAs. Standard data normalization means that a hypothetical mRNA representing the averaged elution pattern of all mRNAs would be assigned a spot ratio of ~1 for each array in the series. To select mRNA candidates, profiles may then be sorted into classes that deviate from the average elution pattern in an interpretable way. For example, those having a higher proportion of long tails than average and those having a higher proportion of short tails than average (22). As an alternative, the ratios of each mRNA from the five elution fractions may be transformed into one relative poly(A) tail length measure, which is then ranked (23), *see* legend to **Fig. 9.2c** for details. In our hands, polyadenylation state data displayed good correlation between the two different experimental designs (**Fig. 9.2c**).

Our PASTA surveys on exponentially growing yeast cells uncovered a widespread coordination of mRNA polyadenylation state among cytotopically and functionally related mRNAs that exhibits evolutionary conservation. It further revealed extensive correlation between tail length and other physical and functional mRNA characteristics (**Fig. 9.2d**). PASTA surveys could be done in other growth conditions, mutant cells, or be carried out in more complex eukaryotic models and cellular contexts such as early embryonic development, in which tail length control is deemed particularly relevant. Indeed, poly(A) tail length control in different organisms has already been studied using related approaches (36–38). Finally, RNA eluates prepared as in **Section 3.1** should be suitable for analysis on any single- or dual-color microarray or alternatively next-generation sequencing platform. This will simply require some adjustments to the experimental design and processing of the RNA samples.

4. Notes

1. The buffers for poly(U) chromatography are best made from stocks directly before use. This is particularly important for EB and HSBB since they contain deionized formamide, which may become ionized with time. To rule

out any variance in water quality we use Sigma's molecular biology-grade water for preparation of all small-scale buffers and the suspension of RNA and cDNA samples.

2. The fine bore size of insulin syringes results in very few beads being inadvertently removed at buffer change and elution steps and allows the beads to be substantially depleted of liquid minimizing the mixing of fractions.

3. We find that elution steps at 12, 25, 30, 35, 45°C work well to achieve a good size separation between fractions, each 5°C increment resulting in approximately 10 nt tail length shift. Ideally, the higher temperature elution steps are carried out in a second thermomixer placed at room temperature. Transferring the first one from the cold room to a lab bench and reconnecting it immediately might cause electrical faults due to condensation; however, we still do this once the first elution (12°C) has started, the block maintains the incubation temperature for ~5 min.

4. Our large format electrophoresis system is 35 cm wide, 45 cm long, and the gel has a width of 0.2 mm (Owl S3S).

5. We have used standard procedures outlined in the manufacturer's instructions for PNK to generate ^{32}P-labeled 10 bp DNA ladders that can be run alongside RNA samples to approximate the length of poly(A) tracts associated with each fraction.

6. 16% urea polyacrylamide sequencing gels tend to be quite dry and their thinness makes handling difficult. To transfer the gel onto blotting paper remove the top plate, place blotting paper (pre-cut to size) onto the gel surface (it will not stick efficiently like lower percentage gels), and invert onto a clean bench-top, allowing ~4 cm overhang of the gel relative to the edge of the bench. Pry the overhanging gel from the glass plate allowing it to fall onto the blotting paper. Next lift the glass plate up allowing the rest of the gel to fall onto the blotting paper. The gel is then dried onto the paper using a vacuum gel drier (2 h at 80°C).

7. In practice, the dilution of the oligo(dT)$_{12–18}$ primers linker is empirically determined and varies according to the activity of the T4 ligase and other parameters that are difficult to quantify. Readers are urged to refer also to Sallés and coworkers (39) and references therein for further discussion.

8. The TVN-PAT reaction shows the minimal length that can be resolved by the LM-PAT assay and can unveil the occurrence of alternate poly(A) cleavage sites. Typically this reaction is slightly more efficient than the LM-PAT reaction, and a PCR product of defined size will be more readily visible than a smear of amplicons. To compensate for this,

TVN-PAT samples undergo a higher dilution (or lower PCR cycle number) than equivalent LM-PAT samples.

9. The gene-specific primers for *Saccharomyces cerevisiae* transcripts are generally designed at or near the stop codon to give a primer of 18–22 nt length with a melting temperature of >60°C and a PCR product size of between 100 and 300 base pairs. To design primers for LM-PAT assays on *Schizosaccharomyces pombe* and mammalian mRNAs where 3′ UTRs tend to be longer, a suitable primer sequence is chosen within 200 base pairs from the cleavage and polyadenylation site. If unknown, the site can be narrowed down experimentally starting with a primer based around the sequence at the stop codon. In all cases, the amplified LM-PAT products encompass a region of the mRNA 3′ UTR in addition to the stretch of poly(A) tail.
10. By using low cycle numbers and a heat-activated polymerase (Fast-Start Taq by Roche) we avoid most PCR bias (22). However, it is important to note that bias toward 'apparently' shorter poly(A)-tails can occur when using too high a concentration of the oligo $(dT)_{12-18}$ linker during the ligation or over-cycling in the PCR step. It is useful to analyze total RNA from wild-type cells (*PAN2/CCR4*) and an isogenic strain mutant for both yeast deadenylases (*Δpan2/ccr4-1*) in parallel to ensure saturation of the poly(A) tail during cDNA synthesis.

Acknowledgments

We thank Jürg Bähler and Daniel Lackner for their input to PASTA method development and application. This work was supported by the Victor Chang Cardiac Research Institute, the Australian Research Council, the National Health and Medical Research Council of Australia, and the Sylvia and Charles Viertel Charitable Foundation.

References

1. Gebauer, F., and Hentze, M. W. (2004) Molecular mechanisms of translational control. *Nat. Rev. Mol. Cell. Biol.* **5**, 827–835.
2. Beach, D. L., and Keene, J. D. (2008) Ribotrap: targeted purification of RNA-specific RNPs from cell lysates through immunoaffinity precipitation to identify regulatory proteins and RNAs. *Methods Mol. Biol.* **419**, 69–91.
3. Keene, J. D. (2007) RNA regulons: coordination of post-transcriptional events. *Nat. Rev. Genet.* **8**, 533–543.

4. Mathews, M. B., Sonenberg, N., and Hershey, J. W. (2007) Origins and principles of translational control. In: Mathews, M. B., Sonenberg N., and Hershey J. B. W. (eds.), *Translational Control in Biology and Medicine* (pp. 1–40). Cold Spring Harbor, NY: Cold Spring Harbor Laboratory Press.
5. Gallie, D. R. (1991) The cap and poly(A) tail function synergistically to regulate mRNA translational efficiency. *Genes Dev.* **5**, 2108–2116.
6. Tarun, S. Z., Jr., and Sachs, A. B. (1995) A common function for mRNA 5′ and 3′ ends in translation initiation in yeast. *Genes Dev.* **9**, 2997–3007.
7. Preiss, T., and Hentze, M. W. (1998) Dual function of the messenger RNA cap structure in poly(A)-tail-promoted translation in yeast. *Nature* **392**, 516–520.
8. Gebauer, F., Corona, D. F., Preiss, T., Becker, P. B., and Hentze, M. W. (1999) Translational control of dosage compensation in Drosophila by Sex- lethal: cooperative silencing via the 5′ and 3′ UTRs of msl-2 mRNA is independent of the poly(A) tail. *EMBO J.* **18**, 6146–6154.
9. Bergamini, G., Preiss, T., and Hentze, M. W. (2000) Picornavirus IRESes and the poly(A) tail jointly promote cap- independent translation in a mammalian cell-free system. *RNA* **6**, 1781–1790.
10. Jacobson, A., and Favreau, M. (1983) Possible involvement of poly(A) in protein synthesis. *Nucleic Acids Res.* **11**, 6353–6368.
11. Amrani, N., Ghosh, S., Mangus, D. A., and Jacobson, A. (2008) Translation factors promote the formation of two states of the closed-loop mRNP. *Nature* **453**, 1276–1280.
12. Tarun, S. Z., Jr., and Sachs, A. B. (1996) Association of the yeast poly(A) tail binding protein with translation initiation factor eIF-4G. *EMBO J.* **15**, 7168–7177.
13. Imataka, H., Gradi, A., and Sonenberg, N. (1998) A newly identified N-terminal amino acid sequence of human eIF4G binds poly(A)-binding protein and functions in poly(A)-dependent translation. *EMBO J.* **17**, 7480–7489.
14. Richter, J. D., and Sonenberg, N. (2005) Regulation of cap-dependent translation by eIF4E inhibitory proteins. *Nature* **433**, 477–480.
15. Hentze, M. W., Gebauer, F., and Preiss, T. (2007) Cis-regulatory sequences and trans-acting factors in translational control. In: Mathews M. B., Sonenberg N., and Hershey J. W. B. (eds.), *Translational Control in Biology and Medicine* (pp. 269–295). Cold Spring Harbor, NY: Cold Spring Harbor Laboratory Press.
16. Goldstrohm, A. C., and Wickens, M. (2008) Multifunctional deadenylase complexes diversify mRNA control. *Nat. Rev. Mol. Cell. Biol.* **9**, 337–344.
17. Humphreys, D. T., Westman, B. J., Martin, D. I., and Preiss, T. (2005) MicroRNAs control translation initiation by inhibiting eukaryotic initiation factor 4E/cap and poly(A) tail function. *Proc. Natl. Acad. Sci. USA* **102**, 16961–16966.
18. Standart, N., and Jackson, R. J. (2007) MicroRNAs repress translation of m7Gppp-capped target mRNAs in vitro by inhibiting initiation and promoting deadenylation. *Genes Dev.* **21**, 1975–1982.
19. Beilharz, T. H., Humphreys, D. T., and Preiss, T. (2009) miRNA effects on mRNA closed-loop formation during translation initiation. In: Rhoads R. E. (ed.), *miRNA Regulation of the Translational Machinery* (pp. 99–112). Berlin: Springer.
20. Eulalio, A., Huntzinger, E., Nishihara, T., Rehwinkel, J., Fauser, M., and Izaurralde, E. (2009) Deadenylation is a widespread effect of miRNA regulation. *RNA* **15**, 21–32.
21. Beilharz, T. H., and Preiss, T. (2009) Transcriptome-wide measurement of mRNA polyadenylation state. *Methods* **48**, 294–300.
22. Beilharz, T. H., and Preiss, T. (2007) Widespread use of poly(A) tail length control to accentuate expression of the yeast transcriptome. *RNA* **13**, 982–997.
23. Lackner, D. H., Beilharz, T. H., Marguerat, S., et al. (2007) A network of multiple regulatory layers shapes gene expression in fission yeast. *Mol. Cell* **26**, 145–155.
24. Beilharz, T. H., and Preiss, T. (2004) Translational profiling: the genome-wide measure of the nascent proteome. *Brief Funct. Genomic Proteomic* **3**, 103–111.
25. Mata, J., Marguerat, S., and Bahler, J. (2005) Post-transcriptional control of gene expression: a genome-wide perspective. *Trends Biochem. Sci.* **30**, 506–514.
26. Thermann, R., and Hentze, M. W. (2007) Drosophila miR2 induces pseudo-polysomes and inhibits translation initiation. *Nature* **447**, 875–878.
27. Binder, R., Horowitz, J. A., Basilion, J. P., Koeller, D. M., Klausner, R. D., and Harford, J. B. (1994) Evidence that the pathway of transferrin receptor mRNA degradation involves an endonucleolytic cleavage within the 3′ UTR and does not involve poly(A) tail shortening. *EMBO J.* **13**, 1969–1980.

28. Palatnik, C. M., Storti, R. V., and Jacobson, A. (1979) Fractionation and functional analysis of newly synthesized and decaying messenger RNAs from vegetative cells of Dictyostelium discoideum. *J. Mol. Biol.* **128**, 371–395.
29. Minvielle-Sebastia, L., Winsor, B., Bonneaud, N., and Lacroute, F. (1991) Mutations in the yeast RNA14 and RNA15 genes result in an abnormal mRNA decay rate; sequence analysis reveals an RNA-binding domain in the RNA15 protein. *Mol. Cell. Biol.* **11**, 3075–3087.
30. Steiger, M. A., and Parker, R. (2002) Analyzing mRNA decay in Saccharomyces cerevisiae. *Methods Enzymol.* **351**, 648–660.
31. Sallés, F. J., and Strickland, S. (1995) Rapid and sensitive analysis of mRNA polyadenylation states by PCR. *PCR Methods Appl.* **4**, 317–321.
32. Clancy, J. L., Nousch, M., Humphreys, D. T., Westman, B. J., Beilharz, T. H., and Preiss, T. (2007) Methods to analyze microRNA-mediated control of mRNA translation. *Methods Enzymol.* **431**, 83–111.
33. Woolstencroft, R. N., Beilharz, T. H., Cook, M. A., Preiss, T., Durocher, D., and Tyers, M. (2006) Ccr4 contributes to tolerance of replication stress through control of CRT1 mRNA poly(A) tail length. *J. Cell. Sci.* **119**, 5178–5192.
34. Tucker, M., Valencia-Sanchez, M. A., Staples, R. R., Chen, J., Denis, C. L., and Parker, R. (2001) The transcription factor associated Ccr4 and Caf1 proteins are components of the major cytoplasmic mRNA deadenylase in Saccharomyces cerevisiae. *Cell* **104**, 377–386.
35. Tucker, M., Staples, R. R., Valencia-Sanchez, M. A., Muhlrad, D., and Parker, R. (2002) Ccr4p is the catalytic subunit of a Ccr4p/Pop2p/Notp mRNA deadenylase complex in *Saccharomyces cerevisiae. EMBO J.* **21**, 1427–1436.
36. Du, L., and Richter, J. D. (2005) Activity-dependent polyadenylation in neurons. *RNA* **11**, 1340–1347.
37. Graindorge, A., Thuret, R., Pollet, N., Osborne, H. B., and Audic, Y. (2006) Identification of post-transcriptionally regulated Xenopus tropicalis maternal mRNAs by microarray. *Nucleic Acids Res.* **34**, 986–995.
38. Meijer, H. A., Bushell, M., Hill, K., et al. (2007) A novel method for poly(A) fractionation reveals a large population of mRNAs with a short poly(A) tail in mammalian cells. *Nucleic Acids Res.* **35**, e132.
39. Salles, F. J., Richards, W. G., and Strickland, S. (1999) Assaying the polyadenylation state of mRNAs. *Methods* **17**, 38–45.

Chapter 10

Enabling Technologies for Yeast Proteome Analysis

Johanna Rees and Kathryn Lilley

Abstract

Whilst the study of yeast genomes and transcriptomes is in an advanced state, there is still much to learn about the resulting proteins in terms of cataloging, characterization of post-translational modifications, turnover, and the dynamics of sub-cellular localization and interactions. Analysis of the transcripts gives little insight into function or diversity as changes in RNA levels do not always correlate with the resulting protein abundance. A number of global and targeted attempts have been made to catalog and characterize the yeast proteome and we describe here the methods used to gain a greater understanding of the yeast proteome. This comprehensive review also describes future approaches that will aid completion in identifying and characterizing the remaining 20% of the undetermined yeast proteome as well as giving new insight into protein dynamics.

Key words: Proteome, quantitation, post-translational modifications (PTMs), interactions, sub-cellular localization, protein turnover, protein microarrays, bioinformatics and data repositories.

1. Introduction

1.1. Why Study at the Proteome Level?

Over the last decade or so, a large body of data describing RNA abundances in cells and tissues have been collected. Our knowledge of cellular mechanisms has been greatly enhanced by the study of RNA expression patterns upon perturbations such as mutation. Characterizing RNA expression does not always lead to mechanistic insight, arguably transcripts can be thought of as surrogates for the proteins they encode, the proteins are the work horses, and their study directly couples expression and function. Furthermore, the function of many proteins remains to be determined, for example, up to 25% of the predicted *Saccharomyces cerevisiae* open reading frames have an unassigned molecular function (http://www.yeastgenome.org). Analysis of the transcripts

J.I. Castrillo, S.G. Oliver (eds.), *Yeast Systems Biology*, Methods in Molecular Biology 759,
DOI 10.1007/978-1-61779-173-4_10,

of these proteins will give little insight into their function, as since changes in RNA levels do not always correlate with their relevant protein product abundance the credible prediction of protein abundance (1, 2). Therefore for a systems-wide knowledge of an organism, characterization of the proteome is crucial. The global study of proteins within a sample has given rise to the field of proteomics.

In general, much less is known about the protein abundances in the cell as methods to measure abundances and changes in expression levels are technically very challenging for a number of reasons. First, the chemical characteristics of proteins are far more varied and complex than are those of nucleic acids, making them more challenging to work with. Second, the range of protein concentrations in a given cell or biofluid is large and is generally over many orders of magnitude in any given sample. Unfortunately, there is no technology yet developed for the amplification of the rarer abundance proteins in contrast to PCR amplification of nucleic acids. Low abundance protein species may be detected only after protracted enrichment methodologies have been applied. Third, an additional layer of complexity is possible as many isoforms of proteins exist that are translated from a common gene but have different functions and different distributions, which make interpretation much more involved. These may arise not only from differential splicing or initiation but also as a result of the plethora of post-translational modifications (PTMs) that are possible such as phosphorylation, methylation, acetylation, ubiquitination, glycosylation, and proteolytic processing amongst others.

An additional complexity when studying the proteome is the spatial arrangement of proteins in terms of recruitment to multiprotein complexes and sub-cellular localization. Different protein–protein interaction combinations give rise to an enormous diversity of protein complexes and functions. Moreover, in order to fulfill their function, proteins must be present at their correct sub-cellular location. In eukaryotes, there are numerous complex membrane-delineated sub-cellular structures to which many proteins are trafficked in order to carry out their correct physiological function. The assignment of the sub-cellular localization of proteins is of great importance if biologists are to elucidate the role of any given protein and fully comprehend cellular processes by tracing certain activities to specific organelles.

Traditionally the characterization of proteins involves identification of proteins within a sample using mass spectrometric technologies, some details of which are given in **Section 2** of this chapter. An additional challenge in the study of proteins is the reliance on well-characterized genomes which enable interpretation of protein and peptide fragmentation patterns generated by the use of such technologies. The *S. cerevisiae* genome is one of the best annotated eukaryote genomes and hence the application of proteomics technologies to this organism is relatively mature and historically it has been a test bed for the development of many proteomics methodologies. Data from yeast studies have aided the mapping of many multicellular proteomes. In fact, as the result of recent heroic efforts by the proteomics community, it is reasonable to suggest that *S. cerevisiae* is one of the species with the best characterized proteome (3) (http://www.peptideatlas.org). However, despite this, 20% of the predicted proteome has yet to be detected despite the numerous experimental approaches.

2. Characterization of the Proteome

The simplest form of proteomics is to catalog the presence of proteins in any given sample. There are numerous approaches to achieve the part list of proteins within a yeast cell. Several years ago an extensive study from the O'Shea and Weissman laboratories led to the creation of a *S. cerevisiae* fusion library where each open reading frame was tagged with a high-affinity epitope and expressed from its natural chromosomal location (4). Using immunodetection of the common tag, they obtained a census of proteins expressed during log-phase growth and measurements of their absolute levels, discovering that approximately 80% of the proteome is expressed during normal growth conditions. The results of this study are of great utility to the community but are not without some caveats. Protein fusions were created which may have perturbed expression levels, trafficking, the ability to form protein–protein complexes, and physiological function of the protein.

The technological development which has aided the study of the global protein content of a sample, without resorting to piecemeal creation of tagged fusion proteins, has been the advent of soft ionization methods coupled to mass spectrometry. Two technologies, matrix-assisted laser desorption ionization time-of-flight mass spectrometry (MALDI-TOF) and nanoelectrospray ionization, earned their inventors, Koichi Tanaka and John Fenn, respectively, Nobel prizes in 2002 (5, 6). Both approaches result in stable products of ionization of proteins and peptides within mass spectrometry. These stable ions can then be characterized in terms of their mass and also quantity. Good reviews of the mass spectrometry used in conjunction with analysis of the proteome can be found in (7, 8). Mass spectrometry methods have advanced considerably over the last few years and now typically proteins are identified by performing tandem mass spectrometry experiments on peptides derived from proteins using specific proteases, typically trypsin. Peptides ionize more readily than whole proteins and can be fragmented using collision-induced dissociation within a tandem mass spectrometer to produce informative fragment ions whose masses can be used *in silico* to reconstruct the sequences which would give rise such fragmentation patterns using a variety of algorithms (MASCOT, SEQUEST, Phenyx, and X!Tandem). For such approaches to be of any use in the global analysis of the proteome, protein separation methods have made these methodologies employable with complex samples such as whole cell lysates.

Traditionally, two-dimensional polyacrylamide gel electrophoresis (2D PAGE) has been the protein separation technique most associated with proteomics studies (9) and remains one of the key methodologies. Here proteins are separated first as a function of their isoelectric point (p*I*) and second by their denatured molecular weight (Mw). A major drawback of using 2D gels is their incompatibility with hydrophobic membrane proteins which make up a mechanistically important subset of proteins and are likely to play crucial roles within organelle proteins (10).

Non-2D gel-based technologies therefore are more attractive methodologies to employ in a global study of the sub-cellular proteomes as they do not suffer the same bias toward analysis of the soluble proteome as do 2D gels, provided that the proteins are successfully solubilized prior to analysis. Typically in non-2D gel-based approaches, the proteome undergoes simplification before mass spectrometric analysis in order to maximize the amount of information about the protein content of the sample. These methods are often referred to as shot-gun proteomics. One such approach, MudPIT (multi-dimensional protein identification technology) (11), involves a solution-phase digestion of proteins to peptides and then multi-dimensional chromatographic separation of peptides before mass spectrometric analysis typically based on separation using strong cation exchange chromatography followed by orthogonal separation using reverse-phase chromatography. Washburn and co-workers first demonstrated the utility of this approach using a *S. cerevisiae* cell lysate, identifying 1,484 proteins. At the time this catalog of the yeast proteome was significantly larger than any previous attempt using gel-based methods or single-dimensional peptide chromatography. Subsequent refinement of this approach means that it is a method of choice for cataloging the proteome in an unbiased manner, sampling proteins from all sub-cellular portions of the cell with extremes in p*I*, Mw, abundance, and hydrophobicity.

The above approaches are good for gaining part list of proteins within a given proteomics, but more edifying datasets also need to contain information about protein abundances.

3. Quantitative Proteomics

There are two approaches to interrogating the proteome in a quantitative manner. Relative quantification methods have been the most widely used method to date. More recently, complementary methods have been developed to give absolute quantification values for any given protein. This latter approach tends to be used in more focused studies where prior knowledge of the proteins is used to design quantification experiments. Relative quantification methods are generally used for hypothesis-generating studies.

3.1. Relative Quantitation

The technique historically associated with quantitative proteomics studies is 2D PAGE (12). Relative quantification utilizing 2D gels is achieved by comparing the protein spot intensities preferably with fluorescent dyes which bind to proteins stoichiometrically (Sypro Ruby, Deep Purple, and CyDyes) and there are many software packages to facilitate quantification (13).

With more traditional approaches to 2D PAGE, problems with sensitivity, reproducibility, and normalization across the gels have been overcome by introducing difference in-gel electrophoresis (DIGE) technology (14, 15) in which proteins from different samples to be compared quantitatively are labeled with fluorescent dyes based on cyanine and pooled before being applied onto a gel. An internal standard is typically used to correct for the gel-to-gel variation. Within the yeast community, Valerius and co-workers (16) used this approach to study changes in cell interactions of *S. cerevisiae* during nutrient starvation. Utilizing the DIGE methodology, they determined a set of post-transcriptionally regulated proteins which were regulated proteins in response to amino acid starvation which included the ribosomal protein Cpc2p/Asc1p. They went on to show that the deletion of CPC2/ASC1 abolished amino acid starvation-induced adhesive growth and impaired basal expression of FLO11 and its activation upon starvation in haploid cells.

A weakness of all 2D PAGE-based methodologies, however, is that it is not always possible to relate the signal of a spot to a particular protein when multiple proteins are detected by more sensitive mass spectrometers.

Alternatively, non-2D gel-based methods can be used in a quantitative manner by coupling them with either the incorporation of differential stable isotopes or label-free quantification. The latter is achieved by one of two methods: either by comparing the peptide peak intensities or by counting the number of MS/MS spectra (spectral counts) collected per protein (17, 18). Within the field of yeast proteomics, both approaches have been used to yield significant datasets. Mosley and co-workers (19) used the spectral counting approach to characterize an enriched nuclear fraction using sucrose density gradients followed by MudPIT. They identified 2,674 proteins and estimated their abundance, some of which has been shown to be in abundances less than 100 copies per cell by Ghaemmaghami and co-workers (4). These researchers also assessed the co-fractionation of nuclear complexes including transcriptional regulation complexes.

Peptide ion peak intensity measurement as a method to gain quantitative proteomics data has been challenging because it heavily relies on being able to align MS matrices between different LC-MS runs and hence accurately match the intensity of the same ion between successive runs. Foss and colleagues (20) measured protein levels in total unfractionated cellular proteins which involved using graph theory to enable alignment of 408 LC-MS datasets. They went on to apply this approach to elucidate the genetic basis of variation in protein abundance in a cross between two diverse strains of yeast, BY4716 and a vineyard isolate. They discovered that loci which were found to influence protein abundance were different from those that influenced transcript levels. This study again emphasized the importance of direct analysis of the proteome. One of the advantages of these approaches is the lack of requirement for labeling of peptides. A disadvantage of these methods, however, is the potentially semi-quantitative character of the data as accurate quantitation is highly dependent on reproducible peptide generation, chromatography, and ionization efficiency (21). Another level of technical challenge within label-free quantitation methods comes from the fact that samples to be compared are not combined at any point in the experimental protocol and hence technical variability, particularly in the case of studies with protracted workflows, will have influence on the reproducibility of the data. This in turn may lead to the necessity to run many replicate experiments to achieve statistically relevant data (22).

Alternative non-gel-based methods which overcome some elements of technical irreproducibility involve the differential incorporation of stable isotopes. Incorporation of stable isotopes into a protein or peptides can be achieved in vivo, or in vitro by the incorporation of chemical tags.

In the case of yeast, in vivo incorporation of stable isotopes during growth is possible in two ways. The first entails the replacement of an amino acid by on containing a heavy isotope in growth media (23). This approach is known as SILAC (stable isotope labeling of amino acids in cell culture). The second method, metabolic labeling, involves complete replacement of an element used to synthesize amino acids, typically ^{15}N, by using growth media which contain only the stable isotopic form of the element (24). Samples grown in the presence of the heavy isotope are pooled with samples grown in the presence of the light isotope and are subsequently digested to peptides and analyzed by LC-MS/MS. For both SILAC and metabolic labeling, the isotope label is uniformly incorporated into all proteins after several rounds of cell division. The relative abundance of the light peptide is then compared to that of the corresponding heavy peptide, giving information on the relative protein quantities in addition to protein identity. The strength of this approach is that samples can be combined early on in an experimental protocol and thus from the point of combination share experimental variance. A weakness of the technique, however, is the limited multiplexing capabilities of the system, with typically pair-wise comparisons being made within experiments. The SILAC technology has been used by several researchers to investigate the yeast proteome. de Godoy and co-workers (25) recently employed this technology to compare protein levels of 4,399 endogenous proteins in haploid yeast cells to their diploid counterparts spanning four orders of magnitude in protein abundance. They achieved this admirable level of protein coverage (approx. 75% of the yeast proteome) by using a combination of multi-dimensional separations at the protein and peptide levels, and repeat mass spectrometric analysis at finite, but overlapping mass ranges. They demonstrated one-to-one ratios of most yeast proteins but noted that key members of the pheromone pathway were specific to haploid yeast where other members of the same pathway were unaltered. From this they postulated an efficient control mechanism of the mating response. Additionally they noted several retrotransposon-associated proteins which were specific to haploid yeast and also highlighted a significant change for cell wall components consistent with the fact that diploid cells have twice the volume but not twice the surface area of haploid cells. The large coverage of the yeast proteome demonstrated the utility of this approach to gain very information-rich datasets.

The alternative in vivo stable isotope labeling strategy, where replacement of an element with a stable isotopic variant occurs during growth, has also been used to great effect in yeast. de Groot and colleagues (26) replaced ^{14}N with ^{15}N in yeast culture to reveal post-transcriptional regulation of key cellular processes during anaerobiosis.

More widely applicable differential stable isotopic labeling can be achieved by labeling extracted proteins in vitro with one or more isotopic variants of a tag. There are many such tagging methods currently available. One of the most simple approaches involves proteolysis of proteins in the presence of $H_2{}^{16}O$ or $H_2{}^{18}O$, leading to the incorporation of two isotopic variants of oxygen at the C terminus of every peptide generated (except the original C terminus) (27). Currently, however, the most commonly used peptide tagging system is the iTRAQ (amine-reactive reagents for relative and absolute quantitation), developed by

Ross and co-workers at Applied Biosystems (28). The tagging system is a multiplexed set of four or eight isotope tags which label via amino groups of peptides typically generated by proteolysis of proteins using trypsin. iTRAQ tags are isobaric and differentially labeled versions of a peptide appear as a single precursor ion peak. When an iTRAQ-labeled peptide is subjected to collision-induced dissociation in MS/MS mode, however, the iTRAQ tags release diagnostic, low-mass ions (reporter ions) that are used for quantitation. This technology thus allows for significant levels of multiplexing within experimental designs. iTRAQ labeling to achieve relative quantification of the yeast proteome has been used by many groups (29–31).

In a set of collaborating laboratories including the authors', this method was used for a systems-wide study of the transcriptional, proteomic, and metabolic levels under defined controlled conditions where either nitrogen, carbon, sulfur, or phosphorus was limiting (32). The studies demonstrated that characteristic signatures at the transcriptomic, proteomic, and exo- and endometabolomic levels reveal groups of growth-regulated genes, proteins, and biological processes controlled at different levels, with specific outliers characteristic of each nutrient-limiting condition.

Although the authors used iTRAQ data in the publication describing this study, earlier datasets were collected on the same material using DIGE (*see* **Fig. 10.1**), demonstrating that complementary datasets can be achieved using different quantitative methods.

Each approach listed above has strengths and weaknesses, and such technologies should not be used without careful consideration of appropriate experimental designs and data analyses. A full description of these is outside the scope of this manuscript and explored in (10) and many other publications.

3.2. Absolute Quantitation

Methods to define differences between two or more proteomes in a relative manner have been used with great effect by the community, but such methods have several limitations.

The relative nature results in comparison possible only for the same protein and not between different protein species, which would be very useful in terms of determining stoichiometries of proteins within multiprotein complexes. Moreover, in relative proteomics experiments carried out on a global scale, a sample of interest is usually compared against a "normal" sample acting as a reference. Much care must be taken into consideration when determining and preparing what can be considered to be the normal sample, otherwise serious errors in quantitative measurements will arise. Large amounts of this normal sample also have to be prepared for larger scale experiments, which often means that storage of the sample is challenging if its quality is not to

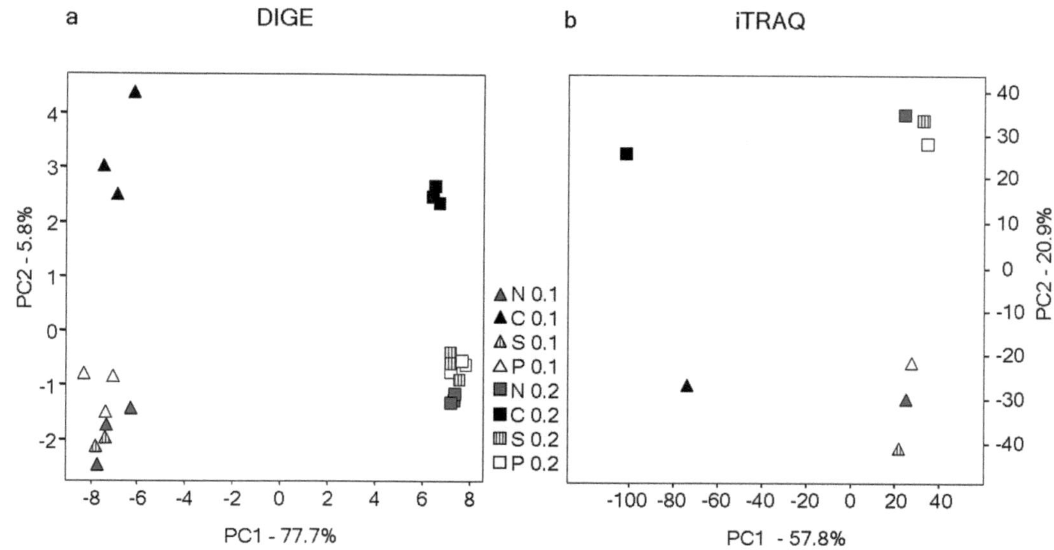

Fig. 10.1. Comparison of DIGE and iTRAQ proteomics data. The *x*- and *y*-axes represent the two principal components (PC1, PC2), the groups responsible for the majority of the variance in (**a**) proteomics dataset collected using DIGE technology on cell lysates from replicate samples grown with either carbon (C), nitrogen (N), phosphorus (P), or sulfur (S) limitation and (**b**) dataset using the same samples but quantifying the proteome using the iTRAQ technology. Growth rates μ=0.1 and 0.2 h^{-1}. Reproduced from (32), with permission from BioMed Central (copyright retained by the authors). For more details *see* (32).

be compromised. Moreover, in label-free quantification methods, the experimental conditions need to be kept constant at all times to assure pattern reproducibility. Additionally, in relative quantification experiments involving mass spectrometric measurements of peptides, it is often impossible to sample the set of peptides in different experiments and different peptide surrogates for a protein may be measured in successive experiments. This can cause issue with respect to peptides which may be common to different protein isoforms. Finally, there is a limit of detectability in most relative quantification experiments as the peptides are sampled in a data-dependent manner such that only the most abundant peptides are analyzed, leading to a reduction in the abundance range sampled (33).

Methods to interrogate the proteome in a manner which returns absolute quantitation data for peptide species acting as surrogates for the proteins from which they were proteolytically generated have become more widely used in recent years. Broadly such methods fall into four categories:

1. spiking a known concentration of a synthetic stable isotope-labeled target peptides which serve as internal standards (AQUA) (34);
2. creation of a concatenated stable isotope-labeled protein (QconCAT) which when digested liberates target peptides which serve as internal standards (35);

3. spiking of a known concentration of a stable isotope-labeled version of the protein whose absolute concentration is to be determined. The sample is then usually digested to peptides after addition of the labeled internal standard protein and quantification is carried out at the peptide level (36);
4. spiking of a known concentration of an abundance calibration protein and using peak intensity measurements of peptides generated from this protein after digestion to infer the absolute concentration of peptides within the same mass spectrometric run (37).

The first three of the above approaches use the relative intensities of the spiked internal standard peptides (generated upon digestion of their parent protein) and the unlabelled peptides of the same sequence generated from the native proteins within a sample to extrapolate the absolute amount of the latter in the sample.

These three approaches are often used in conjunction with selective reaction monitoring (SRM) approaches. Here, the only selected ions corresponding to the mass/charge ratio of the target peptides (native peptide and the internal standard surrogate peptide) are selected in the mass spectrometer and relative measurements are made via a set of discriminatory fragment ions generated as part of the tandem mass spectrometry process (38).

All four approaches have strengths and weaknesses. The first two methods assume that the stoichiometric amounts of the native peptide are released upon digestion but are suitable for multiplexing such that absolute abundance of several proteins can be recorded per experiment. The third approach is in many respects the most robust as internal standard and native peptides are generated simultaneously as part of the same proteolytic digest, but this method requires production and purification of recombinant labeled proteins and thus does not lend itself to high levels of multiplexing. The fourth method assumes that each peptide has the same ionization efficiency as the internal standard which is not likely.

In the case of the first three approaches is of utmost importance to design internal standard peptide surrogates such that they are specific (proteotypic) for the isoform of protein of interest and are not likely to undergo differential chemical modification compared with the native peptide during sample preparation, for example, oxidation or deamidation (39).

A recent study, of great significance to the yeast community, has come from the Aebersold group and makes use of the above methodologies (3). They applied a targeted proteomics approach using SRM to detect *S. cerevisiae* proteins expressed at concentrations below 50 copies/cell from total cellular digests (*see* **Fig. 10.2**). They achieved this by carrying out SRM analysis on five

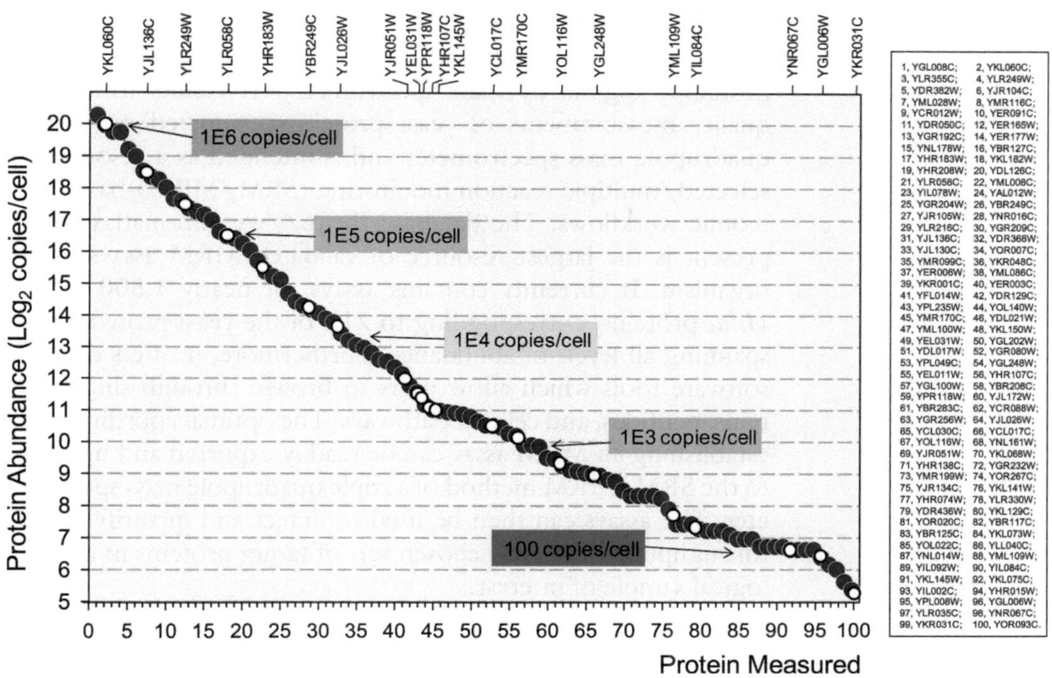

Fig. 10.2. Cellular concentrations of the set of measured proteins. Protein abundances are derived from (4). Yeast proteins detected by SRM assays are sorted by abundance to show the even distribution across the whole range of concentration (*filled circles*). Proteins for which the absolute abundance was measured using isotopically labeled standards are indicated on top of the graph (*open circles*). Reproduced from Picotti et al., (2009) (3) with permission from Elsevier.

proteotypic peptides per 100 proteins which spanned the abundance range of the *S. cerevisiae* proteome as determined in the antibody-based study of the O'Shea and Weissman laboratories (4). They demonstrated a linear correlation between the abundance of proteins and the SRM signal intensity of the most intense proteotypic peptide and confirmed their findings by using stable isotope-labeled reference peptides to absolutely quantify 21 selected proteins distributed across all levels of cellular abundances. They went on to show the power of this approach by displaying rapid and reproducible analysis of a network of proteins spanning the entire abundance range over a growth time course of *S. cerevisiae* involving passage through a series of metabolic phases. The great power of this approach is that as only a set number of proteins are assayed in any experiment, sensitivity and speed of analysis are greatly improved.

Focused quantitative proteomics approaches hold huge promise in situations where the proteome is well annotated and suitable proteotypic peptides can be selected. Entire signaling pathways can be targeted and, because of the fast mass spectrometric-based assays associated with this approach, it lends itself to temporal studies or experimental schema where multiple treatments are measured with high replication (40).

The MRMAtlas contains a compendium of targeted proteomics assays to detect and quantify yeast proteins in complex proteome digests by mass spectrometry. It results from high-quality measurements of yeast proteins conducted on a triple quadrupole mass spectrometer and is intended as a resource for selected/multiple reaction monitoring (SRM/MRM)-based proteomic workflows. The database (http://www.mrmatlas.org) at present is the largest resource of validated MRM assays of any organism. It currently contains assays for nearly 1,500 *S. cerevisiae* proteins, corresponding to 21% of the yeast proteome and spanning all levels of abundance. Furthermore, it offers different software tools which allow users to browse through single proteins, peptides, and cellular pathways. The optimal coordinates for establishing an MRM assay can be readily exported and uploaded to the SRM/MRM method of a triple quadrupole mass spectrometer. The assays can then be used to detect and quantify at high throughput deliberately chosen sets of target proteins in any biological sample of interest.

4. Post-translational Modifications

The previous section dealt with characterization of the proteome simply in terms of abundance. A whole added layer of complexity is possible when one considers chemical events which may occur to proteins after their translation.

Post-translational modifications (PTMs) are paramount for cell life. PTMs dictate progression of cells through the cell cycle and metabolites through the various metabolic pathways, synthesis of phospholipids, and cell wall assembly amongst other processes. Most PTMs can now be detected by protein and peptide analysis by mass spectrometry (MS), either as a mass increment or a mass deficit relative to the nascent unmodified protein (41). In addition, the modified amino acid may liberate a unique diagnostic ion in MS/MS. Most PTMs can be searched for by selecting such variable modifications in parameters in software for example, Mascot (Matrix Science) such that the additional mass of a peptides associated with the modification is utilized within the search (42).

4.1. Phosphorylation

Many pathways and protein interactions are regulated by reversible phosphorylation and studies date back to the 1950s (43–45). More recently, MS has become the method of choice for the analysis of protein phosphorylation with thousands of phosphopeptides and phosphorylation sites being routinely identified (46). The first yeast phosphopeptides were discovered in 1987 (47) and to date there are 145 published MS-based studies of which 39 were in 2009 (PubMed).

Phosphoprotein and phosphopeptide detection is still a rapidly developing field in MS with several complementary methods for enriching and detecting specific phosphopeptides from complex mixtures. Such enrichment methods included titanium dioxide (TiO_2), iron (Fe^{3+}), and gallium (Ga^{3+}) that bind phosphate groups in an acidic environment. A spectral shift of 80 Da is indicative of a phosphorylated peptide with a neutral loss mass difference of 98 Da (including 18 Da for a water molecule) for phosphothreonines and -serines.

Comparisons of phosphoproteomes of three yeast species (*S. cerevisiae, Candida albicans,* and *Schizosaccharomyces pombe*) have been performed to quantify the evolutionary change in phosphorylation and resulted in the proposal that protein kinases are an important source of phenotypic diversity (48). Phosphoproteins are conserved within these three species with predominantly phosphoserines (72–82%), phosphothreonines (15–25%), and few phosphotyrosines (0.8–1.9%), with an average of 20% of each proteome comprising phosphoproteins. In the same study, in an analysis of 6,333 *S. cerevisiae* proteins, 1,185 were phosphoproteins with 3,486 phosphosites. The conserved functions assigned to the phosphoprotein groups were cytokinesis, cell budding, morphogenesis, and signal transduction. Functions that were not conserved across the three species included RNA polymerase I, respiratory chain proteins, and outer kinetochore proteins. Global analyses, in particular of Cdk1 substrate phosphorylation sites, have revealed that the position of most phosphorylation sites in *S. cerevisiae* is not conserved in evolution and that the clusters shift position in rapidly evolving disordered regions (49).

Many databases exist to identify potential/predicted phosphorylation sites such as Phosida and more recently PhosphoPep (http://www.phosphopep.org), a database of protein phosphorylation sites in four model organisms including yeast (46). This paper reported 9,554 phosphopeptides with confidence (p value>0.8) and a total of 8,901 phosphorylation sites (of which 5,890 are uniquely assigned) in *S. cerevisiae*, covering one-third of the predicted yeast proteome. In addition, PhosphoPep has the ability to support cross-species comparisons and viewing and aligning of orthologous phosphoproteins in whole signaling pathways at different cellular states that could be useful in targeted experiments.

MS technologies are frequently improving in order to preserve PTMs and the introduction of electron transfer dissociation (ETD) has offered a complementary approach to determining sites of phosphorylation by MS detection of phosphoproteins. ETD employs radical anions to bring about fragmentation of the peptide backbone during tandem mass spectrometry. This fragmentation generates c and z fragment ions from the original peptide but largely leaves side chain modifications such as phosphorylation intact (50). In collision-induced dissociation (CID), the method typically employed in tandem mass spectrometry results in the loss of the labile phosphate groups on phosphoserines and phosphothreonines, hence use of ETD is a good approach to confirm the existence of phosphorylation at these two residues. Using ETD, Chi et al. identified 1,252 phosphorylation sites, including a phosphohistidine (pHis) site, on 629 proteins with phosphoprotein expression levels ranging from <50 to 1,200,000 copies per cell. In addition, they were able to identify a consensus site representing a motif for uncharacterized kinases and that yeast kinases contain a disproportionately large number of phosphorylation sites (51). More recent large-scale phosphoproteome analysis of *S. cerevisiae* and human embryonic stem cells using a decision tree-driven algorithm resulted in 5,874 phosphopeptides of 39,507 total identifications using ETD (52).

4.2. Glycosylation

The most complex and energetically most costly PTM of yeast proteins is glycosylation. Although yeast possesses only two types of glycolsylation, N-glycosylation of asparagine residues and O-mannosylation of serine and threonine residues, further modifications of the primary sites of modification lead to mannosyl extensions of various lengths but, unlike eukaryotes, these are relatively simple and not as highly branched as oligo- and polysaccharide groups (53). The biosynthesis of these PTMs, however, is highly conserved in eukaryotes and studies in yeast and yeast mutants have been invaluable in elucidating altered cellular functions and human developmental diseases such as CDG (congenital disorders of glycosylation) syndrome, congenital muscular dystrophy, and neuronal cell migration defects (54, 55). Detection of glycosylated peptides or proteins is more complex due to the varieties of groups, N-linked glycans result in mass increments of 162, 203, 291, and 365 Da and O-linked glycans result in many increments >203 Da (55). Glycoproteins or proteolytically derived glycopeptides can be enriched for using lectins. Following enrichment, glycopeptides can be treated with enzymes to remove the core sugar moieties to aid identification of the modified peptide. Studies have used ^{18}O labeling to enable site- specific assignment of glycosylation sites by ion signals (56). There is little yeast glycoproteome data reported at a global scale. Most studies to date have been more focussed such as the characterization the oligosaccharyltransferases (57), and glycosylated membrane proteins (58), and the functional effects of perturbing glycosylation (54). There is much to investigate in this area and new developments in refining the detection techniques will greatly improve the high-throughput identification and mapping of these complex PTMs.

4.3. Acetylation

Almost as abundant as phosphorylation, acetylation is a reversible modification, neutralizing the positive charge on lysine residues thereby changing protein function, and has many roles within a cell. The most extensively studied acetylations in all organisms are those involving lysine residues within histone proteins. Acetylation and deacetylation of histone lysine residues are among the best of all characterized post-translational modifications in yeast. There are over 20 reported histone acetyltransferases and deacetylases in *S. cerevisiae* (59). Acetylation is also crucial in extranuclear metabolic proteins such as the gluconeogenesis enzymatic Pck1p. Tandem MS identified multiple essential Lys residues crucial for Pck1p activity which allow yeast growth on non-fermentable carbon sources and deacetylation of a single residue resulted in a loss of Pck1p activity blocking the yeast life span (60). Other important roles of acetylated proteins in yeast are in the separation of sister chromatids by SMC3 and the regulation of kinase activity for CDC28, CDK9, and CDK6 (61). These authors concluded that lysine acetylation appears to contribute to the regulation of all nuclear functions and that there appears to be extensive cross talk between different PTMs, notably phosphorylation and ubiquitination.

Acetylation of proteins or protein complexes can be easily detected by MS especially as acetyl–lysine is a very stable PTM during tandem MS analysis. Acetylation of lysine residues can be observed by a spectral peak shift of 42 Da corresponding to an acetyl group. An alternative or a complementary method to identify acetylated proteins is by immunoaffinity purification using an antibody specific for acetyl-lysine followed by further enrichment with IEF and high-resolution MS. This technique used by Choudhary and colleagues identified 3,600 acetylation sites in 1,750 human proteins many of which were involved in complexes and are homologous to yeast. Unlike phosphorylation, acetylation appears to target structurally ordered regions and is not only restricted to the nucleus. Analysis of amino acids surrounding acetylated lysines in cytoplasmic and mitochondrial acetylated proteins was also mapped, providing potential motifs for mapping other acetylomes including yeast (62).

The only global attempt to map the yeast acetylome are protein acetylation microarrays that have characterized acetylated yeast substrates revealing 91 proteins acetylated by the NuA4 complex that were validated in vitro using GST fusions (60).

4.4. Methylation

Methylation of proteins is a critical process in gene transcription, both activation and repression (63), and is abundant especially in histones and ribosomes occurring most commonly on lysine and arginine residues but also aspartate and glutamate residues where different diverse functions have been proposed in yeast (64).

Methylation detection is possible by MS but the mass shift of trimethylation is similar to that of acetylation (approx. 42 Da); however, high-resolution MS is able to distinguish these as well as establish the site specificity. In addition, the retention times of acetylated or trimethylated version of the same peptide differ (63) and the diagnostic ions are unique for each (tri-)(di-)methylated lysine or arginine peptide, whereas acetylated peptides generate ammonium ions at *m/z* 126 and 143 (41). To date, no global methylation proteome in yeast has been reported.

4.5. Ubiquitination

Ubiquitination (Ub) is a factor in protein turnover and is implicated in many diseases. Sequential addition of ubiquitin molecule(s) by ubiquitin ligase to lysine residues in proteins, although a rapid and a specifically targeted event, is easily detectable by MS as a 114-Da Gly-Gly tag remains even after tryptic digestion of proteins. Current studies are adding to the knowledge of ubiquitination sites and tools are now available to predict such sites (UbPred) (65). This chapter also describes how mutant yeast strains have been invaluable in targeting short-lived proteins for ubiquitination and the sequence and structural preferences of ubiquitination sites. To date there are hundreds of confirmed and predicted Ub sites in *S. cerevisiae* (65–67).

4.6. Sumoylation

Sumoylation, like ubiquitination, is reversible modification of a lysine residue, which is a rapid event, and often protein targets are at very low abundance. A number of proteomics approaches have addressed this modification at a quantitative and global scale and SILAC has been useful in studying sumoylation dynamics in response to stimuli that initiate sumoylation (68, 69). Also yeast-two-hybrid approaches have been used to identify SUMO substrates and distinguish the covalent from non-covalent modifications (70). A comprehensive review documents many other protocols to identify and characterize sumoylation in yeast and other organisms (71). As far back as 2004, Panse and co-workers identified 139 sumoylated substrates by affinity purification followed by MS with most being nuclear proteins or nuclear pore complexes. The following year, a further >150 sumoylated substrates were characterized by Hannich et al. by alternative approaches. No single method, or large-scale approach, has been able to unravel the entire SUMO proteome and known targets often remain undetected due to low abundance, reiterating the fact that orthogonal approaches are required for such analyses.

5. Protein–Protein Interactions

Establishing abundance levels and post-translational states gives insight into only fraction of a protein's function. Proteins rarely exist in isolation and usually interact with other biomolecules as part of their function. Characterization of protein interactions is essential to further our understanding of other complex pathways underpinning biological functions. There are a number of approaches to establish multi-component protein complexes using a variety of methodologies. A specific chapter in this volume details methods for analysis of yeast protein–protein interactions and networks (*see* **Chapter 12**).

5.1. Yeast-Two-Hybrid

Yeast-two-hybrid (Y2H) approaches have been used widely in the last two decades and have generated vast proteome datasets in many organisms as a result of its primary application in yeast (72). This technique is well documented (73, 74).

5.2. Tandem Affinity Purification (TAP)

Tandem affinity purification (TAP) has been adopted as a method of isolating not only target proteins but also their binding partners at an individual and global scale. In addition, TAP tagging combined with Y2H methods and mutant studies have identified PTMs in proteins as described previously.

At a more global scale, in 2005 a collection of C-terminal TAP- and GFP-tagged yeast strains were generated, tagging approximately 75% of the yeast proteome (75). These can be, and have been, used in many studies to determine interactions, PTMs, and sub-cellular localizations and are commercially available (http://www.openbiosystems.com).

5.3. Interactions by Parallel Affinity Capture (iPAC)

Another recent methodology for identifying protein–protein interactions is iPAC, interactions by parallel affinity capture (76). This procedure involves isolating in vivo-tagged endogenous proteins from lysates and purifying baits and binding partners in native conditions using two or more affinity capture techniques in parallel, rather than the TAP sequential approach. The resulting eluted protein complexes are analyzed by MS and the datasets compared for overlap, with proteins in common being assigned higher confidence scores. Interaction lists are compared to public repositories such as those listed below. This parallel approach overcomes the low yields of interacting partners often recovered at the end of two or more sequential elution steps during TAP affinity protocols.

5.4. Interaction Repositories

There are a number of public and private repositories for yeast protein–protein interactions described below:

- Yeast protein complexes and interactions (http://yeast-complexes.embl.de/).
- BioGRID (http://www.thebiogrid.org); 5,483 interactions to date.
- PIMRider Yeast (http://pim.hybrigenics.com/) (account required although site no longer maintained) has yeast protein interactions and resulting networks are available in the PSI downloadable format.

- The *Saccharomyces* Genome Database (SGD) (http://www.yeastgenome.org) documents literature relating to 65 large-scale protein interaction studies.

6. Sub-cellular Localization

A further complexity when studying the proteome is the spatial arrangement of proteins. In eukaryotes, there are numerous complex sub-cellular structures which are generally delineated by membranes. Proteins traffic to the correct sub-cellular position in order to carry out their correct physiological function. Assigning the sub-cellular location of proteins is of paramount importance to biologists for two reasons: first, in the elucidation of their role and second in the refinement of knowledge of cellular processes by tracing certain activities to specific organelles.

There are many technologies which have been developed over the years to characterize the sub-cellular location of proteins.

The first large-scale yeast protein sub-cellular localization study was conducted by the O'Shea laboratory. This study involved the construction and analysis of a collection of yeast strains expressing full-length, chromosomally tagged green fluorescent protein fusion proteins representing 75% of the yeast proteome (77). These were then classified into 22 distinct sub-cellular localization categories and provide localization information for 70% of previously unlocalized proteins. Analysis of this high-resolution, high-coverage localization dataset in the context of transcriptional, genetic, and protein–protein interaction data helped reveal the logic of transcriptional co-regulation and provided a comprehensive view of interactions within and between organelles in eukaryotic cells.

A criticism of this approach is that while it performs well with soluble proteins, problems may arise when applying the same strategy to localize integral membrane proteins. Artifactual results arise as GFP attached to a protein may result in the protein trafficking in an aberrant manner.

Organelle proteomics methods have largely relied to date on the ability to purify an organelle to homogeneity and then catalog the proteins present generally using shot-gun proteomics strategies. Such approaches have recently been reviewed by Premsler and colleagues (78). In the case of certain organelles, such as the mitochondria, this approach works very well and gives high-quality datasets of mitochondrial resident proteins (79). For other organelles, purification to near homogeneity is not possible. Many compartments, such as those associated with the endomembrane system, have similar physicochemical properties and are therefore impossible to obtain without the presence of contaminating organelles. Also, homogenization steps often used to release the content of the cells may damage the organelles, increasing heterogeneity in their size, shape, and density. Moreover, given that many proteins traffic from one compartment to another, it can be a challenge to distinguish cargo from "full-time" residents. For example, residents of the endoplasmic reticulum (ER) continuously cycle between the ER and Golgi via COPI and COPII vesicles (80), and many plasma membrane (PM) proteins are recycled to the PM after endocytosis. Additionally, in plant and animal cells, endocytic and exocytic routes are thought to share common intermediates, making it difficult to separate these systems and assign endocytic or exocytic proteins. Finally, alternatively spliced forms of transcripts may lead to protein products, which are targeted to different sub-cellular compartments. If the extent of contamination by other organelles is not assessed in these cases, many false assignments of uncharacterized proteins to specific sub-cellular locations will occur. Overall, these factors necessitate an approach which allows the measurement of the global steady-state distributions within organelles to obtain insight into principal sub-cellular distributions.

Several methods have been developed in the last few years which allow assignment of proteins to sub-cellular locations which do not rely on purification of an organelle to homogeneity. First, subtractive proteomics methods have been used to look at enrichment of proteins within a system where a relatively crude fraction is partitioned from a fraction which contains the organelle of interest. An example of this approach within the *C. albicans* is the strategy of phase partitioning for enrichment of the detergent-rich membranes (81). In this study, the enrichment of proteins associated with lipid rafts was facilitated by the use of aqueous phase partitioning in different detergents. Lipid rafts have been shown to have an important role in mating and hyphal formation and also contain certain transporter proteins such as the arginine permease Can1. Identification of proteins associated with lipid rafts is therefore important to understand function of the rafts. Wiederhold and co-workers (82) used a similar approach employing iTRAQ tagging methods and robust statistical analysis to look at the enrichment of *S. cerevisiae* vacuolar proteins. They reported 148 proteins which were significantly enriched in the "pure" vacuolar fractions when compared with a crude vacuolar preparation, including well-characterized vacuolar proteins such as the subunits of the vacuolar H^+-ATPase. The issue with this approach is that the comparison is pair-wise and therefore reveals only whether a protein is present in the enriched fraction and thus gives no information about proteins with multiple locations. Furthermore, most enrichment methods do not achieve purification of any organelle, and hence data are still peppered with false assignments from contaminating organelles.

Many decades ago, de Duve (83) noted that when analytical centrifugation was applied in order to enrich for organelles, true resident(s) of a spatially distinct organelle(s) were not unique to a single fraction but had a characteristic distribution within the gradient. He postulated that if the distribution pattern could be determined for a marker known to be specific to a particular organelle, then determining the organelle residency of a protein of unknown location could be achieved by matching its distribution to that of a known marker. By applying this approach, several organelles can be investigated at the same time. Two high-throughput proteomics studies have been developed that make use of organelle-specific distribution patterns to identify novel organelle residents. Here distribution patterns of organelle marker proteins along density gradients are matched to patterns of proteins with no known localization to enable the assignment of organelle residency.

The first, protein correlation profiling, employs label-free quantification methodologies to determine protein distribution patterns (84). This method allows the simultaneous assignment of proteins to multiple organelles and has been used by several groups to map proteins to different organelle locations in mammalian tissues (85, 86). The utility of label-free approaches for this type of analysis is still hotly debated in that each sample is processed separately during all steps of the procedure from protein extraction to MS analysis, and thus the variance associated with the data may be too great to allow the subtle discrimination required to assign proteins confidently to organelles such as the ER and Golgi. LOPIT (localization of organelle proteins by isotope tagging) is a method also inspired by de Duve's observations but uses the iTRAQ technology to measure distribution patterns of proteins from self-generating iodixanol density gradients (87). This technique also allows simultaneous assignment of proteins to multiple organelles and has been used in the author's laboratory to assign proteins to organelles in plants, Drosophila, and animal cell culture, respectively (88–90). Four or eight fractions, enriched with different organelles, are selected for comparison and protein distributions are determined by measuring their relative abundance using iTRAQ quantification. Multivariate statistical techniques such as principal component analysis (PCA) and partial least squares discriminant analysis (PLS-DA) are employed to cluster proteins according to the similarities in their gradient distributions and distributions of known organelle markers, and thus to assign proteins to specific organelles. The strength of this method is that iTRAQ tagging and co-analysis of different fractions from the same gradient in a single LC-MSMS experiment reduces experimental variation, and the use of robust statistical tests and the appropriate use of replicates enable the creation of high-quality organelle protein maps.

Yeast cell biology is ripe for having a more thorough organelle proteomics study using some of the more advanced methods described above to characterize more organelles, especially those associated with the secretory pathway. Such datasets offer an excellent complementary to resource to that of Huh and co-workers in cataloging organelle-specific localization of proteins in yeast (77).

7. Protein Turnover

The knowledge of the abundance of a protein and its interaction partners and sub-cellular location still does not give a complete picture of the dynamics of the proteins. For the vast majority of protein species, there is continuous turnover of the protein even in an unstressed steady-state condition. Proteins are synthesized and degraded at relative rates which either maintain the steady-state concentrations of a protein or vary it to allow adjustment of the protein pool which is a vital component of the modulation of response of the cell to changing conditions, such as nutrient availability. Proteins turn over at different rates from one another which may range from a matter of seconds to months, the latter generally being non-regulatory proteins or those whose regulation is modulated by post-translational processes.

The measurement of synthesis and degradation rates of proteins is tricky, especially if they are to be carried out at a global scale.

One approach to study the half-life of yeast proteins was taken by Belle and co-workers (91), who used the TAP-tagged yeast collection (75) to determine the degradation rates of 3,751 yeast proteins using quantitative Western blotting techniques and were able to classify 161 proteins as very unstable with half-lives of less than 4 min. They compared these data with previous measurement of mRNA abundance and concluded that mRNA abundance plays a major role in determining protein abundance but actual abundance is determined by degradation. Additionally, based on their data, they were able to cluster proteins into production and regulation groups in which the former was comprised of mostly cytoplasmic proteins such as ribosomal proteins which are produced in larger quantities and are generally stable compared to low-abundance, rapidly degrading regulatory proteins such as cell cycle and transcriptional proteins. They correlated half-lives with protein sequence and determined that proteins with short half-lives were significantly enriched in serine residues, whereas stable proteins were enriched in valines. They also proposed a model which predicts correlation between mRNA expression and half-lives of unstable proteins. This approach, though useful, is extremely time consuming and open to criticism as to whether the Tap tag fusion had any effect on degradation rates.

A SILAC proteomics-based study to look at the turnover of proteins in *S. cerevisiae* was reported by Pratt and co-workers (92). In this study the *S. cerevisiae* were grown in chemostat cultures in the presence of deuterated leucine for the equivalent of seven doubling times. An excess of unlabeled amino acid was then added instantaneously to the culture vessel and samples taken at time points and the decrease of deuterated peptides and the appearance of undeuterated peptides were measured mass spectrometrically to determine the turnover of each protein. Although they produced a limited dataset in this study, they demonstrated that the average rate of degradation of 50 proteins was 2.2% per hour, although some proteins were turned over at imperceptible rates, and others had degradation rates of almost 10 per hour.

A similar elegant approach has also been recently published by Selbach and co-workers (93, 94) and termed pulsed SILAC. This method allows the direct quantification of protein translation at a proteome-wide scale and is able to differentiate between synthesis rates rather than protein turnover rates of proteins. Cells are first grown in a standard growth medium with the normal light (L) amino acids. The cells are then transferred after differential treatment such as application of an miRNA species, which shuts off production of a subset of proteins, to a culture medium containing either heavy (H) or medium–heavy (M) amino acids. After an incubation phase, all newly synthesized proteins appear in the H or M form. Both samples are then combined and processed and the ratio of H to M peptides measured. The H versus M peptide ratio reflects differences in translation of the corresponding proteins. Since preexisting proteins remain in the L form and can be ignored, this method is independent of differences in protein stability that would otherwise interfere with accurate quantification. Although degradation will also affect newly synthesized proteins, this degradation will normally occur for proteins in both the M and the H samples and thus will not have great effect on the H/M ratios.

8. Bioinformatics

Proteomics technologies are highly reliant on bioinformatics methodologies and resources to enable identification of protein species and addition of context to proteomics datasets in terms of function, known genetic interactions, and collation of data from numerous studies.

8.1. Data Repositories

The *Saccharomyces* Genome Database (SGD) (http://www.yeastgenome.org) hosts a number of public repositories for protein detection, protein interactions, protein modifications, and protein localizations, reporting a plethora of experimental and computational techniques to add to the increasing collection of yeast data.

In addition the BioGRID (http://www.thebiogrid.org) also reports protein interactions and in 2001 the YEAST protein complex database (http://yeast.cellzome.com) describes kinase-targeted protein complexes and their localization(s). However, the latter is no longer maintained as the data are now located within many general public repositories (e.g., http://yeast-complexes.embl.de/). More repositories for protein–protein interaction data are described later.

Peptide Atlas (http://www.peptideatlas.org/) is a repository reporting all peptides identified from yeast proteins using MS studies. A new build was publicly available in April 2009 reporting 3,110 peptides from 324,658 spectra. This release runs in parallel with the MRM Atlas (http://www.mrmatlas.org/; Yeast_MRM_Atlas_2008-03_P0.9) that is useful for designing targeted approaches to identify specific proteins, in particular low-abundance proteins in complex mixtures.

Much of the above data are compiled into central repositories such as IntAct (http://www.ebi.ac.uk/intact) that reports 1,281 protein interactions in *Saccharomyces* species.

Interaction datasets often report confidence scores; however, currently there is no consistent method or range for reporting these. IntAct hosts a collection of interaction datasets that can only be compared qualitatively. Until such standardization is implemented, any proteomics approach requires orthogonal experimental validation.

In order to standardize all proteomics and interactomics datasets for all organisms, the HUPO Proteomics Standard Initiative Molecular Interaction (PSI-MI) format was established in 2004 by Hermjakob et al. and updated by Kerrien et al. (95, 96). The minimum information required for reporting a molecular interaction experiment (MIMIx) for a proteomics experiment is described in (97). More recently, further developments have been reported, with PSI-M XML2.5 schema allowing easy comparison of consistent large datasets.

8.2. Assessment of False Discoveries

With the ever increasing amount of data being deposited into public repositories, there is a need for assessing the quality of such mass produced datasets. Many groups now report the false discovery rates (FDRs) for protein identification from MS experiments.

This value however must not be confused with the false positive rate, also described in MS experiments.

Protein quantification data may also be polluted with many false discoveries, not least because many quantification datasets have been interrogated with univariate statistical tests in which tests, each carrying a probability that the test will return a false finding, are used in isolation and thus an accumulation of erroneous results, also referred to as the multiple testing issue (98). As with all quantitative datasets, successful quantitative proteomics relies on appropriate experimental design and employment of statistical tests which are suitable for the type of data analyzed. For more details about issues related to experimental design and data analysis in quantitative proteomics, *see* (22).

9. Other Methodologies

9.1. Protein Microarrays

Other approaches to catalog the yeast proteome use protein microarrays and also interrogate its functionality. Protein microarrays are microarray chips printed with proteins as opposed to DNA or RNA and date back to 2000 (99) and 2001 where Zhu and co-workers (100) expressed proteins from 5,800 cloned ORFs to screen diverse biochemical pathways and PTMs. Protein microarrays have also been used for ubiquitination screening (101) where substrates of the ubiquitin E3 ligase were identified at a global scale. Commercially available yeast protein microarrays exist with more than 4,000 GST-and His6-tagged yeast proteins (Invitrogen). Protein acetylation microarrays have been used to characterize acetylated yeast substrates with 91 proteins acetylated by the NuA4 complex that were validated in vitro using GST fusions, a large proportion being enzymes and effectors involved in signaling pathways responsive to nutrient availability and energy status (60).

10. Future Prospects

It is clear that the application of proteomics technologies has greatly enhanced our knowledge of yeast biology. Despite large-scale global studies from several groups, there is still approximately 20% of the yeast proteome which has failed to be identified in proteomics experiments and 25% of the yeast proteome uncharacterized in terms of location and function. The wish list of what proteomics needs to deliver to the yeast community in the future, however, involves global studies which define abundance, sub-cellular localization, protein interaction data, and post-translational status for all proteins expressed by proteome in any given situation. Furthermore, these studies need to be extended to give dynamic information, for example, changes in the above when yeast is grown in different growth conditions. Finally, proteomics studies need to be expanded such that functional information about the proteome is also captured in global experiments.

To come close to fulfilling this wish list, future proteomics experiments need to address the characterization of low copy number proteins and also use methods to interrogate the yeast proteome and to fill in the gaps in our knowledge, which have as yet been applied only in other organisms.

Targeted proteomics approaches such as those using SRM to detect and quantify as pioneered by Aebersold and co-workers, coupled with some of the methodologies allowing robust sub-cellular fractionation, will be necessary to push the detection boundary of proteins. Application of methods such as LOPIT or protein correlation profiling and comparison of resulting datasets with the GFP-tagged sub-cellular localization studies of Huh et al. will give more refined knowledge about the protein sub-cellular localization. Refinement of these methods is needed so that they can be used in a dynamic sense such that correlated re-location of proteins upon perturbation can be tracked at a scale not possible with the use of fluorescent protein fusion tags. Moreover, extending sub-cellular localization data to include differential information about protein isoforms, particularly phospho-isoforms, has the potential to enrich our understanding of signaling pathways and their execution. These extensions to existing technologies will be technically challenging but not beyond current technology, particularly with the availability of new generations of mass spectrometers where resolution, sensitivity, and mass accuracy are significantly improved of their predecessors (102, 103). In terms of protein–protein interaction networks, yeast has historically been the test bed for technologies, but these now require refinement such that high-throughput, robust datasets can be achieved with low frequencies of false interactors and also coupled with quantification technologies more readily such that dynamic changes in protein complex components can be easily monitored.

In the case of both sub-cellular distributions and protein complex makeup, datasets in the past have tended to produce a snap shot of the cell rather than continuum representing the truly dynamic nature of a living cell.

Another dynamic aspect of protein behavior is its synthesis and degradation. In future, coupling of absolute quantification methods to measure the absolute amount of a protein with the SILAC type methods described by Selbach and co-workers

(93, 94) will be accurately able to dissect the degradation and synthesis for any protein. As yet such studies have not been published at a large scale.

The yeast proteome is probably one of the best characterized in terms of post-translational modifications. Our knowledge of some modifications is less complete than others; for example, glycoproteomics studies have a long way to go to give us a full picture of possible glycosylation sites within proteins and how these change in response to given perturbations. The dynamic nature of post-translational modifications thus also needs to be worked upon.

In terms of defining protein function, non-mass spectrometry-based approaches such as protein microarrays offer complementary methods to assign function to the 25% of proteins with no known function such as those described in **Section 9.1**. Coupled with more traditional proteomics approaches characterizing the proteomes of various deletion mutants, functional arrays offer an attractive complementary technology to complete our understanding of what roles proteins play within yeast.

Proteomics datasets to date largely involve studies carried out on *S. cerevisiae*, but application of bioinformatics tools particularly those that allow the investigation of the evolution of yeast strains will give insight also into the functional groups of proteins and their evolution and allow extrapolation of protein biology from the well-studied *S. cerevisiae* to other important yeast species.

The creation of robust proteomics datasets has served and will continue to serve as valuable resources to the yeast community and act as a complement to the wealth of data emerging from transcriptomics studies and characterization of the yeast metabolome. There is still much methodological development required, however, to ensure completeness of such datasets.

References

1. Gygi, S. P., Rochon, Y., Franza, B. R., and Aebersold, R. (1999) Correlation between protein and mRNA abundance in yeast. *Mol. Cell. Biol.* **19**, 1720–1730.
2. Lu, P., Vogel, C., Wang, R., Yao, X., and Marcotte, E. M. (2007) Absolute protein expression profiling estimates the relative contributions of transcriptional and translational regulation. *Nat. Biotechnol.* **25**, 117–124.
3. Picotti, P., Bodenmiller, B., Mueller, L. N., Domon, B., and Aebersold, R. (2009) Full dynamic range proteome analysis of *S. cerevisiae* by targeted proteomics. *Cell* **138**, 795–806.
4. Ghaemmaghami, S., Huh, W.-K., Bower, K., et al. (2003) Global analysis of protein expression in yeast. *Nature* **425**, 737–741.
5. Tanaka, K., Waki, H., Ido, Y., et al. (1988) Protein and polymer analyses up to m/z 100,000 by laser ionization time-of-flight mass spectrometry. *Rapid Comm. Mass Spectrom.* **2**, 151–153.

6. Fenn, J. B., Mann, M., Meng, C. K., Wong, S. F., and Whitehouse, C. M. (1989) Electrospray ionization for mass spectrometry of large biomolecules. *Science* **246**, 64–71.
7. Steen, H., and Mann, M. (2004) The abc's (and xyz's) of peptide sequencing. *Nat. Rev. Mol. Cell. Biol.* **5**, 699–711.
8. Gstaiger, M., and Aebersold, R. (2009) Applying mass spectrometry-based proteomics to genetics, genomics and network biology. *Nat. Rev. Genet.* **10**, 617–627.
9. Klose, J., Nock, C., Herrmann, M., et al. (2002) Genetic analysis of the mouse brain proteome. *Nat. Genet.* **30**, 385–393.
10. Lilley, K. S., and Dupree, P. (2006) Methods of quantitative proteomics and their application to plant organelle characterization. *J. Exp. Bot.* **57**, 1493–1499.
11. Washburn, M. P., Wolters, D., and Yates, J. R. (2001) Large-scale analysis of the yeast proteome by multidimensional protein identification technology. *Nat. Biotechnol.* **19**, 242–247.
12. Lilley, K. S., Razzaq, A., and Dupree, P. (2002) Two-dimensional gel electrophoresis: recent advances in sample preparation, detection and quantitation. *Curr. Opin. Chem. Biol.* **6**, 46–50.
13. Kang, Y., Techanukul, T., Mantalaris, A., and Nagy, J. M. (2009) Comparison of three commercially available DIGE analysis software packages: minimal user intervention in gel-based proteomics. *J. Proteome Res.* **8**, 1077–1084.
14. Lilley, K. S., and Friedman, D. B. (2004) All about DIGE: quantification technology for differential-display 2D-gel proteomics. *Exp. Rev. Proteomics* **1**, 401–409.
15. Unlu, M., Morgan, M. E., and Minden, J. S. (1997) Difference gel electrophoresis: a single gel method for detecting changes in protein extracts. *Electrophoresis* **18**, 2071–2077.
16. Valerius, O., Kleinschmidt, M., Rachfall, N., et al. (2007) The *Saccharomyces* homolog of mammalian RACK1, Cpc2/Asc1p, is required for FLO11-dependent adhesive growth and dimorphism. *Mol. Cell. Proteomics* **6**, 1968–1979.
17. Old, W. M., Meyer-Arendt, K., Aveline-Wolf, L., et al. (2005) Comparison of label-free methods for quantifying human proteins by shotgun proteomics. *Mol. Cell. Proteomics* **4**, 1487–1502.
18. Silva, J. C., Denny, R., Dorschel, C., et al. (2006) Simultaneous qualitative and quantitative analysis of the *Escherichia coli* proteome: a sweet tale. *Mol. Cell. Proteomics* **5**, 589–607.
19. Mosley, A. L., Florens, L., Wen, Z., and Washburn, M. P. (2009) A label free quantitative proteomic analysis of the *Saccharomyces cerevisiae* nucleus. *J. Proteomics* **72**, 110–120.
20. Foss, E. J., Radulovic, D., Shaffer, S. A., et al. (2007) Genetic basis of proteome variation in yeast. *Nat. Genet.* **39**, 1369–1375.
21. Usaite, R., Wohlschlegel, J., Venable, J. D., et al. (2008) Characterization of global yeast quantitative proteome data generated from the wild-type and glucose repression *Saccharomyces cerevisiae* strains: the comparison of two quantitative methods. *J. Proteome Res.* **7**, 266–275.
22. Karp, N. A., and Lilley, K. S. (2007) Design and analysis issues in quantitative proteomics studies. *Proteomics* **7**, 42–50.
23. Ong, S. E., Blagoev, B., Kratchmarova, I., et al. (2002) Stable isotope labeling by amino acids in cell culture, SILAC, as a simple and accurate approach to expression proteomics. *Mol. Cell. Proteomics* **1**, 376–386.
24. Conrads, T. P., Alving, K., Veenstra, T. D., et al. (2001) Quantitative analysis of bacterial and mammalian proteomes using a combination of cysteine affinity tags and N-15- metabolic labeling. *Anal. Chem.* **73**, 2132–2139.
25. de Godoy, L. M. F., Olsen, J. V., Cox, J., et al. (2008) Comprehensive mass-spectrometry-based proteome quantification of haploid versus diploid yeast. *Nature* **455**, 1251–1254.
26. de Groot, M. J. L., Daran-Lapujade, P., van Breukelen, B., et al. (2007) Quantitative proteomics and transcriptomics of anaerobic and aerobic yeast cultures reveals post-transcriptional regulation of key cellular processes. *Microbiology* **153**, 3864–3878.
27. Wang, J., Gutierrez, P., Edwards, N., and Fenselau, C. (2007) Integration of O-18 labeling and solution isoelectric focusing in a shotgun analysis of mitochondrial proteins. *J. Proteome Res.* **6**, 4601–4607.
28. Ross, P. L., Huang, Y. L. N., Marchese, J. N., et al. (2004) Multiplexed protein quantitation in *Saccharomyces cerevisiae* using amine-reactive isobaric tagging reagents. *Mol. Cell. Proteomics* **3**, 1154–1169.
29. Pham, T. K., Chong, P. K., Gan, C. S., and Wright, P. C. (2006) Proteomic analysis of *Saccharomyces cerevisiae* under high gravity fermentation conditions. *J. Proteome Res.* **5**, 3411–3419.
30. Pham, T. K., and Wright, P. C. (2008) The proteomic response of *Saccharomyces*

cerevisiae in very high glucose conditions with amino acid supplementation. *J. Proteome Res.* 7, 4766–4774.
31. Lin, F. M., Tan, Y., and Yuan Y. J. (2009) Temporal quantitative proteomics of *Saccharomyces cerevisiae* in response to a nonlethal concentration of furfural. *Proteomics* 9, 5471–5483.
32. Castrillo, J. I., Zeef, L. A., Hoyle, D. C., et al. (2007) Growth control of the eukaryote cell: a systems biology study in yeast. *J. Biol.* 6, 4.
33. Mirzaei, H., McBee, J. K., Watts, J., and Aebersold, R. (2008) Comparative evaluation of current peptide production platforms used in absolute quantification in proteomics. *Mol. Cell. Proteomics* 7, 813–823.
34. Gerber, S. A., Rush, J., Stemman, O., Kirschner, M. W., and Gygi, S. P. (2003) Absolute quantification of proteins and phosphoproteins from cell lysates by tandem MS. *Proc. Natl. Acad. Sci. USA* **100**, 6940–6945.
35. Rivers, J., Simpson, D. M., Robertson, D. H. L., Gaskell, S. J., and Beynon, R. J. (2007) Absolute multiplexed quantitative analysis of protein expression during muscle development using QconCAT. *Mol. Cell. Proteomics* **6**, 1416–1427.
36. Brun, V., Dupuis, A., Adrait, A., et al. (2007) Isotope-labeled protein standards: toward absolute quantitative proteomics. *Mol. Cell. Proteomics* **6**, 2139–2149.
37. Silva, J. C., Gorenstein, M. V., Li, G.-Z., Vissers, J. P. C., and Geromanos, S. J. (2006) Absolute quantification of proteins by LCMSE: a virtue of parallel MS acquisition. *Mol. Cell. Proteomics* **5**, 144–156.
38. Anderson, L., and Hunter, C. L. (2006) Quantitative mass spectrometric multiple reaction monitoring assays for major plasma proteins. *Mol. Cell. Proteomics* **5**, 573–588.
39. Mead, J. A., Bianco, L., Ottone, V., et al. (2009) MRMaid, the web-based tool for designing multiple reaction monitoring (MRM) transitions. *Mol. Cell. Proteomics* **8**, 696–705.
40. Wolf-Yadlin, A., Hautaniemi, S., Lauffenburger, D. A., and White, F. M. (2007) Multiple reaction monitoring for robust quantitative proteomic analysis of cellular signaling networks. *Proc. Natl. Acad. Sci. USA* **104**, 5860–5865.
41. Larsen, M. R., Trelle, M. B., Thingholm, T. E., and Jensen, O. N. (2006) Analysis of posttranslational modifications of proteins by tandem mass spectrometry. *Biotechniques* **40**, 790–798.
42. Creasy, D. M., and Cottrell, J. S. (2002) Error tolerant searching of uninterpreted tandem mass spectrometry data. *Proteomics* **2**, 1426–1434.
43. Engstrom, L. (1958) Studies of phosphoprotein functions. III. The effect of some inhibitors of oxidative phosphorylation on the incorporation rate of 32P into trichloroacetic acid-soluble nucleotides and protein phosphorylserine of baker's yeast. *Acta Soc. Med. Ups.* **63**, 149–154.
44. Kotyk, A. (1958) Effect of potassium ions on phosphorylation processes in *Saccharomyces cerevisiae*. *Biokhimiia* **23**, 737–750.
45. Stoppani, A. O., and De Favelukes, S. L. (1957) Effect of certain inhibitors of oxidative phosphorylation on the fixation of carbon dioxide by *Saccharomyces cerevisiae*. *Rev. Soc. Argent. Biol.* **33**, 252–256.
46. Bodenmiller, B., Campbell, D., Gerrits, B., et al. (2008) PhosphoPep-a database or protein phosphorylation sites in model organisms. *Nat. Biotechnol.* **26**, 1339–1340.
47. Biemann, K., and Scoble, H. A. (1987) Characterization by tandem mass spectrometry of structural modifications in proteins. *Science* **237**, 992–998.
48. Beltrao, P., Trinidad, J. C., Fiedler, D., et al. (2009) Evolution of phosphoregulation: comparison of phosphorylation patterns across yeast species. *PLoS Biol.* 7, e1000134.
49. Holt, L. J., Tuch, B. B., Villen, J., Johnson, A. D., Gygi, S. P., and Morgan, D. O. (2009) Global analysis of Cdk1 substrate phosphorylation sites provides insights into evolution. *Science* **325**, 1682–1686.
50. Syka, J. E., Coon, J. J., Schroeder, M. J., Shabanowitz, J., and Hunt, D. F. (2004) Peptide and protein sequence analysis by electron transfer dissociation mass spectrometry. *Proc. Natl. Acad. Sci. USA* **101**, 9528–9533.
51. Chi, A., Huttenhower, C., Geer, L. Y., et al. (2007) Analysis of phosphorylation sites on proteins from *Saccharomyces cerevisiae* by electron transfer dissociation (ETD) mass spectrometry. *Proc. Natl. Acad. Sci. USA* **104**, 2193–2198.
52. Swaney, D. L., McAlister, G. C., and Coon, J. J. (2008) Decision tree-driven tandem mass spectrometry for shotgun proteomics. *Nat. Methods* **5**, 959–964.
53. Tanner, W., and Lehle, L. (1987) Protein glycosylation in yeast. *Biochim. Biophys. Acta* **906**, 81–99.
54. Herscovics, A., and Orlean, P. (1993) Glycoprotein biosynthesis in yeast. *FASEB J.* 7, 540–550.
55. Lehle, L., Strahl, S., and Tanner, W. (2006) Protein glycosylation, conserved from yeast

to man: a model organism helps elucidate congenital human diseases. *Angew. Chem. Int. Ed. Engl.* **45**, 6802–6818.
56. Kaji, H., Saito, H., Yamauchi, Y., et al. (2003) Lectin affinity capture, isotope-coded tagging and mass spectrometry to identify N-linked glycoproteins. *Nat. Biotechnol.* **21**, 667–672.
57. Schulz, B. L., and Aebi, M. (2009) Analysis of glycosylation site occupancy reveals a role for Ost3p and Ost6p in site-specific N-glycosylation efficiency. *Mol. Cell. Proteomics* **8**, 357–364.
58. Lee, B. K., Jung, K. S., Son, C., et al. (2007) Affinity purification and characterization of a G-protein coupled receptor, *Saccharomyces cerevisiae* Ste2p. *Protein Expr. Purif.* **56**, 62–71.
59. Lee, K. K., and Workman, J. L. (2007) Histone acetyltransferase complexes: one size doesn't fit all. *Nat. Rev. Mol. Cell Biol.* **8**, 284–295.
60. Lin, Y. Y., Lu, J. Y., Zhang, J., et al. (2009) Protein acetylation microarray reveals that NuA4 controls key metabolic target regulating gluconeogenesis. *Cell* **136**, 1073–1084.
61. Choudhary, C., Kumar, C., Gnad, F., et al. (2009) Lysine acetylation targets protein complexes and co-regulates major cellular functions. *Science* **325**, 834–840.
62. Smith, K. T., and Workman, J. L. (2009) Introducing the acetylome. *Nat. Biotechnol.* **27**, 917–919.
63. Yang, L., Tu, S., Ren, C., et al. (2010) Unambiguous determination of isobaric histone modifications by reversed-phase retention time and high-mass accuracy. *Anal. Biochem.* **396**, 13–22.
64. Sprung, R., Chen, Y., Zhang, K., et al. (2008) Identification and validation of eukaryotic aspartate and glutamate methylation in proteins. *J. Proteome Res.* **7**, 1001–1006.
65. Radivojac, P., Vacic, V., Haynes, C., et al. (2010) Identification, analysis, and prediction of protein ubiquitination sites. *Proteins* **78**, 365–380.
66. Hitchcock, A. L., Auld, K., Gygi, S. P., and Silver, P. A. (2003) A subset of membrane-associated proteins is ubiquitinated in response to mutations in the endoplasmic reticulum degradation machinery. *Proc. Natl. Acad. Sci. USA* **100**, 12735–12740.
67. Peng, J., Schwartz, D., Elias, J. E., et al. (2003) A proteomics approach to understanding protein ubiquitination. *Nat. Biotechnol.* **21**, 921–926.
68. Andersen, J. S., Matic, I., and Vertegaal, A. C. (2009) Identification of SUMO target proteins by quantitative proteomics. *Methods Mol. Biol.* **497**, 19–31.
69. Wohlschlegel, J. A. (2009) Identification of SUMO-conjugated proteins and their SUMO attachment sites using proteomic mass spectrometry. *Methods Mol. Biol.* **497**, 33–49.
70. Kroetz, M. B., and Hochstrasser, M. (2009) Identification of SUMO-interacting proteins by yeast two-hybrid analysis. *Methods Mol. Biol.* **497**, 107–120.
71. Ulrich, H. D. (2009) *SUMO Protocols*, Vol. 497. New York, NY: Humana Press.
72. Fields, S., and Song, O. (1989) A novel genetic system to detect protein-protein interactions. *Nature* **340**, 245–246.
73. Giot, L., Bader, J. S., Brouwer, C., et al. (2003) A protein interaction map of *Drosophila melanogaster*. *Science* **302**, 1727–1736.
74. Melcher, K., and Johnston, S. A. (1995) GAL4 interacts with TATA-binding protein and coactivators. *Mol. Cell. Biol.* **15**, 2839–2848.
75. Howson, R., Huh, W. K., Ghaemmaghami, S., et al. (2005) Construction, verification and experimental use of two epitope-tagged collections of budding yeast strains. *Comp. Funct. Genomics* **6**, 2–16.
76. Rees, J. S., Lowe, N., Armean, I. M., Roote, J., Johnson, G., Drummond, E., Spriggs, H., Ryder, E., Russell, S., Johnston, D. S., and Lilley, K. S. (2011) In vivo analysis of proteomes and interactomes using parallel affinity capture (iPAC) coupled to mass spectrometry. *Mol. Cell. Proteomics* **10**, M110.002386. Epub 2011 Mar 29.
77. Huh, W.-K., Falvo, J. V., Gerke, L. C., et al. (2003) Global analysis of protein localization in budding yeast. *Nature* **425**, 686–691.
78. Premsler, T., Zahedi, R. P., Lewandrowski, U., and Sickmann, A. (2009) Recent advances in yeast organelle and membrane proteomics. *Proteomics* **9**, 4731–4743.
79. Reinders, J., Zahedi, R. P., Pfanner, N., Meisinger, C., and Sickmann, A. (2006) Toward the complete yeast mitochondrial proteome: multidimensional separation techniques for mitochondrial proteomics. *J. Proteome Res.* **5**, 1543–1554.
80. Lee, M. C. S., Miller, E. A., Goldberg, J., Orci, L., and Schekman, R. (2004) Bi-directional protein transport between the ER and Golgi. *Annu. Rev. Cell Dev. Biol.* **20**, 87–123.

81. Insenser, M., Nombela, C., Molero, G., and Gil, C. (2006) Proteomic analysis of detergent-resistant membranes from *Candida albicans*. *Proteomics* **6**, S74–S81.
82. Wiederhold, E., Gandhi, T., Permentier, H. P., Breitling, R., Poolman, B., and Slotboom, D. J. (2009) The yeast vacuolar membrane proteome. *Mol. Cell. Proteomics* **8**, 380–392.
83. de Duve, C. (1971) Tissue fractionation. *J. Cell. Biol.* **50**, 20D–55D.
84. Andersen, J. S., Wilkinson, C. J., Mayor, T., Mortensen, P., Nigg, E. A., and Mann, M. (2003) Proteomic characterization of the human centrosome by protein correlation profiling. *Nature* **426**, 570–574.
85. Foster, L. J., de Hoog, C. L., Zhang, Y. L., et al. (2006) A mammalian organelle map by protein correlation profiling. *Cell* **125**, 187–199.
86. Wiese, S., Gronemeyer, T., Ofman, R., et al. (2007) Proteomics characterization of mouse kidney peroxisomes by tandem mass spectrometry and protein correlation profiling. *Mol. Cell. Proteomics* **6**, 2045–2057.
87. Dunkley, T. P. J., Hester, S., Shadforth, I. P., et al. (2006) Mapping the *Arabidopsis* organelle proteome. *Proc. Natl. Acad. Sci. USA* **103**, 6518–6523.
88. Sadowski, P. G., Groen, A. J., Dupree, P., and Lilley, K., S. (2008) Sub-cellular localization of membrane proteins. *Proteomics* **8**, 3991–4011.
89. Tan, D. J. L., Dvinge, H., Christoforou, A., Bertone, P., Martinez Arias, A., and Lilley, K. S. (2009) Mapping organelle proteins and protein complexes in *Drosophila melanogaster*. *J. Proteome Res.* **8**, 2667–2678.
90. Hall, S. L., Hester, S., Griffin, J. L., Lilley, K. S., and Jackson, A. P. (2009) The organelle proteome of the DT40 lymphocyte cell line. *Mol. Cell. Proteomics* **8**, 1295–1305.
91. Belle, A., Tanay, A., Bitincka, L., Shamir, R., and O'Shea, E. K. (2006) Quantification of protein half-lives in the budding yeast proteome. *Proc. Natl. Acad. Sci. USA* **103**, 13004–13009.
92. Pratt, J. M., Petty, J., Riba-Garcia, I., et al. (2002) Dynamics of protein turnover, a missing dimension in proteomics. *Mol. Cell. Proteomics* **1**, 579–591.
93. Selbach, M., Schwanhausser, B., Thierfelder, N., Fang, Z., Khanin, R., and Rajewsky, N. (2008) Widespread changes in protein synthesis induced by microRNAs. *Nature* **455**, 58–63.
94. Schwanhäusser, B., Gossen, M., Dittmar, G., and Selbach, M. (2009) Global analysis of cellular protein translation by pulsed SILAC. *Proteomics* **9**, 205–209.
95. Hermjakob, H., Montecchi-Palazzi, L., Bader, G., et al. (2004) The HUPO PSI's molecular interaction format – a community standard for the representation of protein interaction data. *Nat. Biotechnol.* **22**, 177–183.
96. Kerrien, S., Orchard, S., Montecchi-Palazzi, L., et al. (2007) Broadening the horizon – level 2.5 of the HUPO-PSI format for molecular interactions. *BMC Biol.* **5**, 44.
97. Orchard, S., Salwinski, L., Kerrien, S., et al. (2007) The minimum information required for reporting a molecular interaction experiment (MIMIx). *Nat. Biotechnol.* **25**, 894–898.
98. Karp, N. A., McCormick, P. S., Russell, M. R., and Lilley, K. S. (2007) Experimental and statistical considerations to avoid false conclusions in proteomics studies using differential in-gel electrophoresis. *Mol. Cell. Proteomics* **6**, 1354–1364.
99. MacBeath, G., and Schreiber, S. L. (2000) Printing proteins as microarrays for high-throughput function determination. *Science* **289**, 1760–1763.
100. Zhu, H., Bilgin, M., Bangham, R., et al. (2001) Global analysis of protein activities using proteome chips. *Science* **293**, 2101–2105.
101. Gupta, R., Kus, B., Fladd, C., et al. (2007) Ubiquitination screen using protein microarrays for comprehensive identification of Rsp5 substrates in yeast. *Mol. Syst. Biol.* **3**, 116.
102. Second, T. P., Blethrow, J. D., Schwartz, J. C., et al. (2009) Dual-pressure linear ion trap mass spectrometer improving the analysis of complex protein mixtures. *Anal. Chem.* **81**, 7757–7765.
103. Wallace, A., Castro-Perez, J., Major, H., et al. (2009) Synapt G2: Breakthrough quantitative and qualitative performance for UPLC/MS and MS/MS (MS^E) applications, p Technical Note, Water Corporation, Milford, USA.

Chapter 11

Protein Turnover Methods in Single-Celled Organisms: Dynamic SILAC

Amy J. Claydon and Robert J. Beynon

Abstract

Early achievements in proteomics were qualitative, typified by the identification of very small quantities of proteins. However, as the subject has developed, there has been a pressure to develop approaches to define the amounts of each protein – whether in a relative or an absolute sense. A further dimension to quantitative proteomics embeds the behavior of each protein in terms of its turnover. Virtually every protein in the cell is in a dynamic state, subject to continuous synthesis and degradation, the relative rates of which control the expansion or the contraction of the protein pool, and the absolute values of which dictate the temporal responsiveness of the protein pool. Strategies must therefore be developed to assess the turnover of individual proteins in the proteome. Because a protein can be turning over rapidly even when the protein pool is in steady state, the only acceptable approach to measure turnover is to use metabolic labels that are incorporated or lost from the protein pool as it is replaced. Using metabolic labeling on a proteome-wide scale in turn requires metabolic labels that contain stable isotopes, the incorporation or loss of which can be assessed by mass spectrometry. A typical turnover experiment is complex. The choice of metabolic label is dictated by several factors, including abundance in the proteome, metabolic redistribution of the label in the precursor pool, and the downstream mass spectrometric analytical protocols. Key issues include the need to control and understand the relative isotope abundance of the precursor, the optimization of label flux into and out of the protein pool, and a sampling strategy that ensures the coverage of the greatest range of turnover rates. Finally, the informatics approaches to data analysis will not be as straightforward as in other areas of proteomics. In this chapter, we will discuss the principles and practice of workflow development for turnover analysis, exemplified by the development of methodologies for turnover analysis in the model eukaryote *Saccharomyces cerevisiae*.

Key words: Metabolic labeling, amino acids, mass spectrometry, stable isotope, cell culture, yeast metabolism, protein synthesis, protein degradation, turnover rate.

1. Introduction

Metabolic incorporation of stable isotope labels is a fundamental methodological approach in comparative proteomics, particularly

J.I. Castrillo, S.G. Oliver (eds.), *Yeast Systems Biology*, Methods in Molecular Biology 759,
DOI 10.1007/978-1-61779-173-4_11, © Springer Science+Business Media, LLC 2011

applied to isolated cells grown in culture and most commonly known as SILAC (stable isotope labeling with amino acids in cell culture) experiments (1). In SILAC, one cell culture is labeled with, for example, an amino acid in which every carbon atom is carbon-13 (^{13}C, "heavy", abbreviated H), and a second culture is labeled with the same amino acid in which every carbon atom is carbon-12 (^{12}C, "light", abbreviated L). Often the labeled cells are grown under normal conditions, and the corresponding unlabeled cells are exposed to a differing physiological, or pathological, environment. This has the advantage of a single control sample for multiple perturbed samples, without the need for additional expensive labeled amino acids. Conversely, the perturbed samples could be labeled and compared to a single unlabeled control. Moreover, different isotopes, such as [$^{13}C_6$]arginine and [$^{13}C_6$][$^{15}N_4$]arginine, allow for a degree of multiplexing (the first is 6 Da heavier than the unlabeled counterpart and the second is 10 Da heavier). The two (or more) cell cultures are combined, post-labeling, and the mixture subjected to downstream processing for proteomics. Subsequently, every peptide that contains at least one instance of that amino acid will appear as a H–L doublet in mass spectrometric analysis. The relative intensities of the H and L ions will indicate the relative expression of the protein when the cell cultures are compared. A fundamental goal that is implicit in the SILAC approach is that the labeling is complete – each label can only report on the entire population of proteins. In practice, this is not formally necessary, but with single cells grown in culture, this is attainable and makes downstream analysis much simpler.

Thus, an assumption with SILAC experiments is the completeness of labeling of the control culture. If any unlabeled proteins remain, their constituent peptides will add to the peak intensity of the peptides representing protein from the unlabeled culture, and therefore peak ratios will not indicate true biological protein abundance ratios without more detailed analysis. Incomplete labeling, whether through the use of partially labeled precursors or as a consequence of sampling cells before they are fully labeled, can compromise a SILAC experiment if not spotted or complicate the data analysis if known. In this regard, a rapid method of checking incorporation is valuable (2).

In marked contrast, proteome-level turnover measurements differ from standard comparative proteomics experiments in one fundamental way – the proteins *cannot* be fully labeled, as it is the time dependence of the process of labeling of each protein that defines the turnover rate. We have referred to such studies as "dynamic SILAC" experiments (3) (for a summary of the few studies of this type, *see* **Table 11.1** and references therein). Although single time point experiments are possible, the preferred experiments are those in which the proteome is sampled

Table 11.1
Summary of current literature detailing turnover experiments using stable isotope labels with cells in culture

Organism	Label	Labeling protocol	Analytical protocol	First author (ref.)	Date
Saccharomyces cerevisiae (BY4743)	DL-$[^{2}H_{10}]$Leucine	Glucose-limited chemostat culture. Once cells were fully labeled with leucine, the growth medium was changed to contain unlabeled leucine. Samples taken regularly during the light chase.	2D SDS-PAGE gel followed by in-gel tryptic digestion and MALDI-ToF MS analysis	Pratt (4)	2002
Escherichia coli (K-12)	$[^{13}C_6]$Glucose	Cells were grown in unlabeled growth medium and harvested 30 min after addition of labeled glucose.	1D SDS-PAGE gel followed by in-gel tryptic digestion. Analysis using MALDI-ToF/ToF and LCQ ion-trap MS	Cargile (5)	2004
HeLa cells	^{15}N-enriched medium	Cells were cultured in natural isotope containing medium for 2 days, then transferred to ^{15}N-enriched growth medium. Samples collected at intervals after transfer. Experiments included two heat-shock exposures of 30 min.	2D SDS-PAGE gel followed by in-gel tryptic digestion and MALDI-ToF/ToF analysis	Gustavsson (6)	2005

(continued)

Table 11.1 (continued)

TAP-tagged yeast	TAP tag	Cells grown to exponential (log) phase. Samples collected following cycloheximide treatment after 0, 15, and 45 min.	SDS-PAGE and Western blotting followed by chemiluminescence at three exposure times using a CCD camera	Belle (7)	2006
Mycobacterium smegmatis (mc2 155)	[^{15}N]Ammonium sulfate	*For acid shock*: cultured at mid-log phase doped with [^{15}N]ammonium sulfate and grown for one doubling time before harvesting. *For iron starvation*: cells grown in heavy medium and harvested at mid-log phase. Cell pellets then resuspended in medium containing ^{14}N and harvested after one doubling time.	Tris–HCl SDS-PAGE gel followed by in-gel digestion and nanoLC/LTQ-FT	Rao (8)	2008
Human adenocarcinoma A549 cells	L-[$^{13}C_6$]Arginine	Cells grown in labeled medium for 13 days, then transferred to unlabeled medium. Samples taken regularly.	GeLC-MS using 1D SDS-PAGE and analysis using LTQ ion-trap	Doherty (3)	2009

along the labeling trajectory, to allow the rate of change in labeling profile to be assessed (3–8).

1.1. Fundamental Structure of a Turnover Experiment

Because turnover can operate even in the absence of any change in the protein pool size, it follows that tracer labels must be used to measure flux through the protein pool. Methods that measure regression of a protein pool after poisoning of protein synthesis with cycloheximide, for example, are of limited value; the toxic effects limit the time frame over which loss of protein in the absence of protein synthesis is physiologically relevant. As such, tracer methods to follow turnover in the steady state are best used with living cells. Turnover experiments invoke an added degree of complexity such that comparative turnover analyzes have yet to become routine, and in the first instance, preliminary studies have concentrated upon building a profile of turnover rates for individual proteins. In all cases, the experimental outline is essentially the same.

Protein turnover can be monitored in two different ways using the same general approach: labeling to completeness followed by a "light" chase using non-labeled amino acid, or alternatively labeling over time from the initial unlabeled culture. In the latter, unlabeled (L) cells are exposed to the heavy labeled (H) precursor, and the incorporation of the heavy isotope reflects the rate of synthesis. Alternatively, the cells can be pre-labeled with the H precursor, and at the start of the turnover experiment, the precursor is swapped to the L counterpart. Although the two experiments are formally equivalent, there are some advantages of taking the latter approach, related to the need for control of the relative isotope abundance (RIA) of the precursor (9, 10). In a turnover study, the transition of the precursor pool from labeled to unlabeled, or vice versa, is an obligatory component. Ideally, this transition will be from an RIA of 1 to an RIA of 0 (a "light chase"), or from an RIA of 0 to an RIA of 1 (a "heavy chase"). Whilst not essential, management of the transition such that it is between 0 and 1, whether in a light or a heavy chase experiment, simplifies the analysis and maximizes the range of labeling. One incidental benefit of a light chase experiment is cost; an excess of amino acid is required to ensure that any labeled amino acid returned to the precursor pool from protein degradation and then reincorporated is negligible, thus rapidly fixing the RIA at 0. If this excess is prepared from unlabeled precursor, the costs will be dramatically lower.

1.2. Overall Experimental Strategy

As metabolic labeling involves the incorporation of a stable isotope label into newly synthesized proteins in vivo, it also enables the rate of protein synthesis, not just change in abundance, to be monitored by assessing the peak intensity ratio between corresponding labeled and unlabeled peptides over time. Any peptide that could contain the label can be used as a surrogate to measure turnover for the entire protein, as turnover kinetics are the same for all peptides from the same protein (3).

1.3. Choice of Label

As stable isotopes are chemically identical to their non-labeled counterparts, they respond identically in a mass spectrometer so that if peptides containing either the labeled or non-labeled forms are present in the same quantity in a sample, they will have the same intensity in a mass spectrum. The most common stable isotope labels used are ^{13}C, ^{15}N, and ^{2}H, which are often integrated into amino acids or other compounds such as glucose (11). The benefit of using a stable isotope-labeled amino acid, such as [$^{13}C_6$]arginine or [$^{2}H_{10}$]leucine, instead of glucose, is that the mass offset created by the label is constant and therefore H–L pairs are easy to identify. When [^{15}N]H_4Cl is used as the sole source of nitrogen, all amino acids incorporate the label at the α-nitrogen position, and those with nitrogen-containing side chains (Asn, Gln, His, Trp, Lys, and Arg) will also incorporate different numbers of heavy nitrogen atoms into the side chain. The outcome is that peptides of differing chain length and amino acid composition have different mass offsets from the unlabeled counterpart. Compare this complexity of labeling to the use of [$^{13}C_6$]arginine, where a peptide containing a single arginine residue will always show a 6 Da separation between peaks representing the labeled and unlabeled forms. A di-arginine peptide will show a 12 Da separation and so on.

Similar is true of deuterated amino acids, but transamination of the α-carbon deuteron can lead to a mass shift less than that expected. For example, [$^{2}H_{10}$]leucine can be transaminated and lose the label at the α-carbon atom, generating a 9 Da difference between labeled and unlabeled peptides. Additionally, hydrogen is more hydrophilic than deuterium, which can lead to elution of peptides labeled with [^{2}H]amino acids from reversed-phase chromatography columns before their unlabeled counterpart. This could result in inaccurate intensity ratios between H–L pairs, although the deuterium interaction with the column is dependent on the properties of the other amino acids in the sequence of the peptide (12).

Stable isotope-labeled precursors are not isotopically pure, and isotope purities of 98.5–99%, often referred to as 98.5 or 99APE (atom percent excess), are typical. For the incorporation of a single labeled amino acid, such as [$^{13}C_6$]arginine, every peptide containing a single arginine residue will, even if "fully" labeled with the heavy isotope, demonstrate about 1% ion intensity from the unlabeled contaminant. In most circumstances this is not a major issue, although it could introduce quantitative errors when only a few percent of unlabeled peptide is present in the analyte. By contrast, nitrogen labeling is more complex and requires analysis and correction. To illustrate, the peptide LVFHSASTEDNNQLIMEGR has the elemental composition $C_{91}H_{145}N_{27}O_{32}S_{1}$ and thus at an isotopic purity of 99% ^{15}N (therefore 1% ^{14}N), there are 27 different ways to construct a peptide with a single "light" nitrogen atom, all of which have a mass 1 Da less than the monoisotopic true heavy labeled peptide. This peak is therefore about 27% of the height of the fully labeled peptide, representing a significant portion of the heavy synthesized material, and has to be factored into calculations.

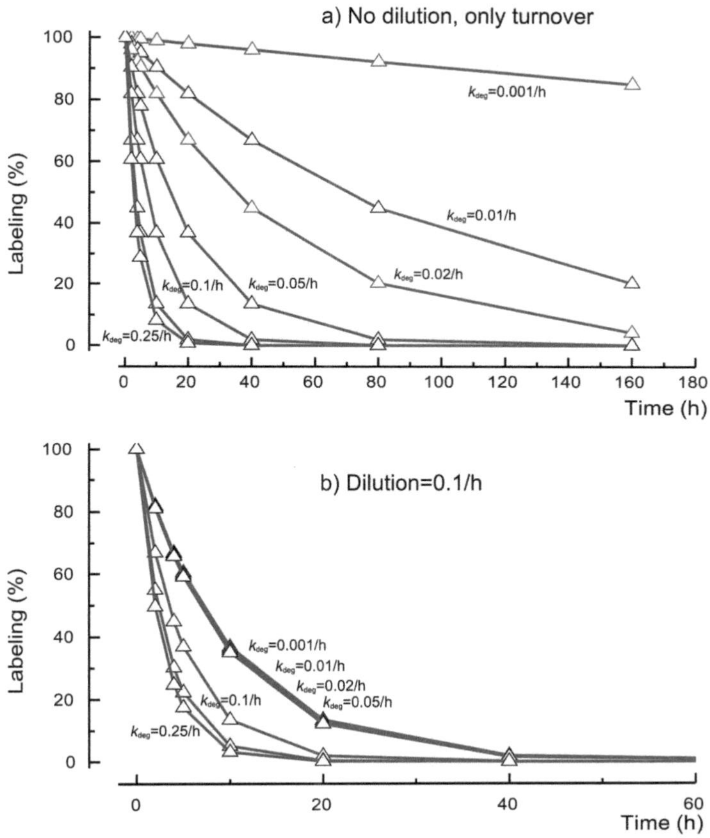

Fig. 11.1. Accurate determination of degradation rate requires multiple points to define the rate of loss of material, and the range of degradation rates that are anticipated defines the sampling strategy (**Panel a** – degradation in the absence of any pool expansion). However, if the system is growing rapidly (e.g., cells in exponential growth or a chemostat operating at a dilution rate of 0.1/h), then the change of labeling within any protein is a combination of degradation and dilution (**Panel b**). The outcome of this complication is that lower degradation rates are not readily discerned, yielding similar rates of loss of labeling irrespective of degradation rate, and therefore placing extra demands on the quality of isotopic data. In this example, it is likely that only those proteins with degradation rates more than two times the dilution rate would yield satisfactory degradation rates.

1.4. Limitations of Sampling Frequency

To access quantitative data on proteins that are turned over within minutes, samples must be taken rapidly and at very early time points in the chase. How rapidly this can occur is determined mostly by the practicalities of sampling and of arresting the cells metabolically. Moreover, sampling usually disrupts the mixing and aeration in batch culture, and thus the cells are changed. A continuous chemostat culture has the advantage that the eluted cells can be collected, but large samples will be collected over a significant time window. To illustrate, a 1 L chemostat culture operating at a dilution rate (flow/volume ratio; $D=F/V$) of 0.1/h will produce 100 mL of cell suspension/h, or 1.6 mL/min. If biomass equivalent to 20 mL of cells were required, cells would have to be collected for 15 min, during which time some high turnover proteins could be fully replaced. Sampling frequencies should be geometrically distributed in order to cover the broadest range of degradation rates. For example, a sampling regimen of 0, 2, 5, 10, 20, 40, 80, 160 min will yield at least five data points containing between 100 and 5% undegraded material for all rates of degradation from 0/min (no degradation) to 0.25/min: a half life of 2.7 min. A second complication derives from growth. If cells are actively growing (such as in exponential phase), then proteins are lost from cells in part by degradation and in part by dilution into daughter cells. This sets a limit on the range of degradation rates that can be reliably assessed (**Fig. 11.1**). Low turnover proteins will transition from labeled to unlabeled (or vice versa) mostly by dilution, and the closer the degradation rate is to zero, the more difficult it will to be to resolve losses due to degradation from those due to dilution.

2. Materials

2.1. Labeling of Cells in Culture

1. This chapter aims to give details on the analysis of proteomic data to determine protein turnover rates in yeast; for a more comprehensive guide to culture methodology, please refer to other chemostat culture methods, also (3, 8, 13).
2. Downstream analysis will be simplified if a yeast strain auxotrophic for the amino acid to be labeled is used, e.g., BY4743, a leucine auxotroph. This is not critical however, especially if an essential amino acid is chosen, as the organisms may be able to use the amino acid provided in the growth medium (in labeled form) (8) and not dilute the pool with biosynthesis de novo.
3. Additional materials required include the stable isotope-labeled amino acids, for example, $[^2H_{10}]$leucine or

[$^{13}C_6$]lysine and [$^{13}C_6$]arginine (Cambridge Isotope Laboratories) and the corresponding unlabeled amino acid (Sigma-Aldrich) (*see* **Notes 1** and **2**):

i. For a "light" chase experiment, 100 mg stable isotope-labeled amino acid is added to the culture for the initial labeling phase.

ii. The unlabeled amino acid is then added (1 g/50 mL medium) to the culture and the inflowing medium changed to also contain unlabeled amino acid (50 mg/L).

2.2. Harvesting Soluble Proteins

1. To suspend protein synthesis at the instant of collection, cycloheximide (Sigma-Aldrich, freshly prepared in water or buffer at neutral pH) is added to each sample taken at a final concentration of 100 μg/mL.
2. Before lysis, 20 mM HEPES (pH 7.5) (Sigma-Aldrich) containing 1 protease inhibitor cocktail tablet (EDTA-free) (Roche Diagnostics) per 10 mL is required.
3. Glass beads (Sigma-Aldrich) are used to lyse the cells.
4. Make solutions of DNase and RNase (Sigma-Aldrich) at 1 mg/mL.

2.3. Sample Preparation

1. Ammonium bicarbonate (50 mM; Analar grade) is required as the buffer for tryptic proteolysis. For best results, all solutions used here should be made fresh.
2. For in-solution proteolysis, a detergent is added to the sample to denature the proteins present. RapiGest™ surfactant (Waters) is made up to 1% (w/v) using 50 mM ammonium bicarbonate.
3. Reduction of disulfide bonds between cysteine residues during in-gel and in-solution proteolysis is achieved using a final concentration of 10 and 3 mM dithiothreitol (DTT) (Sigma-Aldrich), respectively, in 50 mM ammonium bicarbonate. For the in-solution protocol outlined below, this is achieved with a 9.2 mg/mL DTT starting solution.
4. Alkylation with final concentration of 60 and 9 mM iodoacetamide (IAA) (Sigma-Aldrich) in 50 mM ammonium bicarbonate is also required. The 9 mM IAA for in-solution proteolysis requires a 33 mg/mL starting solution.
5. For in-gel proteolysis, 25 mg lyophilized trypsin (Roche Diagnostics) is resuspended in 250 μL of 50 mM acetic acid. This is then diluted 1:9 with 25 mM ammonium bicarbonate to give a final trypsin concentration of 0.01 μg/μL.
6. For the in-solution digestion protocol, 125 mL of 10 mM acetic acid is used to resuspend the trypsin, which is then

added to the sample for proteolysis at 0.2 μg/μL concentration at a 1:50 trypsin:protein ratio.

7. All other solutions (acetonitrile and trifluoroacetic acid) are obtained from VWR.

2.4. Mass Spectrometry (MS)

1. For matrix-assisted laser desorption ionization time-of-flight (MALDI-ToF) MS, the matrix α-cyano-4-hydroxycinnamic acid should be made up to 8 mg/mL using 60% acetonitrile and 1% trifluoroacetic acid. Best results are obtained when 1 μL sample is dried onto the MALDI target and then 1 μL matrix is added and allowed to crystallize.

The buffers used for reversed-phase liquid chromatography (RP-LC) are made from HPLC-grade acetonitrile and HPLC-grade water (VWR) plus 0.1% formic acid. A linear gradient is most commonly used, for example, 0–50% buffer B over 30 min, when using a 75 μm capillary reversed-phase (C18) column.

3. Methods

3.1. Labeling of Cells in Culture

The main details covered in this section will not be the growing of the yeast but the subsequent mass spectrometric analysis of labeled peptides following protein isolation from the yeast cells. However, the basic method for a protein turnover experiment, using deuterium-labeled leucine as the stable isotope-labeled amino acid, is detailed below (3, 13) (*see* **Note 3**):

1. Grow cells in a glucose-limited chemostat culture using complete synthetic medium, supplemented with 100 mg DL-[$^2H_{10}$]leucine (98.5APE) at a dilution rate of 0.1/h (*see* **Note 4**):
   - i. To ensure cells are fully labeled, maintain cells in this growth medium for at least seven doubling times (*see* **Note 5**).
   - ii. Following this, add the 50 mL solution of unlabeled L-leucine to the culture and change the inflowing medium to also contain unlabeled L-leucine (50 mg/L).
   - iii. Begin sampling of the yeast cells at appropriate time points, e.g., 0, 10, 40 min, and then continue with fairly regular sampling (e.g., seven samples every 1–2 h for the first 12 h) reducing frequency until the two final samples at approximately 24 and 51 h.
   - iv. If the chemostat is sampled directly, and large volumes are removed, the apparent dilution rate will be modified (*see* **Note 6**).

3.2. Harvesting Soluble Proteins

1. When cells (40 mL) reach an A_{600} of about 1.6, collect into ice-cold Falcon tubes containing cycloheximide (100 μg/mL final concentration).
2. Centrifuge at 4°C for 5 min at 7,000 × *g* and discard the resulting supernatant.
3. Resuspend the pellet in 1 mL ice-cold double distilled water and transfer to 1.5 mL microcentrifuge tube.
4. Centrifuge again at 16,000 × *g* and discard the supernatant:
   i. Pellets can be stored at –80°C at this point.
5. Resuspend the pelleted cells in 300 μL of 20 mM HEPES, pH 7.5, plus protease inhibitor solution.
6. Vortex with glass beads to lyse cells in six bursts of 45 s (allowing 45 s cooling).
7. Add 6 and 2 μL of the DNase and RNase solutions, respectively.
8. Incubate at 4°C for 1 h.
9. Centrifuge the cell lysate at 4°C for 10 min at 2,500 × *g* and collect the supernatant.

3.3. Sample Preparation

For ease of analysis, and to increase information gained, the sample must be separated before analysis using mass spectrometry. Gel-based methods can be used, with separation of the proteins by 1D or 2D SDS-PAGE before in-gel proteolysis; if a gel-free approach is taken, an in-solution proteolysis step usually precedes separation.

3.3.1. In-Gel Proteolysis Following Gel-Based Separation

1. Excise spots corresponding to the same protein from each gel and perform standard in-gel digestion technique:
   i. De-stain gel plug of protein spot with 25 μL of 1:1 acetonitrile:50 mM ammonium bicarbonate ammonium bicarbonate for 15 min or until all color removed.
   ii. Reduce disulfide bonds between cysteine residues using 25 μL of 10 mM DTT. Incubate at 60°C for 60 min, then discard the liquid.
   iii. Alkylate with 25 μL of 60 mM IAA. Incubate in the dark at room temperature for 45 min, then discard the liquid.
   iv. Dehydrate the gel plug by incubating at 37°C in 10 μL acetonitrile, leaving the microcentrifuge tube lid open. After 15 min, if the gel plug is white, remove the remaining liquid and leave for a further 10 min.
   v. Rehydrate the gel plug in 10 μL of 0.01 μg/μL trypsin solution and incubate overnight at 37°C.

3.3.2. In-Solution Proteolysis for Gel-Free Separation

1. To 100 μg protein in 160 μL (final volume) of 50 mM ammonium bicarbonate, add 10 μL of 1% (w/v) RapiGest^{TM} surfactant. Incubate at 80°C for 10 min.
2. Add 10 μL of 3 mM (final concentration) DTT and incubate at 60°C for 15 min.
3. Alkylate with 10 μL of 9 mM (final concentration) IAA for 30 min in the dark.
4. To this add 10 μL of 0.2 μg/μL trypsin and incubate overnight at 37°C.
5. Inactivate the detergent by the addition of 1 μL trifluoroacetic acid (0.5%, v/v, final concentration) and incubate for 45 min at 37°C.
6. Before analysis, remove all precipitation formed at this stage by centrifugation.

3.4. Mass Spectrometry (MS)

If a gel-based strategy is used, analysis of individual protein spots by MALDI-ToF MS is sufficient to acquire data for individual proteins, which also avoids any possible complications introduced by the differential elution of deuterium-labeled peptides. An in-solution tryptic digest of the whole-cell lysate is however too complex to analyze by MALDI-ToF MS alone. Online liquid chromatography (LC) separation into an electrospray (ESI) mass spectrometer, such as a Q-ToF or linear ion-trap instrument, is preferred. Alternatively, offline LC followed by MALDI-ToF MS can be used, taking care to combine all spectra for labeled and unlabeled peptides:

1. Acquire protein identification data:
   i. For MALDI-ToF MS analysis, peptide mass fingerprinting (PMF) will identify the protein present in the spot.
   ii. Analysis using ESI-MS can take advantage of MS/MS capabilities and sequence information can also be obtained for database searching (*see* **Note 7** and **Fig. 11.2**).
2. The protein identifications can then be used to validate H–L peptide pairs, separated by a mass difference corresponding to the number of labeled amino acids present (*see* **Note 8**). The number of labeled amino acids present can also be used to confirm and support the protein identification (14).
3. Record isotopomer peak intensities for MS peaks representing L and H peptides or the corresponding peak areas from extracted ion chromatograms (*see* **Notes 9** and **10**).
4. The RIA of the label in each specific protein at each time (RIA_t) is calculated as follows:

$$RIA_t = \frac{\text{Labeled isotopomer}}{\text{Labeled isotopomer} + \text{Unlabeled isotopomer}} = \frac{H}{H+L}$$

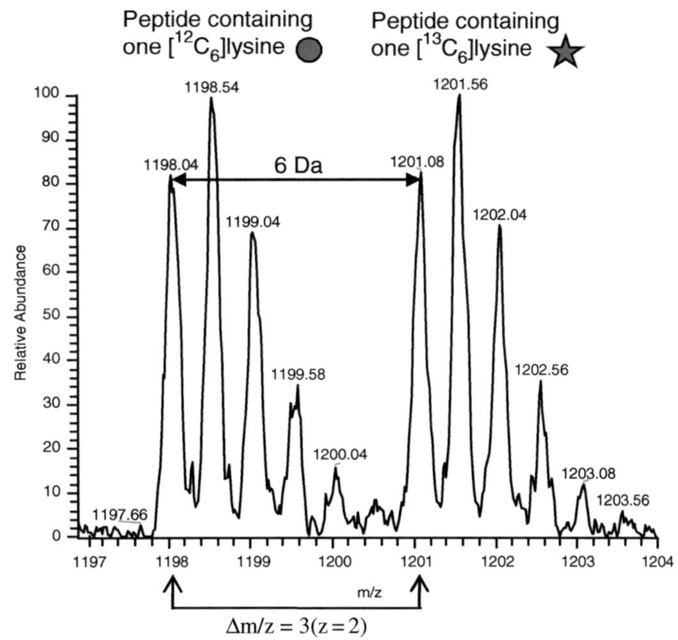

1. P32497|EIF3C_YEAST Mass: 93375 Score: 365 Queries matched: 5 emPAI: 0.12

Eukaryotic translation initiation factor 3 subunit C OS=Saccharomyces cerevisiae GN=NIP1 PE=1 SV=2

☐ Check to include this hit in error tolerant search or archive report

	Query	Observed	Mr(expt)	Mr(calc)	Delta	Miss	Score	Expect	Rank	Peptide
☑	228	1198.04	2394.07	2394.18	-0.11	0	(110)	5.4e-010	1	K.DQLDSADYVDNLIDGLSTILSK.Q ●
☑	233	1201.53	2401.05	2400.20	0.85	0	118	7.4e-011	1	K.DQLDSADYVDNLIDGLSTILSK.Q + (PFG)_6HeavyLys (K) ★

Fig. 11.2. An example of protein identification from MS/MS data searched using the MASCOT search engine with [$^{13}C_6$]lysine set as a variable modification in the search parameters. The data available from the search enables an extracted ion chromatogram to be obtained to assess peak area or a zoom scan (shown) to obtain peak intensity values from the isotopomer envelope.

5. Plot these values over time for each protein.
6. Using non-linear curve fitting, the rate of loss of the label (k_{loss}) can be calculated from the gradient of the line when the RIA at the beginning (RIA_0) and the end (RIA_∞) of the experiment are taken into consideration:
   i. RIA_0 can be determined using a random selection of proteins taken before the start of the light chase. Although this RIA should be 1, it is possible that the value may be slightly less due to the purity of the label used and degradation of unlabeled proteins releasing amino acids into the precursor pool.
   ii. The RIA of the culture at the end of the labeling period can be taken as 0 as long as there has been more than seven doubling times (but this should be directly observable).
   iii. When $RIA_\infty = 0$, $RIA_t = RIA_0 \times \exp^{-kt}$, where $k = k_{loss}$.
7. From this it is possible to determine the rate of protein degradation (k_{deg}) by factoring in the effect of the rate of dilution (k_{dil}) into the value of k_{loss}, i.e., $k_{deg} = k_{loss} - k_{dil}$:
   i. As k_{dil} is constant, it does not affect the error in the parameter estimates.
8. It is also possible to measure the rate of loss, and thence, turnover from a single time point t using the equation $k_{loss} = (-\log_e(RIA_t/RIA_0))/t$.

   If single time point experiments are to be used, an increased number of peptides per protein would be recommended to decrease the error. However, a true time course is valuable as it defines the trajectory of the loss of label from the proteins and, for example, allows it to be assessed as formally first order.

4. Notes

1. Stable isotopically labeled compounds are often named informally, according to the nomenclature such as $[^{13}C_6]$lysine. In this instance, there are only six carbon atoms and therefore, all six atoms must be replaced – the compound is referred to as (U-$^{13}C_6$ lysine) where "U" refers to "uniformly labeled". However, it is also possible to purchase labeled lysine variants such as "2-^{13}C, 99%; ε-^{15}N, 99%" that is only labeled at the α-carbon atom and the side chain nitrogen atom. Care must be taken in defining exactly the precursor form being used.

2. Trypsin is the endopeptidase most commonly used to generate peptides for mass spectrometric analysis and has a clearly defined specificity (cleavage at Arg-X and Lys-X, Arg-Pro and Lys-Pro excepted). Labeling with [$^{13}C_6$]arginine and [$^{13}C_6$]lysine has the advantage that almost all tryptic peptides are identically labeled with the +6 Da variant. The exceptions are peptides that contain internal arginine or lysine residues, either because of missed cleavages or because the peptide contains the uncleavable Arg-Pro or Lys-Pro bonds.
3. Experiments, especially in chemostat culture, that use stable isotope-labeled amino acids can be costly. There is merit in performing an entire experiment as a "dry run" using unlabeled amino acids throughout. In a perfect experiment, the switch from heavy to light amino acid (or vice versa) will be near instantaneous, complete and have no effect on the growth rate or gene expression profile of the organism. In particular, a dry run would identify the bottlenecks in the period of rapid sampling and allow appropriate measures to be taken to avoid these bottlenecks.
4. L-amino acids are the forms that are metabolically active and the precursors for protein synthesis. To reduce costs further, it is possible to use the DL racemic mixture of a labeled amino acid, and to assume that label from the D-amino acid does not make a significant contribution to the labeling pattern.
5. It is possible that during exponential growth, some proteins have zero degradation, and thus, label can be incorporated into the protein only during pool expansion, i.e., growth. After each doubling time, half of that protein in each cell would have been replaced by newly synthesized proteins containing the stable isotope-labeled amino acid from the growth medium. Therefore, after one generation, at least 50% of the proteins will be labeled, after two generations 75% etc., until after seven generations over 99% of proteins with the slowest rate of turnover will contain a heavy amino acid.
6. The changes in dilution rate can be caused by the removal of cells from a chemostat culture during sampling. When a set volume is removed from the culture, the rate of cell loss from the outflow is reduced, while the volume is replenished by new medium. The remaining cells therefore remain in the chemostat system for an extended period of time. This can be avoided by sampling directly from the culture outflow, as long as the flow rate is fast enough

to deliver sufficient cells for analysis, or by using a batch culture.

7. Proteomic search engines (MASCOT, SEQUEST, etc.) allow the use of variable modifications when analyzing mass spectrometric data. If the chosen labeled amino acid is set as a variable modification when identifying proteins using these search tools, a comprehensive list indicating labeled and unlabeled peptides will be created which helps to reduce analysis time (**Fig. 11.2**). In addition, some software, such as ProteinLynx Global Server (Waters), will also give the intensity values of the identified peaks.
8. Fully deuterated amino acids are relatively inexpensive but will be labeled at the α-carbon atom. This atom is metabolically labile, and the deuteron would be lost by the reversible process of transamination. In our experience, the α-deuteron of [$^{10}H_2$]leucine was completely lost in a yeast labeling experiment, and turnover was assessed by a 9 Da separation between unlabeled and single labeled peptides.
9. A critical step in any turnover study is the need for an instantaneous transition as the experiment is switched between the two labeled variants. Two things can go wrong here; the changeover might be slow or the changeover might be incomplete. If the changeover is incomplete, the precursor pool will possess an RIA that is neither zero nor unity. In this instance, some of the precursor pool will be labeled and some unlabeled. This is detectable by the inability of the labeling curve to move between 0 and 1, or vice versa, and by the appearance of intermediate labeled forms of peptides where the peptide contains two instances of the amino acid. It is advisable to deliberately look at such peptides to ensure that there is complete replacement of the precursor pool.
10. There are a number of bioinformatic tools that have been created to automatically acquire information regarding peak intensities and relative abundances in stable isotope-labeled samples (not discussed in this review), each with their advantages and disadvantages (15).

References

1. Ong, S. E., Blagoev, B., Kratchmarova, I., et al. (2002) Stable isotope labeling by amino acids in cell culture, SILAC, as a simple and accurate approach to expression proteomics. *Mol. Cell. Proteomics* **1**, 376–386.
2. Schmidt, F., Strozynski, M., Salus, S. S., Nilsen, H., and Thiede, B. (2007) Rapid determination of amino acid incorporation by stable isotope labeling with amino acids in cell culture (SILAC). *Rapid Commun. Mass Spectrom.* **21**, 3919–3926.
3. Doherty, M. K., Hammond, D. E., Clague, M. J., Gaskell, S. J., and Beynon, R. J. (2009) Turnover of the human proteome: determination of protein intracellular stability

by dynamic SILAC. *J. Proteome Res.* **8**, 104–112.

4. Cargile, B. J., Bundy, J. L., Grunden, A. M., and Stephenson, J. L., Jr. (2004) Synthesis/degradation ratio mass spectrometry for measuring relative dynamic protein turnover. *Anal. Chem.* **76**, 86–97.
5. Gustavsson, N., Greber, B., Kreitler, T., Himmelbauer, H., Lehrach, H., and Gobom J. (2005) A proteomic method for the analysis of changes in protein concentrations in response to systemic perturbations using metabolic incorporation of stable isotopes and mass spectrometry. *Proteomics* **5**, 3563–3570.
6. Belle, A., Tanay, A., Bitincka, L., Shamir, R., and O'Shea, E. K. (2006) Quantification of protein half-lives in the budding yeast proteome. *Proc. Natl. Acad. Sci. USA* **103**, 13004–13009.
7. Rao, P. K., Roxas, B. A., and Li, Q. (2008) Determination of global protein turnover in stressed mycobacterium cells using hybrid-linear ion trap-Fourier transform mass spectrometry. *Anal. Chem.* **80**, 396–406.
8. Pratt, J. M., Petty, J., Riba-Garcia, I., et al. (2002) Dynamics of protein turnover, a missing dimension in proteomics. *Mol. Cell. Proteomics* **1**, 579–591.
9. Beynon, R. J. (2005) The dynamics of the proteome: strategies for measuring protein turnover on a proteome-wide scale. *Brief Funct. Genomic Proteomic* **3**, 382–390.
10. Doherty, M. K., and Beynon, R. J. (2006) Protein turnover on the scale of the proteome. *Expert Rev. Proteomics* **3**, 97–110.
11. Beynon, R. J., and Pratt, J. M. (2005) Metabolic labeling of proteins for proteomics. *Mol. Cell. Proteomics* **4**, 857–872.
12. Zhang, R., Sioma, C. S., Thompson, R. A., Xiong, L., and Regnier, F. E. (2002) Controlling deuterium isotope effects in comparative proteomics. *Anal. Chem.* **74**, 3662–3669.
13. Baganz, F., Hayes, A., Farquhar, R., Butler, P. R., Gardner, D. C., and Oliver, S. G. (1998) Quantitative analysis of yeast gene function using competition experiments in continuous culture. *Yeast* **14**, 1417–1427.
14. Pratt, J. M., Robertson, D. H., Gaskell, S. J., et al. (2002) Stable isotope labeling in vivo as an aid to protein identification in peptide mass fingerprinting. *Proteomics* **2**, 157–163.
15. Panchaud, A., Affolter, M., Moreillon, P., and Kussmann, M. (2008) Experimental and computational approaches to quantitative proteomics: status quo and outlook. *J. Proteomics* **71**, 19–33.

Chapter 12

Protein–Protein Interactions and Networks: Forward and Reverse Edgetics

Benoit Charloteaux, Quan Zhong, Matija Dreze, Michael E. Cusick, David E. Hill, and Marc Vidal

Abstract

Phenotypic variations of an organism may arise from alterations of cellular networks, ranging from the complete loss of a gene product to the specific perturbation of a single molecular interaction. In interactome networks that are modeled as nodes (macromolecules) connected by edges (interactions), these alterations can be thought of as node removal and edge-specific or "edgetic" perturbations, respectively. Here we present two complementary strategies, forward and reverse edgetics, to investigate the phenotypic outcomes of edgetic perturbations of binary protein–protein interaction networks. Both approaches are based on the yeast two-hybrid system (Y2H). The first allows the determination of the interaction profile of proteins encoded by alleles with known phenotypes to identify edgetic alleles. The second is used to directly isolate edgetic alleles for subsequent in vivo characterization.

Key words: Protein–protein interactions, interactome networks, interactome perturbations, interaction-defective alleles, edgetic alleles, yeast two-hybrid, reverse yeast two-hybrid, disease-associated alleles, gateway cloning.

1. Introduction

Genes and gene products do not work in isolation but instead interact with each other and with other cellular components within complex and dynamic "interactome" networks modeled as graphs of nodes and edges representing individual molecules

B. Charloteaux, Q. Zhong, and M. Dreze contributed equally to this work.

J.I. Castrillo, S.G. Oliver (eds.), *Yeast Systems Biology*, Methods in Molecular Biology 759, DOI 10.1007/978-1-61779-173-4_12, © Springer Science+Business Media, LLC 2011

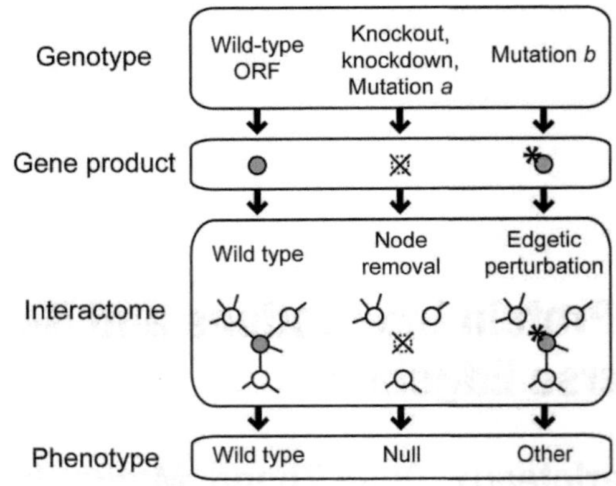

Fig. 12.1. Impact of various genotypic alterations on the interactome and on the resulting phenotype. Mutation *a*: mutation resulting in a protein with drastic molecular defects (*dashed circle*) such as a major truncation and/or an unstable fold. Mutation *b*: mutation (*star*) preserving the folding and some functional properties of at least one domain of a protein.

and the interactions between them, respectively (1, 2). Ongoing binary protein–protein interaction mapping efforts have identified a substantial fraction of interactome networks for human (3, 4) as well as for model organisms (5–7). Phenotypic changes arise from alteration of these networks, due to either complete loss of gene products ("node removal") or perturbations of specific interactions (edge-specific or "edgetic perturbations") (8, 9) (**Fig. 12.1**). For example, nonsense mutations and out-of-frame insertions or deletions resulting in a major truncation of a gene product as well as gene knockouts and RNAi-mediated gene expression knockdowns can be considered as "node removal" in the context of interactome network models. Alternatively, truncations that preserve specific autonomous protein domains or single amino acid substitutions in protein-binding sites likely cause what can be considered as edgetic perturbations (8, 9).

As distinct types of network perturbations can result in diverse phenotypes (**Fig. 12.1**), these perturbations and their impact on the underlying interactome have to be taken into account to improve our understanding of genotype-to-phenotype relationships (8, 9). Importantly, distinguishing node removal from edgetic perturbations can explain complex genotype-to-phenotype relationships associated with human diseases (9). As demonstrated with the *Caenorhabditis elegans* antiapoptotic protein CED-9, systematic isolation of edgetic alleles and their characterization in vivo represents a promising strategy to investigate gene function(s) (8). This strategy complements gene knockout and gene knockdown approaches by investigating the biological relevance of specific interaction network edges instead of concentrating on nodes.

This chapter describes two complementary strategies, "forward and reverse edgetics," to systematically investigate the phenotypic outcomes of specific perturbations of binary protein–protein interaction in interactome networks. The use of Gateway (10) vectors in combination with the yeast two-hybrid (Y2H) system (11) as a binary protein–protein interaction assay ensures compatibility with high-throughput manipulations. Since interaction defects are considered with respect to the wild-type protein, the identification of wild-type protein interaction partners by Y2H is a prerequisite. These interaction partners can be retrieved from available Y2H interactome maps (3–6, 12) or identified by performing proteome-wide Y2H screens (3).

Starting from a set of mutations in a gene associated with particular phenotypes (i.e., disease-associated mutations), the forward edgetics approach uses Y2H to determine the interaction defects of proteins encoded by open reading frames (ORFs) where the corresponding mutations have been introduced by site-directed mutagenesis (**Fig. 12.2**). Starting from a set of Y2H interactions for a protein of interest, the reverse edgetics strategy aims to systematically isolate alleles encoding proteins with desired specific interaction defects (**Fig. 12.2**). Reverse yeast two-hybrid (R-Y2H) selections (13, 14) are first used to isolate interaction-defective alleles from a random library enriched for full-length ORFs (15). The specificity of the interaction defects is then determined by Y2H screens to identify edgetic alleles. Such

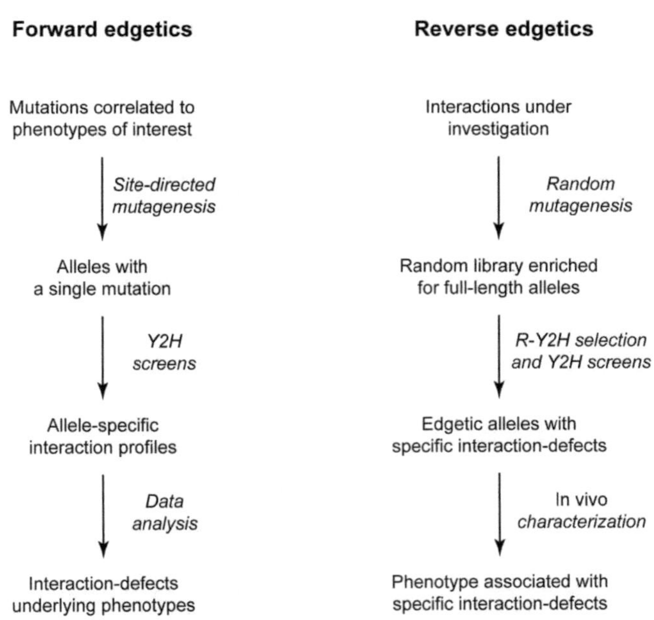

Fig. 12.2. Forward and reverse edgetics. Y2H: yeast two-hybrid and R-Y2H: reverse yeast two-hybrid.

edgetic alleles can subsequently be reintroduced in vivo to investigate the phenotypic consequences of specifically altering the corresponding molecular interaction(s).

2. Materials

2.1. Forward Edgetics

2.1.1. Gateway-Compatible PCR-Based Site-Directed Mutagenesis

1. KOD Hot Start DNA polymerase (Novagen).
2. Donor vector pDONR223 to generate entry clones of mutant alleles (Invitrogen). Destination vectors pDEST-DB and pDEST-*CYH2*^{S}-AD for Y2H analysis.
3. BP clonase and LR clonase (Invitrogen).
4. Z-competent DH5α cells (Zymo Research).
5. LB medium: dissolve 10 g of bacto-tryptone, 5 g of yeast extract, and 5 g of NaCl in 1 L of water. Adjust to pH 7.0 with NaOH if necessary. Autoclave and store at room temperature. Add needed antibiotics before use. For solid medium, add agar to a final 2% (w/v) concentration prior to autoclaving and add appropriate antibiotics before pouring media into 15-cm Petri dishes.
6. SOC medium: dissolve 20 g of bacto-tryptone, 5 g of yeast extract, 0.5 g of NaCl, 0.186 g of KCl in 1 L of water. Adjust to pH 7.0 with NaOH if necessary. Autoclave and store at room temperature. Add 5 mL each of sterilized 2 M $MgCl_2$, sterilized 2 M $MgSO_4$, and sterilized 40% (w/v) glucose per liter of medium before use.

2.1.2. Generation of Yeast Strains Expressing DB-X and AD-Y Hybrid Constructs

1. MaV103 yeast strain: *MAT***a** *leu2-3,112 trp-901 his3-200 ade2-1 gal4Δ gal80Δ SPAL10-URA3 GAL1-lacZ LYS2::GAL1-HIS3 can1*^{R} *cyh2*^{R}.
2. MaV203 yeast strain: *MATα leu2-3,112 trp-901 his3-200 ade2-1 gal4Δ gal80Δ SPAL10-URA3 GAL1-lacZ LYS2::GAL1-HIS3 can1*^{R} *cyh2*^{R}.
3. Non-selective rich medium (YEPD): mix 10 g of yeast extract, 20 g of bacto-peptone, and 950 mL of water, autoclave and store at room temperature. Add 50 mL of sterilized 40% (w/v) glucose per liter of medium before use. For YEPD agar plates, add 20 g of yeast extract and 40 g of bacto-peptone to 950 mL of water. At the same time, in a separate flask, add 40 g of agar to 950 mL of water. Autoclave both flasks separately and mix together afterward. Cool to 55°C in a water bath. Add 100 mL of sterilized 40% (w/v) glucose before pouring media into 15-cm Petri dishes.

4. Synthetic complete (SC) medium amino acid supplement: mix an equal weight of each of the following amino acids: alanine, arginine, aspartic acid, asparagine, cysteine, glutamic acid, glutamine, glycine, isoleucine, lysine, methionine, phenylalanine, proline, serine, threonine, tyrosine, and valine. Tryptophan (Trp), histidine (His), leucine (Leu), and uracil (Ura) are omitted from the mix and are added individually to SC-selective medium from stock solutions.
5. Stock solutions for selective SC medium: prepare solutions at the following concentrations: 100 mM His, 20 mM Ura, 100 mM Leu, and 40 mM Trp. Filter sterilize all solutions. Store the Ura and Leu solutions at room temperature, store the His solution light protected at room temperature, and store the Trp solution at 4°C light protected.
6. SC liquid medium: dissolve 1.4 g of yeast SC medium amino acid supplement, 1.7 g of yeast nitrogen base (without amino acids and without ammonium sulfate), and 5 g of ammonium sulfate into 925 mL of water. Adjust pH to 5.9 with 10 M NaOH. Autoclave and store at room temperature. Add 50 mL of sterile 40% (w/v) glucose before use. This media is SC-Leu-Trp-Ura-His. To supplement with a missing amino acid or nucleotide, add 8 mL of the appropriate stock solution (e.g., to make SC-Leu-Trp, add 8 mL of 20 mM Ura stock solution and 8 mL of 100 mM His stock solution).
7. SC agar medium and SC medium agar plates: to prepare 2 L of SC medium agar plates, dissolve 2.8 g of yeast synthetic drop-out medium amino acid supplement, 3.4 g of yeast nitrogen base (without amino acids and without ammonium sulfate), and 10 g of ammonium sulfate into 925 mL of water. Adjust pH to 5.9 with 10 M NaOH. In a separate 2 L flask, add 40 g of agar into 925 mL of water. Autoclave both flasks, transfer their content into the agar flask, and mix. Cool to 55°C in a water bath. Add 100 mL of sterilized 40% (w/v) glucose and 16 mL of the appropriate stock solutions of amino acid or nucleotide before pouring into 15 cm Petri dishes (e.g., to make SC-Leu-Trp, add 16 mL of 20 mM Ura stock solution and 16 mL of 100 mM His stock solution).
8. SC-Leu-Trp-His + 3-AT plates ("3-AT plates"): add 8 mL of 20 mM Ura stock solution and 1.68 g of 3-amino-1,2,4-triazole (3-AT, final concentration of 20 mM) per 1 L of SC agar medium before pouring media into 15 cm Petri dishes.
9. Cycloheximide stock solution (10 mg/mL): dissolve cycloheximide in 100% ethanol and filter sterilize (store light protected at –20°C).

10. SC-Leu-His + 3-AT + cycloheximide plates ("CHX plates"): add 8 mL of 20 mM Ura stock solution, 8 mL of 40 mM Trp stock solution, 100 μL of 10 mg/mL cycloheximide stock solution, and 1.68 g of 3-AT powder per 1 L of SC agar medium before pouring media into 15 cm Petri dishes.
11. SC-Leu-Trp-Ura ("-Ura plates"): add 8 mL of 100 mM His stock solution per 1 L of SC agar medium before pouring media into 15 cm Petri dishes.
12. For yeast transformation: Salmon sperm DNA (Sigma D9156), "TE/LiAc" solution, 10 mM Tris-HCl (pH 8.0), 1 mM EDTA (pH 8.0), 100 mM LiAc prepared from 10X TE, and 1 M LiAc. "TE/LiAc/PEG" solution, prepared by adding 50% (w/v) polyethylene glycol 3350 to TE/LiAc. Both solutions need to be freshly prepared from stock solutions before use.

2.1.3. Determination of Allele Interaction Profiles

1. Y2H controls: MaV203 co-transformed with pDEST-DB and pDEST-AD (no interaction control); pDEST-AD-*CYH2*^{S}-E2F1 and pDEST-DB-pRB (weak interaction control and no growth on CHX plates); pDEST-AD-Jun and pDEST-DB-Fos (moderately strong interaction control); pDEST-AD and pCL1 (positive control of Y2H readout); pDEST-AD-dE2F1 and pDEST-DB-dDP (strong interaction control); and pDEST-AD-*CYH2*^{S}-dE2F1 and pDEST-DB-dDP (strong interaction control and no growth on CHX plates) (3, 16).
2. Buffer for β-galactosidase assay: Z-buffer, add 16.1 g $Na_2HPO_4 \cdot 7H_2O$, 5.5 g $NaH_2PO_4 \cdot H_2O$, 0.75 g KCl, and 0.246 g $MgSO_4 \cdot 7H_2O$ to 1 L of water. Autoclave to sterilize and store at room temperature. About 4% (w/v) bromo-chloro-indolyl-galactopyranoside (X-gal), dissolve 40 mg X-gal in 1 mL of *N*,*N*-dimethylformamide. Store at –20°C light protected. β-Gal solution: for each β-gal assay plate, mix 5 mL of Z-buffer with 120 μL of 4% (w/v) X-gal, 13 μL of 2-mercaptoethanol. Prepare fresh each time.
3. Nitrocellulose filter for β-galactosidase assay (Osmonics) and Whatman paper filters (Whatman).

2.2. Reverse Edgetics

2.2.1. Generation of a Random Library Enriched for Full-Length Alleles

1. Platinum Taq DNA polymerase (Invitrogen).
2. Donor vector pDONR-Express to generate entry clones and select for full-length mutants (Invitrogen). Destination vectors pDEST-DB and pDEST-*CYH2*^{S}-AD for R-Y2H and Y2H analyses.
3. TOP10 electrocompetent bacteria (Invitrogen).
4. BP clonase and LR clonase (Invitrogen).

5. S.N.A.P. DNA purification kit (Invitrogen).
6. Midiprep kit (Qiagen).
7. LB plates with kanamycin (10–100 μg/mL) with or without 1 mM isopropyl β-D-thiogalactopyranoside (IPTG) and LB plates with 100 μg/mL ampicillin.

2.2.2. Isolation of Edgetic Alleles

1. MaV203 yeast strain (*see* **Section 2.1.2**, step 2).
2. YEPD plates (*see* **Section 2.1.2**, step 3).
3. SC-Leu-Trp plates (*see* **Section 2.1.2**, step 7).
4. 3-AT plates (*see* **Section 2.1.2**, step 8).
5. -Ura plates (*see* **Section 2.1.2**, step 11).
6. SC-Leu-Trp + 0.2% (w/v) 5-fluoroorotic acid (5-FOA): add 8 mL of 20 mM Ura, 8 mL of 100 mM His, and 2 g of 5-FOA per 1 L of SC agar medium before pouring into 15 cm Petri dishes.
7. Y2H controls (*see* **Section 2.1.3**, step 1).

3. Methods

3.1. Forward Edgetics

The forward edgetics approach has three steps: (i) generate the mutant alleles of interest, (ii) introduce each mutant in appropriate yeast cells for Y2H analysis, and (iii) determine the interaction profile of the mutant alleles.

Mutations are introduced in the wild-type ORF by Gateway-compatible PCR-based site-directed mutagenesis (17), enabling high-throughput manipulations. The mutants are next characterized in the Y2H system, which detects interactions between two proteins through the functional reconstitution of the yeast Gal4 transcription factor (11). The two proteins (X and Y) are fused to the Gal4 DNA-binding domain (DB) and the Gal4 activation domain (AD), respectively (*see* **Note 1**). The DB-X and AD-Y hybrid constructs are expressed from the pDEST-DB and pDEST-*CYH2*^{S}-AD vectors that harbor the *LEU2* and *TRP1* selectable markers for selection of transformants on SC-Leu and SC-Trp, respectively. The *CYH2*^{S} counter-selectable marker on pDEST-*CYH2*^{S}-AD allows for detection of auto-activating DB-X hybrid constructs (detectable activation of Y2H reporter genes in the absence of an interacting partner) by selecting yeast cells without an AD-Y hybrid construct plasmid (16). MaV103 and MaV203 are two non-isogenic strains of opposite mating type **a** and α, respectively. They both have three Gal4-inducible reporter genes integrated into their genome: *GAL1-lacZ*, *GAL1-HIS3*, and *SPAL10-URA3*. By convention, MaV103

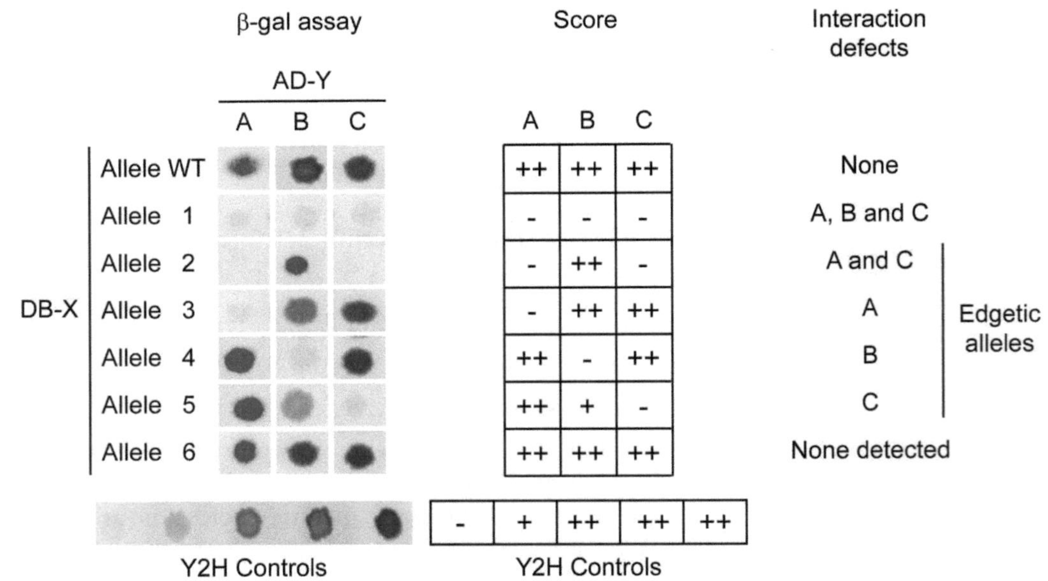

Fig. 12.3. Examples of Y2H readout, scoring and interaction defects observed for six alleles with respect to the wild-type (WT) allele. A, B, and C: Y2H interaction partners of the WT protein. DB: Gal4 DNA-binding domain. AD: Gal4 activation domain. Proteins corresponding to the alleles of the gene of interest are expressed as DB-X hybrid constructs. A, B, and C are expressed as AD-Y hybrid constructs. β-gal: β-galactosidase. Y2H controls from left to right: MaV203 co-transformed with pDEST-AD and pDEST-DB; pDEST-AD-*CYH2*^{S}-E2F1 and pDEST-DB-pRB; pDEST-AD-Jun and pDEST-DB-Fos; pDEST-AD and pCL1; pDEST-AD-dE2F1 and pDEST-DB-dDP.

and MaV203 are transformed with AD-Y and DB-X hybrid constructs, respectively.

All mutants are individually tested for interaction against each interactor known for the wild-type protein. For each interaction, the expression level of each Y2H reporter is scored by comparison to a set of Y2H control strains (**Fig. 12.3**). Expression of the *HIS3* and *URA3* reporters are determined by growth on 3-AT plates and -Ura plates, respectively. The expression of the *lacZ* reporter is determined by a filter colorimetric assay. Exclude any auto-activating DB-X hybrid construct showing growth on CHX plates.

The forward edgetics strategy presented here does not apply to the identification of gain of interaction alleles that can also underlie phenotypic alterations. Such alleles can be identified by performing proteome-wide Y2H screens (**3**).

3.1.1. Gateway-Compatible PCR-Based Site-Directed Mutagenesis

1. Design four specific primers to use in two separate PCR amplifications: conventional Gateway *att*B1.1 forward and *att*B2.1 reverse end primers plus forward and reverse internal primers straddling the desired mutation. Gateway *att*B1.1 forward and *att*B2.1 reverse end primers contain Gateway *att*B1.1 or *att*B2.1 sites followed by ORF-specific

sequence. The ORF-specific sequences for the *att*B1.1 forward primer start from ATG start codon (shown in bold below) and the ORF-specific sequence for the *att*B2.1 reverse primer may or may not contain a stop codon:

*att*B1.1 forward primer: 5′-G GGG ACA ACT TTG TAC AAA AAA GTT GGC ACC **ATG** (ORF 5′ end sequence)-3′.

*att*B2.1 reverse primer: 5′-GGG GAC AAC TTT GTA CAA GAA AGT TGG CAA (ORF 3′ end sequence)-3′.

Design forward and reverse internal primers to create 40 bp homologous regions between the forward and reverse internal primers (*see* **Note 2**).

2. Use fully sequenced wild-type DNA as template for PCR (*see* **Note 3**).
3. Carry out two PCR reactions for each mutant allele to be cloned. Each PCR reaction in 50 μL volume, containing 1 unit of KOD Hot Start DNA polymerase according to the recommendations from the manufacturer (Novagen). Carry out the first PCR with the Gateway attB1.1 primer and the reverse internal primer to generate fragment I. Carry out the second PCR with the forward internal primer and the Gateway attB2.1 primer to generate fragment II.
4. Mix equal volumes of the two PCR products containing fragments I and II (2–3 μL of each) with 150 ng pDONR223 in a standard 10 μL BP reaction mixture according to the instructions from the manufacturer (Invitrogen) and incubate at 25°C for 16 h.
5. Combine 2.5 μL of BP reaction mixture with 25 μL DH5α Z-competent cells (Zymo Research). Incubate on ice for 20 min. Spot the cell suspension on LB agar plates containing 50 μg/mL spectinomycin and incubate overnight at 37°C. There should be no growth from bacteria transformed with a BP reaction negative control (pDONR223 vector only, no PCR products).
6. With a sterile toothpick pick four isolated colonies from each transformation reaction into individual wells of a 96-well deep-well plate containing 1 mL liquid LB media with 50 μg/mL spectinomycin. Incubate the deep-well plate on a shaker at 37°C for 20 h. Combine 80 μL of the culture and 80 μL of 40% (w/v) sterile glycerol to prepare an archival stock (store at –80°C). Use the remainder of the culture for plasmid isolation.
7. To control for clone sequence and correct recombination, PCR-amplify transferred ORF from DNA isolated from individual colonies, using KOD polymerase with M13-based universal primers to generate templates for sequencing (*see* **Note 4**):

M13GF: 5′-CCCAGTCACGACGTTGTAAAACG-3′.

M13GR: 5′-GTAACATCAGAGATTTTGAGACAC-3′.

8. Full-length, sequence-confirmed mutant alleles are transferred to Y2H vectors pDEST-DB and pDEST-*CYH2*^{S}-AD via a Gateway LR reaction (Invitrogen). Carry out LR reactions in 10 μL according to the instructions of the manufacturer (Invitrogen) (*see* **Note 5**). Transform 10 μL of chemically competent DH5α cells with 5 μL of the LR reaction. There should be no growth from bacteria transformed with an LR reaction negative control (destination vector only, no entry clone).

9. Grow transformants in liquid LB media containing 100 μg/mL ampicillin in a 96-well deep-well plate on a shaker at 37°C for 20 h. Remove 80 μL of the overnight culture, mix with 80 μL of 40% (w/v) sterile glycerol to generate archival stocks, and store at –80°C. Use the remainder of the culture for plasmid isolation.

10. Confirm successful recombination of mutant alleles into the two Y2H vectors by PCR amplification using 5′-primers (AD and DB) specific to pDEST-*CYH2*^{S}-AD and pDEST-DB destination vectors, respectively, and 3′-primer (Term) common for both vectors. End-read sequencing can be done using the same primers:
    AD: 5′-CGCGTTTGGAATCACTACAGGG-3′.
    DB: 5′-GGCTTCAGTGGAGACTGATATGCCTC-3′.
    Term: 5′-GGAGACTTGACCAAACCTCTGGCG-3′.

3.1.2. Generation of Yeast Strains Expressing DB-X and AD-Y Hybrid Constructs

1. Streak MaV103 and MaV203 on YEPD plates and incubate for 48 h at 30°C to obtain isolated colonies. Prepare fresh cultures for each strain by inoculating a single colony into 20 mL of YEPD media. Incubate the cultures overnight at 30°C on a shaker. Measure and record the OD_{600} for the overnight cultures of MaV103 and MaV203, and then dilute cells into fresh YEPD media to obtain a final OD_{600} of 0.1. Incubate diluted cultures at 30°C on a shaker until OD_{600} reaches 0.6–0.8 (4–6 h). Use 100 mL of YEPD culture per 96-well plate of transformations.

2. Harvest cells by centrifugation at 800×*g* for 5 min. Wash cells first with 10 mL of sterile water, centrifuge, and discard the supernatant. Then wash with 10 mL of TE/LiAc solution, centrifuge, and discard the supernatant.

3. Resuspend cells in 2 mL of TE/LiAc solution, and then add 10 mL of TE/LiAc/PEG solution and 200 μL of carrier DNA (salmon sperm DNA boiled and kept on ice). Mix the solution by inversion. Aliquot 120 μL of this mix to

each well in a 96-well round bottom plate. Add 5 μL of DNA (miniprep of AD-Y or DB-X hybrid construct mutant clones) to the competent cells and mix. Seal the plate tightly with adhesive aluminum sealing foil. Incubate at 30°C for 30 min. Heat shock in a 42°C water bath for 15 min.

4. Centrifuge the 96-well plate for 5 min at 800×*g* and carefully remove the supernatant using a multi-channel pipette. To each well add 100 μL of sterile water and resuspend the cell pellet. Centrifuge the 96-well plate for 5 min at 800×*g* and then carefully remove 90 μL. Resuspend pellet on a 96-well plate vortex. Spot 5 μL of the cell suspension onto appropriate selective plates (Sc-Trp plates for AD-Y hybrid constructs and Sc-Leu plates for DB-X hybrid constructs). Incubate selective plates at 30°C for 3 days.
5. Pick transformants into a round bottom 96-well plate containing 160 μL of selective media in each well (Sc-Trp media for AD-Y hybrid construct strains and Sc-Leu media for DB-X hybrid construct strains). Incubate on a shaker at 30°C for 2 days. Prepare archival stocks by combining 80 μL of the yeast culture with 80 μL of 40% (w/v) sterile glycerol and store at –80°C.

3.1.3. Determination of Allele Interaction Profiles

1. Grow fresh cultures of MaV103 strains transformed with AD-Y hybrid construct and MaV203 strains transformed with DB-X hybrid construct in SC-Trp and SC-Leu media, respectively, for 2 days at 30°C in round bottom 96-well plates (*see* **Note 1**). Spot 5 μL of each culture, one on top of the other, on YEPD agar plates and grow overnight at 30°C.
2. Select diploid cells by replica plating of cells from YEPD to SC-Leu-Trp plates and grow for 2 days at 30°C.
3. Replicate diploid cells from SC-Leu-Trp plates on three phenotyping plates, including 3-AT, -Ura, and CHX plates. Clean 3-AT and CHX plates after overnight growth at 30°C by replica plates on fresh velvets (usually three to four times until no cells are readily detectable by eye). All phenotyping plates are grown for 5 days at 30°C before scoring.
4. Resuspend a small amount (200 μL tip end size) of diploid cells from SC-Leu-Trp plates in 20 μL SC-Leu-Trp medium. Spot cell suspension onto YEPD agar plates with a nitrocellulose filter laid on top. Incubate YEPD plate overnight at 30°C for β-gal filter lift assay.
5. Retrieve the YEPD/filter plates. For each plate to be assayed, get one empty 15 cm Petri dish. Place two pieces of Whatman filter paper in the plate. Add 5 mL of β-gal solution to each plate. Let the paper soak up the solution and make sure there are no bubbles under the Whatman paper. Remove the nitrocellulose filter from the YEPD (with yeast on the

filter) and place in liquid nitrogen for at least 30 s to lyse the cells. Remove the filter from liquid nitrogen and allow it to thaw in the air (approximately 30 s). Once the filter is flexible again place it into a Petri dish with β-gal solution soaked Whatman paper. Use forceps to remove any bubbles that may be under the filter. Incubate the β-gal assay plates at 37°C overnight (or until control 2 turns blue) before scoring intensity of readout compared to Y2H controls.

6. Growth on 3-AT plates but not on CHX plates indicates activation of the *HIS3* reporter through Y2H interactions. Growth on -Ura plates indicates activation of the *URA3* reporter. Blue coloration on the β-gal assay plates indicates activation of the *lacZ* reporter. Interactions of each mutant are compared to those of the wild type (**Fig. 12.3**). Wild-type-like, null-like, and edgetic alleles with specific interaction defects may be observed.

3.2. Reverse Edgetics

The first step in reverse edgetics is a PCR-based random mutagenesis of a wild-type ORF by taking advantage of the inherent error rate of Taq DNA polymerase. The resulting library of random alleles is subsequently enriched for full-length alleles by a genetic selection in bacteria using the pDONR-Express vector (15). This selection relies on the expression of a kanamycin resistance-encoding gene (Kan^R) placed in-frame with the ORF cloning site. In this construct mutant alleles with nonsense or out-of-frame mutations are selected against in the presence of kanamycin.

Interaction-defective alleles are then selected by R-Y2H, which relies on a genetic selection against wild-type interactions that have been reconstituted in the yeast two-hybrid (Y2H) system (13, 14). In the R-Y2H system, the activation of the *URA3* reporter gene by an interaction confers 5-fluoroorotic acid (5-FOA) sensitivity (5-FOA^S). Events that dissociate or prevent the interaction and provide 5-FOA resistance (5-FOA^R) can be genetically selected for from a complex random library mutant cloned in pDONR-Express. Only interactions driving the expression of the *URA3* reporter gene can be used in our version of the R-Y2H. The interaction profile of the isolated mutants can subsequently be tested against other partners to determine the specificity of the interaction defects and to identify edgetic alleles.

Here we describe a mutant library cloned into the pDEST-DB vector as an example for R-Y2H selection. The mutant library can be cloned into the pDEST-*CYH2*^{S}-AD (*see* **Note 1**).

3.2.1. Generation of a Random Library Enriched for Full-Length Alleles

1. Design two specific primers that add conventional *att*B1 and *att*B2 Gateway recombination sites to the forward (5′) and reverse (3′) ORF-specific primers, respectively. It is imperative that the reverse primer contains no stop codon:

*att*B1: 5′-ACA AGT TTG TAC AAA AAA GCA GGC nnn-3′.

*att*B2: 5′-AC CAC TTT GTA CAA GAA AGC TGG Gnn-3′.

where "n" represents ORF-specific sequences.

2. Carry out 30 cycles of PCR amplification (*see* **Note 6**), using Platinum Taq DNA polymerase with the designed pair of Gateway primers following the instructions of the manufacturer (Invitrogen). Set up mutagenic PCR reactions in 10 different wells to increase the complexity of the final library (*see* **Note 7**). Pool all PCR products for library construction.
3. Analyze pooled PCR products on a 1% (w/v) agarose gel, then gel-purify using the SNAP system and measure concentration at A_{260}.
4. Mix 4 μL of BP buffer 5X, 300 ng of pDONR-Express, 200 ng of gel-purified PCR product (flanked by *att*B sites), and 4 μL of BP clonase mix. Fill up to 20 μL with filter-sterilized TE buffer 1X, pH 8.0. Incubate at room temperature (25°C) for 16 h.
5. Transform electrocompetent TOP10 cells with BP reaction mix. The transformation protocol is done according to the instructions of the manufacturer (Invitrogen) (*see* **Notes 8** and **9**). First, determine the optimal selection condition (optimal kanamycin concentration of selective plates) with a wild-type ORF in pDONR-Express and subsequently apply this selection condition to the construction of the library of mutant clones. Scrape all clones from the selective plates, collect, and pool all cells before isolation of the DNA library by midiprep (Qiagen).
6. Mix 4 μL of LR buffer 5X, 1 μg of pDEST-DB or pDEST-*CYH2*^{S}-AD vector (*see* **Note 1**), 500 ng of the library, and 4 μL of LR clonase mix. Complete to 20 μL with filter-sterilized TE buffer 1X, pH 8.0. Incubate at room temperature (25°C) for 16 h.
7. Transform electrocompetent TOP10 cells with LR reaction mix (*see* **Note 9**). The bacterial transformation protocol is essentially the same one used for BP reaction transformation, except that the selection is done on ampicillin (100 μg/mL). After transformation, scrape all clones from the plates, collect, and pool all cells before isolating the DNA library by midiprep (Qiagen).

3.2.2. Isolation of Edgetic Alleles

1. Transform MaV203 yeast strain with the plasmids encoding wild-type DB-X and AD-Y hybrid constructs (*see* **Section 3.1.2**). Select co-transformants on Sc-Leu-Trp. Incubate plates for 72 h at 30°C. Test for activation of

the *URA3* reporter gene by plating co-transformed yeast cells on -Ura plates and media with 5-FOA (SC-Leu-Trp + 0.2% 5-FOA). Only interactions driving the expression of the *URA3* reporter gene should be kept for further analysis.

2. Transform MaV203 yeast strain with AD-Y hybrid construct carrying the ORF encoding the interaction partner of the wild type (*see* **Section 3.1.2**).
3. Streak the yeast strain obtained from step 2 on an SC-Trp agar plate to obtain isolated colonies. Incubate 48 h at 30°C. Prepare a fresh culture by inoculating a single colony into 20 mL of Sc-Trp media. Incubate the culture overnight at 30°C on a shaker. Measure and record the OD_{600} for the overnight culture and dilute cells into 100 mL of fresh YEPD media to obtain a final OD_{600} of 0.1. Incubate the diluted culture at 30°C on a shaker until OD_{600} reaches 0.6–0.8 (4–6 h).
4. Harvest cells by centrifugation at 800×*g* for 5 min. Wash cells first with 10 mL of sterile water, centrifuge, and discard the supernatant. Then wash the cell pellet with 10 mL of TE/LiAc solution, centrifuge, and discard the supernatant.
5. Resuspend cells in 2 mL of TE/LiAc solution, and then add 10 mL of TE/LiAc/PEG solution and 200 μL of carrier DNA (salmon sperm DNA boiled and kept on ice). Mix the solution by inversion. Add 10 μg of mutant library DNA to the competent cells and mix by pipetting. Distribute equally into 10 polypropylene tubes. Incubate at 30°C for 30 min. Heat shock in a 42°C water bath for 15 min.
6. Centrifuge all tubes for 5 min at 800×*g* and discard the supernatant. Wash cells in each tube with 1 mL of sterile water. Plate yeast cells on 10 plates of media with 0.2% (w/v) 5-FOA (Sc-Leu-Trp + 0.2% 5-FOA). In parallel, two 100 μL aliquots of transformation mix should be diluted 10^3 and 10^4 times and plated on non-selective medium (Sc-Leu-Trp) to determine the efficiency of transformation. An average of 3–5 million yeast transformants should be obtained in total. Incubate all plates at 30°C for 72 h.
7. Pick 5-FOA^R colonies and streak them on non-selective media (SC-Leu-Trp). Incubate at 30°C overnight, and then replicate on 3-AT plates and YEPD plates to assay *HIS3* and *lacZ* expression, respectively. Only colonies that can no longer activate the expression of these reporter genes should be kept for further characterization.

8. PCR-amplify ORFs directly from yeast cells with pDEST-DB-specific primers to identify potential mutations. Only single non-synonymous missense mutants should be kept for further analysis.
9. Using the protocol given in **Section 3.1.2**, transform the strain obtained from step 2 with both linearized pDEST-DB vector and the PCR product obtained from step 8 (gap repair (16)). Retest Y2H interaction between DB-X hybrid construct of the mutant alleles and AD-Y hybrid construct of the wild-type interactors (*see* **Section 3.1.3**).
10. Characterize the interaction profiles of the selected mutant alleles against other interactors (*see* **Section 3.1.3**) and identify edgetic alleles (**Fig. 12.3**).

4. Notes

1. In the Y2H system, some interactions between proteins X and Y can be detected both as DB-X/AD-Y and as DB-Y/AD-X configuration, whereas others can only be detected in one of the two configurations. The cloned mutant alleles can be transferred into pDEST-DB and/or pDEST-*CYH2*^{S}-AD vectors, depending on the configuration of the reconstituted interaction.
2. An overlap shorter than 40 bp between forward and reverse internal primers may sometimes be sufficient. A larger overlap usually increases the recombination efficiency. Mutation close to the 5′- or 3′-end of the ORF can be introduced using the Gateway-tailed *att*B primers.
3. If the wild-type ORF used as PCR template is carried on a plasmid, it is important to ensure that this plasmid contains an antibiotic marker different from that of pDONR223 (spectinomycin) to avoid contamination in the bacterial transformation.
4. Unconfirmed clones are usually due to primer synthesis error or additional mutations introduced by PCR.
5. The volume of Gateway LR reaction can be reduced to 5 μL.
6. Thirty cycles of PCR amplification are suitable for a 750 base pair (bp) ORF. For ORFs of different lengths, PCR conditions may need to be optimized for number of cycles and/or dNTP concentrations to increase the fraction of amplicons with single nucleotide changes, without drastically increasing the fraction of amplicons with multiple mutations.
7. Since ORFs mutated early in the cycling conditions will be amplified during the subsequent cycles, carrying out

independent PCR reactions in distinct wells increases the probability of getting a final library rich in distinct mutations.

8. The optimal selection condition to enrich the library in full-length alleles is determined by titration of kanamycin concentrations (10–100 μg/mL), which yields the highest number of colonies in the presence of IPTG and minimal (or zero) colonies without IPTG.
9. The appropriate number of transformants required to sufficiently cover the complexity of the library needs to be adjusted depending on the size of the ORFs, with longer ORFs requiring a higher number of clones. In our experience, for a 750 bp ORF, approximately 500,000 clones should be sufficient.

Acknowledgements

Special thanks to R. Vidal for her help with the manuscript. This work was supported by the Ellison Foundation and the WM Keck Foundation (MV); NIH grants R01-HG001715 from NHGRI and NIGMS (MV and DEH), U01-CA105423 (PI, S Orkin, project leader, MV), and U54-CA112952 (PI, J Nevins, subcontract, MV) and R33 CA105405, R33 CA132073, and R21/R33 CA081658 (MV) from NCI, and by Institute Sponsored Research funds from the Dana-Farber Cancer Institute Strategic Initiative awarded to CCSB. M.V. is a "Chercheur Qualifié Honoraire" from the Fonds de la Recherche Scientifique (FRS-FNRS, French Community of Belgium). The content is solely the responsibility of the authors and does not necessarily represent the official views of the NCI, NHGRI, NIGMS or the NIH.

References

1. Barabasi, A. L., and Oltvai, Z. N. (2004) Network biology: understanding the cell's functional organization. *Nat. Rev. Genet.* **5**, 101–113.
2. Vidal, M. (2005) Interactome modeling. *FEBS Lett.* **579**, 1834–1838.
3. Rual, J. F., Venkatesan, K., Hao, T., et al. (2005) Towards a proteome-scale map of the human protein-protein interaction network. *Nature* **437**, 1173–1178.
4. Stelzl, U., Worm, U., Lalowski, M., et al. (2005) A human protein-protein interaction network: a resource for annotating the proteome. *Cell* **122**, 957–968.
5. Yu, H., Braun, P., Yildirim, M. A., et al. (2008) High-quality binary protein interaction map of the yeast interactome network. *Science* **322**, 104–110.
6. Li, S., Armstrong, C. M., Bertin, N., et al. (2004) A map of the interactome network

of the metazoan *C. elegans*. *Science* **303**, 540–543.

7. Giot, L., Bader, J. S., Brouwer, C., et al. (2003) A protein interaction map of Drosophila melanogaster. *Science* **302**, 1727–1736.
8. Dreze, M., Charloteaux, B., Milstein, S., et al. (2009) 'Edgetic' perturbation of a *C. elegans* BCL2 ortholog. *Nat. Methods* **6**, 843–849.
9. Zhong, Q., Simonis, N., Li, Q. R., et al. (2009) Edgetic perturbation models of human inherited disorders. *Mol. Syst. Biol.* **5**, 321.
10. Walhout, A. J., Temple, G. F., Brasch, M. A., et al. (2000) GATEWAY recombinational cloning: application to the cloning of large numbers of open reading frames or ORFeomes. *Methods Enzymol.* **328**, 575–592.
11. Fields, S., and Song, O. (1989) A novel genetic system to detect protein-protein interactions. *Nature* **340**, 245–246.
12. Boxem, M., Maliga, Z., Klitgord, N., et al. (2008) A protein domain-based interactome network for *C. elegans* early embryogenesis. *Cell* **134**, 534–545.
13. Vidal, M., Brachmann, R. K., Fattaey, A., Harlow, E., and Boeke, J. D. (1996) Reverse two-hybrid and one-hybrid systems to detect dissociation of protein-protein and DNA-protein interactions. *Proc. Natl. Acad. Sci. USA* **93**, 10315–10320.
14. Vidal, M., Braun, P., Chen, E., Boeke, J. D., and Harlow, E. (1996) Genetic characterization of a mammalian protein-protein interaction domain by using a yeast reverse two-hybrid system. *Proc. Natl. Acad. Sci. USA* **93**, 10321–10326.
15. Gray, P. N., Busser, K. J., and Chappell, T. G. (2007) A novel approach for generating full-length, high coverage allele libraries for the analysis of protein interactions. *Mol. Cell. Proteomics* **6**, 514–526.
16. Walhout, A. J., and Vidal, M. (2001) High-throughput yeast two-hybrid assays for large-scale protein interaction mapping. *Methods* **24**, 297–306.
17. Suzuki, Y., Kagawa, N., Fujino, T., et al. (2005) A novel high-throughput (HTP) cloning strategy for site-directed designed chimeragenesis and mutation using the Gateway cloning system. *Nucleic Acids Res.* **33**, e109.

Chapter 13

Use of Proteome Arrays to Globally Identify Substrates for E3 Ubiquitin Ligases

Avinash Persaud and Daniela Rotin

Abstract

Ubiquitin-protein ligases (E3s) are responsible for target recognition and subsequent modification of selected substrates within the ubiquitin proteasomal system (UPS). Substrates of this pathway are covalently modified by the attachment of ubiquitin usually onto Lys residues. As a result, these modified proteins can be targeted for degradation, endocytosis, protein sorting, subnuclear trafficking, or other fates. Despite the advancements in understanding the underlying mechanisms of the ubiquitin system, the substrates of most E3 enzymes remain largely unknown. Here, we describe the development of a high-throughput method to identify in vitro substrates for E3 ligases on a global proteomic scale. The enzymatic activity (ubiquitylation) and binding of ubiquitin ligases to their substrates are performed using protein (proteome) microarrays as the experimental platform, and using Nedd4/Rsp5 family members as examples of HECT E3 ligases. The in vitro ubiquitylation and binding substrates identified in these screens can provide invaluable insight into the cellular pathways in which E3 ligases participate.

Key words: E3 ubiquitin ligase, ubiquitin proteasomal system (UPS), microarray, proteomics, HECT, Nedd4, high-throughput.

1. Introduction

Ubiquitin-mediated proteolysis is a major proteolytic pathway in the cell and is essential for normal cell functioning and survival (1). Ubiquitin also targets proteins for other cellular fates, such as endocytosis, vesicular transport, sorting of transmembrane and cargo proteins, histone modification, and others (2–5). The ubiquitylation cascade involves the sequential activity of three enzymes: E1 (ubiquitin-activating enzyme), E2 (ubiquitin-conjugating enzyme), and the E3 (ubiquitin ligase);

J.I. Castrillo, S.G. Oliver (eds.), *Yeast Systems Biology*, Methods in Molecular Biology 759,
DOI 10.1007/978-1-61779-173-4_13, © Springer Science+Business Media, LLC 2011

the latter is responsible for substrate recognition and transfer of ubiquitin onto it. The major classes of E3 ubiquitin ligases are the HECT and RING classes. While HECT E3s can directly transfer ubiquitin onto its target proteins, RING E3s facilitate transfer of ubiquitin from an E2 to the substrate (6). Given the importance of E3s for both substrate recognition and ubiquitin transfer, the identification of their substrates is essential for understanding the cellular pathways in which they participate and their biological function.

Protein microarray technology provides a powerful tool to assess the selectivity of protein–protein interactions and protein modifications on a system-wide or proteome-wide scale, an advantage that is limiting in other high-throughput methods. However, only a few studies to date have used this technology to assess substrate recognition and specificity of ubiquitin ligases.

In a recent study, we have utilized the yeast *Saccharomyces cerevisiae* and proteome microarrays (chips) to globally identify ubiquitylation substrates and binding partners for Rsp5, a HECT E3 that is the ortholog of Nedd4 in yeast (**Fig. 13.1a**); this screen identified both known and novel substrates for this E3 ligase (7). The validity of this approach was further confirmed by recent studies, which demonstrated that some of the novel Rsp5 substrates (hits) identified in our proteome array screen (e.g., Sna3, Ear1P, and the arrestin-related trafficking adaptors (ARTs) Ygr068c, Rod1, Rog3, Aly1, and Aly2 [i.e., Art5, Art4, Art7, Art6, and Art3]) were indeed ubiquitylated in vivo by Rsp5 in yeast cells, a process that was required for their biological function (8–13). In a subsequent study, we performed a similar screen in order to identify substrates for the mammalian Nedd4 (Nedd4-1) and Nedd4-2 (Nedd4L) E3 ligases and decipher their substrate specificity using human proteome microarray (14) (**Fig. 13.1**). In addition, proteome microarrays have been used for the identification of ubiquitylated substrates of E3s in cellular extracts obtained from cells at different stages of the cell cycle (15). Furthermore, earlier studies by Hesselberth and colleagues utilized yeast proteome microarrays to identify Rsp5 WW domain-interacting partners (16).

Thus, the aim of the ubiquitylation-on-chip screen is to identify, on a global proteomic scale, all or most substrates for specific ubiquitin ligases. The method to achieve this is described here, which was utilized to identify Nedd4 family substrates, employing a Human ProtoArray Microarray PATH slide (currently available from Invitrogen). These slides contain purified proteins immobilized at a high spatial density on standard size slides. By performing both ubiquitylation and binding reactions on these slides, proteins that are modified by the attachment of ubiquitin or that bind directly to E3 can be rapidly identified. This method works well also for E3s that are not from the HECT family (e.g., a cullin complex, which includes a RING E3) and others. The

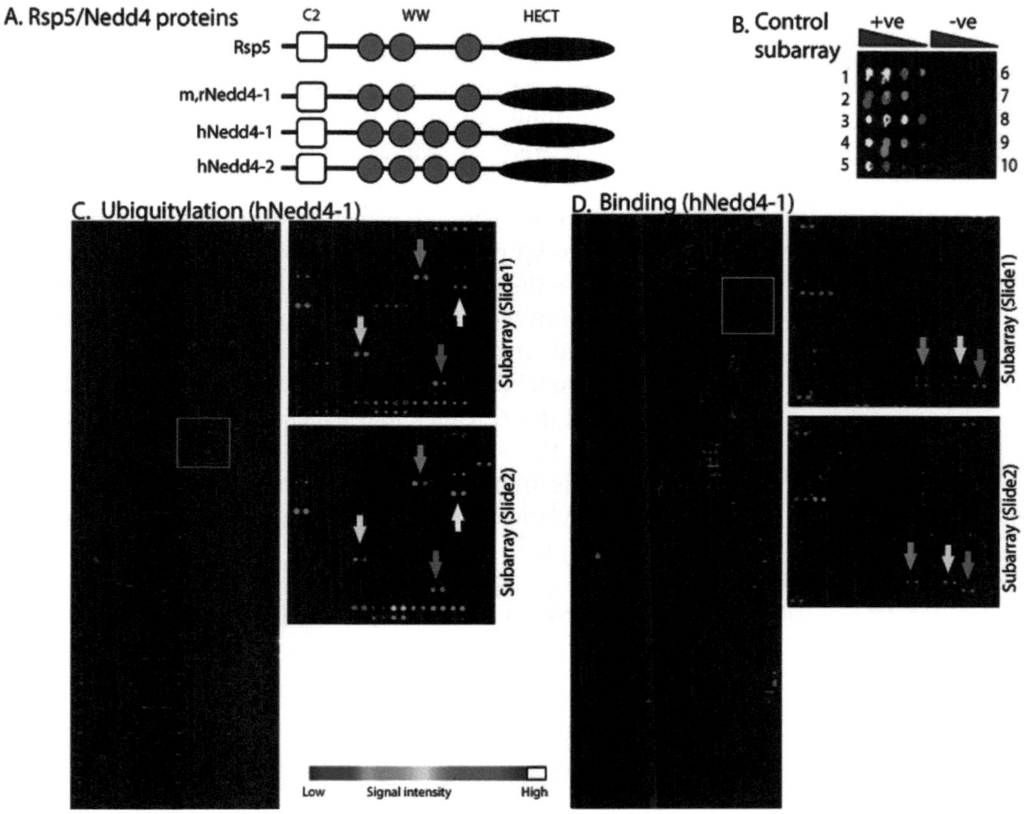

Fig. 13.1. Ubiquitylation and binding assays to identify substrates for E3 ubiquitin ligases. (**a**) Schematic representation of several Nedd4 family members (the yeast Rsp5 and the mammalian mouse (m) and rat (r) Nedd4-1 and human (h) Nedd4-1 and Nedd4-2) showing their domain architecture. Not to scale. (**b**) Control protein microarray spotted with decreasing concentrations of known positive control proteins (containing PY motifs), or negative controls, and probed with hNedd4-1. Positive controls: (1) βENaC-PY motif; (2) Rnf11; (3) LAPTM5-C terminus; (4) *Xenopus* Nedd4-HECT domain; (5) CNrasGEF-C terminus. Negative controls: (6) GST; (7) rNedd-4 –C2 domain; (8) RNF11 (Y to A mutant); (9) Grb10-SH2 domain; (10) CNrasGEF-PDZ domain. (**c**) Ubiquitylation of the proteome array (ProtoArray^{TM}) by hNedd4-1. The array was incubated with E1, E2 (UbcH5b), E3 (hNedd4-1), FITC-Ub and Mg-ATP. Subarrays from duplicate proteome array are shown, to demonstrate reproducibility of the assay. (**d**) Binding of hNedd4-1 to the proteome array (ProtoArray^{TM}). The array was incubated with Alexa647-hNedd4-1. Subarrays from duplicate proteome array are shown, to demonstrate reproducibility of the assay, as in (**c**). Intensity of ubiquitylation and binding to protein on the arrays is revealed by the *color bar*, which is depicted in (14) and in the online version of the volume. Adapted from (14) with permission from Nature Publishing Group (NPG).

improvements made over our previous screen of yeast proteome microarray probed with Rsp5 are described here as well.

2. Materials

The ProtoArray technology used in this study contains purified proteins immobilized at a high spatial density on standard nitrocellulose-coated glass slides and is based on the yeast protein microarray technology developed by Zhu and colleagues (17) to detect molecular interactions of proteins.

The ProtoArray Human Protein Microarray contains thousands of purified human proteins that have been expressed using a baculovirus expression system, purified from insect cells and printed in duplicate on a PATH protein microarray slides (nitrocellulose-coated glass slide). Currently, ProtoArray Human Protein Microarray v5.0 is available from Invitrogen that contains 9,483 proteins, as well as 2,016 controls, all spotted in duplicate. Previously, ProtoArray Yeast Protein Microarrays were printed that contained >4,000 *S. cerevisiae* proteins. However, these yeast proteins were printed on a FAST protein microarray slide and are no longer in production. The PATH slide is coated with an ultra-thin film of nitrocellulose, which confers the advantage of very low background fluorescence and excellent signal-to-noise ratio, as compared to the older FAST slide, thus providing a higher quality data. The methods described in **Section 3** are for performing ubiquitylation screens, as well as parallel binding screens, using ProtoArrays.

2.1. Materials for Printing Control Microarray PATH Slides

1. PATH® Protein Microarray slides (Gentel Biosciences).
2. Biochip Arrayer (Perkin Elmer PiezoArrayer or any standard microarray arrayer) to print the control slides.
3. Purified proteins of interest to be spotted on slide (positive and negative controls; *see* **Note 1**).

2.2. Materials for Ubiquitylation Assay

1. Respective E1, E2, and E3 enzymes required for ubiquitylation reaction (*see* **Note 2**).
2. Ubiquitylation reaction buffer (10X): 250 mM Tris-HCl pH 7.5, 500 mM NaCl, 1 μM DTT, 40 mM $MgCl_2$. Store in aliquots at –20°C for up to 2 months (*see* **Note 3**).
3. Blocking buffer (1X): 50 mM HEPES pH 7.5, 200 mM NaCl, 0.08% (w/v) Triton X-100, 25% glycerol, 20 mM glutathione, 1.0 mM DTT, 10 M NaOH, 1% (w/v) BSA. Store at 4°C for up to 4 months.
4. Wash buffer: 0.5% PBST (phosphate-buffered saline (PBS) pH 7.4 with 0.5% (w/v) Triton X-100. Store at room temperature.
5. Fluorescein-N terminus ubiquitin, FITC (U-580, Boston Biochem). Store in aliquots at –20°C for up to 2 months.
6. 100 mM Mg-ATP dissolved in distilled water.
7. Invitrogen Human ProtoArray® Microarray PATH slide.
8. ProScan Array HT™ (Perkin Elmer) slide scanner.

2.3. Materials for Binding Assay

1. Fluorescently labeled E3 ligase of interest (*see* **Note 2**).
2. Ubiquitylation reaction buffer (10X): 250 mM Tris-HCl pH 7.5, 500 mM NaCl, 1 μM DTT, 40 mM $MgCl_2$. Store in aliquots at –20°C for up to 2 months (*see* **Note 3**).

3. Blocking buffer (1X): 50 mM HEPES pH 7.5, 200 mM NaCl, 0.08% (w/v) Triton X-100, 25% glycerol, 20 mM glutathione, 1.0 mM DTT, 10 M NaOH, 1% (w/v) BSA. Store at 4°C for up to 4 months.
4. Wash buffer: 0.1% PBST (PBS, pH 7.4 with 0.1% (w/v) Triton X-100). Store at room temperature.
5. Invitrogen Human ProtoArray® Microarray PATH slide.
6. Alexa Fluor 647 Microscale protein Labeling Kit (A30009, Molecular Probes).
7. ProScan Array HT™ (Perkin Elmer) slide scanner.

3. Methods

Ubiquitin ligases are required for the targeted recognition of specific substrates within the ubiquitylation system. The E3 must first interact with the substrate before modifying it or facilitating its modification. This provides the rationale for the use of binding screens to complement the ubiquitylation screens. However, it is worthwhile to note that the ubiquitylation assay is more sensitive than the binding assay (**Fig. 13.1c**, **d**) and usually results in a higher quality data set. One possible explanation for this is that the binding of E3 ligases to their targets is relatively weaker and more transient as compared to the covalent attachment of ubiquitin onto these substrates. Furthermore, polyubiquitylation (or multi-monoubiquitylation) of substrates on the proteome microarray may result in fluorescent signal amplification. In addition to being more sensitive, the ubiquitylation screen is also a more direct assay of the E3 ligase's activity (7).

Both the ubiquitylation and binding screen are performed in duplicate on the ProtoArray® Microarray slides in order to ensure reproducibility.

3.1. Printing Control Microarray PATH Slides

Control microarray slides (**Fig. 13.1b**) are printed in order to optimize conditions for both ubiquitylation and binding assays.

1. Printing of control microarray PATH slides can be done using a Biochip arrayer (for example, PiezoArrayer from Perkin Elmer).
2. Purified proteins (epitope tagged) are diluted such that the total amount of control proteins spotted within a subarray range from 0.016 to 0.8 ng per spot (usually 4 serial dilutions per protein, *see* **Note 1**).
3. Each control protein subarray can be reprinted in a six row by two-column format for testing multiple optimization conditions simultaneously on the same microarray

slide using a multiplexing gasket (SIM*plex*™ Multiplexing System, Gentel Biosciences).

4. The quality of the print can be assessed by probing the slide with a primary antibody against the epitope tag on the control proteins followed by a secondary antibody–fluorophore conjugate. The slide can be scanned using a standard fluorescent-based microarray scanner such as ProScan Array HT™ scanner (Perkin Elmer) or Scanarray™ 4000 laser scanner (Perkin Elmer) to visualize protein spots.

3.2. Ubiquitylation Assays

3.2.1. Optimizing Ubiquitylation-On-Chip Assay Using Control Microarray PATH Slides

1. Rinse the control PATH slide briefly with 0.5% PBST.
2. Following rinsing, the slide is blocked at room temperature for 40 min by gentle agitation in ubiquitylation blocking buffer (*see* **Note 3**).
3. Wash slide gently with 1X ubiquitylation reaction buffer (prepared by diluting the 10X stock in water) at room temperature for 5 min. Slide is then placed on moistened filter paper (in a petri dish) on a flat surface following this wash step.
4. Ubiquitylation reaction mixture: To optimize the ubiquitylation screen, it is recommended that you maintain similar concentrations to those used in an in vitro ubiquitylation reaction with your cascade enzymes as the starting platform. If this is not known, the following can be used: 3 μg E1, 6 μg E2, 10 μg E3, 8 μg of FITC-Ub, and 8 mM Mg-ATP in ubiquitylation reaction buffer to a final volume of 300 μL (*see* **Note 4**). The concentration of the contents of the ubiquitylation reaction mixture can be varied to increase the signal-to-noise ratio and to selectively ubiquitylate the positive controls.
5. Incubate ProtoArray slide with ubiquitylation reaction mixture by pipetting the mixture onto the slide surface. Ensure that the entire slide surface is covered with ubiquitylation reaction mixture (*see* **Note 5**). The slide is left in the dark for 1.5 h.
6. Following this incubation, wash slide gently 3×5 min each with 0.5% PBST.
7. Place the slide in a 50 mL Falcon tube and dry it by centrifugation at 1,000×*g* for 5 min at room temperature (*see* **Note 6**).
8. The slide is then air dried by gentle shaking in the dark.
9. Scan slide at 10 μm resolution using a 488 nm laser on ProScan Array HT™ scanner (Perkin Elmer) using Scan Array Express software. Quantitation of fluorescent signal is also accomplished using this software.

3.2.2. Ubiquitylation Assay Using ProtoArray Microarray PATH Slides

To conduct the actual ubiquitylation screen, use the same protocol and stoichiometric ratios of E1, E2, and E3 enzymes, as well as FITC-Ub and Mg-ATP, as those optimized for the control PATH slides, in order to probe the human ProtoArray Microarray (**Fig. 13.1c**).

3.3. Binding-On-Chip Assays

3.3.1. Optimizing Binding-On-Chip Assay Using Control Microarray PATH Slides

1. Follow the instructions in the Alexa Fluor 647 Microscale protein Labeling Kit for labeling the purified E3 of interest.
2. Rinse the control PATH slide briefly with 0.1% PBST.
3. Following rinsing, the slide is blocked at room temperature for 40 min by gentle agitation in ubiquitylation blocking buffer (*see* **Note 3**).
4. Wash slide gently with 1X ubiquitination reaction buffer (prepared by diluting the 10X stock in water) at room temperature for 5 min. Slide is then placed on moistened filter paper (in a petri dish) on a flat surface following this wash step.
5. Binding reaction mixture: Starting amounts of labeled E3 can vary between 2 and 6 μg (or 0.15–0.5 μM of Alexa647-E3 as a function of the ubiquitin ligase) in ubiquitylation reaction buffer to a final volume of 300 μL (*see* **Note 7**). This can be varied subsequently to maximize signal-to-noise ratio and to selectively bind the positive controls.
6. Incubate the control slide with binding reaction mixture by pipetting the mixture on the slide surface. Ensure that entire slide surface is covered by the binding reaction mixture (*see* **Note 5**). The slide is left in the dark for 1 h.
7. Following this incubation, wash slide gently 3×5 min each with 0.1% PBST.
8. Place the slide in a 50 mL Falcon tube and dry it by centrifugation at 600×*g* for 5 min at room temperature (*see* **Note 6**).
9. The slide is then air dried by gentle shaking in the dark.
10. Scan slide at 10 μm resolution using a 633 nm laser on ProScan Array HT^{TM} scanner (Perkin Elmer or any standard fluorescent-based microarray scanner) using Scan Array Express software.

3.3.2. Binding Assay Using ProtoArray Microarray PATH Slides

To conduct the actual binding screen, use the same protocol and concentration of Alexa-647-labeled ubiquitin ligase established in optimizing the binding assay on the control PATH slides to probe the human ProtoArray Microarray (**Fig. 13.1d**).

3.4. Quantitation and Data Analysis

Spot fluorescent intensity is quantitated using ProScan Array HT^{TM} (Perkin Elmer) software. Comparison of hits on replicate slides is accomplished using Protein Prospector Analyzer software (Invitrogen).

Duplicate screens are compared using Protein Prospector Analyzer (Invitrogen) software to align similar protein spots on both slides. Spots in which 50% of the pixels produced a signal greater than 2 standard deviations (SD) above background were identified as 'hits.' These proteins are included in the substrate and interaction data sets only if both duplicate spots in the two replicated slides (i.e., all four spots) meet the criteria. Once the data sets are generated, the spots are normalized and ranked according to their signal intensity per unit protein spotted on slide [signal intensity = mean signal on the spot – background)/RFU (relative fluorescent units) of the protein spotted] (*see* **Note 8**). The RFU of the proteins spotted on the slide is provided upon purchase of the ProtoArray.

4. Notes

1. Control proteins are spotted in specified amounts so as to replicate the range of proteins spotted on the Human Protoarray Microarray PATH slide. These control proteins can be epitope tagged as it allows the visualization of the spotted proteins on the control slide after probing with specific antibody/fluorophore conjugate.
2. The ubiquitin ligase probe should *not* be a GST fusion protein; most of the proteins printed on the Human Protoarray Microarray PATH slide are GST tagged and thus dimerization of GST can lead to false positives. If it is absolutely necessary to generate the E3 protein first as a GST fusion, ensure GST is completely removed prior to the assay (for example, through proteolytic cleavage).
3. DTT is omitted when making the ubiquitylation blocking buffer. This is added to the buffer prior to blocking the ProtoArray slide. It is also important to prevent the slide from drying for most of the duration of the ubiquitylation and binding assays, in order to reduce background noise when scanning. Protocol steps should therefore flow smoothly into each other.
4. The amounts of E1, E2, and E3 needed for the ubiquitylation reaction will vary depending on the activity of your E3. In our assay we used the following concentrations: 0.303

μM E1, 1.96 μM E2 (UbcH5b), 0.49 μM E3 (Nedd4 protein), and 8 mM Mg-ATP. Volume of ubiquitylation reaction mixture pipetted on entire slide surface ranges from 300 to 700 μL.

5. Ensure that reaction mixture is evenly distributed over the slide surface, since uneven distribution would result in inconsistent ubiquitylation (or binding) of targets between replicate experiments.
6. When drying by centrifugation, ensure that the surface of the slide with proteins spotted on it is facing the center of rotation and that slide barcode is facing upward. This orientation reduces background noise considerably.
7. The amounts of Alexa647-E3 needed for the ubiquitylation reaction will depend on the amount used in your binding optimization reaction. Starting amount used can vary between 0.15 and 0.3 μM of Alexa647-E3 as a function of the ubiquitin ligase. Volume of ubiquitylation reaction mixture pipetted on entire slide surface ranges from 300 to 700 μL.
8. It is difficult to measure the amount of protein spotted per spot on the microarray slide. As such, their amounts are given in terms of relative fluorescent units (RFU). These were derived from comparing the fluorescence from these spots (after treatment with an antibody–fluorophore conjugate) to those of standard protein controls with known amounts spotted.

   Previously, we had used the Nedd4 homologue in the yeast *S. cerevisiae*, Rsp5, to probe the yeast ProtoArray in order to identify substrates for this ubiquitin ligase within the yeast proteome (7). However, the yeast proteome was printed on a FAST microarray slide. This slide has a thicker layer of nitrocellulose on its surface in comparison to the PATH slide used for printing the human ProtoArray and results in higher background noise. There are also few differences with regard to the protocol used to conduct the ubiquitylation and binding assays. The major differences for both assays are outlined below:

   (a) Gupta et al. (7) used 5% skim milk to block the microarray slide in their screen while a blocking buffer (recipe given) was used for the new PATH slide.
   (b) Incubation time with the ubiquitylation reaction mixture in the yeast microarray screen was 3 h vs. 1.5 h used in the revised current protocol (for the binding screen, incubation times were 2 h vs. 1 h, respectively).
   (c) Although both methods used centrifugation to dry the slide, the orientation of the slide during this step is crucial to reducing the background noise (*see* **Note 6**).

Acknowledgments

The work described here was supported by the Canadian Institute of Health Research (CIHR) and the Canadian Cancer Society Research Institute (CCSRI/NCIC) to DR and by the CFI-funded SIDNET. AP is supported by RESTRACOMP and DR by a Canada Research Chair (CRC, Tier II).

References

1. Hershko, A., and Ciechanover, A. (1998). The ubiquitin system. *Ann. Rev. Biochem.* **67**, 425–479.
2. Schwartz, A. L., and Ciechanover, A. (2009) Targeting proteins for destruction by the ubiquitin system: implications for human pathobiology. *Ann. Rev. Pharmacol. Toxicol.* **49**, 73–96.
3. Hicke, L. (2001) A new ticket for entry into budding vesicles-ubiquitin. *Cell* **106**, 527–530.
4. Woelk. T., Sigismund, S., Penengo, L., and Polo, S. (2007). The ubiquitination code: a signaling problem. *Cell Div.* **2**, 11.
5. Dikic, I., Wakatsuki, S., and Walters, K. J. (2009) Ubiquitin-binding domains – from structures to functions. *Nat. Rev. Mol. Cell Biol.* **10**, 659–671.
6. Rotin, D., and Kumar, S. (2009) Physiological functions of the HECT family of ubiquitin ligases. *Nat. Rev. Mol. Cell Biol.* **10**, 398–409.
7. Gupta, R., Kus, B., Fladd, C., et al. (2007) Ubiquitination screen using protein microarrays for comprehensive identification of Rsp5 substrates in yeast. *Mol. Syst. Biol.* **3**, 116.
8. McNatt, M. W., McKittrick, I., West, M., and Odorizzi, G. (2007) Direct binding to Rsp5 mediates ubiquitin independent sorting of Sna3 via the multivesicular body pathway. *Mol. Biol. Cell* **18**, 697–706.
9. Oestreich. A, J., Aboian, M., Lee, J., et al. (2007) Characterization of multiple multivesicular body sorting determinants within Sna3: a role for the ubiquitin ligase Rsp5. *Mol. Biol. Cell* **18**, 707–720.
10. Stawiecka-Mirota, M., Pokrzywa, W., Morvan, J., et al. (2007) Targeting of Sna3p to the endosomal pathway depends on its interaction with Rsp5p and multivesicular body sorting on its ubiquitylation. *Traffic* **8**, 1280–1296.
11. Watson, H., and Bonifacino, J. S. (2007) Direct binding to Rsp5p regulates ubiquitination-independent vacuolar transport of Sna3p. *Mol. Biol. Cell* **18**, 1781–1789.
12. Leon, S., Erpapazoglou, Z., and Haguenauer-Tsapis, R. (2008) Ear1p and Ssh4p are new adaptors of the ubiquitin ligase Rsp5p for cargo ubiquitylation and sorting at multivesicular bodies. *Mol. Biol. Cell* **19**, 2379–2388.
13. Lin, C. H., MacGurn, J. A., Chu, T., Stefan, C. J., and Emr, S. D. (2008) Arrestin-related ubiquitin-ligase adaptors regulate endocytosis and protein turnover at the cell surface. *Cell* **135**, 714–725.
14. Persaud, A., Alberts, P., Amsen, E. M., et al. (2009) Comparison of substrate specificity of the ubiquitin ligases Nedd4 and Nedd4-2 using proteome arrays. *Mol. Syst. Biol.* **5**, 333.
15. Merbl, Y., and Krischner, M. W. (2008) Large-scale detection of ubiquitination substrates using cell extracts and protein microarrays. *Proc. Natl. Acad. Sci. USA* **106**, 2543–2548.
16. Hesselberth, J. R., Miller, J. P., Golob, A., Stajich, J. E., Michaud, G. A., and Fields, S. (2006) Comparative analysis of *Saccharomyces cerevisiae* WW domains and their interacting proteins. *Genome Biol.* **7**, R30.
17. Zhu, H., Bilgin, M., Bangham, R., et al. (2001) Global analysis of protein activities using proteome chips. *Science* **293**, 2101–2105.

Chapter 14

Fit-for-Purpose Quenching and Extraction Protocols for Metabolic Profiling of Yeast Using Chromatography-Mass Spectrometry Platforms

Catherine L. Winder and Warwick B. Dunn

Abstract

Metabolomics involves the investigation of the intracellular (endometabolome) and extracellular (exometabolome) pools of metabolites in biological systems. Methods to sample the exometabolome and to quench metabolism and extract intracellular metabolites for the model eukaryote *Saccharomyces cerevisiae* are presented here. These methods have been developed and validated to provide a fit-for-purpose protocol for global analyses of the *S. cerevisiae* metabolome. The protocol allows the extraction of a wide variety of metabolite classes and provides reproducible results to allow relative and semi-quantitative comparisons between samples of different origin. For exometabolome studies, fast sampling and separation of cells by syringe filtration is recommended. For endometabolome studies, fast quenching of intracellular metabolism is performed using a 60:40 (v/v) methanol:aqueous ammonium hydrogen carbonate solution at –48°C. Extraction of intracellular metabolites is performed using multiple freeze/thaw cycles in a 60:40 (v/v) methanol:water solution at temperatures lower than 0°C.

Key words: Metabolomics, endometabolome, exometabolome, metabolite quenching, metabolite extraction, mass spectrometry, metabolic footprinting.

1. Introduction

Metabolomic studies involve the qualitative and quantitative investigation of low molecular weight chemicals present in biological systems (typically lower than 1,500 Da), at the intracellular and extracellular level (1–3). The aim is to analyze a representative picture of the whole pool of metabolites present in a biological system, the "metabolome," whose definition originates from the microbial research community (4, 5). Metabolomics is

J.I. Castrillo, S.G. Oliver (eds.), *Yeast Systems Biology*, Methods in Molecular Biology 759,
DOI 10.1007/978-1-61779-173-4_14,

playing an important role in microbial-focused research (6, 7), biotechnological and clinical studies, and systems biology (8, 9).

Two experimental strategies are applied in metabolomics, "targeted" and "non-targeted." Targeted investigation of a limited number of metabolites (typically less than 20) entails sample preparation and the use of analytical protocols for absolute quantification, with high specificity and accuracy. These investigations require the previous knowledge of the metabolites to be studied and the availability of chemical standards (10). Non-targeted metabolomic strategies (i.e., "metabolic profiling" studies), do not require previous knowledge of the metabolites of interest (11) and can provide a global (holistic) picture of metabolic patterns. These methods can be used for hypothesis-generation strategies (12) to investigate, for example, the effect of a gene deletion or an environmental perturbation on the metabolome, or relative changes in metabolite concentrations, without the need for absolute quantification. In this case, sample preparation and analytical protocols are developed to provide a representative unbiased profile of all metabolites in the biological system, with suitable accuracy and precision. Since different metabolites exhibit different physicochemical properties there is no single method that can be applied for the analysis for all metabolites (2, 13). The selected "fit-for-purpose" method should be able to extract a high number of compounds, prevent loss of metabolites and should be non-destructive (13).

When developing non-targeted methods a validation comparing the technical variability (from multiple analyses of biologically identical samples) against the biological variability (single analysis of biologically discrete samples) can be performed, to confirm low technical variability (this should be lower than between biological replicates). In our validation protocol we used a continuous culture system, with cellular metabolism in steady state (substrate, product concentrations and metabolic fluxes are constant at a constant growth rate). A batch culture in exponential phase, with a higher number of replicates can also be used to validate the methods (14).

The sampling, quenching, and extraction procedures to be applied are defined by the strategy and biological system to be studied. In microbial systems, two metabolic profiles can be investigated (1, 15): the intracellular (endometabolome) and the extracellular (exometabolome) pools, this is also defined as the "metabolic footprint" in cultures where metabolite-rich medium (medium with a high number of metabolites) is used (16). A fit-for-purpose protocol to enable the sampling and untargeted metabolic profiling of both metabolomes in *Saccharomyces cerevisiae* will be described in this chapter. This is summarized in **Fig. 14.1**.

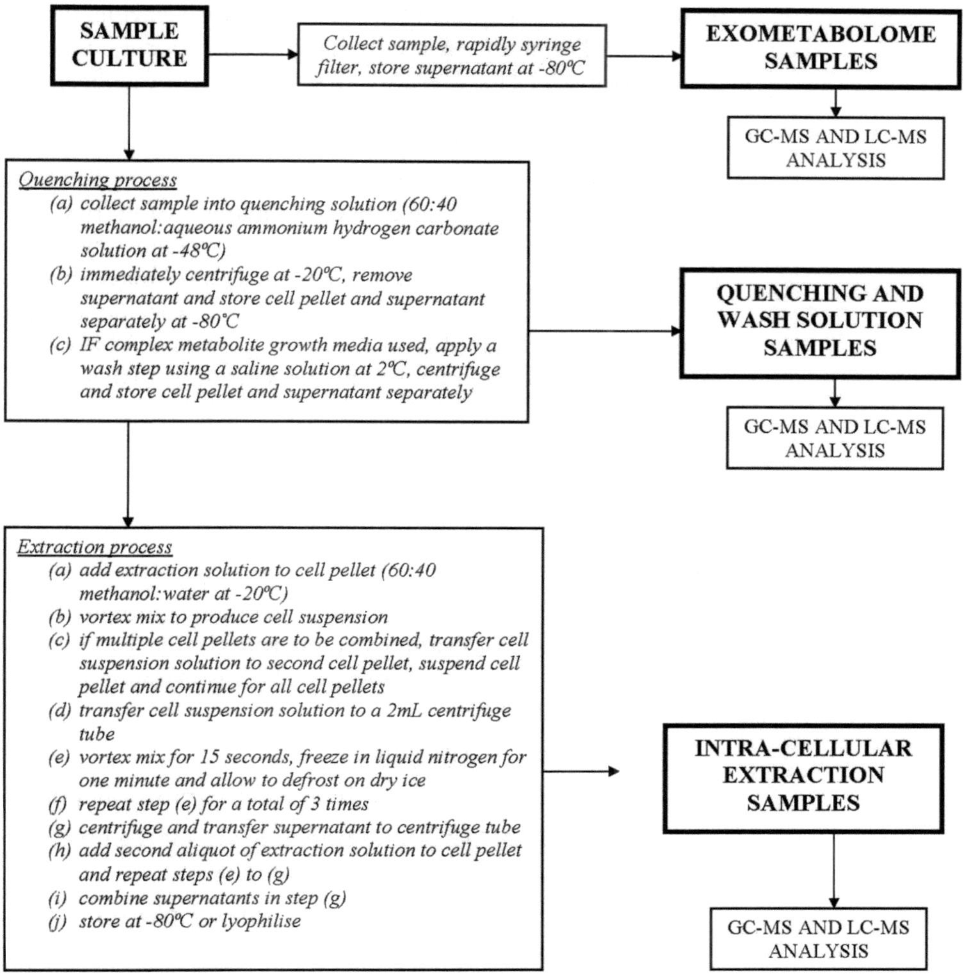

Fig. 14.1. Workflow describing the collection of exometabolome samples and the quenching and extraction method for the analysis of intracellular metabolites (endometabolome).

The collection of samples for exometabolome studies requires the separation of the biomass from the growth medium, which can be performed by filtration (17, 18) or centrifugation (13). Endometabolome sampling is technically more demanding due to the high rates of conversion of metabolites (high rates of intracellular reactions; e.g., the rate of ethanol synthesis from acetaldehyde can be as high as 2 mmol/L/s (19)). Therefore, a two-step process is applied where step 1 involves the fast quenching of enzymatic and chemical reactions, followed by permeabilization or lysis of cells and the release of intracellular metabolites into a suitable solvent as the second step. Quenching can be performed by a decrease or increase in temperature. Permeabilization or cell lysis can be performed with a number of mechanical or non-mechanical approaches including temperature increase (20) or rupturing of cell walls by freezing or sonication (21). Different protocols have been reported for each of these processes (21). Here we describe the protocols applied to acquire a sample representative of the whole metabolome where multiple metabolite classes are detected, with methods which are reproducible and robust across a wide range of concentrations. These methods have been validated using two analytical platforms: gas chromatography-mass spectrometry (GC-MS) and ultra-performance liquid chromatography-mass spectrometry (UPLC-MS). More targeted protocols specific for certain metabolites or metabolite classes (for example, ATP extraction using perchloric acid) may be applied separately (10, 13).

2. Materials

2.1. Collection of Exometabolome Samples

1. Sample bottles for bioreactors, accurately graduated in 5 mL increments. For example, screw top glass vials, universal fitting, depending on sample port fitting (Moulded Pathology Media Vials, LSL).
2. 5 and 10 mL sterile syringes.
3. Syringe filters 0.22 μm (Millex Filter Units, Millipore).
4. Pipette capable of handling 5 mL sample volume, for fast sampling and manipulation of samples.
5. 15 mL centrifuge tubes (good quality plastics, previously tested to ensure no chemical leakage into solutions) (e.g., polypropylene tubes, Sterilin).
6. 2 mL centrifuge tubes for sample storage (good quality plastics previously tested to ensure no chemical leakage into solutions; e.g., see above).
7. Liquid nitrogen, appropriate gloves, and material for sample manipulation (e.g., metal ladle).
8. For samples to be analyzed by GC-MS, succinic d_4 acid and d_{15} octanoic acid (Sigma-Aldrich) solutions can be added as internal standards (final concentration, 0.17 mg/mL in water). For samples to be analyzed by UPLC-MS no internal standards are added.

2.2. Endometabolome Studies: Sampling and Quenching of Intracellular Metabolism

1. Quenching solution: 60:40% (v/v) solution of methanol/water containing ammonium hydrogen carbonate at a final concentration of 0.85% (w/v), adjusted to pH 5.5. Use analytical or higher grade solvents and chemicals. The quenching solution should be cooled to –48°C (for example, in a cryostat, in a dry ice/ethanol bath with controlled addition

of dry ice or in a –80°C freezer for a minimum of 2 h). The temperature can be measured using a digital thermometer.

2. Sample bottles for bioreactor, graduated in 5 mL increments. For example, screw top glass vials, universal fitting, depending on sample port fitting, minimum volume of 50 mL (e.g., Moulded Pathology Media Vials, LSL, UK).
3. Digital thermometer (to operate at temperature between –45 and +150°C).
4. Dry ice, frozen carbon dioxide.
5. 50 mL centrifuge tubes (good quality plastics, previously tested to ensure no chemical leakage into solutions) (e.g., polypropylene tubes, Sterilin).
6. Pipette capable of handling 10 mL sample volumes, for fast sampling and manipulation of samples.
7. Centrifuge, able to cool to below –20°C and provide a minimum of 3,000×*g*.
8. Saline solution: 9 g/L aqueous solution of sodium chloride at 4°C.

2.3. Endometabolome Studies. Extraction of Intracellular Metabolites

1. Extraction solution. 60:40% (v/v) solution of methanol: water using analytical or higher grade solvents. The extraction solution should be cooled to a temperature lower than –20°C for a minimum of 2 h in a –80°C freezer or on dry ice for a minimum of 2 h.
2. Centrifuge tubes, 2 mL (good quality plastics, previously tested to ensure no chemical leakage into solutions).
3. Centrifuge, able to be cooled to –20°C and operate at 13,000×*g*.
4. Liquid nitrogen, gloves, and material for sample manipulation (e.g., metal ladle).
5. Dry ice (frozen carbon dioxide).
6. Vortex mixer.
7. For samples to be analyzed by GC-MS, succinic d_4 acid and d_{15} octanoic acid (Sigma-Aldrich) solutions can be added as internal standards (final concentration, 0.17 mg/mL in water). For samples to be analyzed by UPLC-MS no internal standards are added.

2.4. GC-MS and UPLC-MS Analysis

1. Solvents, HPLC grade methanol, water, pyridine, and hexane. Pyridine and hexane should be anhydrous.
2. *O*-Methylhydroxylamine (Sigma-Aldrich).
3. *N*-Acetyl-*N*-(trimethylsilyl)-trifluoroacetamide, MSTFA (Acros Chemicals).

4. Appropriate chromatography vials and inserts.
5. Heater and heating blocks for 2 mL centrifuge tubes, with efficient heat transfer.
6. Retention index solution: 0.6 mg/mL $C_{10}/C_{12}/C_{15}/C_{19}/C_{22}/C_{25}$ *n*-alkanes in pyridine/hexane 50:50% (v/v) (Sigma-Aldrich).
7. Vacuum-based lyophilization system (e.g., Lyopro 3000 Freeze dryer, ThermoFisher Scientific).
8. GC-MS instrument (e.g., the authors use an Agilent 6890 N gas chromatograph coupled to a Leco Pegasus III electron impact time-of-flight mass spectrometer).
9. GC column (e.g., SPB50 column, Supelco).
10. UPLC-MS instrument (e.g., Waters Acquity UPLC system coupled to a ThermoFisher hybrid electrospray LTQ-Orbitrap XL mass spectrometer).
11. UPLC column (e.g., Acquity BEH C_{18} chromatography column; 1.7 μm, 2.1 × 100 mm).
12. Centrifuge tubes, 2 mL (good quality plastics, previously tested to ensure no chemical leakage into solutions) (e.g., Eppendorf brand tubes).
13. Vortex mixer.

3. Methods

3.1. Collection of Exometabolome Samples

3.1.1. Sample Withdrawal from a Bioreactor

1. Connect sample bottle to the bioreactor.
2. Transfer 5 mL of culture into the empty sample container. This can be performed using the relatively high pressure in the reactor to expel the sample or with a syringe connected to the sampling system to pull sample into the container.
3. Immediately transfer the culture sample to a 5 mL sterile syringe and filter to separate the exometabolome (filtrate) from the cell biomass using a 0.22 μm syringe filter.
4. Snap-freeze the exometabolome samples in liquid nitrogen.
5. Store the sample at –20 or –80°C for further analysis (*see* **Note 1**).

3.1.2. Sample Withdrawal from a Flask

1. Fill a sterile 5 mL syringe with 5 mL of culture.
2. Immediately filter the sample through a 0.22 μm syringe filter to separate the exometabolome (filtrate) from the cell biomass.

3. Snap-freeze the exometabolome samples in liquid nitrogen.
4. Store the sample at –20 or –80°C for further analysis (*see* **Note 1**).

3.2. Endometabolome Studies. Sampling and Quenching of Intracellular Metabolism

For quenching and extraction of intracellular metabolites an amount of 30 mg of dry weight biomass is recommended. With a 10 mL sampling volume this corresponds to a *S. cerevisiae* culture with a biomass concentration of 3 mg/mL (dry weight). For cultures with lower biomass concentrations, strategies to combine cell pellets for optimum quenching and extraction are presented below.

The use of organic solvents (methanol) and low temperatures (–48°C) may compromise the integrity of the cell membrane resulting in leakage of intracellular metabolites during the quenching (13, 14, 22). A quantitative method can be used to determine the extent of the leakage in the quenching step. It has been shown that the level of leakage is metabolite specific and some metabolites may be differentially affected (23). It is recommended that the extent of the leakage is assessed by analysis of the quenching and washing solutions together with an exometabolome sample using identical analytical methods. Ensure that the residence times of cells in the quenching solution are identical as it has been shown that longer residence times may provide greater leakage (14). The aim is to assess, qualitatively and quantitatively, which metabolites were present in the medium before quenching and which appear from leakage.

After quenching, the cells need to be rapidly separated from the external environment. This can be performed by centrifugation or rapid filtration. In our experience, for the recommended amount of biomass, a centrifugation step is the most appropriate method.

A range of quenching methods have been reported (21). These can be separated in two classes: (a) those applying water:organic solvent solutions at low (e.g., –48°C) (13, 14, 22) or high (90°C) (21) temperatures and (b) those which apply an aqueous solution containing either glycerol, a saline solution (24) or buffer (25) at 4°C.

When high temperatures are used to quench metabolism, this is generally followed by extraction at high temperatures, in a combined quenching and extraction protocol (21). This combined method has some limitations when a metabolite-rich medium is used. When many metabolites are already present in the external medium these metabolites cannot be assayed in the intracellular metabolome and should be removed from any subsequent data analysis. Alternatively, a defined medium with a minimum set of known metabolites can be used.

When low temperatures are used to effectively quench metabolism, an organic solution (e.g., methanol/water) to keep the temperature below –20°C is necessary (20, 26).

3.2.1. Quenching Metabolism in Samples Taken from a Bioreactor

1. Add 40 mL of quenching solution at –48°C to the sample bottle, connect it to the bioreactor.
2. Fast sampling. Quickly transfer 10 mL of culture into the quenching solution (*see* **Note 2**). If a vacuum system coupled to the sampling port is not available, fast sampling can be achieved using the pressure in the reactor to expel the sample or a syringe for fast sampling into the quenching solution (*see* **Notes 3** and **4**).
3. Transfer the solution rapidly by pouring into a 50 mL centrifuge tube.
4. Centrifuge for 5 min at a recommended temperature of –20°C and at a maximum of 4,000×*g* (*see* **Notes 5**, **6**, and 7).
5. Remove all supernatant (quenching solution) and store supernatant and cell pellet at –80°C for subsequent analysis.
6. If a "metabolite-rich" medium is used, after removal of the supernatant in step 5 perform steps 7–9 (*see* **Note 8**).
7. Add 10 mL of saline solution (4°C) to the cell pellet, vortex mix for 15 s to resuspend the biomass.
8. Centrifuge for 5 min at 4°C and at 4,000×*g*.
9. Remove all supernatant and store cell pellet and supernatant for subsequent analysis at –80°C.

3.2.2. Quenching Metabolism in Samples from Flasks

1. Add 40 mL of quenching solution at –48°C to a 50 mL centrifuge.
2. Using a pipette rapidly transfer 10 mL of culture to the quenching solution (*see* **Note 2**). Add the sample to the central part to ensure cell pellets do not freeze on container surfaces (*see* **Notes 3** and **4**).
3. Centrifuge for 5 min at –20°C and at a maximum of 4,000×*g* (*see* **Notes 5**, **6** and 7).
4. Remove all supernatant (quenching solution) and store supernatant and cell pellet at –80°C for subsequent analysis.
5. If a "metabolite-rich" medium is used, after removal of the supernatant in step 5 perform steps 7–9 (*see* **Note 8**).
6. Add 10 mL of saline solution (4°C) to the cell pellet, vortex mix for 15 s to resuspend the cells.

7. Centrifuge for 5 min at 4°C and at 4,000×*g*.
8. Remove all supernatant and store cell pellet and supernatant for subsequent analysis at –80°C.

3.3. Endometabolome Studies: Extraction of Intracellular Metabolites

Once metabolism has been arrested the metabolites need to be extracted from the cells. At this point the endogenous enzymes should be permanently deactivated and permeabilization of the cells should be performed to release the intracellular metabolites into a suitable solvent with the highest possible recovery. To perform global metabolic profiling we recommend that the equivalent of 30 mg dry weight is used per profile (*see* **Note 9**). A minimum of three replicates is recommended for samples from steady-state continuous cultures. For batch cultures whose metabolic pattern can be less reproducible a minimum of six replicates is recommended:

1. Add 500 μL of extraction solution (*see* **Notes 10** and **11**) to the 50 mL centrifuge tube containing the frozen cell pellet (*see* **Note 2**).
2. Vortex mix for 15 s to resuspend the cell pellet.
3. Transfer cell suspension solution into a 2 mL centrifuge tube. Store on dry ice.
4. Add 250 μL of extraction solution to the 50 mL centrifuge tube, vortex mix for 15 s and transfer the cell suspension to the 2 mL Eppendorf tube. This results in a cell suspension of 750 μL.
5. Vortex mix the solution for 15 s.
6. Place the sealed Eppendorf tube in liquid nitrogen for 1 min (or until frozen) and then remove and store on dry ice.
7. Allow the frozen solution to thaw on dry ice and then vortex mix for 15 s.
8. Repeat steps 5–7 three times.
9. Centrifuge for 5 min at 13,000×*g* at –20°C.
10. Transfer the supernatant to a new 2 mL centrifuge tube and store on dry ice.
11. Add 500 μL of extraction solution to the cell pellet and repeat steps 5–9.
12. Pool both volumes of supernatant to produce a solution of ca. 1250 μL.
13. The combined extraction solution is suitable for analysis by GC-MS or UPLC-MS (*see* **Note 12**).

3.4. GC-MS and UPLC-MS Analysis

1. For GC-MS and UPLC-MS analysis, lyophilize the exometabolome, extraction, quenching and washing solutions

in 2 mL centrifuge tubes with a vacuum-based lyophilization system. Typically 50–400 μL of exometabolome sample, 500–1,250 μL of intracellular extraction and 1,000–3,000 μL of quenching and washing solutions are lyophilized (*see* **Notes 13**, **14**, **15**, **16**, and **17**).

2. For reversed-phase UPLC-MS analysis no internal standards are used. Reconstitute the lyophilized sample in water and transfer to a suitable high recovery LC vial and seal with a screw cap. A reconstitution volume of 100–200 μL is recommended. The samples can then be analyzed using a range of reversed-phase methods. The authors apply a method which entails analysis on a Waters Acquity UPLC system coupled to a ThermoFisher hybrid electrospray LTQ-Orbitrap XL mass spectrometer using an Acquity BEH C_{18} chromatography column (1.7 μm, 2.1 × 100 mm). Elution is performed over 22 min with a non-linear gradient from 100% A (water + 0.1% (w/v) formic acid) to 100% B (methanol + 0.1% (w/v) formic acid). An injection volume of 10 μL is applied and samples are analyzed in positive and negative ion modes (28). Other methods are perfectly valid. All samples should be stored in an autosampler at 4°C and analyzed within 48 h of reconstitution.

3. For GC-MS analysis, two internal standards can be used. The authors use succinic d_4 acid (0.17 mg/mL) in the extraction solution to compensate for physical variations during the extraction process and d_{15} octanoic acid (0.17 mg/mL) in the solution before lyophilization, to compensate for physical variations during derivatization and analysis. The use of two internal standards allows the researcher to differentiate the experimental variability from the two separate processes of sample extraction and sample derivatization/analysis. The next step is the chemical derivatization of the lyophilized sample, to reduce the boiling point of the metabolites (11). A range of protocols can be performed with different derivatization schemes. The authors apply a two-step process of (a) oximation and (b) trimethylsilylation. An aliquot (50 μL) of 20 mg/mL *O*-methylhydroxylamine solution in pyridine is added, vortex mixed, and heated at 40°C for 80 min followed by addition of 50 μL of MSTFA (*N*-acetyl-*N*-(trimethylsilyl)-trifluoroacetamide), vortex mixed and heated at 40°C for 80 min. A retention index solution is added for chromatographic alignment (20 μL, 0.6 mg/mL $C_{10}/C_{12}/C_{15}/C_{19}/C_{22}$ *n*-alkanes in pyridine/hexane 50:50% v/v). From here, the samples can be analyzed with a range of different methods available. The authors use an Agilent 6890 N gas chromatograph coupled to a Leco Pegasus III electron impact time-of-flight mass spectrometer (11). About 1 μL of derivatized sample is

injected onto an SPB50 (Supelco) column at an initial temperature of 70°C and with a split ratio of 1:4. A temperature ramp is operated between 70 and 290°C over a 20 min period with detection provided by electron impact time-of-flight mass spectrometry. Other methods are perfectly valid. All samples should be processed within 24 h of derivatization completion.

4. Notes

1. Unless lysis or high production of extracellular enzymes is expected, the concentration of enzymes in the exometabolome can be considered negligible in the majority of cases (i.e., no need for additional steps to inhibit enzymatic activities).
2. If the biomass is low (i.e., a maximum of 10 mL of sample does not result in a 30 mg of biomass) do not collect more than 10 mL of sample, the use of larger volumes may increase the temperature during the quenching step and may result in longer times to pellet the biomass during the centrifugation step. Instead collect multiple 10 mL sample volumes and combine cell pellets at the extraction stage. If multiple cell pellets are to be combined, suspend the cell pellet in the first tube and transfer the cell suspension solution to tube 2 (containing the second pellet), suspend the cell pellet in tube 2, and continue the procedure for all cell pellets.
3. When sampling from bioreactors do not remove more than 5% of the total working volume, to avoid disruption of the steady-state culture conditions during sampling.
4. The quenching temperature after adding the sample should be less than −20°C. An independent test to ascertain the volume of quenching solution required to maintain this temperature is recommended. Quenching solution volumes can be optimized, though a minimum ratio of 2:1 (v/v) (quenching solution/sample) is recommended.
5. Previous optimization of the centrifugation step is recommended to ensure that the biomass is harvested in the minimum amount of time. However do not exceed 5,000×*g* as higher regimes may result in cell wall permeabilization.
6. The extent of metabolic quenching can be assessed using the adenylates energy charge ratio (EC) (27). This is a quantitative measurement of the halting of metabolism by the comparison of the concentrations of ATP with those

of ADP and AMP. For yeast an EC of 0.8–0.9 is typical. Quenching methods which do not halt metabolism rapidly can be expected to have lower EC values.

7. Minimize the residence time of cells in organic solvents as this will minimize the leakage of metabolites from cells.
8. When a metabolite-rich medium is used (17) a washing step to remove external metabolites is required. Use saline washing solution at 4°C and analyze washing solutions to assess metabolite leakage.
9. An amount of biomass lower than 30 mg dry weight may compromise the detection of specific metabolites (low sensitivity).
10. The extraction solution should always remain at a temperature lower than 0°C during the extraction process.
11. Other solvent systems can be used for quenching and extraction of specific metabolites or metabolite classes. For example, perchloric acid for adenosine nucleotides (10). The complexity of metabolites present in the medium influence the sample composition, as the final sample may contain intracellular and external metabolites. Cultures in defined media with a limited number of metabolites are recommended.
12. Part of the extract can be analyzed by GC-MS and the rest by UPLC-MS. However, for high sensitivity purposes the extract should be analyzed by a single technique only.
13. The volume of solution to be lyophilized varies depending on the metabolite concentration and the medium composition. For exometabolome studies, large volumes can only be dried if the concentrations of sugars and salts are low (typically less than 1 mM). Higher sugar and salt concentrations result in incomplete lyophilization of large volumes, which is detrimental to the precision and robustness of the analytical steps.
14. Volumes to be lyophilized in endometabolome studies. Determination of biomass levels by optical density or dry weight in cultures allows normalization of raw analytical data to the amount of biomass extracted. Thus, for samples with different amounts of biomass the volumes to be lyophilized can be adjusted to account for the differences. For example, if the extracted biomass for samples A and B were 30 and 35 mg the volumes to lyophilize would be 1,000 and 857 μL, respectively.
15. Ensure samples are fully dried before analysis. The presence of water or methanol can influence the efficiency of the derivatization reactions for GC-MS analyses.

16. Some reported methods have been applied and tested in one analytical platform only, typically GC-MS (14). We have assessed and validated the described methods on both GC-MS and UPLC-MS platforms.
17. The periodic injection of quality control (QC) samples is recommended to assess instrumental reproducibility. For GC-MS and UPLC-MS applications, respectively, the authors inject five and ten QC samples at the start of the analytical batch and then five analytical samples followed by a new QC sample. The procedure is repeated until completion of the analysis. All metabolite peaks with a relative standard deviation (RSD) of less than 20% for GC-MS and 30% for UPLC-MS compared to all QC samples from injection 5 to the end of the analytical batch are filtered and retained for further data analysis whereas those metabolite peaks with an RSD greater than 20% are removed. Only biological samples for which repeated analyses result in reproducible results are retained. Further information on the use of quality control samples can be found elsewhere (28, 29).

Acknowledgments

CLW would like to thank the BBSRC and Unilever for financial assistance. CLW and WBD would like to thank the BBSRC and EPSRC for financial support of The Manchester Centre for Integrative Systems Biology (BBSRC/EPSRC Grant BB/C008219/1).

References

1. Dunn, W. B. (2008) Current trends and future requirements for the mass spectrometric investigation of microbial, mammalian and plant metabolomes. *Phys. Biol.* **5**, 11001.
2. Dunn, W. B., Bailey, N. J. C., and Johnson, H. E. (2005) Measuring the metabolome: current analytical technologies. *Analyst* **130**, 606–625.
3. Goodacre, R., Vaidyanathan, S., Dunn, W. B., Harrigan, G. G., and Kell, D. B. (2004) Metabolomics by numbers: acquiring and understanding global metabolite data. *Trends Biotechnol.* **22**, 245–252.
4. Oliver, S. G., Winson, M. K., Kell, D. B., and Baganz, F. (1998) Systematic functional analysis of the yeast genome. *Trends Biotechnol.* **16**, 373–378.
5. Tweeddale, H., Notley-McRobb, L., and Ferenci, T. (1998) Effect of slow growth on metabolism of Escherichia coli, as revealed by global metabolite pool ("Metabolome") analysis. *J. Bacteriol.* **180**, 5109–5116.
6. van der Werf, M. J., Overkamp, K. M., Muilwijk, B., et al. (2008) Comprehensive analysis of the metabolome of *Pseudomonas putida* S12 grown on different carbon sources. *Mol. Biosyst.* **4**, 315–327.
7. Oldiges, M., Lutz, S., Pflug, S., Schroer, K., Stein, N., and Wiendahl, C. (2007)

Metabolomics: current state and evolving methodologies and tools. *Appl. Microbiol. Biotechnol.* **76**, 495–511.
8. Kell, D. B. (2006) Metabolomics, modelling and machine learning in systems biology – towards an understanding of the languages of cells. *FEBS J.* **273**, 873–894.
9. Castrillo, J. I., Zeef, L. A., Hoyle, D. C., et al. (2007) Growth control of the eukaryote cell: a systems biology study in yeast. *J. Biol.* **6**, 4.
10. Klawitter, J., Schmitz, V., Klawitter, J., Leibfritz, D., and Christians, U. (2007) Development and validation of an assay for the quantification of 11 nucleotides using LC/LC-electro spray ionization-MS. *Anal. Biochem.* **365**, 230–239.
11. Pope, G. A., MacKenzie, D. A., Defernez, M., et al. (2007) Metabolic footprinting as a tool for discriminating between brewing yeasts. *Yeast* **24**, 667–679.
12. Kell, D. B., and Oliver, S. G. (2004) Here is the evidence, now what is the hypothesis? The complementary roles of inductive and hypothesis-driven science in the post-genomic era. *Bioessays* **26**, 99–105.
13. Winder, C. L., Dunn, W. B., Schuler, S., et al. (2008) Global metabolic profiling of *Escherichia coli* cultures: an evaluation of methods for quenching and extraction of intracellular metabolites. *Anal. Chem.* **80**, 2939–2948.
14. Villas-Boas, S. G., Hojer-Pedersen, J., Akesson, M., Smedsgaard, J., and Nielsen, J. (2005) Global metabolite analysis of yeast: evaluation of sample preparation methods. *Yeast* **22**, 1155–1169.
15. Mashego, M. R., Rumbold, K., De Mey, M., Vandamme, E., Soetaert, W., and Heijnen, J. J. (2007) Microbial metabolomics: past, present and future methodologies. *Biotechnol. Lett.* **29**, 1–16.
16. Kell, D. B., Brown, M., Davey, H. M., Dunn, W. B., Spasic, I., and Oliver, S. G. (2005) Metabolic footprinting and systems biology: the medium is the message. *Nat. Rev. Microbiol.* **3**, 557–565.
17. Allen, J., Davey, H. M., Broadhurst, D., et al. (2003) High-throughput classification of yeast mutants for functional genomics using metabolic footprinting. *Nat. Biotechnol.* **21**, 692–696.
18. Taymaz-Nikerel, H., de Mey, M., Ras, C., et al. (2009) Development and application of a differential method for reliable metabolome analysis in *Escherichia coli. Anal. Biochem.* **386**, 9–19.
19. Pritchard, L., and Kell, D. B. (2002) Schemes of flux control in a model of *Saccharomyces cerevisiae* glycolysis. *Eur. J. Biochem.* **269**, 3894–3904.
20. Castrillo, J. I., Hayes, A., Mohammed, S., Gaskell, S. J., and Oliver, S. G. (2003) An optimized protocol for metabolome analysis in yeast using direct infusion electrospray mass spectrometry. *Phytochemistry* **62**, 929–937.
21. Villas-Boas, S. G., Roessner, U., Hansen, M. A. E., Smedsgaard, J., and Nielsen, J. (2007) *Metabolome Analysis. An Introduction*. Hoboken, NJ: Wiley.
22. Wittmann, C., Kromer, J. O., Kiefer, P., Binz, T., and Heinzle, E. (2004) Impact of the cold shock phenomenon on quantification of intracellular metabolites in bacteria. *Anal. Biochem.* **327**, 135–139.
23. Canelas, A. B., Ras, C., ten Pierick, A., van Dam, J. C., Heijnen, J. J., and Van Gulik, W. M. (2008) Leakage-free rapid quenching technique for yeast metabolomics. *Metabolomics* **4**, 226–239.
24. Villas-Boas, S. G., and Bruheim, P. (2007) Cold glycerol-saline: the promising quenching solution for accurate intracellular metabolite analysis of microbial cells. *Anal. Biochem.* **370**, 87–97.
25. Faijes, M., Mars, A. E., and Smid, E. J. (2007) Comparison of quenching and extraction methodologies for metabolome analysis of *Lactobacillus plantarum. Microb. Cell Fact.* **6**, 27.
26. Lange, H. C., Eman, M., van Zuijlen, G., et al. (2001) Improved rapid sampling for in vivo kinetics of intracellular metabolites in *Saccharomyces cerevisiae. Biotechnol. Bioeng.* **75**, 406–415.
27. Link, H., Anselment, B., and Weuster-Botz, D. (2008) Leakage of adenylates during cold methanol/glycerol quenching of *Escherichia coli. Metabolomics* **4**, 240–247.
28. Zelena, E., Dunn, W. B., Broadhurst, D., et al. (2009) Development of a robust and repeatable UPLC-MS method for the long-term metabolomic study of human serum. *Anal. Chem.* **81**,1357–1364.
29. van der Greef, J., Martin, S., Juhasz, P., et al. (2007) The art and practice of systems biology in medicine: mapping patterns of relationships. *J. Proteome Res.* **6**, 1540–1559.

Chapter 15

The Automated Cell: Compound and Environment Screening System (ACCESS) for Chemogenomic Screening

Michael Proctor, Malene L. Urbanus, Eula L. Fung, Daniel F. Jaramillo, Ronald W. Davis, Corey Nislow, and Guri Giaever

Abstract

The automated cell, compound and environment screening system (ACCESS) was developed as an automated platform for chemogenomic research. In the yeast *Saccharomyces cerevisiae*, a number of genomic screens rely on the modulation of gene dose to determine the mode of action of bioactive compounds or the effects of environmental/compound perturbations. These and other phenotypic experiments have been shown to benefit from high-resolution growth curves and a highly automated controlled environment system that enables a wide range of multi-well assays that can be run over many days without any manual intervention. Furthermore, precise control of drug dosing, timing of drug exposure, and precise timing of cell harvesting at specific generation times are important for optimal results. Some of these benefits include the ability to derive fine distinctions between growth rates of mutant strains (1) and the discovery of novel compounds and drug targets (2). The automation has also enabled large-scale screening projects with over 100,000 unique compounds screened to date including a thousand genome-wide screens (3). The ACCESS system also has a diverse set of software tools to enable users to set up, run, annotate, and evaluate complex screens with minimal training.

Key words: Robotics, phenotypic screening, genome-wide screening, yeast, bacteria, drug discovery.

1. Introduction

This chapter describes an automated cell, compound and environment screening system (ACCESS) that was developed as a platform for chemogenomic research. In *Saccharomyces cerevisiae*

M. Proctor and M.L. Urbanus contributed equally to this work.
Supporting web site: http://chemogenomics.med.utoronto.ca/supplemental/ACCESS/

J.I. Castrillo, S.G. Oliver (eds.), *Yeast Systems Biology*, Methods in Molecular Biology 759,
DOI 10.1007/978-1-61779-173-4_15, © Springer Science+Business Media, LLC 2011

a number of genomic screens rely on the modulation of gene dose to determine the mode of action of bioactive compounds or the effects of environmental/compound perturbations. These genomic screens operate by either reducing the gene dose of a drug target (and thereby increasing drug sensitivity) or increasing gene dose and conferring resistance to a compound. For example, the haploinsufficiency profiling (HIP) (4) assay uses a pool of heterozygous yeast knockout (YKO) mutants (5, 6). Each individual strain is barcoded with two unique 20 base-pair sequences to determine the gene dose reduction that leads to a growth defect. Briefly, the strains are pooled at approximately equal volume and grown competitively in a single culture. At an optimum time (empirically determined), cells are collected, genomic DNA is isolated, and the barcodes, following amplification using common primers incorporated into every barcode, are hybridized to a microarray carrying the barcode complements. The multicopy suppression profiling (MSP) assay is performed in a converse assay, based on a gene dose increase using a pool of overexpression strains carrying a random genomic DNA library on a 2 μm plasmid under the endogenous promoter. Such a collection of overexpression clones can be treated with a high dose of drug and screened for surviving strains (2). The ensuing growth defect or relative resistance is assayed using a commercial microarray (TAG4, Affymetrix Item No. 511331), which has been designed to carry all molecular barcodes of the YKO collection as well as two probes for each predicted open reading frame (ORF) (7). ACCESS was designed to streamline and automate such assays and thereby increase their reproducibility.

Although these experiments can be performed by growing the mutant pools in flasks as described in Pierce et al. (8) this approach is labor intensive and requires large (and often prohibitive) volumes of drug. Furthermore, precise control of drug dosing, timing of drug exposure, and precise timing of cell harvesting at specific generation times are important for optimal results. To address these needs for reproducibility and precision, we developed the automated cell, compound and environment screening system (ACCESS). ACCESS miniaturizes the culture volume, increases the throughput, and automates the assays to minimize user-induced variation. The ACCESS has been instrumental in several large-scale genome-wide screening efforts (1–3, 8–13). The genome-wide screens using the pooled YKO collection have been miniaturized to 48-well plates, using 700 μl of medium (plus compound) per well for HIP and MSP screens (2, 4). Currently, an ACCESS equipped with a commercial liquid handler can process 48–72 HIP assays and 192–288 MSP assays per week taking into account typical assay run times and the microplate capacity of ACCESS. All the ACCESS versions can run a minimum of 4,032 single-strain growth assays per week.

In this chapter and its accompanying web site, we describe the ACCESS platform, different versions of ACCESS robot installations, and the variety of applications they can serve. The ACCESS platform is an extensive system that includes robot control of instruments, user interfaces, and laboratory information management system (LIMS)-type environment for sample tracking and data analysis. On the supplementary web site we provide a step-by-step user manual and suggestions for assembling and operating an ACCESS in your own laboratory.

2. Materials

2.1. Microplates AND Plate Seals

1. Microplates format (*see* **Note 1**).
2. Microplates from Greiner Bio One, Monroe, NC, USA; CELLSTAR 48-well microplates, Cat. No. 667 102; and CELLSTAR 96-well microplates, Cat. No. 655 180.
3. Microplates from Nunc/Thermo Fisher Scientific, Rochester, NY, USA. 96-well microplates, Cat. No. 269787 and 384-well microplates, Cat. No. 242757.
4. Microplates from Falcon/BD Biosciences, Franklin Lakes, NJ, USA. 384-well microplates, Cat. No. 353233.
5. Plate seals, Cat. No. AB-0580, ABgene/Thermo Fisher Scientific, Rockford, IL, USA (*see* **Note 2**).

2.2. Manual Liquid Handling ACCESS Robot Equipment

1. Computer with multiple serial ports (*see* **Note 3**) and Internet access (*see* **Note 4**).
2. File server.
3. Web server configured to run Perl scripts.
4. MySQL database server, co-hosted with web server or remotely hosted.
5. National Instruments LabVIEW version 8.5 or later, National Instruments, Austin, TX, USA.
6. ACCESS software. The software, supporting manual and additional documents can be found on the supplemental web site http://chemogenomics.med.utoronto.ca/supplemental/ACCESS/.
7. RdrOLE reader server (Tecan Group Ltd., Männedorf, Switzerland) version 4.62, install with XFluor4.
8. Any of the following Tecan readers that communicates via RdrOLE (*see* **Note 5**). The Sunrise absorbance reader with temperature control module, firmware version 3.31. The GENios reader, firmware version 4.5 or 6.0.2 or later. This

instrument is no longer produced and has been replaced by the Infinite series (*see* **Note 6**). The Safire[2] reader, Firmware v1.90, software version XSafire2 2.4.

9. Fan and portable air conditioner (*see* **Note 7**).

2.3. ACCESS Version 1 Liquid Handling Equipment

1. Computer hardware and operating system recommended by PerkinElmer (*see* **Note 8**) expanded with multiple serial ports and PCI bus for Packard robot communication card.

2–7. As in **Section 2.2**.

8. GENios reader (Tecan) with plate tray clip modification (*see* **Note 9**).
9. Packard Multiprobe IIEx (Perkin Elmer Life and Analytical Sciences, Waltham, MA, USA) – without physical extended deck. The extended deck is configured in software to enable access to the plates in the readers. This platform has been replaced by the JANUS automated platform, which should perform comparably (*see* **Note 10**). Disposable tips, supplies, and accessories are still available.
10. 200 μl disposable conductive tips Cat. No. 6000683.
11. Winprep version 1.22.0.252 or later and Winprep scripts for the Packard robot (*see* supporting web site).
12. Positioning hardware to fix the readers in the correct position; machinist drawings are provided in the online manual.
13. Plate tray support stands – provides support for plate tray during pipetting; machinist drawings are provided in the online manual.
14. 4°C save plate coolers; machinist drawings are provided in the online manual.

2.4. ACCESS Version 2 Liquid Handling Equipment

1. Computer – specified and supplied by Tecan.

2–7. As in **Section 2.2**.

8. Freedom EVO 150 (or 200) robot (Tecan) with Freedom EVOware version 2.2 SP1 (Tecan) and EVOware scripts written for ACCESS plate movement and liquid handling steps (*see* supporting web site). The Freedom EVO deck contains the following Tecan hardware: two cooling plate decks (three plate positions each) for saved samples, a wash stand for pipet tips, and a 3-position plate deck for transient plate rest – i.e., for save/inoculation execution.
9. Thermoshakers (Inheco, Martinsried/Munich, Germany).
10. Reader with a robot friendly tray (Safire[2] recommended (Tecan)).

11. Circulating water bath that can cool to –2°C.
12. Incubator enclosure for the Thermoshakers with a heating element and fan above the plate (*see* **Note 11**).

3. Methods

3.1. ACCESS Platform Overview

The ACCESS platform consists of three major components assembled from a mix of custom-designed and commercially available hardware and software. These components consist of (1) a hardware robot component, (2) a robot control and user interface software component, and (3) a data management/database server and software component. The custom design and implementation of all components was performed in-house. The layout of each component and communication between them is shown in **Fig. 15.1**.

The details of the different ACCESS robot versions are described in **Sections 3.2**, **3.3**, and **3.4**. Briefly, each ACCESS robot comprises a computer and multiple microplate shakers and a microplate reader for optical measurements. In most versions of ACCESS the microplate reader also acts as the shaker and temperature-controlled incubator. For screens requiring sample manipulation, the ACCESS robot is integrated with a liquid handling robot (**Figs. 15.1a** and **15.2**). In **Fig. 15.2** the two different versions of the liquid handling ACCESS are shown. One version is based upon a Packard robot (**Fig. 15.2a**) and the other on a Tecan Freedom EVO robot (**Fig. 15.2b**), described in detail later.

Advanced protocol execution is possible using comprehensive, in-house developed user interfaces and robot control software written in LabVIEW (National Instruments). The ACCESS software is the master application that continuously monitors and controls up to six microplate readers and, optionally, a liquid handling robot. In addition, the ACCESS software also has powerful experiment configuration, data visualization, and analysis capabilities (**Sections 3.6**, **3.7**, **3.8**, **3.9**, **3.10**, **3.11**, **3.12**, and **3.13**).

The software structure has a modular design, i.e., each robotic part and user interaction service of the ACCESS program work independently and in parallel (**Fig. 15.1b**). Each microplate reader or shaker is controlled by its own copy of the reader software module and runs independent from other readers or shakers. The reader software module can autonomously request services from the liquid handling, the database, or data analysis modules. The ACCESS software communicates with hardware either directly or through the manufacturer's ActiveX software driver libraries such as RdrOLE.ddl and Evoware.exe for

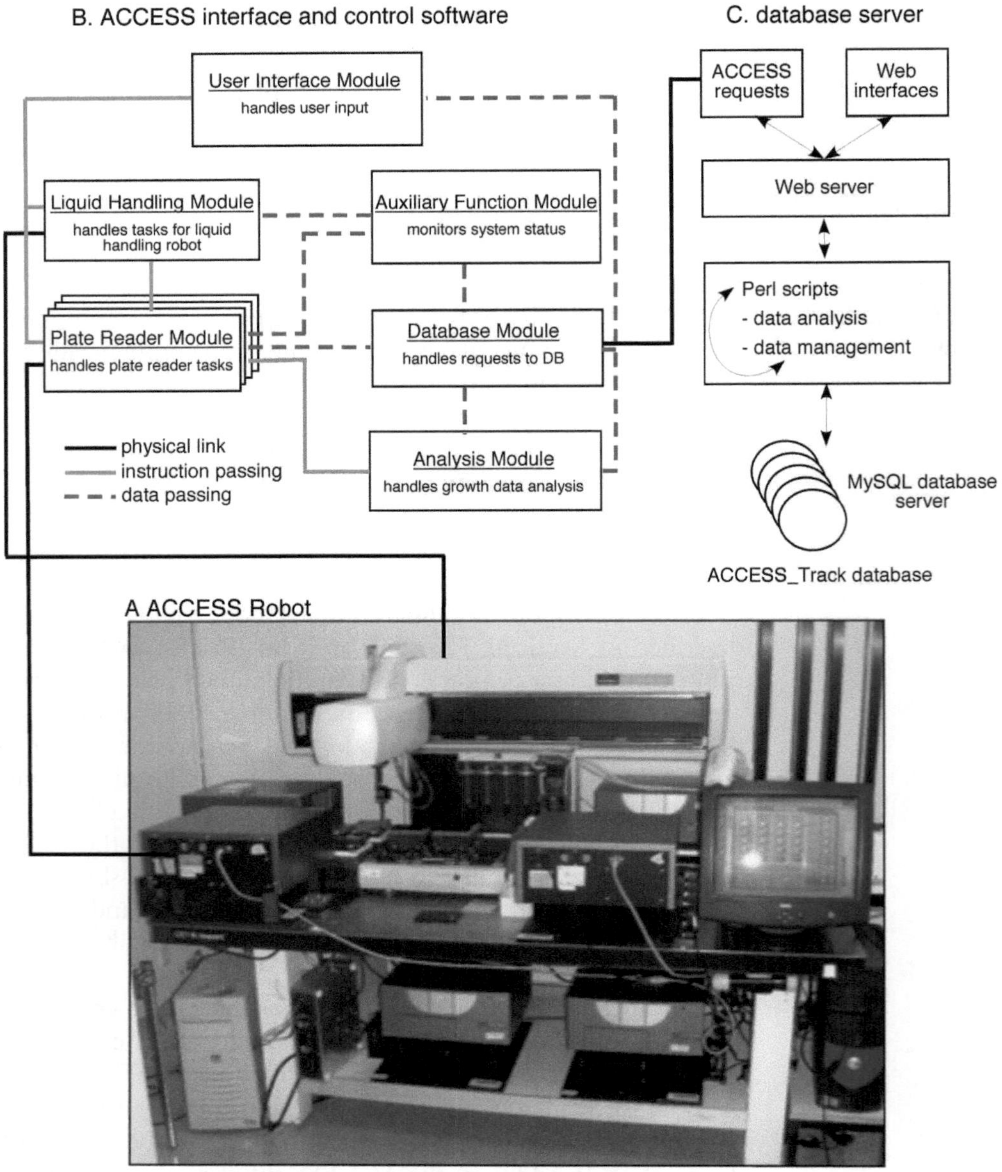

Fig. 15.1. ACCESS platform overview. This figure shows the major components of the ACCESS robot. (**a**) shows a picture of the version 1 liquid handling ACCESS composed of a Packard Multiprobe IIEx robot with four GENios readers positioned for direct access by the Packard robot and two GENios readers below deck (*see* **Fig. 15.2** and **Section 3.2** for a more detailed description). (**b**) shows the modular structure of the ACCESS command and control software that coordinates the physical robot modules. It also hosts the user interface windows that allow the users to configure, view, analyze, and annotate the experiments (*see* **Section 3** for more details on ACCESS software). The ACCESS modules run independently and coordinate and communicate between themselves using message queues that send command requests either autonomously or upon request from the users. Each robot part has its own physical link and copy of its control module to enable it to run independently of the other parts. The Database module communicates with the data management and database server (**c**) and handles data uploads, status updates, and many of the ACCESS configuration parameters that enable remote administration. (**c**) shows the database and data management server that uses a web server to handle the requests from the ACCESS robot or web interfaces that can also be used to interface with the robot and the experiment configuration, annotations, or data (*see* **Sections 3.6**, **3.7**, **3.8**, **3.9**, **3.10**, **3.11**, **3.12**, and **3.13**). The web server executes Perl scripts that manipulate the data before upload to the database or after retrieval from the ACCESS track database before passing the data back to the ACCESS robot or web interface.

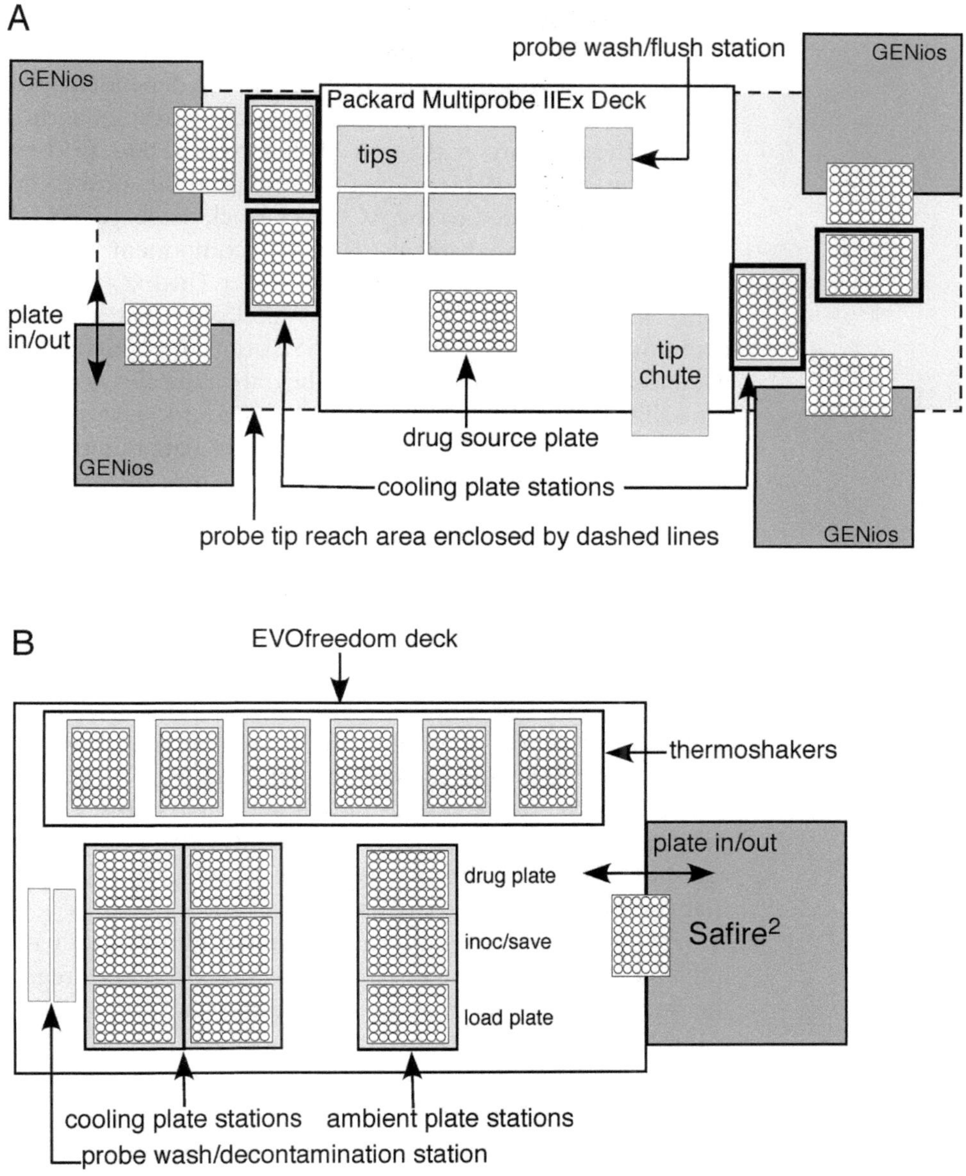

Fig. 15.2. ACCESS robot layout. This figure shows a detailed schematic layout of the version 1 Packard based (**a**) and version 2 Freedom EVO based (**b**) ACCESS robots. In (**a**) the four GENios readers are placed beside the Packard deck and the microplates are accessible to the Packard probes when the reader plate tray is in the open position. The probe reach area is indicated by the *gray dashed lines*. A peltier-cooled plate station for saved cell culture storage is placed in close proximity to each reader. The Packard uses disposable tips that pierce a transparent plate seal when there is a need to aspirate or dispense to a well. The positions of tip boxes, drug source plate, probe wash/flush station, and tip chute position on the physical Packard robot deck are indicated. (**b**) shows the layout of the ACCESS version 2 based on the Tecan Freedom EVO robot in combination with the Safire^2 reader. The Freedom EVO is configured with eight fixed-tip liquid handling probes and a plate gripper arm. Tip probes are washed and decontaminated in the probe wash and decontamination station. The EVO robot deck contains two cooled plate stations (three plate positions each) that are used to store saved cell culture samples. An ambient plate deck provides stations for a drug source plate, inoculation and save actions, and plate loading. Arrayed along the back of the deck (out of range for the liquid handling probes) are the Thermoshakers plate shakers/incubators. The Safire^2 is placed to the side of the EVO deck. The reader plate tray is accessible to the robot plate gripper arm in the open position.

Tecan instruments and MPIIX.exe for the Packard. Data collection integrity is protected against mechanical or software failures. During an experiment, each reader receives independent instructions for each shake–read cycle and reports the results of the read back to the software at the end of that cycle. These read results are written immediately to a text file stored on the local computer and are also uploaded to the ACCESS Track database (ATDB).

The data management/database component is custom designed using a MySQL database server (http://www.mysql.com) and Perl scripts (http://www.perl.org) that are used to communicate between the ACCESS robot and ACCESS Track database (ATDB) (**Fig. 15.1c**). They are typically hosted on a centralized server. The ATDB is designed to service multiple installations of ACCESS robots. The robots' data requests sent to the ATDB are handled by a web server, such as Apache (Apache Foundation, http://www.apache.org). The web server executes Perl scripts to perform the desired data management or database request tasks. The advantage of this design is that it uses public available software that is flexible to configure and scales easily with increasing system use and not tied to any computer system architecture. A further advantage of the web server/Perl script design is that the same Perl scripts can also serve web interface requests. This allows for the development of web interfaces that users can remotely view and manipulate data and to set up, monitor, and modify the robot or experiment and their statuses. We have designed other web interfaces (**Sections 3.12** and **3.13**) to manage and analyze multi-plate experiments or large-scale projects that run over many experiments.

A comprehensive set of system monitoring measures has been implemented, including e-mail notifications to users regarding the status of their experiments as well as the status of the ACCESS robot. All ACCESS operations and reader errors reported by the RdrOle software (Tecan) are written to a log file. Notifications of errors, run startup, or run completion are relayed to the user and administrator by e-mail. At each read cycle, the status of the reader (with timestamp) is written to the ATDB. Every 30 min, each ACCESS installation checks that all attached readers and liquid handling robots have a valid timestamp. A server script, that runs remotely, also monitors the status and timestamps stored in the ATDB for each ACCESS installation. If any status or timestamp shows that either the ACCESS or any of its components has an error, the error is reported both to the user and to the system administrator by e-mail. In addition, the quality of the growth curves is also monitored. This quality control feature is particularly useful when testing new compounds, conditions, or strains, whose behavior cannot be predicted beforehand. When the difference between any two reads is higher than the predetermined threshold, the user will receive an e-mail reporting a data jump.

Data jumps can be caused by a number of biological and technical reasons, e.g., flocculating cells, air bubbles on the surface of the medium, condensation on the film, a precipitate in the medium, or if the growth plate is no longer secured correctly in the reader's plate tray.

3.2. ACCESS Robot Version: Manual Liquid Handling

The simplest version of the ACCESS robot is composed of a computer that controls up to six microplate readers that shake and incubate the experiment plates. This version of ACCESS is used for experiments that do not require pipetting, such as prescreening compounds, phenotypic analysis of mutants, or fluorescence experiments. However, an experiment can be configured to e-mail the user when the culture in any well of a plate has reached its predefined target OD or other inoculation trigger. At this point, the user can pause the experiment, take out the microplate, inoculate the next well, add drug or save the samples as instructed in the e-mail, and then continue the experiment (*see* supplementary manual for instructions). This basic version of the software permits one to prototype different assays and is also a useful way for users to familiarize themselves with the different capabilities of the software prior to integrating one or more liquid handling robots.

3.3. Liquid Handling ACCESS Version 1

The full ACCESS installation includes a liquid handling robot. These robots are used (i) to inoculate from one well to another well to keep cultures in logarithmic growth for many generations, (ii) to save cultures on a cooled destination plate when a predefined target OD is reached (or after a predetermined time interval), or (iii) to add· drug at the beginning of an experiment or when a new well is inoculated.

The original liquid handling version of ACCESS comprised one computer controlling up to six microplate readers and a Packard Multiprobe IIEx liquid handling robot (PerkinElmer) (**Fig. 15.2a**). Four of the microplate readers are positioned such that the plate trays in the open position overlap with the Packard deck and are directly accessible for liquid handling.

The Packard robot is controlled by Winprep software. The Winprep environment contains the deck layout of experiment plates, e.g., cooled sample saving plates, drug plates, and tip boxes. The Winprep scripts are written to perform the liquid handling operations such as saving and inoculating samples and adding drug. The details, e.g., which wells to pipet, well source position, volume, and destination well, are all read from pipetting map files written by the ACCESS software just before the Winprep script is activated. When a liquid handling experiment reaches the predetermined target OD (or other parameter), ACCESS opens the microplate reader plate tray, activates the Winprep script, and provides instructions to the robot with the well positions and the volumes to add drug, inoculate next well, or save the sample. After the Packard robot performs the liquid handling action, ACCESS checks for errors reported by the Packard robot, the plate tray is closed, and the experiment continues on to the next shake–read cycle.

3.4. Liquid Handling ACCESS Version 2

The ACCESS robot versions described in the earlier sections use the reader as a shaking incubator. This is particularly hard on the shaking mechanism of the microplate readers and limits the scalability of the robot due to size constraints of the robot deck. The latest version of the liquid handling ACCESS uses temperature-controlled plate shakers (Thermoshakers, Inheco) that are more compact and intended for continuous operation. This configuration allows increased liquid handling throughput. This version 2 liquid handling ACCESS consists of a higher performance plate reader, six Thermoshakers, and one Tecan EVO robot (**Fig. 15.2b**). The EVO robot has a liquid handling arm with fixed pipet tips and a robot gripper arm for microplate manipulation.

On the EVO_ACCESS robot, the experiment plates need to be moved to and from different deck positions such as Thermoshakers, the reader, and pipet positions. Furthermore, the experiment plates are competing for (i) the liquid handling arm, (ii) the plate moving robot arm, and (iii) the plate reader time. To coordinate these demands, the EVO_ACCESS software has a Growth Control Center (GCC) module that receives requests from the plate or user for all tasks that need coordination and puts them in the GCC waiting list queue. Most tasks are grouped into action sets that perform a sequential series of operations with the target experiment plate. The next action set is performed only after the action set for the preceding target plate has been completed.

A typical shake–read cycle in the EVO_ACCESS robot version is as follows. After the set shake time has elapsed, ACCESS appends a task action set (stop shake, move plate to the reader, read OD, etc.) to the GCC queue. Once the EVO robot arm and the reader become available, ACCESS stops the Thermoshaker and activates the EVOware script to move the experiment plate to the reader. The reader is then configured to the measurement parameters defined for that experiment plate, such as OD and fluorescence reads. If after the evaluation of the OD reads a liquid handling event is needed, ACCESS inserts a new task into the GCC queue that instructs the EVO robot to move the plate to an inoculation/save position on the deck that is accessible by the liquid handling component of the EVO robot. An EVO inoculation-save script then executes a sample manipulation worklist written by ACCESS for those samples that need manipulation. Any errors from the Freedom EVO robot are reported to the user or administrator by e-mail. If an error requires user intervention, the robot pauses all operations except plate shaking until the error has been cleared.

The advantage of the EVO_ACCESS is that it can be integrated with a significantly more capable plate reader that is available to all experiments. This is especially advantageous for fluorescence assays. With the Tecan Safire^2 installed, the monochromator-based wavelength selection and detection allows for fine-tuning to optimize the light detection at multiple wavelengths in one assay. Another advantage is the added flexibility in the growth environment, e.g., one can grow light-requiring photosynthetic organisms and operate at an expanded range of growth temperatures. To date we have successfully used temperatures ranging from 16 to 42°C. This robot has a higher throughput (six shakers vs. four readers) for assays that require liquid handling. However, its total throughput is limited by the need to channel all experiment plates through one robotic arm and one reader. Incorporation of a second reader and/or second arm would alleviate this bottleneck and increase capacity. If all six shakers are active the shortest possible shake–read cycle interval is 15 min. A long inoculation/save step will hold up the queue, but as the plates are kept shaking until it is their turn to be read, there is no detriment to the quality of the data. The EVO_ACCESS uses fixed tips that are cleaned with a bleach wash and high-speed flush. This results in a considerable savings compared to the cost of disposable tips needed in the Packard-based ACCESS.

One constraint of the EVO_ACCESS setup is that the plates are manipulated by one arm and read using one reader. If any of the robot components fail then all the experiments stall. In contrast to the ACCESS version where the plates never leave the plate reader, when plates are moved about the system error recovery is more complex. Accordingly, cameras and remote login capabilities for the computer are recommended to aid in error recovery.

3.5. Post Hoc Analysis

For reader models that do not yet communicate directly with the ACCESS software, it is possible to create an ACCESS format data file and import it into the system after the experiment is completed (*see* manual for details). The user would still have all the advantages of output of all growth metrics, annotation, data storage, data retrieval, and data views. A Perl Script to convert measurements from the Infinite reader (Tecan) using I-Control software (Tecan) is already available (*see* supplementary web site).

3.6. ACCESS Software User Interaction Features

The main user interface to the ACCESS robot is shown in **Fig. 15.3a**. Each shaker/incubator has its own separate column of controls for the user to quickly access the experiment configuration interfaces, view the data, and manipulate the run if necessary. When appropriate, individual controls are grayed out to prevent mistaken operation. None of the allowed user operations can inadvertently interfere with the running experiments. The

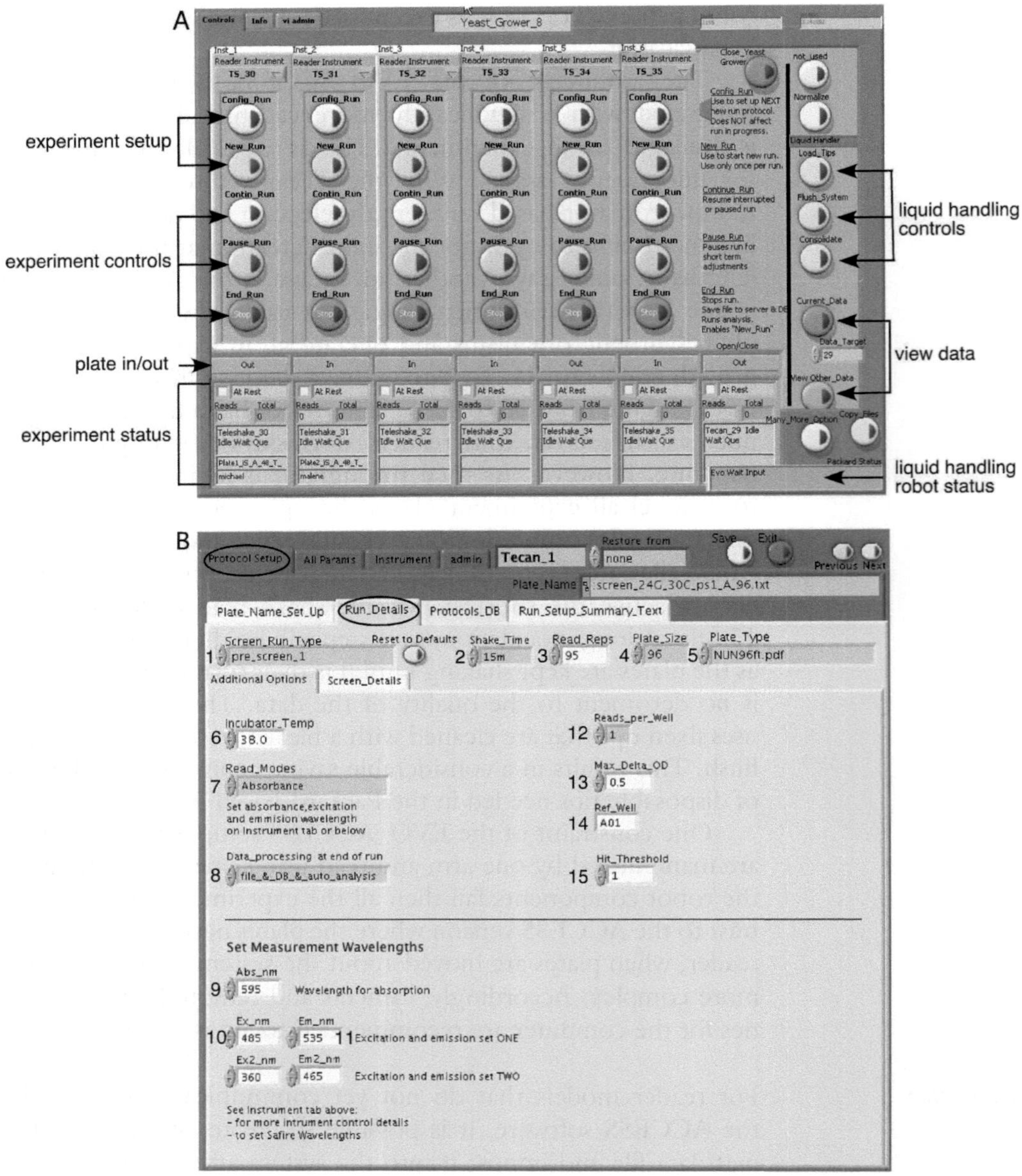

Fig. 15.3. Main user and configuration interface windows. (**a**) shows the main user interface window. Each shaker/incubator has its own separate column of controls to allow the user to quickly access experiment configuration interfaces, view the data, and manipulate the run if necessary. To load the plates into the readers there is a control to open or close the reader plate tray (or for the EVO ACCESS to move a plate from the load position to the Thermoshaker). The experiment status section shows the status of the reader/shaker, number of reads completed, shake time left, and experiment name and user. On the *right* are liquid handler controls, status and controls to access the plate data view, analysis and annotation interfaces. (**b**) shows the Run_Details tab of the configuration interface used to set the experiment parameters such as temperature (**6**), duration (**2, 3**) and read modes (**7**). **Table 15.1** lists a detailed description of each numbered item.

experiment status section gives an overview of the present ACCESS tasks. The status of the reader/shaker, number of reads completed, shake time left, and experiment name and user are listed for each instrument. On the right are liquid handler controls, status and controls to access the plate data view, and analysis and annotation interfaces.

3.7. Experiment Configuration

Each ACCESS screen begins by defining the experiment configuration. This will set the parameters for the screens and ensure that all subsequent screen steps are properly associated. Experiment parameters are set in the configure interface (**Fig. 15.3b**) and accessed from the main interface (**Fig. 15.3a** – experiment setup). Pull-down menus are available for the most often used parameters. This simplifies and helps manage the many available options (**Fig. 15.3b**, **Table 15.1**). Parameters that are rarely changed are edited in the All_Params tab. Reader-specific parameters (such as shaking intensity, shaking mode, and measurement wavelength settings) can be changed in the Instrument tab. A step-by-step instruction for experiment configuration and setup and a list of all experiment parameters can be found in the supplementary manual.

Table 15.1 lists a selection of parameters that can be set in the configuration screen. For each parameter, options and comments are listed and the most commonly used option is highlighted in bold italic. All parameters in the configuration can be changed during the course of an experiment (although currently not all from the configuration user interface; for exceptions and instructions *see* supplementary manual). This feature of the ACCESS platform is particularly helpful if an experiment is taking longer than expected or if a mistake was made during the initial configuration and a modification can rescue the screen. To help with the setup of complex screens, experiment templates can be saved to the DB and retrieved for quick configuration of subsequent experiments.

3.8. Data Read Configuration

The ACCESS measures high-resolution growth curves with data reads taken every 1–15 min. This allows for very precise phenotypic analysis required in the drug-discovery experiments for which ACCESS was initially developed. In addition, ACCESS can be configured for any of the available readers' detection mode, such as optical density (OD) or fluorescence (**Fig. 15.3b.7**).

A typically read cycle is as follows:

- shake 15 min (*see* **Note 12**).
- read OD.
- read fluorescence (optional) (*see* **Note 13**).
- liquid handling intervention if needed.

This read cycle is then repeated 95–500 times (**Fig. 15.3b.3**, **Table 15.1**). If the experiment is configured to include a fluorescence measurement, it is read immediately after the OD

Table 15.1
ACCESS configurable run parameters

Parameter name	Values	Comments	Fig. reference
Screen_Run_Type	***pre_screens_1-3*** ***innoc_well_to_well*** dilute_&_replenish _media template	For simple growth assays For liquid handling runs (manual and robotic) For removing and replacing with fresh medium For storage and retrieval of save protocols in the DB	**15.3b.1**
Shake_time	10 s, 1, 5, 10, or *15* min		**15.3b.2**
Number of read repetitions	Number of shake–read cycles (1–∞)	For absorbance + fluorescence assays this value needs to be doubled as one shake–read cycle takes two reads	**15.3b.3**
Plate_Size	6, 12, 24, ***48***, ***96*** and 384	Number of wells in microplate	**15.3b.4**
Plate_Type	Physical plate dimensions file Drop down menu populated from DB	Used by reader; supplied by manufacturer, changing Plate_Size auto selects plate_type and vice versa	**15.3b.5**
Temperature	Temperature to run assay *30° C*	Needs to be ~5°C above ambient, typically not lower than 28°C in a well-controlled temperature environment	**15.3b.6**
Read_Modes	**Absorbance,** fluorescence, or luminescence Abs_&_Flor Abs_&_Flor_Flor2 Abs_&_Lumin Abs_Barcode Abs_Flor_Barcode	Combinations of read modes: absorbance and fluorescence absorbance and two fluorescence wavelengths absorbance and luminescence multi-plate end point absorbance assay multi-plate end point absorbance fluorescence assay	**15.3b.7**

Table 15.1 (continued)

Parameter name	Values	Comments	Fig. reference
Data_processing at end of run	None file_to_server ***file_&_DB_&_auto_analysis***	Do nothing at end of run Copy data file to file server Copy data file to server, run data analysis, and upload result and data file to DB	**15.3b.8**
Measurement filter	***595 nm***	Absorption measurement wavelength (nm) of installed excitation filter or setting of the monochromator (Safire^2)	**15.3b.9**
Excitation filter	Populated by installed reader filters	Excitation wavelength (nm) of installed excitation filter or setting of the monochromator (Safire^2) Up to two wavelengths possible, separate with comma	**15.3b.10**
Emission filter	Populated by installed reader filters	Emission or luminescence wavelength (nm) of installed emission filter or setting of the monochromator (Safire^2) Up to two wavelengths possible, separate with comma	**15.3b.11**
Reads_per_Well	*1*, 4	Number of reads per well best kept to 1	**15.3b.12**
Max_Delta_OD	0.25	ΔOD threshold between reads. Changes exceeding ΔOD are flagged as data jumps and e-mails are sent to the user	**15.3b.13**
Ref_Well	A01…H12	Well that contains a wild-type or non-drug growth	**15.3b.14**

(continued)

Table 15.1 (continued)

Parameter name	Values	Comments	Fig. reference
BaseLine_OD	atstart	Use first read of the run as baseline OD	**15.5a.1**
	atpipet	Uses the first read after a pipetting event	
	atstart_minimum	Uses first reads but tracks OD to use minimum OD of run – best option to use for pre-screens	
	atpipet_minimum	For screens: uses the minimum values in the first reads after a pipet – best option to use for add drug screens	
MultiProbe_Run_File	Robot script filenames, populated from the DB	Use to select liquid handling script to run inoculations, save, and add drug events	**15.5a.2**
Pipet_Trigger	***OD_Save_Innoc***	Use OD to trigger both save and inoculation	**15.5a.3**
	OD_Save_Time_Innoc	Use OD for saves and number of reads for inoculations	
	Time_Save_Innoc	Use the number of reads to trigger inoculation and saves	
	Time_Save_OD_Innoc	Use number of reads for saves and OD for inoculation	
	OD_Save_Innoc_Dynamic	Use OD calculated from first read to trigger inoculation or save – not recommended	
Pipet_Volume_Adj and Add_Drug	***Con_OD_Save***	Constant inoculation and save volumes	**15.5a.4**
	Con_Save_Adj_Innoc		
	Adj_Save_Con_Innoc		
	Adj_Innoc_Save		
	Normalize_Save_w_Dilute		
	Add_Drug_at_Innoc	Add drug just prior to every inoculation	
	Add_Drug_at_Start_Innoc	Add drug at start of run and just prior to every inoculation	
	no_add_drug	Same as Con_OD_Save	

Table 15.1 (continued)

Parameter name	Values	Comments	Fig. reference
Read_to_1st_Add_Drug	0–16	Read cycle at which to add first drug. Thereafter drug is added to wells at time of inoculation	**15.5a.5**
Innoc_Direction	***down***	Screens run down in columns	**15.5b.1**
	across	Screens run across in rows	
	use_innoc_map	Uses a map to determine where to inoculate. Can inoculate from one well to multiple wells	
	save_only	No well-to-well inoculation – only pipet to the save plate	
Save_Mode	***one_to_one***	Save from well position in growth plate to same well position in save plate	**15.5b.2**
	use_save_map	Uses map to determine what well to use on save plate. Use for consolidation of saved cultures	

to allow for the precise normalization of the fluorescence signal by the OD. **Figure 15.4** shows an example of quantitative fluorescence measurements, which could potentially be used to quantify and monitor changes in expression levels during different growth phases. Using green fluorescent protein (GFP) the promoter strength of the histone gene *HTA1* was measured in yeast *swi4*, *hir1*, and *asf1* deletion mutants. Hir1 and Asf1 are known repressors of *HTA1* transcription (14, 15), and deletion of these genes would cause de-repression of the GFP expression. In contrast, Swi4 plays a role in activation of HTA1 transcription, and deletion of *SWI4* should repress GFP signal (16). **Figure 15.4** shows the growth curves (a), fluorescence curves (b), and the fluorescence/OD (c) of the GFP constructs in the different deletion backgrounds. In agreement with previously published data, the GFP fluorescence is increased in the *hir1* and *asf1* deletion

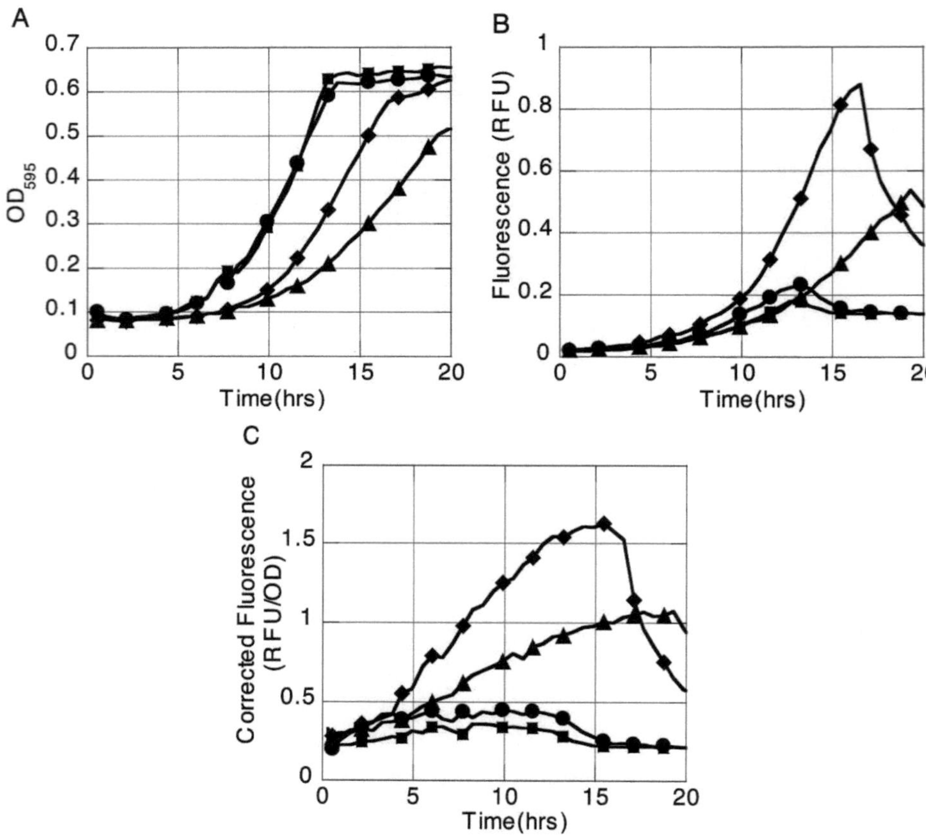

Fig. 15.4. Near-simultaneous OD–fluorescence measurements. Yeast deletion strains *swi4::kan, hir1::kan, asf1::kan*, and a wild-type control in the Y7092 genetic background containing HTA1 promoter upstream of the reporter gene GFP were grown in low fluorescence medium (18) in a 96-well plate in the ACCESS version 2 robot using the Safire^2 reader. OD_{595} and fluorescence measurements (excitation wavelength 485 nm bandwidth 9 nm, emission wavelength 535 nm bandwidth 9 nm, and PMT gain 90 V) were taken every 15 min. (**a**) shows the OD curves, (**b**) the fluorescence curves, and (**c**) the fluorescence corrected for culture OD. *Curves* were plotted by interpolation between the 15 min measurement points, and the strains are indicated as follows: wild type (●), *swi4::kan* (■), *hir1::kan* (◆) and *asf1::kan* (▲).

strains and decreased in the *swi4* deletion strain (**Fig. 15.4b, c**). The application of high-resolution correlated OD and fluorescence measurements can be seen when the cultures reach stationary phase, and GFP fluorescence decreases due to downregulation of *HTA1* expression in the stationary phase (17).

3.9. Liquid Handling Screen Configuration

A large number of screen configurations are possible and are discussed in the supplementary manual. As an example, **Fig. 15.5**

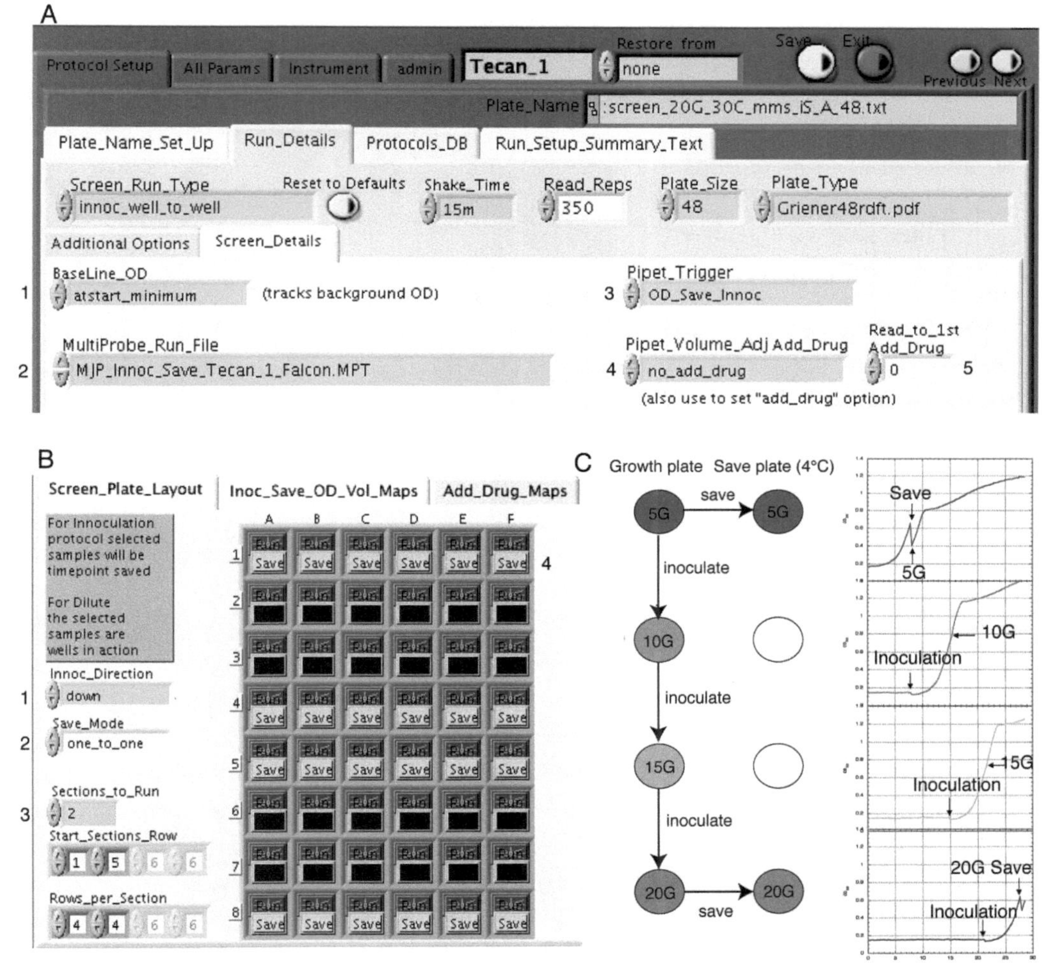

Fig. 15.5. A "down" plate liquid handling screen configuration and execution. This figure shows configuration settings and a cartoon of a liquid handling screen that inoculates wells down a column. (**a**) shows the Run_Details and the Screen_Details for a Screen_Run_Type = "innoc_well_to_well" set when the experiment needs liquid handling (**a1–a5** are described in **Table 15.1**). (**b**) shows the tab configuring the sample manipulation during the experiment. In the example shown (**b, c**) the inoculations are set to move down (**b1**) the columns and the plate is divided into two sections (**b3**) each with four wells. Samples selected for saving are moved to the equivalent well on the save plate (**b2**, one-to-one) and the save button (**b4**) is set for that well. (**c**) shows a cartoon and resulting growth curves of the "down" plate experiment. For every five generations of growth as determined by the target OD (set in another window), an inoculant of sample is moved to the following well. Samples moved to the save plate have a corresponding dip in the OD of the adjacent growth curve.

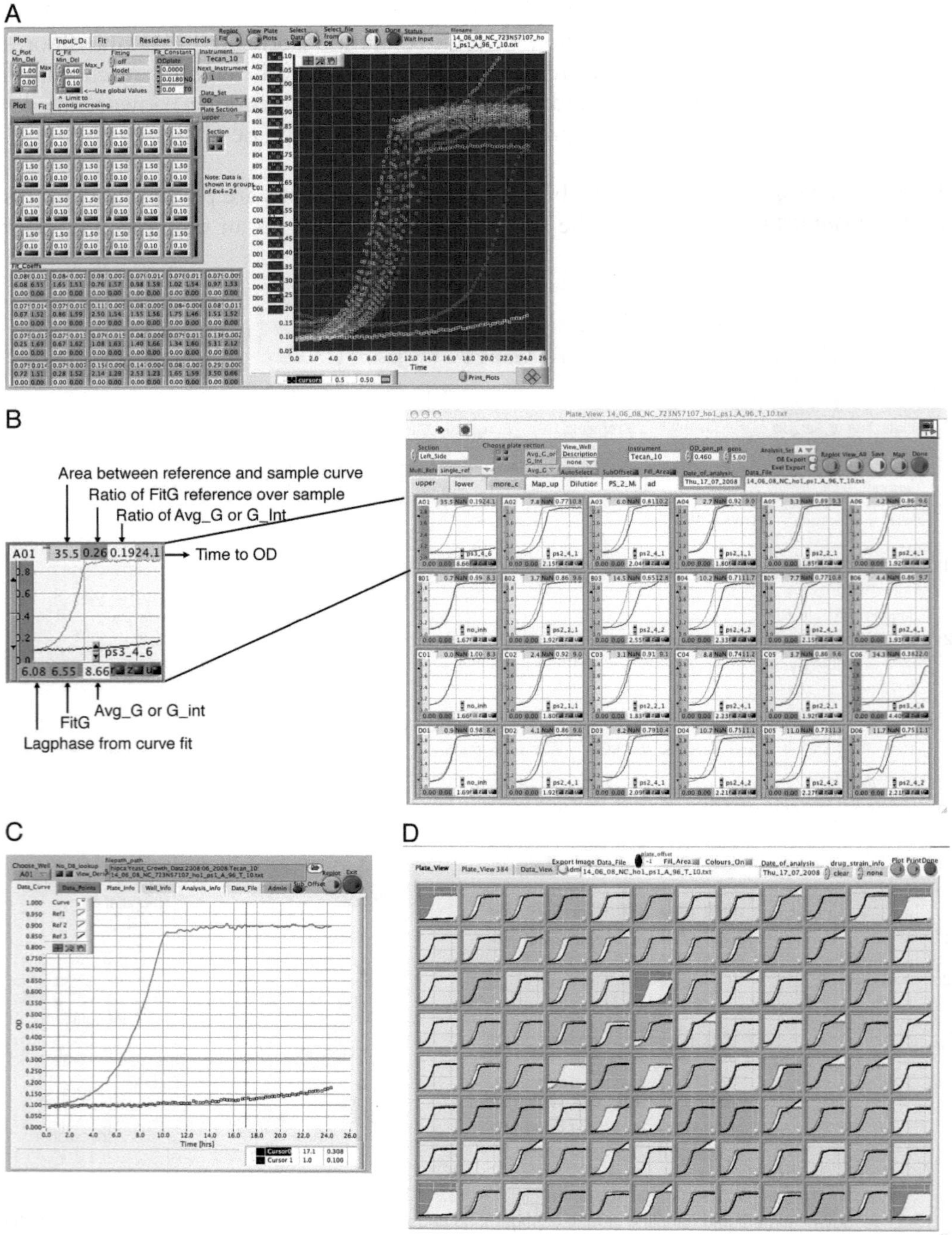

Fig. 15.6. ACCESS data views. The ACCESS software has a variety of interfaces used to view the growth data and analysis metrics. These interfaces also contain controls to manually perform curve fits, set reference wells and access growth condition, and annotate interfaces and update the ACCESS Track database (ATDB). The Analyze_Data interface (**a**) displays up to 24 curves in one plot and contains the curve fitting controls. Other data sets such as fluorescence, luminescence, or temperature can also be selected and plotted singularly or combined with the OD data set. Buttons at the top of the interface are used to move to other data views, select another data file, and save any data analysis results to the data files. The Plate_View interface (**b**) plots the growth curves on independent axes for each well, viewed in plate sections of 24 wells. The controls along the top give access to other viewing options and to save or export the analysis

shows configuration and experiment in which the wells are inoculated down a column. In addition to the basic configuration parameters such as temperature and shaking time (**Fig. 15.3b.6** and **15.3b.2**, **Table 15.1**) the setup of a liquid handling screen requires setting parameters, such as the robot script that executes the liquid handling events (**Fig. 15.5a.2**, **Table 15.1**) and pipet trigger target type (**Fig. 15.5a.3**, **Table 15.1**). In this example the plate is divided into two sections (**Fig. 15.5b.3**). Samples that are to be saved are selected using the save button (**Fig. 15.5b.4**). At the start of the experiment, rows 1 and 5 are inoculated with cells. When the culture reaches its target OD for five generations, the inoculant is pipetted into the following well as illustrated in **Fig. 15.5c**. As the sample is moved down the plate (e.g., from well F01 to F04) the culture maintains growth in the exponential phase over 20 generations. Note the dip in the growth curves where a sample has been moved from the growth well to the save plate (**Fig. 15.5c**). If an add drug step is specified (**Fig. 15.5a.4**, **Table 15.1**), the drug is aspirated from a drug source plate and dispensed into the well just before the inoculant is added. This could reduce the loss of efficacy of unstable drugs due to exposure to the growth medium or condition.

The two primary limitations on the miniaturization of liquid handling screens are (i) to have enough representative of each strain in the pool (i.e., at least 150–300 cells of each strain (8)) and (ii) to use inoculation volumes that are large enough so as not to induce errors from the limitation of the liquid handling robot (e.g., ±1 μl plus any inoculant residue on the outside of the tip).

3.10. On-The-Fly Data Analysis and Visuals

During the course of an experiment, the data can be viewed in a variety of plate view formats. As each new data point is collected, the plate views and data analysis screens are updated. The user can monitor the results in real time, make adjustments to the configurations, and decide when to end the experiment.

There are three main plate view formats. Format 1 shows the growth curves from a plate section overlaid in one plot (**Fig. 15.6a**). Individual curves or sub-groups of up to 24 curves can be selected (*see* supplementary manual). This format is best for viewing many curves in detail, to perform curve fitting, and for viewing the overall progress of the experiment. Other data

Fig. 15.6. (continued) metrics to the ATDB. Reference curves can be selected plate wide or separately for each well. Each well plot displays (*inset* in **b**) selected data analysis metrics, such as average generation time (Avg_G, ratio_Avg_G) and total time to five generations (time_to_OD), and the area between the curves (for detailed description, *see* **Table 15.2**). (**c**) shows the zoom view used to view details of the growth curve and its reference curves and the complete listing of growth conditions, annotations, and data analysis metrics stored in the ATDB. (**d**) shows the complete plate wide view of the growth data shown in this example with the area between the curve and the reference curve *highlighted*.

modes such as fluorescence, luminescence, or temperature can also be selected and plotted singularly or combined with the OD data set.

Format 2 plots the growth curves on independent axes for each well, viewed in plate sections of 24 wells (**Fig. 15.6b**). Reference curves can be selected across the entire plate or individually for each well. Selected data analysis metrics such as average generation time (Avg_G), the Avg_G ratio of the reference to the condition (ratio_Avg_G), total time to five generations (time_to_OD), and the area between reference and sample curve are shown in the inset of **Fig. 15.6b**. Details on the data analysis metrics are provided in **Section 3.11**. For each well, one can zoom in to view the details of the growth curve, and its references, and the complete listing of growth conditions, annotations, and data analysis metrics stored in the ATDB (**Fig. 15.6c**).

Format 3 displays all the growth curves plotted on separate axes in a single window (**Fig. 15.6d**) laid out in the growth plate format. This View_All option is useful for a quick visual inspection of all the growth curves or, after the experiment is finished, to prepare a paper record of the experiment. For screen plates where the samples are collected for genomic analysis, this view also includes the option to print barcodes for each well to allow for sample tracking in downstream sample analysis.

3.11. Data Analysis

Several growth metrics are automatically calculated during an experiment and after an experiment is finished. **Table 15.2** describes a number of growth metrics available from the ACCESS data analysis module. For quick and reproducible calculation of the generation times, we initialize our experiments at 0.0625 OD_{595}. The metrics Avg_G and G_by_Interval (illustrated in **Fig. 15.7**) are calculated using a culture OD that is equivalent to five doublings from the start OD. This "OD generation time" was chosen empirically as the point at which most cultures are still growing exponentially. It is derived from a calibration curve used in the microplate experiments (typically 100 μl in 96-well plates and 700 μl in 48-well plates). The calibration curve correlates OD_{595} measured in a cuvette of 1 cm path length in a spectrophotometer to that of the reader OD_{595} in the microplate. Each plate type and volume of medium needs to be calibrated in this manner. Avg_G is the average generation time from start to the OD generation point and includes both the lag and exponential phases in its calculation. G_by_Interval calculates the doubling time of the exponential growth phase by excluding the first doubling from the calculation and estimates the lag phase.

For precise calculation of Avg_G and G_by_Interval, high-quality growth curves are important. Shaky growth curves or cultures that undergo only one or a few doublings diminish the precision of these growth metrics. Sum, which records the area

Table 15.2
Description of ACCESS growth metrics

Growth metric	Description
Term definitions used to define the growth metrics	nG is the number of generations, OD_{nG} is the calibrated OD after nG doublings calibrated using wild type using a standard starting OD. T_0=0 is the time at start of run, T_{0_G0} is the estimated time that exponential growth begins if there is any lag phase from a dormant culture. T_{nG} is time to reach OD_{nG}. Typically we use OD_5, i.e., nG=5. OD_5 is determined by calibration. We use the reader OD equivalent to OD=2 for 10 mm path length measured using a conventional spectrometer. This is equivalent to five generations if we start our cultures with ODt_0=0.0625. If the data curve under examination does not reach OD5, then a new OD_{nG} and nG are calculated as outlined in Generations_at_time_G
Generations_at_time_G	nG – the number of generations nG used to calculate Average_G, i.e., nG used for T_{nG} and OD_{nG}. For slow-growing cultures that do not reach the OD_{nG} a new end point nG is established by fractional halvings of OD_{nG} until the end point OD is reached. The new nG is then calculated by the number of halvings needed, i.e., if OD_5=2 is the calibrated OD for five generations then an OD=1.414 is the calibrated $OD_{4.5}$ using OD_g*exp(ln(2)*x) to calculate an increase or decrease in OD where x is the generation increase and OD_g is the OD of the present generation
Time_to_Each_n_Gens	T_{nG} time required to reach OD_{nG}
Average_G (hours, h) (Avg_G)	Average generation time, i.e., growth rate; calculated by $(T_{nG} - T_0)/nG$ Average_G also includes any lag or dormant phase
Ratio_Avg_G	Ratio of Average_G to that of Average_G of a reference condition

Table 15.2 (continued)

Growth metric	Description
G_by_interval (h) (G_int)	Growth rate calculated over an interval, usually $(T_{nG} - T_{nG-4})/4$ where T_{nG-4} is the time to OD_{nG-4} where OD_{nG-4} is calculated by four halvings of OD_{nG} where OD_{nG} is usually the calibrated OD for five generations. By comparing G_int*nG to Avg_G, we can estimate a lag phase time for T_0
Ratio_G_by_Int	Ratio of G_by_interval to that of G_by_interval of the selected reference condition for that well
t0_from_interval	$T_{0\_go} = T_{nG} - nG$*G_by_interval, estimate of the time that exponential growth begins
Max_slope	Better described as "max_growth" (max slope)*ln 2/2, where max slope is maximum slope of the curve calculated using a moving averaged over four data points
Sum_20 Also 12, 24	Sums of the $\Delta OD/\Delta t$, i.e., increase in OD for an increase in time up to t=12, 20, or 24 h
Sum_Delta_Ref	Difference between the sum under the curve and the sum under the curve of a reference condition
Sum_del_20 Also 12, 24	Difference between the sum under the curve and the sum under the curve of a reference condition time up to t=12, 20, or 24
Time_saturation	Time that growth saturates can also be labeled as the stationary phase After T_{sat} the curve is either flat or with a linear slope. Determined by examining the first and second derivatives
OD_saturation	OD at which there is no longer any growth
Time_inflex	Time at which growth is no longer exponential base 2. Determined by examining the first and second derivatives
OD_inflex	OD at which the growth is no longer exponential base 2

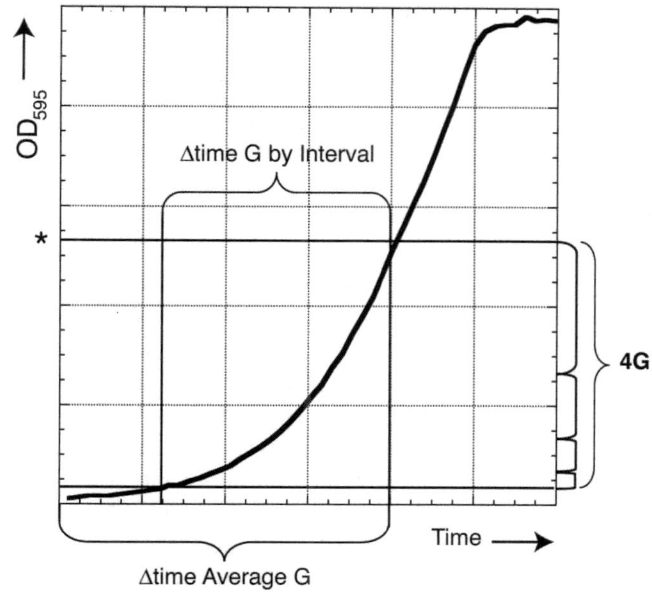

Fig. 15.7. Data analysis metrics. This graph shows the time intervals used to calculate the growth rates Average_G and G_by_Interval (**Table 15.2**). Time interval for Average_G is calculated from the start of the experiment to the time that the culture has grown for five generations. OD_5 generations, the calibrate OD equivalent to five generations, also called the OD generation point, is indicated by (*). The time interval for G_by_Interval is calculated back four doublings (**4G**) from OD_5 generations (*).

under the curve, is less affected by these limitations but is less precise in detecting small differences between two growth curves. Therefore, the exact choice of metric will depend on both the experiment and particular conditions, and it is useful to explore several of these automatically generated metrics.

In the case of large data sets, where not every growth curve is visually inspected, metrics such as saturation OD and sum are used to check the quality of the growth curves and verify the validity of the Avg_G and G_by_Interval values.

3.12. The ACCESS Track Database (ATDB): Data Annotation and Tracking

One ACCESS installation can generate 288 (48-well plates) to 2,304 (384-well plates) growth curves per day with six readers. This large amount of data produced must be properly tracked and stored. To avoid data loss, each experiment must be saved along with its experimental conditions, strains, and compounds/conditions used. At the start of the experiment, the user is prompted to select a username, project name, growth medium, and other well annotations such as compound, concentration, and strain. This information, together with all the experiment configuration parameters (such as temperature and plate type), the growth data, and analysis metrics, is automatically uploaded to the ATDB at the end of the experiment. In addition, the data file, which also contains all experiment parameters and well annotations, is copied to a file server that stores all data files organized by year, month, and plate reader ID.

For large-scale projects, or for experiments that require many variables to be tracked, we established a pipeline using barcodes to minimize operator errors and increase tracking capabilities. First, source plates containing strains or compounds are labeled with a unique barcode. These source plates can be supplied pre-barcoded from a vendor. Plate layouts that define the strains or compounds on the source plates are loaded into the ATDB and linked to the source plate's unique barcode. Each experiment plate receives a unique barcode, which is entered into the software and associated with the source plate via the barcodes. An experiment protocol is assigned to the experiment plate by selecting from a menu of appropriate protocols supplied by the ATDB. At the start of an ACCESS run, the user scans the reader's barcode and the barcode of the experiment plate. With no further intervention from the user, the robot verifies the validity of barcode, retrieves the plate annotations and experiment parameters from the ATDB, and automatically configures and starts the run.

If the source and experiment plates are defined with barcodes, several web interfaces can be used to generate scripts to robotically prepare the growth plates. For example, determining the optimal drug dose is essential for experimental success. Toward this end, the Titrate_Drugs web interface (**Fig. 15.8a**) analyses the growth rates from multi-plate sets of data, for each compound listed in the master source plate to compute the ideal concentration for uniform growth inhibition. The user can review the growth curves and analysis for quality (**Fig. 15.8b**) and determine if any of the compounds need further dilution. The script then creates a daughter source plate in the ATDB and generates a robotic script that is used to dilute the compounds to, e.g., IC20 or IC10 concentrations (the preferred concentrations for sensitivity assays). These daughter plates then become the source plates for further experiments using barcode tracking. An example is the full automation of multi-generational compound screening using the liquid handing robot capability. With only the need to scan three barcodes, reader, source drug plate, and growth plate, a just-in-time add drug screen is configured, annotated, and run to completion with no user intervention, including the end of run analysis and linking to the ATDB that tracks downstream use of the saved samples such as the HIP assay microarray analysis. This dramatically reduces operator workload and setup errors, improving accuracy in annotation, sample tracking, and data analysis.

3.13. Data Recall and Database Tools

A non-robotic ACCESS version of the software (called DB_Interface) exists for both Mac and PC platforms, so users can pre-configure experiments, annotate, view, and analyze data at any

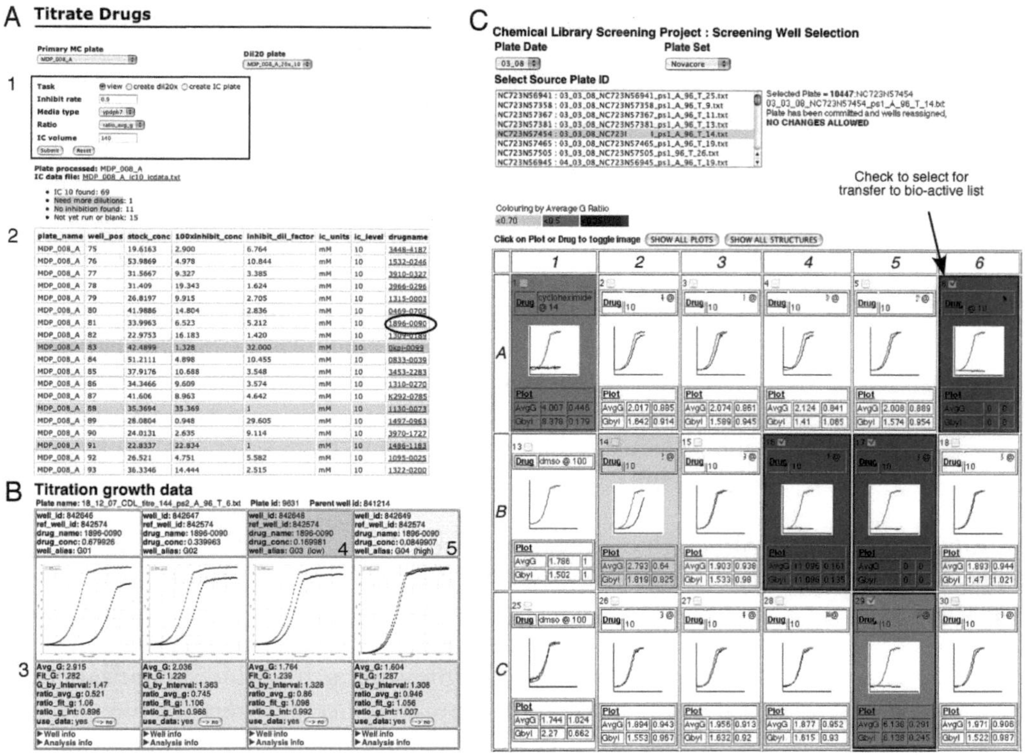

Fig. 15.8. ACCESS web interfaces. This figure shows examples of ACCESS web data interfaces. (**a**) shows the Titrate Drug Interface that tracks titration experiments used to find the desired inhibition rate for compounds for downstream studies. The titration data management script retrieves and analyses data stored in the ATDB to collate growth data from nearly 600 wells. Using the controls (**1**) the user can create liquid handling scripts to automate the compound dilution and reformatting to create daughter compound plates with uniform inhibition rates. Item **2** shows the review of the titration analysis to determine the optimum dilution and if compounds need further dilution. Clicking the compound names (circled) brings up the interface in (**b**) that shows the growth curves and analysis results (**3**) of the serial dilutions of that compound. Only compounds that are flagged as problematic need user review. Items **4** and **5** show the curves that straddle the desired inhibition. An interpolation between their inhibition rate and drug concentrations is used to calculate the dilution volumes needed to create the daughter plate. (**c**) shows the web interface to review experimental data from a large-scale study of a small-molecule compound library. The growth data and analysis are retrieved from the ATDB for on-the-fly presentation. Compounds that give a reduced viability phenotype are colored with different levels of *gray* that indicate the level of growth inhibition. The user can quickly review those compounds and mark them to be added to the bioactive list. Bioactive compounds are then reformatted into "hit plates" to be used in downstream studies.

computer. This version of the ACCESS software can perform all tasks except running the experiments. Both the ACCESS robot and DB_Interface software contain several database tools. These tools query the compounds and strains entered in the ATDB, view each experiment plate, as well as view the layouts of source plates listed by experiment type, project, or user. Another powerful ATDB query is the ability to list all experiments associated with a specified drug or strain. From the returned list, one can select the growth curve data and the full annotations of the plate of interest.

The ATDB is also easily integrated with custom and project-specific web-based tools, such as the example for visualization and well selection shown in **Fig. 15.8c**. In this particular example, project-specific tools were developed for a large-scale 50,000 compound library screening project. Scripts query the ATDB to retrieve data belonging to the screening project. Growth curves, colored by ratio_avg_G cutoffs, create an online plate view. Each plate is visually inspected and active compounds are selected/confirmed by clicking the on-screen well.

Several freeware tools exist to directly query MySQL databases. This allows users to run custom queries that are not integrated in the ACCESS software or part of a web-based tool. For instance, users can retrieve specific data analysis metrics for one particular strain or compound from several different growth experiments and export this data.

3.14. Future Directions and Perspectives

Although the majority of screens performed to date on ACCESS have used *S. cerevisiae*, ACCESS is very versatile and the only limitation for screening is that the microbes grow planktonically and aerobically. Furthermore, ACCESS can incorporate almost any standard plate reader measurement, including optical density (OD), fluorescence, or luminescence in a temperature range of 16–42°C, every 1–15 min. If necessary, for anaerobic conditions or lower temperatures, the entire reader can be placed in an environmentally controlled enclosure. This flexibility allows screens of temperature-sensitive mutants or, for example, GFP-tagged strains. Besides *S. cerevisiae*, ACCESS has been used successfully to culture a variety of microbes including *Schizosaccharomyces pombe*, *Cryptococcus neoformans*, *Candida albicans*, *Escherichia coli*, *Bacillus subtilis*, and *Chlamydomonas reinhardtii*. The versatility and high-throughput nature of this platform makes it a promising tool for the systematic use in screening microorganisms for drug discovery and for understanding cell physiology. We are continually improving on the ACCESS and have established a web site to support these efforts: http://chemogenomics.med.utoronto.ca/supplemental/ACCESS/

4. Notes

1. The Tecan microplate readers can read almost any plate format that fits in the plate tray. We generally use sterile, clear polystyrene microplates with flat bottom wells. Some plate types, such as the Nunc 48-well plates, have wells too close to the edge of the plate and the measurement light path is obstructed by the reader's plate tray.

2. Use a pierceable plate seal such as the plate seals from ABgene. So-called breathable seals are too elastic and will only deform and not pierce, whereas plate seals that are too stiff will resist piercing.
3. Multiple USB to serial port converters do not work well with the Tecan RdrOLE server, so it is best to provide each computer with dedicated serial port cards.
4. At the cost of reduced data security from computer failure, it is possible to run ACCESS using a stand-alone computer (i.e., without Internet access). In this case the MySQL database server, web server, and Perl script engine would be installed on the same computer as the ACCESS software.
5. Other microplate readers can be integrated with ACCESS. Readers that communicate via a serial port require moderate additions to the software.
6. A comparable reader is the Infinite F200, option PMT enhanced, heating, filter fluorescence top, and filter absorbance. In the future, the ACCESS software will be able to communicate with the Infinite readers.
7. For accurate temperature control, the Tecan microplate readers are limited to assay temperatures that are ~5°C above ambient. During use, the continuous shaking cycles cause the readers to heat up by a few degrees. To ensure that assay temperatures are maintained, it is recommended to use a fan to circulate air underneath the reader and to cool the room to ~7°C below the desired assay temperature. This can be accomplished with an inexpensive, portable room air conditioner.
8. The Packard software Winprep was not designed to work with computers with dual core processers; it is therefore recommended to switch this option off in the computer configuration.
9. It is important that the microplate is fixed in the reader plate tray while the robot is pipetting. In this version of ACCESS, the reader plate tray clamps are modified to ensure that they continue to clamp the plate even when the plate tray is in the open position.
10. ACCESS can be modified to use other liquid handling robots with suitable deck access and remotely controlled software.
11. Because Thermoshakers heat only from the bottom, condensation can occur on the plate seal and interfere with the reader measurements. This is solved by building a heating element above the Thermoshaker that keeps the

temperature ~5–8°C above the Thermoshaker setting (*see* online manual and supporting web site).

12. For cultures at medium or high density it is recommended to use longer (>10 min) shaking times, especially if the plate has to be at rest, as shorter times may not be enough to completely re-suspend the cells for accurate OD measurements.

13. For fluorescence measurements, grow yeast in low fluorescent medium such as described in Sheff and Thorn (18).

Acknowledgments

The authors would like to thank Bob St. Onge (Stanford HIPHOP lab) for suggestions for the continued development of the ACCESS robots; Larry Heisler (Toronto HIPHOP lab) for development of the Infinite conversion scripts and web site tools; Pinay Kainth from the University of Toronto for the generous gift of the HT1A-GFP strains; Mark Torresan (Tecan USA) for many years of helpful advice and support; and Susan Dutcher (Washington University School of Medicine) for *C. reinhardtii* strains and advice. Finally, we would like to thank all past and present members of the Stanford and Toronto HIPHOP laboratories for their feedback, support, and innovative use of the ACCESS robots. This work was supported by grants from NIH (NCI, NBIB, and NHGRI) and CIHR to R.W.D., C.N., and G.G, and from the Canadian Cancer Society (020380) to G.G.

References

1. Deutschbauer, A. M., Jaramillo, D. F., Proctor, M., et al. (2005) Mechanisms of haploinsufficiency revealed by genome-wide profiling in yeast. *Genetics* **169**, 1915–1925.
2. Hoon, S., Smith, A. M., Wallace, I. M., et al. (2008) An integrated platform of genomic assays reveals small-molecule bioactivities. *Nat. Chem. Biol.* **4**, 498–506.
3. Hillenmeyer, M. E., Fung, E., Wildenhain, J., et al. (2008) The chemical genomic portrait of yeast: uncovering a phenotype for all genes. *Science* **320**, 362–365.
4. Giaever, G., Shoemaker, D. D., Jones, T. W., et al. (1999) Genomic profiling of drug sensitivities via induced haploinsufficiency. *Nat. Genet.* **21**, 278–283.
5. Winzeler, E. A., Shoemaker, D. D., Astromoff, A., et al. (1999) Functional characterization of the *S. cerevisiae* genome by gene deletion and parallel analysis. *Science* **285**, 901–906.
6. Giaever, G., Chu, A. M., Ni, L., et al. (2002) Functional profiling of the *Saccharomyces cerevisiae* genome. *Nature* **418**, 387–391.
7. Pierce, S. E., Fung, E. L., Jaramillo, D. F., et al. (2006) A unique and universal molecular barcode array. *Nat. Methods* **3**, 601–603.
8. Pierce, S. E., Davis, R. W., Nislow, C., and Giaever, G. (2007) Genome-wide analysis of barcoded *Saccharomyces cerevisiae* gene-deletion mutants in pooled cultures. *Nat. Protoc.* **2**, 2958–2974.
9. Giaever, G., Flaherty, P., Kumm, J., et al. (2004) Chemogenomic profiling: identifying

the functional interactions of small molecules in yeast. *Proc. Natl. Acad. Sci. USA* **101**, 793–798.

10. Lee, W., St. Onge, R. P., Proctor, M., et al. (2005) Genome-wide requirements for resistance to functionally distinct DNA-damaging agents. *PLoS Genet.* **1**, e24.
11. St. Onge, R. P., Mani, R., Oh, J., et al. (2007) Systematic pathway analysis using high-resolution fitness profiling of combinatorial gene deletions. *Nat. Genet.* **39**, 199–206.
12. Ericson, E., Gebbia, M., Heisler, L. E., et al. (2008) Off-target effects of psychoactive drugs revealed by genome-wide assays in yeast. *PLoS Genet.* **4**, e1000151.
13. Yan, Z., Costanzo, M., Heisler, L. E., et al. (2008) Yeast Barcoders: a chemogenomic application of a universal donor-strain collection carrying bar-code identifiers. *Nat. Methods* **5**, 719–725.
14. Spector, M. S., Raff, A., DeSilva, H., Lee, K., and Osley, M. A. (1997) Hir1p and Hir2p function as transcriptional corepressors to regulate histone gene transcription in the *Saccharomyces cerevisiae* cell cycle. *Mol. Cell. Biol.* **17**, 545–552.
15. Sutton, A., Bucaria, J., Osley, M. A., and Sternglanz, R. (2001) Yeast ASF1 protein is required for cell cycle regulation of histone gene transcription. *Genetics* **158**, 587–596.
16. Simon, I., Barnett, J., Hannett, N., et al. (2001) Serial regulation of transcriptional regulators in the yeast cell cycle. *Cell* **106**, 697–708.
17. Gasch, A. P., Spellman, P. T., Kao, C. M., et al. (2000) Genomic expression programs in the response of yeast cells to environmental changes. *Mol. Biol. Cell* **11**, 4241–4257.
18. Sheff, M. A., and Thorn, K. S. (2004) Optimized cassettes for fluorescent protein tagging in *Saccharomyces cerevisiae. Yeast* **21**, 661–670.

Chapter 16

Competition Experiments Coupled with High-Throughput Analyses for Functional Genomics Studies in Yeast

Daniela Delneri

Abstract

Competition experiments are an effective way to provide a measurement of the fitness of yeast strains. The availability of the *Saccharomyces cerevisiae* yeast knock-out (YKO) deletion collection allows scientists to retrieve fitness data for the ~6,000 *S. cerevisiae* genes at the same time in a given environment. The molecular barcodes, characterizing each yeast mutant, serve as strain identifiers, which can be detected in a single microarray analysis. Competition experiments in continuous culture using chemically defined media allow a more specific discrimination of the strains based on their fitness profile. With this high-throughput approach, a series of genes that, when one allele is missing, result in either defective (haplo-insufficient) or favored (haplo-proficient) growth phenotype have been discovered, for each nutrient-limiting condition tested. While haplo-insufficient genes seemed to overlap largely across all the media used, the haplo-proficient ones seem to be more environment specific. For example, genes involved in the protein secretion pathway were highly haplo-insufficient in all the contexts, whereas most of the genes encoding for proteasome components showed a haplo-proficient phenotype specific to nitrogen-limiting conditions. In this chapter, the method used for implementation of competition experiments for high-throughput studies in yeast is presented.

Key words: Competition experiments, continuous culture, chemostat, molecular barcode, haplo-insufficient, haplo-proficient, fitness, genome-profiling.

1. Introduction

Biological fitness is defined by the relationships between genotype, phenotype, and the environment. In higher eukaryotes, the fitness value is usually difficult to estimate due to the multitude of variables associated with this concept (i.e., mating behavior, survival and reproductive rates, and seasonal mating). However, in single-celled asexual organisms such as yeast and bacteria the

J.I. Castrillo, S.G. Oliver (eds.), *Yeast Systems Biology*, Methods in Molecular Biology 759, DOI 10.1007/978-1-61779-173-4_16, relative fitness is simply given by their growth rate as they compete for a pool of resources. Competition experiments between microorganisms are always in relation to the environment, and so is their output fitness, which is the "relative fitness" of the strains in that particular environment context.

Typically, two yeast strains with different auxotrophic or drug markers were grown together for a number of generations, in a one-to-one competition. The final frequencies of the strains were then identified via quantitative PCR analysis or via plate essays using different selective media (1–3). However, in the last 15 years, scientists tried to develop large-scale quantitative methods to detect differences in fitness.

A major step forward in yeast molecular biology and genomic study happened about a decade ago when an international consortium of yeast laboratories created an almost complete collection of yeast mutants carrying two unique barcodes (4). This collection was unique in the sense that every gene that was deleted was also marked by two unique 20 bp sequences (UP-TAG and DOWN-TAG) flanking the selectable marker (KanMX). So each and every mutant strains created was "barcoded" and could be identified from the population via its TAG sequences. Furthermore, all these mutation-specific oligonucleotides were flanked by two universal sequences that could be used to amplify, in only two PCR reactions, all the up-TAGs and down-TAGs present in the population. DNA arrays (i.e., Affymetrix TAG3) were created with oligonucleotides complementary to the TAGs, so that these could be discriminated upon hybridization, since the intensity of the signal from each individual spot directly relates to the amount of TAG and consequently to the abundance of that particular strain in the population. So, if a gene is important for growth in a particular condition, the corresponding mutant strain will be disadvantaged during the competition experiment, and the associated molecular tags will diminish in the population. This technological advance meant that competition experiments could be carried out using simultaneously the entire yeast mutant collection and functional data could be collected for all the genes in one single experiment.

This methodology, pioneered in Ronald Davis's laboratory in 1996 (5), has been extensively applied over the years to study gene function, drug target, and the adaptive response to altered environmental conditions.

The homozygous yeast deletion collection was used to determine gene function of all non-essential genes in different conditions such as lack of amino acids, osmolarity, variation of pH, growth on non-fermentable source, or plasma membrane maintenance, (4, 6, 7). Genes involved in the biogenesis of the endoplasmic reticulum (8) and in maintaining plasma membrane integrity (9) were identified using this approach.

The heterozygous yeast deletion collection, with strains in which only one of the two copies of the gene is deleted, has been originally exploited to uncover molecular mechanisms of action of drugs, and several comprehensive genome surveys in the presence of chemical compounds have been carried out over the years (10–12). Moreover the heterozygous collection has been used to study the effect of the copy number variation on the fitness of the mutant strains in rich media (13), in nutrient-limited environments (14), and in the presence of toxic metals (15).

Most of the barcode experiments had been carried in batch (discontinuous) cultures (i.e., cultures without addition and/or removal of medium) in which the nutritional and chemical context in which the mutants were grown cannot be kept constant. Over time, the environmental changes, such as the level of sugars, vitamins, salts, minerals, secondary metabolites, and pH, could favor different mutants. Thus, a strain that could be in disadvantage at the beginning of the competition may regain fitness as the conditions change, confusing the outcome of the competition experiment. Serial dilutions in batch can partially alleviate this problem by resuspending the cell population in new fresh medium every few generations, so that the original nutritional conditions are recovered (13).

Although it is difficult to discriminate small changes in the growth rate by direct measurement in "batch" conditions, competition experiments in continuous (chemostat) culture provide a powerful approach to the analysis of the growth rate. Keeping the environment constant (*see* **Section 3.1**), the background variation is minimized allowing a sharp discrimination between strains with different fitness. Moreover, using chemically defined medium, such as F1, over rich (YPD) or minimal (SD) medium also allows the detection of smaller differences among strains (14). In particular, by using continuous cultures, both haplo-insufficient and haplo-proficient genes could be identified in a given context (14).

2. Materials

1. Strains and oligonucleotides
   a. The heterozygous yeast deletion collection deletion strains, in the diploid BY4743 background (MAT**a**/MATαhis3Δ1/his3Δ1 leu2Δ0/leu2Δ0 met15Δ0/MET15 LYS2/lys2Δ0 ura3Δ0/ura3Δ0), were obtained from the Saccharomyces Deletion Consortium (http://www-sequence.stanford.edu/group/yeast_deletion_project/deletions3.html).

b. The sequences of the biotinylated primers used to amplify the TAGs are the following: 5′ biotin-GATGTCCACGA GGTCTCT-3′ (BU1) and 5′ biotin-GTCGACCTGCAG CGTACG-3′ (BU2-comp) for amplifying the UP-TAGs; 5′ biotin-CGGTGTCGGTCTCGTAG-3′ (BD1) and 5′-biotin-CGAGCTCGAATTCATCG-3′ (BD2-comp) to amplify the DOWN-TAGs.

c. The sequences of the eight oligonucleotides used in the hybridization mixture are U1 5′-GATGTCCACGAGGT CTCT-3′ (U1), D1 5′-CGGTGTCGGTCTCGTAG-3′ (D1), 5′-CGTACGCTGCAGGTCGAC-3′ (U2), 5′-CG ATGAATTCGAGCTCG-3′ (D2); 5′-AGAGACCTCGT GGACATC-3′ (U1comp), 5′-CTACGAGACCGACAC CG-3′ (D1comp), 5′-GTCGACCTGCAGCGTACG-3′ (U2comp), and 5′-CGAGCTCGAATTCATCG-3′ (D2-comp).

2. Medium

a. The F1 medium (2, 14) can be limited for glucose (carbon limitation), nitrogen, phosphorus, or sulfur (*see* **Table 16.1**). YPD medium contains 1% (w/v) yeast extract, 2% (w/v) peptone, and 2% (w/v) glucose. YPD/glycerol contains same amounts of nutrients in 15% (v/v) glycerol.

b. First, the stock and final solutions of mineral salts, trace elements, and vitamins are prepared, according to whether the medium is C-, N-, P-, or S-, limited (*see* **Table 16.1**). The mineral salts are individually dissolved in 1 L of H_2O and then poured in a 20 L vessel (Nalgene) containing 18.5 L of water.

c. All the trace elements, except for $FeCl_3$, were dissolved in 1 L of sterile water (trace element mix). The $FeCl_3$ solution was dissolved separately in 1 L of sterile water. These solutions are autoclaved.

d. The vitamin stock solution is prepared as described in **Table 16.1** and filter-sterilized (0.22 μm Millex-LG filter, Millipore, # SLGV013SL). 33 mL aliquots are made in 50 mL Falcon tubes and stored at –20°C.

e. Filter-sterilized uracil, histidine, and leucine are then added to the medium (0.4 g Ura, 0.4 g His, and 2 g Leu for a 20 L final volume).

f. The sugar is dissolved in the medium: 400 g glucose is added except for the C-limited medium where only 50 g of glucose is included.

g. Sartopore 2 150 filters, 0.2 μm (Sartorius) for filter-sterilization of the medium (*see* **Section 3.2**).

Table 16.1
F1 nutrient-limited medium

	C-lim	N-lim	P-lim	S-lim
Mineral salts solution				
(Final concentration, g/L)				
$(NH_4)2SO_4$	3.13	0.46	3.13	0.024
NH_4CI				2.5
KH_2PO_4	2.0	2.0	0.054	2.0
KCl			1.10	
$MgSO_4{\cdot}7H_2O$	0.55	0.55	0.55	
$MgCl_2{\cdot}2H_2O$				0.45
NaCl	0.10	0.10	0.10	0.10
$CaCl_2{\cdot}2H_2O$	0.09	0.09	0.09	0.09
Trace elements solutions				
Trace elements mix 1	10,000X stock (g/L)		Final concentration (mg/L)	
$ZnSO_4{\cdot}7H_2O$	0.7		0.07	
$CuSO_4{\cdot}5H_2O$	0.1		0.01	
H_3BO_3	0.1		0.01	
KI	0.1		0.01	
–				
Trace elements mix 2[a]				
$FeCl_3{\cdot}6H_2O$	0.5		0.05	
Vitamins solution[b]				
Vitamins stock solution	600X stock (g/L)		Final concentration (mg/L)	
Inositol	37.2		62	
Thiamine/HCl	8.4		14	
Pyridoxine	2.4		4.0	
Ca-pantothenate	2.4		4.0	
Biotin	0.18		0.3	

[a]The 10,000X $FeCl_3{\cdot}6H_2O$ stock solution is prepared and kept separately.
[b]The 600X vitamin stock solution is filter-sterilized.

h. Air filters (Acro® 50 Vent with 0.2 μm PTFE membrane, Pall).

i. The feeding and the pH lines of the chemostat are cleaned and sterilized with 1 M NaOH solution and 70% (v/v) ethanol solution.

3. Hybridization

   a. The Genomic DNA is extracted using the DNA tissue kit (Qiagen). The concentration of DNA in the extract was determined using a Nanodrop spectrophotometer (Thermo Scientific).

b. Platinum PCR Supermix (Invitrogen) has been used to amplify the TAGs and Microcon YM10 columns (Millipore) to purify them.

c. Hybridization buffer composition: 100 mM MES, 1 M MES-Na^+, 20 mM EDTA, and 0.01% (v/v) Tween (*see* **Note 1**).

d. TAG3 (or TAG4) chips, 213B oligonucleotide control, and eukaryotic control (Affymetrix). Streptavidin/R-phycoerythrin conjugate (SAPE) (Invitrogen).

e. Non-stringent wash buffer: 6X SSPE, 0.01% (v/v) Tween 20.

f. Stringent wash buffer: 100 mM MES, 0.1 M MES-[Na^+], 0.01% (v/v) Tween 20.

g. Stain buffer: 100 mM MES, 1 M MES-[Na^+], 0.05% (v/v) Tween 20 (*see* **Note 2**).

h. SAPE solution: 1X Stain buffer, 2 mg/mL acetylated BSA, and 10 μg/mL SAPE. The stock solutions are 2X SAPE solution, 50 mg/mL acetylated BSA, and 1 mg/mL SAPE (*see* **Note 3**).

i. Standard Goat IgG antibodies (Sigma-Aldrich). A 10 mg/mL stock solution is prepared by resuspending 50 mg of IgG in 5 mL PBS.

j. Anti-streptavidin (goat) and biotinylated antibodies (final concentration 3 μg/mL) (Vector laboratories). Prepare a 0.5 mg/mL stock solution.

k. Antibody solution: 1X Stain buffer, 2 mg/mL acetylated BSA, 0.1 mg/mL goat IgG, 3 μg/mL biotinylated anti-streptavidin antibodies (*see* **Note 3**).

l. GeneChip® Hybridization Oven 645, GeneChip® Fluidics Station 450, and GeneChip® Scanner 300 7G (Affymetrix) are used to hybridize, wash, and scan the TAG3 microarrays (Affymetrix).

3. Methods

3.1. The Chemostat System

The chemostat culture is an open system in which the input medium is fed at a fixed dilution rate (D) with ($D = F/V$), where F is the input flow (liter/hour) and V the volume of the fermenter (in liters). The volume is controlled and kept constant by pumping out medium from the system. The pH is also kept constant by addition of acid or base (using ammonium as nitrogen source the pH tends to decrease and main pH control can be achieved

Fig. 16.1. Small-scale parallel fermentation system Fedbatch-pro (DASGIP Technology). Up to 16 flasks can be set up for continuous cultures inside a shacking incubator (Infors HT, UK); For optimum performance, a maximum number of 12 vessels per fermentation series was used.

by addition of base, such as NaOH). The population is initially grown in batch to early stationary phase and then the system is switched to continuous culture. After a number of generation times the population reaches a steady state, characterized by constant conditions (e.g., pH and temperature) and concentrations of substrates and products. In order to study several conditions at the same time a small-scale parallel fermentation system, such as Fedbatch-pro (Das Gip AG), is ideal in order to monitor up to 16 flasks (**Fig. 16.1**).

3.2. Preparation of the Medium and the Pool of Mutant Strains

The mineral salts, amino acids, and glucose are dissolved in water. 1 mL of trace element mix, 1 mL of $FeCl_3$, and 33 mL of vitamin stock are then added to the 20 L vessel and water is added to reach 20 L final volume. The medium is then filter-sterilized into another sterile 20 L vessel.

For the construction of the pool of yeast mutants, the strains were grown in YPD with 15% (v/v) glycerol using 96-well plates at 30°C until they reached stationary phase (48 h). Using a multichannel pipette, 10 μL of each heterozygous strain was put together in a sterile Petri dish. The pool was then divided in 1 mL aliquots, which were stored at –80°C in YPD with 15% (v/v) glycerol.

3.3. Preparation of the Fermenters

1. First, the calibration of the pH probes is carried out using the relative Fedbatch-pro software. The caps, labeled with numbers 1–16, which are connected to the machine, are screwed onto the relative probes and the calibration is carried out by placing the probes in a standard buffer at pH 7 and subsequently at pH 4. The temperature probe is also placed in the same buffers.

2. Once the pH probes are calibrated, they are inserted into the flask fermenters and prepared for autoclaving. Two air filters are screwed on the flasks and covered with aluminum foil. Medium inlet and outlet and base inlet are closed with their respective caps. The medium outlet is controlled mechanically, with the metal rods connected to the output feeding lines accurately positioned to maintain the desired constant volume, typically 100 mL (*see* **Note 4**).
3. Cleaning of the medium and pH buffer feeding lines: The pH buffer lines require two cleaning procedures, one with 70% (v/v) ethanol and the other with 2% (w/v) NaOH. This is done through the FedBatch software and takes approximately 2 h for 16 flasks. The medium feeding lines are cleaned manually by pumping through the lines first 70% (/v) ethanol, then 2% (w/v) NaOH, and finally rinsing them with sterile water and the sterilized medium to be used in the experiment (*see* **Note 5**).
4. The flasks are now ready to be connected to their feeding lines (*see* **Note 6**).

3.4. Batch and Continuous Culture

1. The medium is pumped in the flasks till it reaches the desired volume, typically 100 mL. Then the pumps are switched off and the yeast inoculum is added (ca. 500 μL of a stationary phase culture mix) by injecting it directly into the flask.
2. The batch phase will last ca. 24 h, where the microorganisms are grown at 30°C with shaking at 170 rpm. A sample is then collected, and the cultures are switched to continuous culture by turning on the pumps and the pH control.
3. The cultures are switched to continuous culture with a dilution rate, $D = 0.1/\text{h}$, and a constant pH of 4.5. A dilution rate of 0.1/h corresponded to about 3.5 generations per day. After ca. 36–40 h of growth the culture reaches a steady state. Each competition experiment was conducted in four replicates for about 35–42 generations (10–12 days) to avoid the incidence of secondary mutations, usually occurring after 50 generations. Samples of 10–15 mL were collected every day (*see* **Note** 7) and stored at –20°C. For each collection, OD_{600} values were measured and the genomic DNA extracted.

3.5. TAGs Amplification and Hybridization

To monitor the changes in the population profile, four samples were chosen to carry out the genomic hybridization using the TAG3 array: the initial population of cells before the inoculum, the sample after the batch culture phase, the culture at beginning of the steady state, and after 18–22 generations and at the end of the competition. The batch sample was useful for the identification of mutations that severely hamper the fitness of the strains resulting in their complete disappearance from the culture after the first 24 h of growth.

In this experiment, we have followed the standard protocol for the discrimination of the strains via their barcodes. This has been exhaustively described in several reviews (16–18); however, for the purpose of this chapter the major steps of the hybridization procedure and data analysis can be summarized as follows:

1. The genomic DNA is extracted from the samples and the amplification of all the barcodes is carried out using the universal primers. The UP-TAGs and DOWN-TAGs are amplified separately in two different PCR reactions using universal primers. Typically, the PCR mix contains 200 ng of genomic DNA as template, 0.2 μM of biotinylated BU1 and BU2-comp primers (for amplification of the UP-TAGs) or 0.2 μM of biotinylated BD1 and BD2-comp (for amplification of the DOWN-TAGs), 5 U of Platinum Taq DNA polymerase and 1X Buffer (Platinum PCR Supermix). The total volume of the PCR is 100 μL. The PCR cycle is as follows: 1 cycle at 94°C for 5 min, 35 cycles of 94°C for 20 s, 56°C for 20 s, 72°C for 30 s, and the last cycle at 72°C for 5 min. The PCR products are cleaned using Microcon YM10 columns and run on a 2% (w/v) agarose gel (5 μL). The total amount of the UP-TAGs and DOWN-TAGs is estimated by densitometry using a low DNA mass ladder as reference.

2. The PCR products are then hybridized to the TAG3 array. The chip is pre-washed with 180 μL of hybridization buffer before the hybridization mix is added. Hybridization mix: 500 ng of UP-TAGs, 500 ng of DOWN-TAGs, hybridization buffer, 1.8 μL of 50 mg/mL of BSA solution, 1.8 μL of 5 nM 213B control oligonucleotide solution (hybridize to the rim of the microarray), eight oligonucleotides solutions (25 μM final concentration each) corresponding to the four universal primers used for the tag amplification and their four complements (these primers serve to block the communal regions of the amplified TAGs before hybridization), and 7.2 μL of eukaryotic control. The hybridization mix (total volume 180 μL) is boiled for 2–5 min and cooled in ice, before being added to the chip. The hybridization is carried out for 18 h at 42°C with constant rotation in the GeneChip® Hybridization Oven 645. The chips are then mechanically washed with stringent and non-stringent buffers before being stained with 600 μL of streptavidin–phycoerythrin (SAPE) solution, followed by 600 μL of antibody solution and again 600 μL of SAPE solution in the GeneChip® Fluidics Station 450, according to manufacturer instructions. The SAPE solution is then removed and the chips washed with non-stringent buffer and scanned using the GeneChip® Scanner 300 7G, *see* **Note 8**.

3. The intensity of the hybridization signal from each individual spot reflects the amount of that specific barcode and consequently of that particular strain in the population.

The arrays are normalized by median centering intensity values from TAGs corresponding to the yeast strains, while the intensity values from the TAGs not corresponding to deletion mutants were taken as being representative of the background intensity and used to determine the presence or absence of deletion strains in the initial pool.

The log ratios are calculated from two biological replicates taken at three time points during the steady state: t0 (beginning of the steady state), t1 (mid-point), and t2 (end of the competition). Growth rates are estimated by robust linear regression on the normalized log ratios. Type I error rates (p values) are estimated by model-based resampling, with suitably rescaled residuals. To account for multiple testing, the false-discovery rate (FDR, q values) is estimated, according to the method of Benjamini and Hochberg, using the R statistical software environment (http://www.r-project.org/). Growth rates with $q < 0.01$ are considered as statistically significant and the corresponding ORFs selected for further bioinformatics analysis.

3.6. Conclusions and Future Directions

Several competition studies have now been carried out in either batch or chemostat cultures, using the TAG3 chip. Continuous cultures are capable of uncovering more haplo-insufficient genes than batch cultures due to their higher discriminatory power.

Quality control study of the yeast deletion collection was carried out by re-sequencing the molecular barcode (19). It was found that 31% of the TAGs contained differences from those designed originally. However, only a small subset of those caused decreased hybridization signal, below the background.

A new generation of array, TAG 4, was designed to improve the detection of the strains with sequence errors in their barcodes (20). In this array five replicates of each barcode are present and a larger set of control tags were added.

High-density microarrays provided so far an excellent tool for genome-wide assays such as the one described; however, the rapid advances in the next generation sequencing technology is now revolutionizing most of the array-based applications. A new method, barcode analysis by sequencing (Bar-seq), has been recently developed where the TAGs are discriminated by sequencing on the Illumina/Solexa platform (21). The Bar-seq method outperformed the classical hybridization array in terms of sensitivity, dynamic range, and detection. Moreover, using a DNA sequencer allows multiplexing (sequencing of several samples on the same flow cell) and eliminates the need of designing new microarray for different types of TAGs in other organisms.

4. Notes

1. A 12X MES stock (1.22 M MES and 0.89 M [Na^+]) is prepared, and the pH is checked in order to be between 6.5 and 6.7. The stock solution is filter-sterilized and stored at 4°C and protected from light. If the solution turns yellow it should be discarded.

   A 2X hybridization buffer is prepared to be added to the hybridization mixture to reach the final 1X final concentration. The 2X hybridization buffer should be stored at 4°C.
2. A 2X stain buffer is prepared to be added to the staining solution (so that the final concentration is 1X).
3. For each array, prepare a total volume of 1,200 μL of SAPE solution and divide in two 600 μL aliquots. Prepare 600 μL of antibody solution.
4. It is advisable that the diameter of the outlet tubing is at least 1.4 mm. Smaller tube sections may cause problems because the yeast cells can clog them up, provoking overspilling of the culture from the flasks that can damage the incubator.
5. When not in use, all feeding lines (pH buffer and medium input and output lines) should be kept in ethanol.
6. When connecting the lines, special care should be taken in order to avoid contaminations. Usually a 70% (v/v) ethanol solution is sprayed on the caps, connectors, and gloves, before linking the lines to the flasks.
7. The samples are collected from the outflow to keep the volume in the flask constant and to avoid oscillations of the steady state.
8. The array should be checked for air bubbles. If present, they need to be removed before scanning, by replacing the washing solution in the chip.

References

1. Dykhuizen, D. E., and Hartl, D. L. (1983) Selection in chemostats. *Microbiol. Rev.* **47**, 150–168.
2. Baganz, F., Hayes, A., Farquhar, R., Butler, P. R., Gardner, D. C. J., and Oliver S. G. (1998) Quantitative analysis of yeast gene function using competition experiments in continuous culture. *Yeast* **14**, 1417–1427.
3. Colson, I., Delneri, D., and Oliver, S. G. (2004) Effects of reciprocal translocations on the fitness of *Saccharomyces cerevisiae*. *EMBO Rep.* **5**, 392–398.
4. Winzeler, E. A., Shoemaker, D. D., Astromoff, A., et al. (1999) Functional characterization of the *S. cerevisiae* genome by gene deletion and parallel analysis. *Science* **285**, 901–906.
5. Shoemaker, D. D., Lashkari, D. A., Morris, D., Mittmann M., and Davis, R. W. (1996) Quantitative phenotypic analysis of yeast deletion mutants using a highly parallel

molecular bar-coding strategy. *Nat. Genet.* **14**, 450–456.
6. Giaever, G., Chu, A. M., Ni, L., et al. (2002) Functional profiling of the *Saccharomyces cerevisiae* genome. *Nature* **418**, 387–391.
7. Steinmetz, L. M., Scharfe, C., Deutschbauer, A. M., et al. (2002) Systematic screen for human disease genes in yeast. *Nat. Genet.* **31**, 400–404.
8. Wright R., Parrish, M. L., Cadera, E., et al. (2003) Parallel analysis of tagged deletion mutants efficiently identifies genes involved in endoplasmic reticulum biogenesis. *Yeast* **20**, 881–892.
9. Zakrzewska, A., Boorsma, A., Delneri, D., Brul, S., Oliver, S. G., and Klis, F. M. (2007) Cellular processes and pathways that protect *Saccharomyces cerevisiae* cells against the plasma membrane-perturbing compound chitosan. *Eukaryot. Cell* **6**, 600–608.
10. Giaever, G., Shoemaker, D. D., Jones, T. W., et al. (1999) Genomic profiling of drug sensitivities via induced haploinsufficiency. *Nat. Genet.* **21**, 278–283.
11. Lum, P. Y., Armour, C. D., Stepaniants, S. B., et al. (2004) Discovering modes of action for therapeutic compounds using a genome-wide screen of yeast heterozygotes. *Cell* **116**, 121–137.
12. Bivi, N., Romanello, M., Harrison, R., et al. (2009) Identification of secondary targets of N-containing bisphosphonates in mammalian cells via parallel competition analysis of the barcoded yeast deletion collection. *Genome Biol.* **10**, R93.
13. Deutschbauer, A. M., Jaramillo, D. F., Proctor, M., et al. (2005) Mechanisms of haploinsufficiency revealed by genome-wide profiling in yeast. *Genetics* **169**, 1915–1925.
14. Delneri, D., Hoyle, D. C., Gkargkas, K., et al. (2008) Identification and characterization of high-flux-control genes of yeast through competition analyses in continuous cultures. *Nat. Genet.* **40**, 113–117.
15. Holland, S., Lodwig, E., Sideri, T., et al. (2007) Application of the comprehensive set of heterozygous yeast deletion mutants to elucidate the molecular basis of cellular chromium toxicity. *Genome Biol.* **8**, R268.
16. Nislow, C., and Giaever, G. (2007) Chemical genomic tools for understanding gene function and drug action. *Methods Microbiol.* **36**, 387–414.
17. Pierce, S. E., Davis, R. W., Nislow, C., and Giaever, G. (2007) Genome-wide analysis of barcoded *Saccharomyces cerevisiae* gene-deletion mutants in pooled cultures. *Nat. Protoc.* **2**, 2958–2974.
18. Pierce, S. E., Davis, R. W., Nislow, C., and Giaever, G. (2009) Chemogenomic approaches to elucidation of gene function and genetic pathways. *Methods Mol. Biol.* **548**, 115–143.
19. Eason, R. G., Pourmand, N., Tongprasit, W., et al. (2004) Characterization of synthetic DNA bar codes in *Saccharomyces cerevisiae* gene-deletion strains. *Proc. Natl. Acad. Sci. USA* **101**, 11046–11051.
20. Pierce, S. E., Fung, E. L., Jaramillo, D. F., et al. (2006) A unique and universal molecular barcode array. *Nat. Methods* **3**, 601–603.
21. Smith, A. M., Heisler, L. E., Mellor, J., et al. (2009) Quantitative phenotyping via deep barcode sequencing. *Genome Res.* **19**, 1836–1842.

Chapter 17

Fluorescence Fluctuation Spectroscopy and Imaging Methods for Examination of Dynamic Protein Interactions in Yeast

Brian D. Slaughter, Jay R. Unruh, and Rong Li

Abstract

Protein interactions are inherently dynamic. In no system is this more true and important than in signaling pathways, where spatial and temporal control of specific protein interactions is key to signaling specificity and timing. While genetic and biochemical interactions form a necessary and important starting point for deciphering interactions among signaling components, they struggle to provide precise information of where and when interactions occur in a live cell setting. In contrast, live cell fluorescence studies such as those outlined below are able to provide quantitative information on the strength, nature, timing, and location of homotypic and heterotypic protein interactions.

Key words: Fluorescence fluctuation spectroscopy (FFS), fluorescence correlation spectroscopy (FCS), fluorescence cross-correlation spectroscopy (FCCS), number and brightness analysis (N&B), photon counting histogram (PCH), 2D PCH, fluorescence resonance energy transfer (FRET), signaling pathway.

1. Introduction

Signaling pathways consist of numerous components, each with a specific role or function in facilitating the general goal: a cellular response to a particular stimulus (1). In many cases, signaling molecules have many potential downstream targets that can lead to different responses. Therefore, mechanisms must be in place to control when and where the regulator–target interactions occur, as well as the strength and duration of these interactions. A well-characterized example is the dual targets of the mitogen-activated protein (MAP) kinase – kinase Ste7. It is able

J.I. Castrillo, S.G. Oliver (eds.), *Yeast Systems Biology*, Methods in Molecular Biology 759,
DOI 10.1007/978-1-61779-173-4_17, © Springer Science+Business Media, LLC 2011

to bind and phosphorylate the MAP kinases Fus3 and Kss1 – yet specificity is important for proper response to different stimuli, such as mating pheromone or nutrient conditions (2). In this case and others, in order to better understand how specific signaling responses are generated, it is necessary to work toward an understanding of when and where in a cellular context interactions between components of these highly intertwined signaling pathways occur.

Dynamic interactions in live cells may be characterized by a variety of fluorescence methods. In this chapter, we define dynamic interactions as those that occur with appreciable dissociation rates, and those that may occur transiently at a particular stage of the cell physiology or at a specific location. These interactions may occur between molecules with varying degrees of mobility, which is an important factor in determining the suitability of some of the detection methods. For example, in fluorescence resonance energy transfer (FRET), the interaction between two molecules, a donor and an acceptor, can be observed provided a close spatial relationship ($< \sim 10$ nm) exists. While this technique does not address the mobility of the molecules per se, it is able to diagnose where in the cell, or at what stage of the cell cycle, an interaction occurs. In contrast, fluorescence fluctuation spectroscopy (FFS) techniques such as fluorescence correlation spectroscopy (FCS), cross-correlation spectroscopy (FCCS), photon counting histogram (PCH), and number and brightness analysis (N&B) probe mobile particles only and reveal hetero-oligomeric interactions (FCCS) or homo-oligomeric interactions (PCH, N&B).

Below, we describe in detail methods for applying FRET and FFS techniques to examine dynamic protein interactions in live yeast cells. We do not include any discussion on fluorescence recovery after photobleaching (FRAP), reviewed elsewhere (3), which is a valuable tool for examining dynamics of protein interaction with the membrane or other structures but does not necessarily directly examine the interaction of specific protein pairs. For new work employing FFS techniques to examine fluctuation in time or space of pixels in a series of images, combining the examination of mobility and interactions of single-point FFS with spatial information, we point the reader to the following subset of references (4–8).

2. Materials

2.1. Media and Cell Culture

1. Yeast agar plates: For 500 mL, which makes 20 plates, mix 5 g Bacto-yeast extract, 10 g Bacto-peptone, 10 g

Bacto-agar, and 475 mL ddH_2O. Add 25 mL of sterile 40% (w/v) glucose after autoclaving.

Pour onto sterile plates (e.g., 25384-302, VWR) and allow plates to cool and harden.

For synthetic plates, which allow for selection in the absence (drop-out) of an essential amino acid, we use the following recipe: 3.35 g Bacto-yeast nitrogen w/o amino acids (Becton, Dickinson and Company, Franklin Lakes, NJ), 1 g synthetic complete drop-out mix (Sunrise Science, San Diego, CA), 10 g Bacto-agar (Becton, Dickinson and Company, Franklin Lakes, NJ), and 475 mL H_2O.

Autoclave, add 25 mL of separately autoclaved 40% (w/v) glucose. Pour onto sterile plates.

2. Synthetic liquid medium: (YPD medium is not used for liquid cultures used for microscopy observations due to its high level of autofluorescence): 3.35 g Bacto-yeast nitrogen w/o amino acids (Becton, Dickinson and Company, Franklin Lakes, NJ), 1 g Synthetic complete of drop-out mix (Sunrise Science, San Diego, CA), and 475 mL H_2O.

   Autoclave: 25 mL of sterile 40% (w/v) glucose (or other sugar as desired) will be added before use for a final concentration of 2% (w/v) (20 g/L). Keep in closed cabinet or wrap in foil. Light exposure over long times will lead to increased autofluorescence.

2.2. Microscopy

When choosing a microscope for fluctuation spectroscopy, there are several points to be noted:

(1) The microscope must be confocal or two photon with the capability to stably record intensity in a single spot for long periods of time. It is important that the microscope objective be as free of color and spherical aberrations as possible to minimize the size of the focal volume and maximize the overlap between red and green detection channels. These requirements are often easier to fulfill on a two-photon microscope with non-descanned detection. The microscope must also have the appropriate filters to separate colors well and eliminate scattered excitation light.

(2) The microscope must be equipped with detectors that have true photon counting detection with the highest sensitivity possible. It has become popular in recent years to perform gain calibration and back-calculate the number of "photons" from the observed intensity. These systems are advertized as photon counting, but will not work as effectively as true photon counting systems for correlation analysis and will fail completely for any PCH analysis. Typically, FCS measurements are made on systems with highly sensitive single photon counting avalanche photodiodes

(spAPDs) or gallium arsenide phosphide photomultiplier tubes (GaAsP-PMTs). Less sensitive detectors may work but with significant losses in sensitivity and the ability to detect single-molecule fluctuations. The data recording system should have the speed necessary to record the fluctuations of interest. Fluorescein, which is typically used as a calibration dye for FCS measurements, diffuses through a confocal focal volume in ~30 μs. Therefore, it is wise to have at least 10 μs time resolution for detection.

(3) It is advantageous to have a microscope with the ability to record images of live cells and select positions for FCS measurements. There are several systems commercially available. These include the Confocor module available with the Zeiss LSM microscopes (Jena, Germany), the Leica TCS SMD microscope (with electronics from PicoQuant (Berlin, Germany)), the MicroTime 200 system from PicoQuant, the Alba module from ISS (Champaign, IL), and the DCS 120 system from Becker and Hickl (Berlin, Germany). It is also possible to build an FCS system. This is much easier for two-photon systems than confocal systems. Several of the above products are available for user-built systems to make acquisition easier.

(4) In terms of software for data analysis, all of the above systems are shipped with software to acquire data, calculate correlation curves, and histogram data in at least one dimension. The Globals suite of software from the Laboratory for Fluorescence Dynamics (UC Irvine, CA) can also aid in acquisition for user-built systems. Most software packages also allow for fitting of correlation curves to complex models including multiple components and triplet blinking. All software packages support data export for analysis in other nonlinear least squares fitting programs. A few of the software packages allow data import from other acquisition systems. We have found the OriginPro (OriginLab Corp., Northampton, MA) software package especially useful for nonlinear least squares fitting. Only the Confocor, Alba, and Globals packages allow for fitting of PCH data. To our knowledge there are no software packages available for 2D PCH fitting. To this end, we have created a 2D PCH software plugin for the free image processing program, ImageJ. This plugin is available at http://research.stowers.org/imagejplugins. It is also worth noting that the Alba and Globals software packages as well as our software allow for simulation of FCS data. This can be useful for testing fitting models and learning how data analysis works.

(5) For FRET using acceptor bleaching, the system requirements are less stringent. The system must have capabilities for multichannel imaging. It must be possible to acquire image series, either pausing in the middle to scan with high red laser intensity to bleach the acceptor or acquiring separate donor time series before and after bleaching the acceptor (*see* **Section 3.3**).

3. Methods

The methods for detecting whether, when, and where interactions among signaling components occur in live yeast include FRET, FCS, FCCS, PCH, and number and brightness (N&B) analysis. There are a large number of technical references and protocols and research papers on these methods; here is a subset (5, 6, 9–13). In this chapter, we discuss protocols for applying these methods in live yeast cells, where in general the expression of fluorescent proteins can be kept at their low, native levels. This provides a huge advantage in yeast over systems where non-native overexpression is the main method for visualization of autofluorescent proteins (AFPs)-tagged proteins. Therefore, these protocols likely would need to be modified to work effectively in other systems.

The method that is best suited for observing interacting components depends on the nature of the interaction and protein mobility. We recommend first analyzing what is known about a group of proteins before deciding which technique is most likely to yield beneficial information on its interactions. For example, while FRET is limited to detection of direct interactions between proteins because it requires close proximity ($< \sim 10$ nm), it can detect interactions among not only highly mobile populations but also immobile populations. FRET also can provide highly specific information on the location of the interaction, for example, outer membrane vs. cytosol or internal membrane. In addition, under some conditions, distance information may provide useful information on the structural layout of a large protein complex both in vivo and in vitro (14, 15). Techniques based on fluorescence fluctuation spectroscopy (FFS), such as FCCS and PCH, are useful for detecting heterotypic or homotypic interactions, respectively, regardless of whether the interaction occurs directly or rather is on opposite ends of a large complex, because they simply measure co-diffusion. In contrast to FRET, FFS techniques require mobile particles. An attempt to apply FFS to immobile particles will result in rapid photobleaching and will not give meaningful results. Therefore, FFS is most easily applied toward rapidly moving cytosolic proteins, and we will limit our discussion to these scenarios. For recent work on combining scanning and fluctuation measurements to observe dynamics and interactions among slowly moving proteins, including membrane proteins, see the following subset of references (5, 7, 8, 16, 17).

3.1. Yeast Growth and Culture

(1) Yeast strains expressing one or more AFPs are grown in synthetic medium between 23 and 30°C for 8–10 h. Check optical density (OD) at 600 nm. Imaging experiments are best performed when the OD is between 0.5 and 0.8 (*see* **Note 1**). If OD is above 1.0, dilute cells to an OD of 0.2–0.3 and grow for 2–4 h to an OD between 0.5 and 0.8 before measurements are made.

(2) When possible, we highly recommend tagging genes with AFPs using homologous recombination. Direct C-term tagging, if not affecting protein function, is convenient and allows for observation of tagged proteins that are under their endogenous promoter and that represent the only copy present in the cell. Cassettes are available as templates for PCR to accomplish this (18, 19). In the case of proteins that are not functional with a C-term tag, necessitating N-term tagging, or if a mutation must be made, the gene may be cloned into a centromeric plasmid, such as the pRS313-pRS316 series (20). Alternatively, the N-terminus may be tagged with the AFP through a multi-step process in the endogenous locus under the control of the native promoter or a single-step process if a different promoter is sufficient (19, 21, 22). For a general protocol for gene tagging in yeast, *see* (23).

(3) A variety of AFPs are available for use in yeast experiments. In our experience, virtually any AFP used in other systems can be examined in yeast. However, we note that there may be large differences in expression level depending on whether the codon usage of the AFP has been optimized for yeast. For single-channel measurements, green (GFP) and yellow (YFP) fluorescence proteins and their variants are very effective. For multi-color applications, the cyan–yellow (CFP-YFP) pair will work, though there is significant spectral cross-talk, and we find cellular autofluorescence is a factor at the wavelengths used to examine CFP. The best pair of monomeric AFPs for two-color measurements to date, in our opinion, is GFP paired with mCherry (24) for both FRET and FFS.

(4) When a yeast strain is first established on a yeast agar plate (2% w/v glucose), grow the yeast strain to OD ~1.0 and mix 50:50 with 50% glycerol. Freeze the glycerol stock and keep at –80°C. Keep fresh yeast cells on plates for 3–4 days at 23–30°C, and use a small amount of cells to inoculate

liquid culture. Plates may be kept up to 10 days at 4°C. After 10 days, it is recommended to discard old plates and plate fresh cells from the frozen glycerol stock (*see* **Note 2**).

3.2. Slide Preparation

Yeast cells are first grown to the optimal OD range for fluorescence imaging experiments.

(1) (Optional) 1.0 mL cells may be spun down and re-suspended in 50–100 μL prior to pipetting onto a glass slide to increase cell density. For imaging experiments, if cells are spotted too densely such that cells are touching or are stacked on top of one another, dilute cells and repeat until the desired plate density is reached.

(2) For short-time experiments, or experiments that do not include time-lapse series longer than 20 min, cells may be pipetted directly on glass cover slips (*see* **Note 3**). Pipette 3–4 μL onto a microscope slide and gently cover with a 22 × 22 mm micro cover glass. To help immobilize cells on the slide, turn slide over and press downward very gently onto lens paper. If pressed too roughly, it may cause cells to rupture (*see* **Note 4**).

(3) For experiments longer than 20 min, i.e., long time-lapse movies, an agarose or gelatin pad can be used to supply nutrients to the yeast to allow for survival. We find this results in a slight increase in background fluorescence. Details for this protocol can be found elsewhere (25).

3.3. FRET Measurements in Live Yeast Cells Using Acceptor Photobleaching

There are a number of methods for measuring FRET, including sensitized emission, acceptor photobleaching, and quenching of the fluorescence lifetime of the donor. In our experience, acceptor bleaching is the most straightforward and easily quantified method for detecting FRET in live yeast cells and will be the method discussed here. In addition, acceptor bleaching does not require quantitative measurement of the fluorescence of the acceptor, but it has to be only ensured that it is bleached. As red AFPs are generally less bright than GFP, this provides an advantage for acceptor bleaching over sensitized emission, which relies on quantitative measurements of the intensity of the red probe. For a general review on FRET, *see* (26).

(1) For FRET experiments, it is necessary to have one protein tagged with a donor, such as GFP, and a separate protein tagged with an acceptor, such as mCherry or mStrawberry (24). Great care must be taken to avoid extended imaging prior to measurement of FRET due to the possibility of AFP bleaching during normal image acquisition, especially for the red AFPs, which are less photostable than GFP (24, 27) (*see* **Note 5**). Photobleaching, especially of the red AFP, during regular confocal imaging prior to

FRET measurement will lead to underestimation of FRET. If possible, for medium to low expressing proteins, we recommend avalanche photodiodes (spAPDs) as opposed to PMTs for greater detection sensitivity, which correspondingly allows for use of lower excitation powers (*see* **Section 2.2** and **Note 5**). If spatial resolution in the *z*-axis is not required, open the pinholes on a confocal microscope to allow greater signal at low excitation powers.

(2) Verify that the chosen cell is expressing both donor and acceptor probes by imaging individual channels.

(3) Choose a region of interest for acceptor bleaching. This may include a certain area in the cell or may be the entire cell. If the entire cell is not bleached, rather only a certain area, mobility of the species must be accounted for. We generally recommend to bleach the acceptor in the entire cell.

(4) Acquire 3–5 images, exciting the donor and collecting only donor fluorescence, then bleach the acceptor with a red laser line (544 or 561 nm, for example) with 5–10 scans at high laser power. After acceptor bleaching, acquire 3–5 images of the donor.

(5) Verify that the red AFP is bleached by scanning once with the red laser and comparing with the initial red image before bleaching.

(6) Separately sum the scans of the green channel before and after bleaching of the red probe and compare. Divide the summed after bleach image by the before bleach image. An increase represents FRET. For an example of FRET between the MAPK signaling pathway scaffold Ste5 and the kinase Fus3, *see* **Fig. 17.1**.

(7) Verify that FRET is not underrepresented or missed altogether due to donor photobleaching during regular image acquisition, and that donor is not bleached by the repetitions with the red laser (*see* **Note 6**). As a control, repeat the above protocol with a yeast strain expressing only donor, and not acceptor, using the exactly same image acquisition protocol and laser powers. Verify that donor fluorescence does not change during the time-course of the control experiment. If there is donor bleaching during normal acquisition, if possible, adjust image acquisition parameters to reduce or eliminate acquisition photobleaching. If a small amount of image photobleaching occurs at excitation powers necessary for sufficient signal to noise to quantitate the data, a correction factor may be employed from the control, donor only measurements to apply to the experimental measurement.

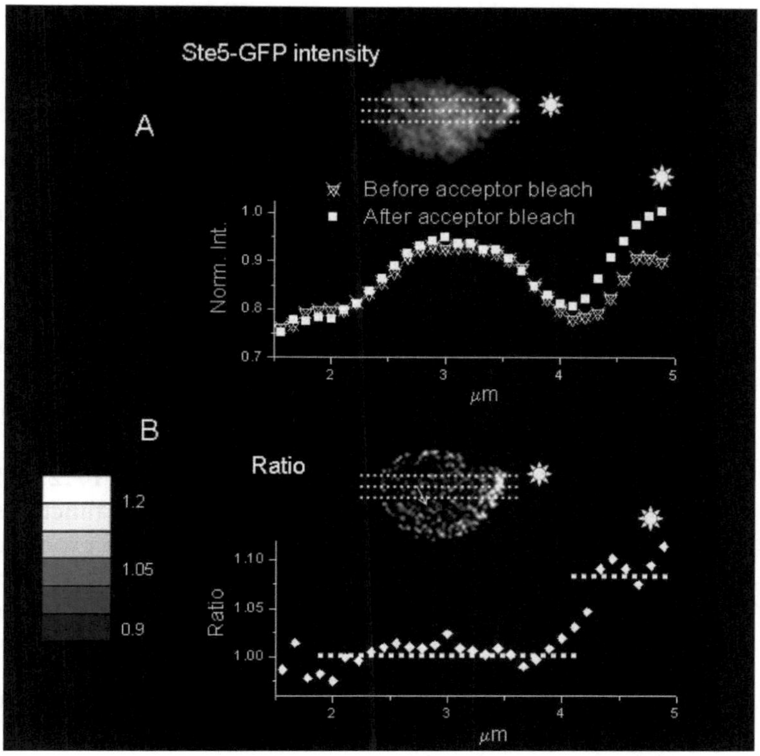

Fig. 17.1. Example of a FRET acceptor bleaching experiment between Ste5-GFP and Fus3-mStrawberry in pheromone-treated cells (adapted from Ref. (28), with permission). (**a**) Intensity line trace of a yeast cell expressing Ste5-GFP. Fus3-mStrawberry has a similar profile (not shown). Ste5-GFP intensity along the line trace is shown before and after acceptor bleach. (**b**) Ratio of Ste5-GFP intensity after acceptor bleaching relative to before. A ratio above 1.0 is observed at the tip, indicative of interaction between Ste5 and Fus3.

(8) FRET efficiency is calculated as $E = 1 - (I_B/I_A)$, where I_B and I_A are the donor intensity before and after acceptor bleach. FRET efficiency is converted to a distance using the following formula, though for FRET between AFPs, there are very significant assumptions that must be made to convert FRET efficiency to a quantitative distance (*see* **Note 7**). For this reason, often simply a FRET efficiency is given as qualitative evidence of an interaction

$$R = \left(\frac{1-E}{E}\right)^{1/6} \cdot R_0. \quad [1]$$

In addition, we note that this method does not distinguish % FRET of interacting proteins from the percentage of interacting proteins.

(9) In theory, any AFP pair can be used where there is an overlap of the donor emission with the acceptor absorption. For

acceptor photobleaching studies, we find that bleaching of CFP at wavelengths necessary to photobleach the acceptor YFP or its variants complicates analysis. We recommend pairing GFP with a monomeric red AFP, such as mStrawberry or mCherry (24).

3.4. Fluorescence Correlation Spectroscopy (FCS)

FCS, FCCS, and PCH are based on statistical analysis of fluctuations in emission intensity (**Fig. 17.2**). The mathematics of FCS and FCCS was developed many years ago (29, 30), but it is only recently that these techniques have gained popularity for live cell measurements as improvements in commercial microscopes have made them accessible to scientists with a wide range of expertise. The base form of the data in fluctuation studies is a single or double fluorescence trace of photon counts at a given location as a function of time (**Fig. 17.2a**). These traces are analyzed with auto- or cross-correlation functions or number and brightness analysis (N&B) or PCH to expose underlying dynamics and interactions of the AFP-tagged components. We start here with a general protocol for acquiring fluctuation data and then diverge into the different methods of data analysis to examine homotypic or heterotypic interactions.

(1) The use of photon counting avalanche photodiodes spAPDs is recommended for FCS and FCCS experiments (*see* **Section 2.2.2**).

(2) Once a cell is located, orient the beam at the chosen location in the cell. On commercial microscopes, this is often included as an option in the software (*see* **Section 2.2.3**) If desired, a separate color AFP can be used to mark locations for analysis, for example, by tagging a protein that localizes to that structure with mCherry. However, analysis inside organelles is limited in yeast due to the small size of many of these structures. The focal volume on most standard confocal scopes is approximately 0.4–0.5 μm in diameter in the radial dimension, and approximately 1–2 μm in the axial dimension. Attempts to take measurements in yeast organelles smaller than this will not truly probe that particular region. In addition, small structures will likely move out of the focal volume during the acquisition time. Thus, measurements are typically made either in the cytosol or in the nucleus.

Immediately following the start of fluorescence data collection, roll the focus up, and then down through the cell before settling in the center. This center may or may not correspond to maximum counts, and likely will not for proteins with a large membrane pool. However, the center point between the focal volumes outside the cell on either end (as photon counts drop sharply) should correspond

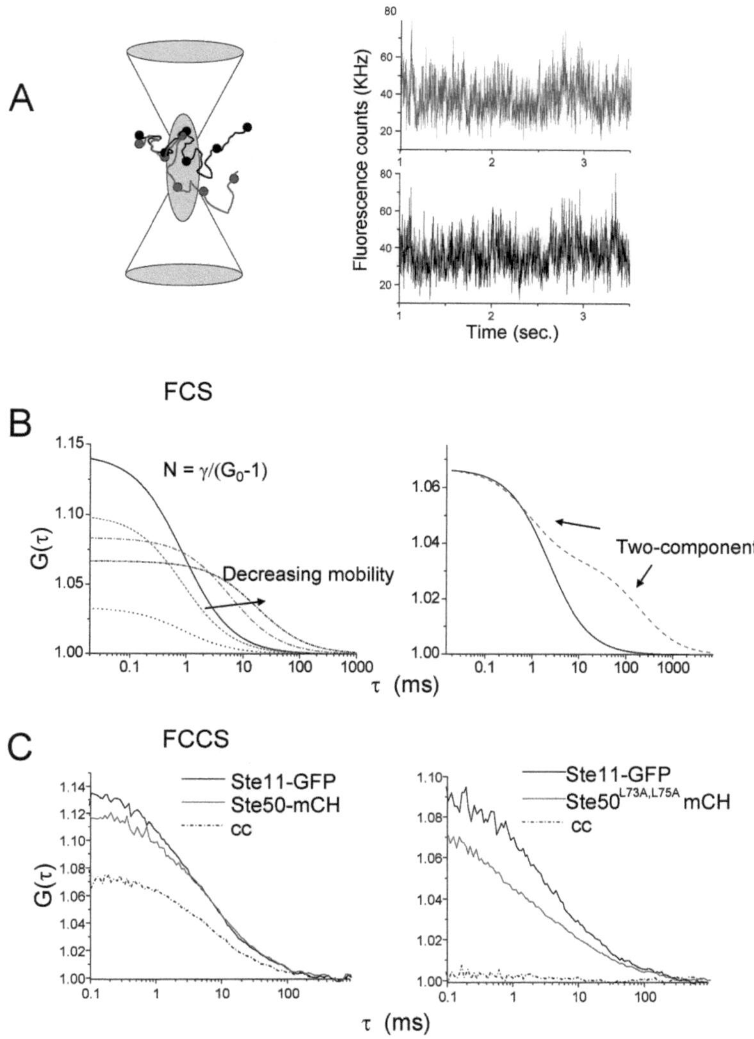

Fig. 17.2. Examples of FCS and FCCS analyses. (**a**) Fluctuations in emission intensity are obtained as molecules diffuse through the focal volume (adapted from Ref. (34), with permission). (**b**) Fluctuations in individual channels may be autocorrelated to reveal information on concentration, mobility, and heterogeneity. Simulated results are shown. (**c**) In two-channel experiments, the two channels may be cross-correlated. A high cross-correlation amplitude relative to the amplitude of the autocorrelation of the individual channels indicated co-diffusion of the two species. Example curves from live yeast are shown for cross-correlation between MAPK pathways proteins Ste11 and Ste50 (adapted from Ref. (34), with permission). The strong cytosolic interaction of these two proteins is disrupted by mutation of the SAM domain of Ste50.

to the middle. This avoids measurements where a significant part of the focal volume is outside of the cell (due to the small size of yeast cells, in some cases the axial resolution of the confocal measurement may approach the axial dimension of the cell).

(3) Collect 3–5 measurements, each 3–7 s in length, per cell (*see* **Note 8**).

(4) The laser power necessary to achieve sufficient signal to noise will vary depending on the overall detection efficiency of the system and the objective. For reference, using a commercially available Zeiss Confocor 3, and a 40x 1.2NA C-Apochromat, we typically use a laser power of ~3.0 μW or 0.15% laser power at the back aperture.

(5) Once the fluorescence fluctuation trace is obtained, it is analyzed by autocorrelation (equation [2]) (*see* **Fig. 17.2**).

$$G(\tau) = \frac{\langle I(t) \cdot I(t+\tau) \rangle}{\langle I(t) \rangle^2} \quad [2]$$

(6) For free 3D diffusion, data can be fit to equation [3], where N is the average number of molecules in the focal volume, τ_D is the transit time through the focal volume, and r_0 and z_0 represent the ~½ radial and axial dimensions of the focal volume, respectively. *See* **Note 9** for discussion of γ. This equation can be expanded to include multiple components.

$$G(\tau_{ac}) = \frac{\gamma}{N} \frac{1}{1 + \left(\frac{\tau}{\tau_D}\right)} \frac{1}{\left(1 + \frac{r_0^2}{z_0^2}\left(\frac{\tau}{\tau_D}\right)\right)^{1/2}} \quad [3]$$

It is sometimes difficult to freely fit the "structure parameter," r_0/z_0. The value of z_0 is often restricted to three to five times larger than the value of r_0.

(7) A least squares fitting algorithm is used to fit the correlation decay to the chosen function.

(8) The most reliable way to find the size of focal volume is to acquire FCS data on a sample of known concentration and diffusion coefficient. Fluorescein in 0.1 M NaOH works well. The size of the confocal volume is found by comparison of the concentration of the sample with the average number of molecules in the focal volume (31).

(9) We recommend fitting fluorescein correlation data to equation [3] to find τ_D and the structure parameter (r_0/z_0), then using the diffusion published coefficient of fluorescein (32, 33) and equation [4] to find the ½ radial dimension of the focal volume. A known radial dimension of the focal volume will then allow for calculation of diffusion coefficient for complexes of interest, using the τ_D from the fit of their correlation curves.

$$D = \frac{{r_0}^2}{4\tau_{\mathrm{D}}}. \qquad [4]$$

(10) Diffusion is related to molecular size by the Stokes-Einstein equation

$$D = \frac{kT}{6\pi \eta r}, \qquad [5]$$

where k is the Boltzmann constant, T is temperature in Kelvin, η is viscosity, and r is molecular radius. For practical applications in yeast, we recommend constructing a control strain to observe the transit time τ_{D} of monomeric GFP in the yeast cytosol. From combination of equations [4] and [5], we see that change in τ_D is proportional to change in molecular radius, which goes as the cube root of molecular weight (assuming spherical geometry). By comparison of τ_D of the protein of interest to τ_D of GFP, which has a known molecular radius, the size of the diffusing complex can be estimated.

3.5. Fluorescence Cross-Correlation Spectroscopy (FCCS)

While FCS is the starting point for any fluorescence fluctuation study and gives useful information on complex size, mobility, and mobile concentration (**Fig. 17.2**), in reality it must be expanded to truly identify heteroprotein interactions. One channel fluctuation data can be treated with N&B or PCH analysis to observe homotypic protein interactions (*see* **Section 3.6**). To detect the strength of interactions among two different proteins in live cells, some form of two-color analysis must be applied, for example, FCCS or two-color N&B. In FCCS, the basic concept is the same as FCS, but in this case two spectrally distinct AFPs are used. Their individual fluctuations are analyzed and cross-analyzed to detect co-diffusion, and thus interaction, of the signaling pathway components (**Fig. 17.2**). GFP and mCherry form one effective pair of AFPs that can be used together for FCCS measurements in live yeast (28, 34, 35).

(1) Steps 1–3 of **Section 3.4** are followed. Take special care to find cells from imaging with as few scans as possible to minimize photobleaching of AFPs prior to obtaining FCCS data. We typically use 488 and 561 nm excitation for GFP and mCherry, respectively, separate emission with a 565LP emission dichroic and collect GFP and mCherry emission through 505–540 BP and LP580 filters, respectively.

(2) The individual fluorescence traces of the individual channels are autocorrelated and fit as discussed in **Sections 3.4.5** and **3.4.6**.

(3) The cross-correlation of the channels is calculated from equation [6],

$$G(\tau) = \frac{\langle I_G(t) \cdot I_R(t+\tau) \rangle}{\langle I_G(t) \times I_R(t) \rangle} \quad [6]$$

where G and R represent green and red channels, respectively.

(4) From the amplitudes of the curves, the number of green and red particles is used to calculate the number of bound species (equation [7]) (36), where $N_{GT} = \gamma/(G_{0Green} - 1)$, for example. N_{cc} is defined similarly by the inverse amplitude of the cross-correlation curve. Importantly, the factor Q is used to correct for spectral leakage of green fluorescence into the red channel. This parameter must be accurately determined and taken into account to avoid false-positive cross-correlation results due to cross-talk, most notably from the green probe into the red emission channel (*see* **Note 10**).

$$N_{bound} = \frac{N_{GT}(N_{RT} + Q \cdot N_{GT})}{N_{cc}} - N_{GT} \cdot Q \quad [7]$$

(5) We highly recommend employing a negative control to verify that observed cross-correlation in experimental samples is not due to spectral cross-talk. This can be accomplished by measuring cross-correlation in a yeast strain co-expressing cytosolic GFP and mCherry. Better yet, if possible, a biological control can be implemented where a mutation is made to one of the proteins that has been shown biochemically to reduce the interaction. For example, a SAM-domain mutation in Ste50 is known to disrupt the interaction of Ste50 and Ste11 (37). We find that this mutation also disrupts the in-vivo cross-correlation (**Fig. 17.2c**). Likewise, we recommend employing a positive control, such as linked GFP-mCherry, as a comparison point for strength of interactions in other samples. We find that even with a linked construct, we observe much less than 100% cross-correlation, likely due to bleaching or incomplete expression or folding of the AFPs (28, 34).

3.6. Detection of Homotypic Interactions Through One Color Number, Brightness Analysis, and the Photon Counting Histogram

Oligomerization of proteins plays a key role in many cellular systems. Proteins may oligomerize in order to assume an activated or inhibited state or to form structural polymers. While structural and biochemical studies may detect the ability of a protein to oligomerize at relatively high concentrations, it is often difficult to know if oligomer formation occurs in native settings and at native expression levels. The photon counting histogram (PCH) and the more computationally straightforward N&B analysis are biophysical techniques that enable quantitative observation of oligomerization in live cells.

The development of moment-based methods (also known as number and brightness analysis, N&B) for the determination of molecular brightness was begun by Quian and Elson in the early 1990s (38, 39). The basic concept has not changed significantly since that time, though several studies have expanded on the original principles. For a subset of this work, *see* (40–42). The technique discussed here makes use of the first two moments: the average and the variance (*see* **Fig. 17.3a**). Following the protocols discussed above, we start with fluctuation data consisting of a single- or dual-color time trace of photons and their arrival times.

(1) If a photon counting intensity trace contains only molecular diffusion and shot noise, the average molecular brightness is defined as follows:

$$\langle \varepsilon \rangle = \frac{\sigma^2 - \langle I \rangle}{\gamma \langle I \rangle} = \sum_i f_i \varepsilon_i \quad [8]$$

where σ^2 is the variance of the intensity in time, I is the average intensity of that signal, γ is a factor accounting for the shape of the focal volume (*see* **Note 9**), f_i is the fractional intensity of species i, and ε_i is the molecular brightness of that species. The molecular brightness is defined as the counts per sampling time per molecule at the center of the confocal focus. This is usually represented as counts per second per molecule (CPSM). The intensity is simply the product of the molecular brightness and the average number of molecules in the focal volume.

(2) If only a single species is present, the number of molecules is calculated as follows:

$$N = \frac{\langle I \rangle}{\langle \varepsilon \rangle} = \frac{\gamma \langle I \rangle^2}{\sigma^2 - \langle I \rangle}. \quad [9]$$

Concentration can then be calculated from the number of molecules given that the size of the focal volume is known (*see* **Section 3.4.8**) (*see* **Note 11**).

(3) Once the molecular brightness is known, it is straightforward to calculate many biologically relevant parameters. The most straightforward of these is an estimation of oligomeric state and heterogeneity. First, if the average molecular brightness is greater than the brightness of the monomeric species, there must be some oligomerization. In the case of dimerization, one can typically approximate the brightness of the dimer as double the brightness of the monomer. In this case the relative concentrations of

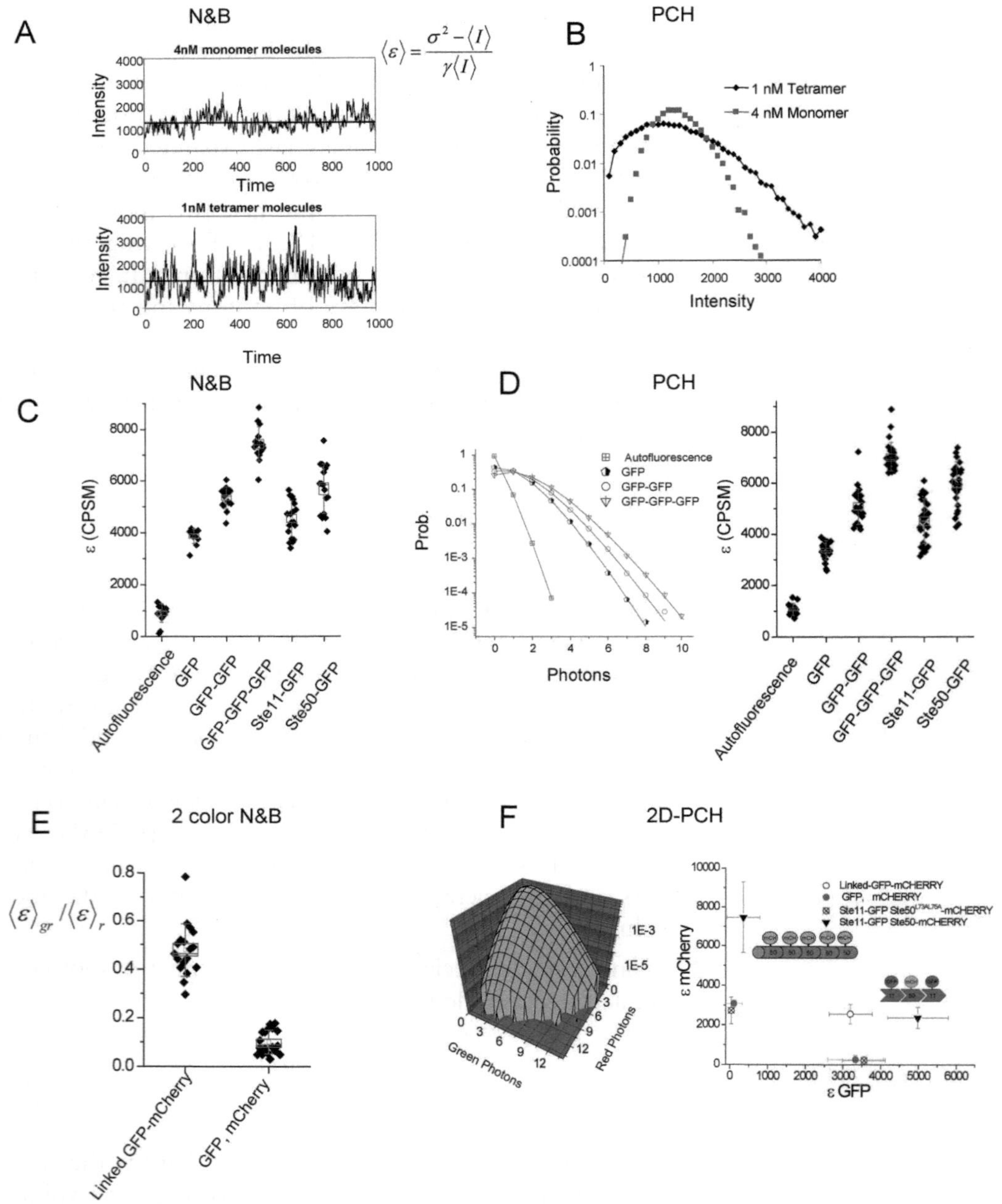

Fig. 17.3. N&B and PCH analyses. (**a**, **b**) Simulated data of monomer and tetramer diffusing particles. The average brightness can be calculated directly from the data using N&B (**a**) or (**b**) by calculating and fitting a photon counting histogram (PCH). (**c**) Fit of data from live yeast cells with N&B to find average brightness of controls, along with Ste11 and Ste50. (**d**) Representative photon histograms and fits for controls, and average brightness values for controls and Ste11 and Ste50 fit using PCH (adapted from Ref. (34), with permission). Note that one-component fits with N&B and PCH are shown here, and these values represent average brightness for Ste11 and Ste50. (**e**) Application of two-color N&B to observe interaction of GFP and mCherry in live yeast. Dual-color brightness divided by brightness of the red species is plotted for linked GFP-mCherry and a yeast strain co-expressing cytosolic GFP and mCherry. $\varepsilon_{\mathrm{gr}}/\varepsilon_{\mathrm{r}}$ is an estimation of % interaction and is consistent with FCCS results between monomeric GFP and mCherry in live yeast cells (28, 34) (*see* **Section 3.5.5**). Note that $\varepsilon_{\mathrm{gr}}/\varepsilon_{\mathrm{r}}$ is not zero for the negative control due to spectral cross-talk. (**f**) Example of 2D PCH histogram, fit, and average results for various yeast strains (adapted from Ref. (34), with permission). Ste11 and Ste50 are present in a complex with 2:1 stoichiometry, while Ste50 also exists as a high order oligomer in a complex without Ste11. A SAM-domain mutant of Ste50 disrupts the interaction and the oligomerization.

the monomer and dimer can easily be determined from the average brightness as follows:

$$N_{\mathrm{m}} = \frac{\langle I \rangle \left(\langle \varepsilon \rangle - \varepsilon_{\mathrm{d}} \right)}{\varepsilon_{\mathrm{m}}^2 - \varepsilon_{\mathrm{m}} \varepsilon_{\mathrm{d}}} \quad [10]$$

$$N_{\mathrm{d}} = \frac{\langle I \rangle \left(\langle \varepsilon \rangle - \varepsilon_{\mathrm{m}} \right)}{\varepsilon_{\mathrm{d}}^2 - \varepsilon_{\mathrm{d}} \varepsilon_{\mathrm{m}}} \quad [11]$$

where the subscripts m and d denote monomer and dimer species. This can be expanded to include higher order oligomers, with some limitations (*see* **Note 12**).

(4) In addition to diagnosis of oligomerization, the average brightness is sensitive to the immobile species as well as the background in a measurement. One can show that the background has zero molecular brightness. Nevertheless, it contributes to the total intensity observed and therefore influences the fractional intensity from equation [8]. The fractional intensity of the immobile and background signal is then given by

$$f_{\mathrm{immobile}} = \frac{\langle \varepsilon \rangle_{\mathrm{mobile}} - \langle \varepsilon \rangle}{\langle \varepsilon \rangle_{\mathrm{mobile}}}. \quad [12]$$

The average molecular brightness of the mobile species must be known beforehand.

Obviously, most biological systems do not fit perfectly into the two state solution above. It would be useful to have an analysis method which contains more flexibility in the measurement of complex equilibrium. Two basic solutions have been proposed in the literature. The first is to fit the intensity histogram (43, 44) and the second is to analyze higher order intensity moments (40). We will describe the first approach because of its historical context and relative simplicity. In 1999, two algorithms were published for intensity histogram analysis: the photon counting histogram (44) and fluorescence intensity distribution analysis FIDA (43). While these algorithms differ in the details of calculation they contain essentially the same basic elements (45). We will describe the PCH method here.

If shot noise did not exist and the confocal volume was sharply bounded, the intensity histogram of photon events in a given bin size would have multiple peaks corresponding to the different oligomers present. Each of these peaks would be distributed according to the distribution in the number of molecules in the focus. This distribution is Poisson for most systems. In reality, this distribution is further broadened by the diffuse nature of the focal volume and shot noise (**Fig. 17.3b**). The final distribution is often referred to as "super-Poisson." Any methodology for analyzing the intensity histogram must take these noise contributions into account. The PCH method begins by calculating the intensity histogram for a single particle based on all of its possible positions within the focal volume, its molecular brightness, and shot noise. Multiple particle histograms are then calculated by convolving the single particle histogram with itself. The final PCH is calculated by weighting the multiple particle histograms by the probability of having each possible number of particles in the focal volume given an average concentration. Multiple species are convolved with one another to obtain a final PCH. This curve is compared with the experimental intensity histogram and then updated using a traditional nonlinear least squares approach to obtain a fit of the data (44).

The mathematics of PCH analysis is complicated (44, 46) and is not presented here. We recommend that beginning users employ a commercially available package (*see* **Section 2.2.4**). We further recommend that a control series of monomer, dimer, and trimer species are examined prior to making conclusions about the oligomerization of the protein of interest (28, 34) (examples – **Fig. 17.3c, d**).

(5) Construct a series of yeast controls, expressing either 1x, 2x, or 3x linked cytosolic GFP. Using identical acquisition settings, acquire FFS data for each.

(6) Fit data to a single component, using a histogram bin time to generate the PCH that is less than ~1/4 the τ_D of the protein. Use of a histogram bin time that approaches the diffusion time of the molecule will lead to underestimation of molecular brightness (47).

(7) Using identical acquisition, the average molecular brightness of the species of interest may be compared to the monomeric, dimeric, and trimeric controls. If desired, the PCH can be fit to multiple species to extract an oligomeric model of the system (*see* **Note 12**).

3.7. Detection of Hetero-oligomeric Interactions Through Two-Color PCH and Number and Brightness Analysis

Given the success of single-channel brightness analysis, it is interesting to extend this to the analysis of dual-color intensity data. The resulting 2D histogram (**Fig. 17.3f**) will contain information about coincidence of single molecules in the two channels as well as molecules that are confined to a single channel (10, 48, 49). Once again, there are simplified, moment-based methods (N&B) (5) as well as distribution fitting methods for this analysis. For this protocol, we assume we are starting with a dual-color time trace of photon events and arrival times. The same data as is used for

FCCS can generally be used for two-color brightness analysis. We present details of two-color moment analysis (5) and a protocol for 2D PCH. The mathematics of 2D PCH is not presented here.

(1) For moment analysis, we define a dual-color brightness based on the covariance of the two signals:

$$\langle \varepsilon \rangle_{\mathrm{gr}} = \frac{\sigma_{\mathrm{gr}}^2}{\gamma \sqrt{\langle I \rangle_{\mathrm{g}} \langle I \rangle_{\mathrm{r}}}} = \sum_i \sqrt{f_{i,\mathrm{g}} f_{i,\mathrm{r}} \varepsilon_{i,\mathrm{g}} \varepsilon_{i,\mathrm{r}}} \quad . \qquad [13]$$

Here g and r denote the green and red channels. Other colors can, of course, be used but we have chosen to use green and red to denote GFP and mCherry. The γ function is the same as in the single-color analysis and σ_{gr} is the covariance of the green and red channels.

In the absence of cross-talk and interaction, the dual-color brightness is zero. Typically cross-talk occurs from the green to the red channel, but not vice versa. We recommend similarly to FCCS, to start with a negative control of non-interacting green and red species, for example, monomeric GFP and mCherry. With no particles interacting, the dual-color brightness due simply to spectral cross-talk is given by

$$\langle \varepsilon \rangle_{\mathrm{gr,independent}} = \langle \varepsilon \rangle_{\mathrm{g}} \beta \sqrt{\frac{\langle I \rangle_{\mathrm{g}}}{\langle I \rangle_{\mathrm{r}}}}, \qquad [14]$$

where β is the ratio of the green species intensity in the red channel to the green species intensity in the green channel. This is the minimum dual-color brightness that can be observed.

(2) Any interaction will present itself as an increase in this parameter. First, independently find the molecular brightness of the red and green species, as discussed in **Section 3.6**. Then it is straightforward to calculate the number of interacting species:

$$N_{\mathrm{bound}} = \frac{\langle \varepsilon \rangle_{\mathrm{gr}} \sqrt{\langle I \rangle_{\mathrm{g}} \langle I \rangle_{\mathrm{r}}}}{\langle \varepsilon \rangle_{\mathrm{g}} \varepsilon_{\mathrm{r}}} - \frac{\beta \langle I \rangle_{\mathrm{g}}}{\varepsilon_{\mathrm{r}}} \qquad [15]$$

where ε_{r} is the brightness of the red species in the red channel and $\langle \varepsilon \rangle_{\mathrm{g}}$ is the average brightness of the green channel.

(3) The number of unbound green species is then given by

$$N_{\mathrm{unbound,g}} = \frac{\langle I \rangle_{\mathrm{g}}}{\langle \varepsilon \rangle_{\mathrm{g}}} - N_{\mathrm{bound}} \qquad [16]$$

Finally, the number of unbound red species is given by

$$N_{\text{unbound,r}} = \frac{\langle\varepsilon\rangle_{\text{r}}\langle I\rangle_{\text{r}} - \beta^2 \langle\varepsilon\rangle_{\text{g}}^2 N_{\text{unbound,g}} - \left(\beta\langle\varepsilon\rangle_{\text{g}} + \varepsilon_{\text{r}}\right)^2 N_{\text{bound}}}{\varepsilon_{\text{r}}^2} \quad [17]$$

Fig. 17.3e shows an example of two-color N&B analysis applied in live yeast.

In its simplest form, two-color N&B is very similar to cross-correlation, in that it reports concentrations of each species and the concentration of bound species. It is more straightforward and more easily applied to images (5) relative to cross-correlation and less dependent on the shape of the focal volume. In addition, if some restrictions are added to the analysis, such as the brightness of the individual green and red species, two-color N&B will report the stoichiometry of the bound complex.

(4) Similar to two-color moment analysis, a 2-dimensional histogram can be generated of coincident photon events in each channel per bin time (*see* **Section 3.6.6** for discussion of bin sizes). An example is shown in **Fig. 17.3f**.

(5) Least squares fitting is used to find the brightness of each species in each channel (**Section 2.2.4**). This analysis will report on the stoichiometry of each individual species and the bound complex.

(6) Similar to multi-species PCH, while 2D PCH is very powerful, statistics can be limited in live yeast cell applications. We highly recommend first examining positive and negative controls to demonstrate the ability to detect interactions and find expected bleedthrough percentages for non-interacting particles. When possible, fitting is markedly improved if restraints can be imposed. For example, if one knows from FCCS analysis of the same data that ~ 50% of the red particles are interacting and one knows from one-color PCH or N&B analysis that the green probe is always monomeric, these restrictions can be added to the fitting procedure.

In summary, there are multiple methods to detect interactions in live yeast cells. We find that acceptor photobleaching is a reliable and relatively simple method for examining if close spatial interactions are present. FCS reports on the concentration and relative mobility of diffusing species, while PCH and single-color N&B report on the stoichiometry of homotypic interactions. FCCS, two-color N&B, and 2D PCH are extensions of these techniques that can be used to examine heterotypic interactions and stoichiometry of interacting complexes. The method most appropriate for a given application depends on the nature of the interaction and the mobility of the species of interest. We are confident that continued work toward applying these techniques in yeast will lead to significant advances in our understanding of protein–protein interactions and will enable testing of hypotheses and models of signaling pathways proposed based on genetic and biochemical results.

4. Notes

1. We find that FFS examination of cells with an OD over 0.8–1.0 results in increased autofluorescence and increased propensity for GFP and mCherry to photobleach.
2. We find that liquid cultures made from cells maintained on plates for a period of time longer than 10 days result in increased autofluorescence and increased photobleaching of GFP and mCherry.
3. Cells may be maintained on glass slides for 20–30 min. We have found from imaging and FFS studies that after that time frame, the growth rate of cells decreases and mobility of proteins inside the yeast cytosol decreases. For studies requiring longer acquisition times, use agarose pads (25).
4. Care must be taken to avoid pressing the glass surfaces together too harshly. This can result in ruptured cells. The presence of ruptured cells can normally be observed through the eye-pieces of a microscope using a high magnification objective. Attempts to use FFS on these cells will be unsuccessful.
5. Red autofluorescent proteins are less photostable than GFP. For two-color studies, including FRET and FCCS, normally 488 nm excitation is used to locate cells for corresponding measurements. Red AFPs also will absorb 488 nm light, albeit at low efficiency. Care must be taken to acquire as few images as possible prior to FFS or FRET acquisition to limit photobleaching.
6. We find that GFP has very little, if any, absorption of 561 nm light. Bleaching of GFP by the red laser should not be an issue.
7. A major assumption of calculating FRET is the value of the orientation factor κ^2 (50). Most FRET calculations assume random orientations of the dipoles of donor and acceptor relative to one another over the time scale of the energy transfer, and a corresponding average κ^2 value of $2/3$. For large AFPs, orientational mobility will be low, causing large uncertainty due to the $2/3$ approximation for κ^2.

8. To limit photobleaching in singe point FCS and FCCS, we recommend acquiring 3–5 measurements, each 3–6 s, per cell. This allows for exclusion of data when the cell moves in the middle of data acquisition, for example. For an examined protein or protein–protein interaction, data can be averaged after acquisition to improve statistics.
9. For a sufficiently small confocal pinhole, the value of γ is typically taken to be 0.3536. For two-photon excitation, the value is ~0.0760 (51). While membranes in principle represent a 2D surface, and therefore should have a higher gamma value than would be for 3D diffusion, recent work has shown that the observed gamma value does not significantly change (Paul Wiseman, personal communication).
10. The value of Q corresponding to donor emission bleedthrough into the acceptor channel can be approximated based on the respective emission profiles, combined with the transmission profiles of the emission filters used. It is highly dependent on the AFPs used and the filters chosen. Using a Zeiss confocor 3, the GFP/mCherry pair, and a LP580 emission filter for mCherry, we estimate Q to be ~5%.
11. The validity of brightness measurement is highly dependent on the shape of the confocal volume. If the pinhole is set to a value larger than 1.0 airy units or the molecules within the excitation are saturated (typically by population of the triplet state), the relative brightness will change anomalously.
12. Equations [10] and [11] and PCH analysis can easily be expanded to include higher order oligomers. However, in our experience we are not able to acquire sufficient statistics in live yeast cells to freely fit to more than two species. Thus, while one-color N&B, two-color N&B, and PCH are valuable for determining if an oligomer exists, in order to analyze heterogeneous populations of oligomers, one must make assumptions and fix the brightness of certain populations prior to fitting. However, we find that it is possible to analyze heterogeneous oligomer populations with 2D-PCH (*see* **Section 3.7**).

Acknowledgments

We thank Veronica Conaway (Stowers Institute for Medical Research, SIMR) for discussion on yeast medium, Winfried Wiegraebe (SIMR) for discussions on fluctuation spectroscopy, and Arupratan Das (SIMR) for critical reading of the manuscript.

References

1. Chen, R. E., and Thorner, J. (2007) Function and regulation in MAPK signaling pathways: lessons learned from the yeast *Saccharomyces cerevisiae*. *Biochim. Biophys. Acta* **1773**, 1311–1340.
2. Schwartz, M. A., and Madhani, H. D. (2004) Principles of MAP kinase signaling specificity in *Saccharomyces cerevisiae*. *Annu. Rev. Genet.* **38**, 725–748.
3. Sprague, B. L., and McNally, J. G. (2005) FRAP analysis of binding: proper and fitting. *Trends Cell Biol.* **15**, 84–91.
4. Unruh, J. R., and Gratton, E. (2008) Analysis of molecular concentration and brightness from fluorescence fluctuation data with an electron multiplied CCD camera. *Biophys. J.* **95**, 5385–5398.
5. Digman, M. A., Wiseman, P. W., Choi, C., Horwitz, A. R., and Gratton, E. (2009) Stoichiometry of molecular complexes at adhesions in living cells. *Proc. Natl. Acad. Sci. USA* **106**, 2170–2175.
6. Kolin, D. L., and Wiseman, P. W. (2007) Advances in image correlation spectroscopy: measuring number densities, aggregation states, and dynamics of fluorescently labeled macromolecules in cells. *Cell Biochem. Biophys.* **49**, 141–164.
7. Digman, M. A., Wiseman, P. W., Horwitz, A. R., and Gratton, E. (2009) Detecting protein complexes in living cells from laser scanning confocal image sequences by the cross correlation raster image spectroscopy method. *Biophys. J.* **96**, 707–716.
8. Ries, J., Chiantia, S., and Schwille, P. (2009) Accurate determination of membrane dynamics with line-scan FCS. *Biophys. J.* **96**, 1999–2008.
9. Schwille, P., Haupts, U., Maiti, S., and Webb, W. W. (1999) Molecular dynamics in living cells observed by fluorescence correlation spectroscopy with one- and two-photon excitation. *Biophys. J.* 77, 2251–2265.
10. Chen, Y., Tekmen, M., Hillesheim, L., Skinner, J., Wu, B., and Muller, J. D. (2005) Dual-color photon-counting histogram. *Biophys. J.* **88**, 2177–2192.
11. Haustein, E., and Schwille, P. (2007) Fluorescence correlation spectroscopy: novel variations of an established technique. *Annu. Rev. Biophys. Biomol. Struct.* **36**, 151–169.
12. Bacia, K., and Schwille, P. (2007) Practical guidelines for dual-color fluorescence cross-correlation spectroscopy. *Nat. Protoc.* **2**, 2842–2856.
13. Cheung, H. C. (1991) Resonance energy transfer. In: Lakowicz, J. R. (ed.), *Topics in Fluorescence Spectroscopy* (pp. 128–176). New York, NY: Plenum Press.
14. Mc Intyre, J., Muller, E. G., Weitzer, S., Snydsman, B. E., Davis, T. N., and Uhlmann, F. (2007) In vivo analysis of cohesin architecture using FRET in the budding yeast *Saccharomyces cerevisiae*. *EMBO J.* **26**, 3783–3793.
15. Goley, E. D., Rodenbusch, S. E., Martin, A. C., and Welch, M. D. (2004) Critical conformational changes in the Arp2/3 complex are induced by nucleotide and nucleation promoting factor. *Mol. Cell* **16**, 269–279.
16. Digman, M. A., Sengupta, P., Wiseman, P. W., Brown, C. M., Horwitz, A. R., and Gratton, E. (2005) Fluctuation correlation spectroscopy with a laser-scanning microscope: exploiting the hidden time structure. *Biophys. J.* **88**, L33–L36.
17. Ries, J., and Schwille, P. (2006) Studying slow membrane dynamics with continuous wave scanning fluorescence correlation spectroscopy. *Biophys. J.* **91**, 1915–1924.
18. Sheff, M. A., and Thorn, K. S. (2004) Optimized cassettes for fluorescent protein tagging in *Saccharomyces cerevisiae*. *Yeast* **21**, 661–670.
19. Janke, C., Magiera, M. M., Rathfelder, N., et al. (2004) A versatile toolbox for PCR-based tagging of yeast genes: new fluorescent proteins, more markers and promoter substitution cassettes. *Yeast* **21**, 947–962.
20. Sikorski, R. S., and Hieter, P. (1989) A system of shuttle vectors and yeast host strains designed for efficient manipulation of DNA in *Saccharomyces cerevisiae*. *Genetics* **122**, 19–27.
21. Prein, B., Natter, K., and Kohlwein, S. D. (2000) A novel strategy for constructing N-terminal chromosomal fusions to green fluorescent protein in the yeast *Saccharomyces cerevisiae*. *FEBS Lett.* **485**, 29–34.
22. Knop, M., Siegers, K., Pereira, G., et al. (1999) Epitope tagging of yeast genes using a PCR-based strategy: more tags and improved practical routines. *Yeast* **15**, 963–972.
23. Maeder, C. I., Maier, P., and Knop, M. (2007) A guided tour to PCR-based genomic manipulations of *S. cerevisiae* (PCR targeting). In: Stansfield, I., and Stark, M. (eds.), *Methods in Microbiology: Yeast gene analysis* (Vol. 36, pp. 55–78). London: Academic.
24. Shaner, N. C., Campbell, R. E., Steinbach, P. A., Giepmans, B. N., Palmer, A. E., and Tsien, R. Y. (2004) Improved monomeric red, orange and yellow fluorescent proteins

derived from *Discosoma sp.* red fluorescent protein. *Nat. Biotechnol.* **22**, 1567–1572.
25. Tran, P. T., Paoletti, A., and Chang, F. (2004) Imaging green fluorescent protein fusions in living fission yeast cells. *Methods* **33**, 220–225.
26. Edidin, M. (2003) Fluorescence resonance energy transfer: techniques for measuring molecular conformation and molecular proximity. *Curr. Protoc. Immunol.* Chapter 18, Unit 18.10.
27. Shaner, N. C., Steinbach, P. A., and Tsien, R. Y. (2005) A guide to choosing fluorescent proteins. *Nat. Methods* **2**, 905–909.
28. Slaughter, B. D., Schwartz, J. W., and Li, R. (2007) Mapping dynamic protein interactions in MAP kinase signaling using live-cell fluorescence fluctuation spectroscopy and imaging. *Proc. Natl. Acad. Sci. USA* **104**, 20320–20325.
29. Elson, E. L., and Magde, D. (1974) Fluorescence correlation spectroscopy. I. Conceptual basis and theory. *Biopolymers* **13**, 1–27.
30. Magde, D., Elson, E. L., and Webb, W. W. (1974) Fluorescence correlation spectroscopy. II. Experimental realization. *Biopolymers* **13**, 29–61.
31. Ruttinger, S., Buschmann, V., Kramer, B., Erdmann, R., Macdonald, R., and Koberling, F. (2008) Comparison and accuracy of methods to determine the confocal volume for quantitative fluorescence correlation spectroscopy. *J. Microsc.* **232**, 343–352.
32. Daly, P. J., Page, D. J., and Compton, R. G. (1983) Mercury-plated rotating ring-disk electrode. *Anal. Chem.* **55**, 1191–1192.
33. Coles, B. A., and Compton, R. G. (1983) Photoelectrochemical ESR. Part I. Experimental. *J. Electroanal. Chem. Intl. Electrochem.* **144**, 87–98.
34. Slaughter, B. D., Huff, J. M., Wiegraebe, W., Schwartz, J. W., and Li, R. (2008) SAM domain-based protein oligomerization observed by live-cell fluorescence fluctuation spectroscopy. *PLoS One* **3**, e1931.
35. Maeder, C. I., Hink, M. A., Kinkhabwala, A., Mayr, R., Bastiaens, P. I., and Knop, M. (2007) Spatial regulation of Fus3 MAP kinase activity through a reaction-diffusion mechanism in yeast pheromone signalling. *Nat. Cell Biol.* **9**, 1319–1326.
36. Rigler, R., Foldes-Papp, Z., Meyer-Almes, F. J., Sammet, C., Volcker, M., and Schnetz, A. (1998) Fluorescence cross-correlation: a new concept for polymerase chain reaction. *J. Biotechnol.* **63**, 97–109.
37. Grimshaw, S. J., Mott, H. R., Stott, K. M., et al. (2004) Structure of the sterile alpha motif (SAM) domain of the *Saccharomyces cerevisiae* mitogen-activated protein kinase pathway-modulating protein STE50 and analysis of its interaction with the STE11 SAM. *J. Biol. Chem.* **279**, 2192–2201.
38. Qian, H., and Elson, E. L. (1990) On the analysis of high order moments of fluorescence fluctuations. *Biophys. J.* **57**, 375–380.
39. Qian, H., and Elson, E. L. (1990) Distribution of molecular aggregation by analysis of fluctuation moments. *Proc. Natl. Acad. Sci. USA* **87**, 5479–5483.
40. Muller, J. D. (2004) Cumulant analysis in fluorescence fluctuation spectroscopy. *Biophys. J.* **86**, 3981–3992.
41. Wu, B., and Muller, J. D. (2005) Time-integrated fluorescence cumulant analysis in fluorescence fluctuation spectroscopy. *Biophys. J.* **89**, 2721–2735.
42. Wu, B., Chen, Y., and Muller, J. D. (2006) Dual-color time-integrated fluorescence cumulant analysis. *Biophys. J.* **91**, 2687–2698.
43. Kask, P., Palo, K., Ullmann, D., and Gall, K. (1999) Fluorescence-intensity distribution analysis and its application in biomolecular detection technology. *Proc. Natl. Acad. Sci. USA* **96**, 13756–13761.
44. Chen, Y., Muller, J. D., So, P. T., and Gratton, E. (1999) The photon counting histogram in fluorescence fluctuation spectroscopy. *Biophys. J.* **77**, 553–567.
45. Meng, F., and Ma, H. (2006) A comparison between photon counting histogram and fluorescence intensity distribution analysis. *J. Phys. Chem. B* **110**, 25716–25720.
46. Chen, Y., Muller, J. D., Ruan, Q., and Gratton, E. (2002) Molecular brightness characterization of EGFP in vivo by fluorescence fluctuation spectroscopy. *Biophys. J.* **82**, 133–144.
47. Perroud, T. D., Huang, B., and Zare, R. N. (2005) Effect of bin time on the photon counting histogram for one-photon excitation. *Chemphyschem.* **6**, 905–912.
48. Hillesheim, L. N., Chen, Y., and Muller, J. D. (2006) Dual-color photon counting histogram analysis of mRFP1 and EGFP in living cells. *Biophys. J.* **91**, 4273–4284.
49. Kask, P., Palo, K., Fay, N., et al. (2000) Two-dimensional fluorescence intensity distribution analysis: theory and applications. *Biophys. J.* **78**, 1703–1713.
50. Van der Meer, B. W. (2002) Kappa-squared: from nuisance to new sense. *Rev. Mol. Biotechnol.* **82**, 181–196.
51. Thompson, N. L. (1991) Fluorescence correlation spectroscopy. In: Lakowicz, J. R. (ed.), *Topics in Fluorescence Spectroscopy* (pp. 337–378). New York, NY: Plenum Press.

Chapter 18

Nutritional Control of Cell Growth via TOR Signaling in Budding Yeast

Yuehua Wei and X.F. Steven Zheng

Abstract

Cell growth is highly regulated and its deregulation is related to many human diseases such as cancer. Nutritional cues stimulate cell growth through modulation of TOR (target of rapamycin) signaling pathway. At the center of this pathway is a large serine/threonine protein kinase TOR, which forms two distinct functional complexes TORC1 and TORC2 in a cell. TORC1 senses the environmental nutrient quality/quantity and transmits the growth signals to multiple effectors to regulate a broad spectrum of biological processes including translation initiation, ribosome biogenesis, autophagy, nutrient uptake, and metabolism. By using budding yeast as a model, recent studies began to elucidate the complexity of the TOR signaling pathway.

Key words: TOR (target of rapamycin), signal transduction, budding yeast, cell growth, nutrient, nutrition, ribosome biogenesis, autophagy, protein translation.

1. Introduction

Cell growth is defined by the increase in cell mass. Cells upregulate macromolecular synthesis to increase cell mass in response to good nutritional conditions while down-regulate it when nutrients are limiting (1, 2). The balance of cell growth is essential for all organisms to adapt to the surrounding environment. Cell growth is also required for cell proliferation because cells divide only when certain cell size is obtained. During development, higher organisms elegantly control cell growth and proliferation for organ formation. Tumors can emerge when cell growth goes uncontrolled (3). Despite the apparent significance, however, the mechanisms underlying growth control remained poorly

J.I. Castrillo, S.G. Oliver (eds.), *Yeast Systems Biology*, Methods in Molecular Biology 759,
DOI 10.1007/978-1-61779-173-4_18, © Springer Science+Business Media, LLC 2011

understood until recently. How does a cell sense and integrate nutritional cues from the surrounding environment to commit cell growth? Specifically, what are the signaling pathways involved in balancing the macromolecular synthesis and degradation in a cell? In the past decade or so, the target of rapamycin (TOR) pathway has been established to be the central regulator of cell growth in all eukaryotic cells. Studies from budding yeast, the simplest eukaryotic model, contribute significantly to the understanding of TOR function and nutrient signaling (4). The TOR signaling cascade is comprised of a large serine/threonine protein kinase called TOR and a broad spectrum of regulators and effectors involving in protein translation, gene transcription, ribosome biogenesis, protein degradation, and autophagy. The versatility of the TOR pathway in regulation of this wide range of cellular responses is in part not only due to its multi-level effectors, but also, importantly, due to the versatile and dynamic regulation of TOR subcellular localization (5).

2. TOR Kinase Is the Central Controller of Cell Growth

TOR proteins are large serine/threonine protein kinase belonging to the kinase family known as phosphatidylinositol kinase-related kinases (PIKK) (6). TOR proteins are highly conserved from budding yeast (Tor1 and Tor2) to mammal (mTOR) (7). Like other members of this kinase family, TOR protein contains a C-terminal kinase domain resembling phosphatidylinositol (PI) kinase, though no lipid substrate has been reported (**Fig. 18.1a**). In the N-terminus, there are about 40 tandem repeats of 37–43 amino acids termed HEAT repeats (named after the four proteins containing the similar sequence: huntingtin, elongation factor 3, the A subunit of PP2A, and Tor). The HEAT repeats are thought to serve as a platform for multiple protein–protein interactions. TOR protein contains a FKBP12-rapamycin binding (FRB) domain in adjacent to its kinase domain, a point mutation (S1972I) in which blocks the inhibition by rapamycin (8). TOR forms two functionally distinct complexes (9). In yeast, TOR complex 1 (TORC1) contains Kog1, Lst8, Tco89, and either Tor1 or Tor2; TOR complex 2 (TORC2) contains Avo1, Avo2, Avo3, Bit6, Lst8, and Tor2 (**Fig. 18.1a**). Interestingly, only TORC1 is sensitive to rapamycin. Rapamycin first binds to its intracellular receptor FKBP12 (FK506-binding protein 12 kDa), then the FKBP12–rapamycin complex binds to FRB domain of Tor1/2, therefore inhibiting TOR kinase activity (8). The reason why FKBP12–rapamycin does not interact with TORC2 is not clear. Two TOR complexes control distinct physiological

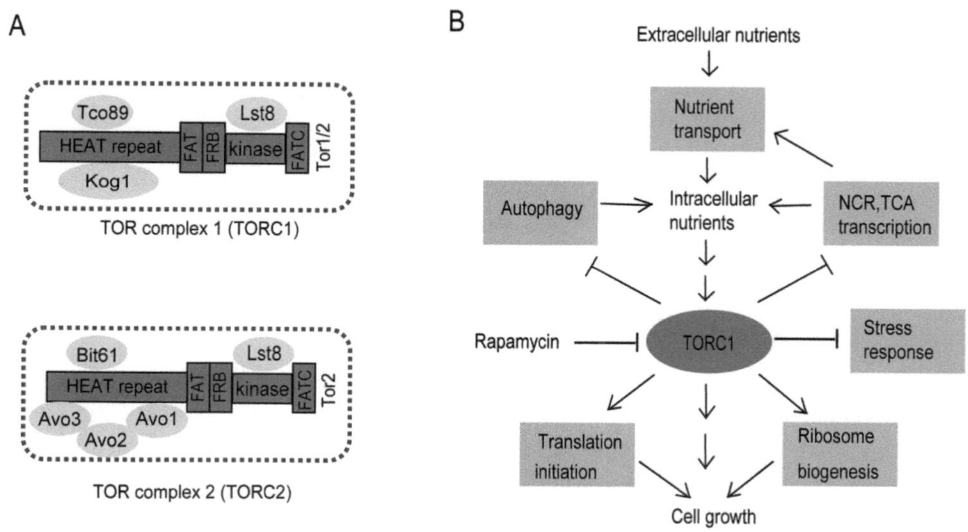

Fig. 18.1. TOR is the central regulator of cell growth. (**a**) The structural organization of two TOR complexes. TOR proteins are large Ser/Thr protein kinases belonging to PI3K-related kinase (PIKK). TOR contains about 40 HEAT repeats at its N terminus and a kinase domain at its C terminus. HEAT repeats mediate TOR interaction with multiple proteins to form distinct two complexes TORC1 and TORC2. Adjacent to the kinase domain is the FKBP12-rapamycin binding (FRB) domain, which mediates rapamycin inhibition of TORC1. Flanking the FRB and kinase domains are FAT (FRAP/mTOR, ATM, TRRAP) and FATC (FRAP/mTOR, ATM, TRRAP C-terminal) domains, which are common to PIKKs. (**b**) TORC1 regulates cell growth by controlling the balance between global biosynthesis and turnover in a cell. TORC1 links environmental nutrients to cell growth by promoting macromolecular biosynthesis such as protein translation and ribosome biogenesis while antagonizing global turnover pathways such as autophagy. TORC1 also represses TCA and NCR transcription and inhibits stress responses.

processes in response to nutrient cues (7). TORC1 regulates growth-related functions including protein synthesis, nutrient transport and utilization, ribosome biogenesis, autophagy, and stress responses (**Fig. 18.1b**). TORC2 is relatively less studied and is thought to control actin polarization in a rapamycin-insensitive manner.

3. How Does TOR Regulate Cell Growth?

Cell growth is a balance between macromolecular biosynthesis and turnover. In response to the extracellular growth stimuli, cells promote growth through global biosynthesis. In contrast, when environmental nutrients become scarce, cells cease growth and assume global turnover. TOR is the central controller of this balance (**Fig. 18.1b**). TORC1 promotes the macromolecular biosynthesis at least in part through translation initiation and ribosome biogenesis. At the same time, TORC1 antagonizes macromolecular turnover such as autophagy, represses NCR (nitrogen-catabolite repression) and TCA (tricarboxylic acid cycle) transcription, and inhibits stress responses. When extracellular nutrients are less available, TORC1 is inhibited. Meanwhile, cells upregulate autophagy and uptake of nutrients and activate NCR and TCA transcription to provide cells with intracellular nutrients to adapt to poor nutrient condition. At the same time, TORC1 activates stress responses to cope with the stresses caused by starvation. Therefore, the balance between macromolecular biosynthesis and turnover is finely tuned by TORC1.

3.1. Regulation of Translation Initiation by TORC1

Protein translation is highly regulated and inhibition of TORC1 results in marked decrease in translational initiation (10). In mammalian cells, mTORC1 robustly regulates this process through two well-studied effectors: S6K1 and 4EBP1 (11). mTORC1 phosphorylates S6K1, which in turn phosphorylates ribosomal protein S6, a subunit of the 40S ribosome, therefore promoting protein translation. mTORC1 also phosphorylates 4EBP1, which sequesters the translation initiation factor eIF4E, thereby preventing the formation of the competent translational initiation complex. In yeast, the AGC (protein kinase A, G, and C) kinase Sch9 is believed to be the functional homolog of S6K1 (12). Sch9 is phosphorylated by TORC1 directly at multiple sites and is required for phosphorylation of ribosomal protein Rps6, the yeast homolog of mammalian S6. Importantly, Sch9 regulates the translational initiation for protein synthesis in addition to its regulation of ribosomal biogenesis at the transcriptional level. The yeast protein Epa1 is probably homologous to 4EBP1, since it also inhibits translational initiation through binding to the highly conserved eIF4E. However, no studies have demonstrated a direct link between Epa1p and TORC1 in yeast. Disruption of *EPA1* gene results in partial resistance to rapamycin, suggesting a conserved role of TORC1 in regulation of translational initiation (13).

TORC1 also regulates several other translational initiation factors, for example, eIF4G and eIF2. Degradation of eIF4G is induced upon rapamycin treatment (10). TORC1 may somehow protect such degradation by an unknown mechanism. But whether such degradation is simply a result of translational inhibition or a regulatory mechanism is not clear. GTP-loaded eIF2 is responsible for recruiting methionyl-tRNA to the first codon of an mRNA. Phosphorylation of the alpha-subunit of eIF2 (eIF2a) by Gcn2 disables such recruitment. Interestingly, Gcn2 is a phosphoprotein whose phosphorylation depends on TORC1. Upon phosphorylation, Gcn2 kinase activity is inhibited, resulting in eIF2 dephosphorylation and activation, leading to translational initiation (14).

3.2. Regulation of Ribosome Biogenesis by TORC1

One major target of TORC1 regulation is ribosome biogenesis (15, 16). Ribosome biogenesis is dedicated to synthesize the translational machinery and is tightly coupled to cell growth. In yeast, TORC1 regulates transcription of 5S, 5.8S, 18S, and 25S ribosomal RNAs (rRNAs) and 137 ribosomal protein (RP) genes by all three major RNA polymerases (Pols) (15, 17–21), which account for approximately 95% of total transcription in a cell (22). Other than transcription of the ribosome components, TORC also controls ribosome maturation by regulating expression of and more than 200 Ribi (ribosome biogenesis) regulons involved in processing and modifying RPs, rRNAs, tRNAs, assembling, and nuclear exporting of ribosomes as well as ribonucleotide metabolism (23). The huge consumption of the cell's resources demands a tight regulation in response to nutrient and energy conditions (20).

Ribosomal protein (RP) genes are transcribed by RNA polymerase II (Pol II). TORC1 promotes transcriptions of RP genes through regulation of several transcription factors. The antagonist interaction of the fork head DNA-binding protein Fhl1 with its co-activator Ifh1 and co-repressor Crf1 appears to be regulated by TORC1: TORC1 promotes Ifh1–Fhl1, whereas suppresses Crf1–Fhl1 complex formation (24). TORC1 likely regulates this interaction switch through inhibiting Yak1-depenent phosphorylation and nuclear translocation of Crf1, though the link between TORC1 and Yak1 is still elusive. Ribosome biogenesis is still regulated by TORC1 in absence of Fhl1, indicating other unknown mechanisms are involved. Sfp1 is a potential transcription factor specifically regulates RP gene and Ribi gene transcription (23). TORC1 interacts with and directly phosphorylates Sfp1, which promotes its nuclear translocation and binding to RP gene promoters (25), resulting in RP gene transcriptional activation. The AGC-family kinase Sch9 is also phosphorylated by TORC1 at multiple sites and regulates RP and Ribi gene transcription independent of Fhl1 and Sfp1 (12, 23). How Sch9 relays TORC1 signals to RP gene and Ribi gene transcription is not known. Remarkably, Sch9 is found at the promoter of several osmo-responsive genes (26), hinting that Sch9 may regulate transcription of RP and Ribi genes at the chromatin level. Other than transcriptional activators, chromatin remodeling factors such as RSC complex, the Rpd3 histone deacetylase, and the Esa1 histone acetylase are also critical for RP gene transcriptional regulation by TORC1 (27, 28). How TORC1 regulates the recruitment of such chromatin remodeling factors to the RP gene promoters are exciting areas to explore.

TORC1 regulates rRNA transcription by directly targeting to rDNA chromatin and activating the transcription factors and/or chromatin remodeling factors (5) (**Fig. 18.2**). TORC1

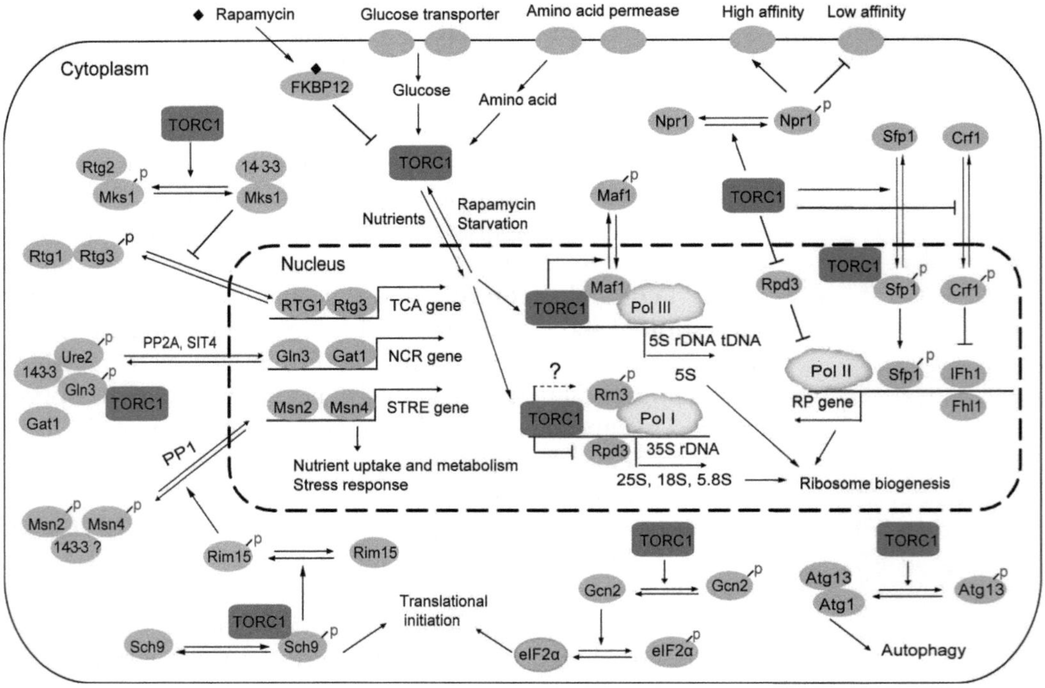

Fig. 18.2. TORC1 regulates diverse effectors in a cell. Nutrients regulate cell growth via TORC1 (*dark rectangle*). In yeast, TORC1 senses nutrients through amino acid and glucose transporters through unknown mechanisms. TORC1 is localized in both cytoplasm and nucleus to regulate a broad range of biological processes including translational initiation, ribosome biogenesis, autophagy, nutrient uptake, and metabolism. One common mechanism of transcriptional regulation by TORC1 is the control of cytoplasm–nucleus shuttling of transcription factors such as Gln3 and Sfp1. Another mechanism is the cytoplasm–nucleus shuttling of TORC1 itself and the binding to regulated-gene promoter (35S rDNA and 5S rDNA), whereby regulating RNA polymerases (*irregular shape*) activity through phosphorylation of transcriptional factors (such as Maf1) or modulating the chromatin remodeling factors (such as Rpd3).

dynamically shuttles between the nucleus and the cytoplasm. In response to good nutrient conditions, TORC1 is localized in the nucleus and binds to 35S rDNA promoter and 5S rDNA gene to promote their transcription by Pol I and Pol III, respectively (17, 18). Under conditions of nutrient starvation or rapamycin treatment, TORC1 delocalizes from the rDNA chromatin and exits from the nucleus, which results in rRNA transcription inhibition. The concomitant binding of TORC1 to 35S rDNA promoter and 5S rDNA gene argues for a role of TORC1 in coregulation of Pol I and Pol III for coordinated rRNA synthesis (17). How does TORC1 regulate Pol I- and Pol III-dependent transcription at the rDNA chromatin? Kinases associated with genes tend to regulate transcription factors through phosphorylation. Maf1 is a phosphoprotein, when dephosphorylated by rapamycin treatment or nutrient starvation, enters the nucleolus and associates with Pol III to repress 5S rDNA and tRNA genes transcription. With good nutrients available, however, TORC1 targets to 5S rDNA gene and phosphorylates Maf1, preventing Maf1 from accumulation in the nucleolus and association with Pol III-transcribed genes (17). Rrn3 is an essential Pol I initiation factor, whose association with and activation of Pol I is regulated by TORC1 (29). It is conceivable that Rrn3 is phosphorylated by TORC1 at the 35S rDNA promoter. Supporting this hypothesis, the mammalian Rrn3 counterpart TIF-IA is also phosphorylated in a TORC1-dependent manner (30). In addition to targeting RNA polymerases for transcriptional regulation, TORC1 is also implicated in modulation of rDNA chromatin structure. Nutrient starvation or rapamycin treatment results in nucleolus contraction and rDNA condensation, occluding Pol I loading on rDNA. Specifically, TORC1 inhibition causes rapid loading of Rpd3-Sin3 histone deacetylase (HDAC) and condensin complex to the rDNA chromatin, which may silence the rDNA transcription (31). It is therefore interesting to investigate whether Rpd3-Sin3 or condensin is targeted for phosphorylation by TORC1.

3.3. TORC1 Antagonizes Macro-autophagy in Growing Cells

In good nutrient condition, TORC1 promotes cell growth by positively regulating several processes for biosynthesis. When shifted to poor nutrient condition, cells rapidly cease macromolecular synthesis and assume global turnover to shrink their transcriptional and translational machineries. One such global turnover pathway is macro-autophagy (briefly autophagy), a process of controlled self-digestion (32). In yeast, autophagy is believed to generate internal supply of nutrients from degrading ribosomes, bulk of proteins, and excessive organelles, allowing cells to adapt to and survive extended periods of starvation. TORC1 negatively regulates induction of autophagy (32, 33). Atg1 is a key kinase required for induction of autophagy when complexed with Atg13. Upon TORC1 inhibition, ATG13 is rapidly dephosphorylated, thereby increasing its interaction with Atg1 kinase. The formation of the Atg1–Atg13 complex then recruits a serial of additional autophagic proteins such as Atg11 and Atg17, resulting in the formation of a double-membrane compartments called autophagosome (34). The autophagosome carrying the engulfed cytosolic mass is finally fused into the lysosome or vacuole for degradation, which generates carbon and nitrogen sources necessary for rapid adaptation to the environment (32). The mammalian Atg1 and Atg13 counterparts have been shown to be directly phosphorylated by mTORC1 (35); however, whether this is true in yeast remains unknown. TORC1 also controls autophagy via transcriptional induction of autophagic genes. One example is the essential gene *APG14*. Inhibition of TORC1 by rapamycin or nutrient limitation results in about 20-fold transcriptional induction of *APG14* via Gln3, a transcriptional factor directly phosphorylated by TORC1 (36). Thus TORC1 likely controls autophagy at multiple levels, but the detailed mechanisms remain poorly understood.

3.4. TORC1 Regulates Nutrient Uptake and Metabolism

Unlike multicellular eukaryotes, yeast cells are unicellular organisms which need to respond directly to a variety of environment cues. When and what kind of nutrients will be transported into the cell are finely regulated processes. Accumulating evidences suggest that TORC1 is the master regulator of nutrient uptake and metabolism (**Fig. 18.2**). Nutrient availability, via TORC1, differentially regulates the degradation of many nutrient transporters (37, 38). With optimal nutrients available, many high-affinity, substrate-selective permeases, such as histidine permease Hip1 and the tryptophan permease Tat2, are expressed and targeted to the plasma membrane to pump in nutrients. At the same time, TORC1 promotes the degradation of some broad-specificity permeases, such as the ammonia permease Mep2 and the general amino acid permease Gap1, both of which are required to import a broad spectrum of nutrients when nutrients become scarce. Rapamycin or nutrient limitation causes the reverse phenomena: degradation of the substrate-selective permeases whereas expression of broad-specificity permeases. Mechanistically, TORC1 positively regulates the phosphorylation of Npr1, which inversely regulates many high-specificity and broad-specificity permeases through the ubiquitin-mediated degradation pathway (39, 40). However, the links between TORC1 and Npr1 and between Npr1 and the permeases are yet to be elucidated.

TORC1 also regulates differential uptake and metabolism of nutrients at the transcriptional level (15, 21). TORC1 inhibition induces expression of several hundred genes involved in nitrogen metabolism, glycolytic pathway, and TCA cycle. The regulation of nitrogen-catabolite repression (NCR) genes is one of the best studied among these processes (5). NCR genes are involved in uptake and metabolism of non-preferred nitrogen sources such as urea and proline. NCR genes are normally repressed and only induced when the environmental nitrogen become scarce. TORC1 regulates the NCR gene transcription through two GATA transcription factors: Gln3 and Gat1. During growth in preferred nitrogen sources, phosphorylation of Gln3 by TORC1 renders Gln3 cytoplasmic retention by anchoring to Ure2. Upon rapamycin treatment or amino acid starvation, however, Gln3 is targeted for dephosphorylation by PP2A or Sit4, resulting in its translocation to the nucleus and induction of NCR-sensitive genes (41, 42). TORC1 also prevents the nuclear translocation of the GATA transcription factor Gat1 (43), but the regulatory details are not well studied. The glycolytic pathway and the TCA cycle produce metabolic intermediates such as alpha-ketoglutarate, required for de novo synthesis of some amino acids, such as glutamine and glutamate. Genes involved in this pathway are induced in response to starvation, which is believed to compensate for the lack of nitrogen source (44). With ample nitrogen source, however, TORC1 represses these genes by antagonizing the nuclear translocalization of the heterodimeric transcription factors Rtg1/Rtg3. Similar to Gln3, Rtg1/Rtg3 complex is sequestered in the cytoplasm when Rtg3 is hyperphosphorylated, while translocates into the nucleus upon dephosphorylation. Regulation of Rtg1/Rtg3 by TORC1 requires both Msk1 and Rtg2: Msk1 inhibits Rtg1/Rtg3 nuclear translocation, while Rtg2 inhibits Msk1 (44). Upon TORC1 inhibition, Rtg2 enhances the association with the Rtg1/Rtg3 inhibitor Msk1, which releases Rtg1/Rtg3 into the nucleus to induce their target gene expression. Both Msk1 and Rtg3 are phosphorylated in good nutrient condition and dephosphorylated in poor nutrient condition. However, whether TORC1 is directly involved in these events and what are the phosphatases involved remain unanswered questions.

3.5. TORC1 Negatively Regulates Stress Signaling

Starvation is an environmental stress to all organisms. In yeast, starvation or TORC1 inhibition induces stress-responsive element (STRE)-dependent transcription, which allows yeast cells to adapt to starvation stress (45, 46). Msn2 and Msn4 are two partially redundant zinc finger transcription factors responsible for the STRE-dependent transcription. TORC1 promotes Msn2 and Msn4 phosphorylation and cytoplasmic localization. Glucose depletion or rapamycin treatment results in their rapid dephosphorylation and nuclear translocation and the subsequent target gene transcription. This transcriptional induction requires Rim15, a protein kinase involved in G_0 entry and life span regulation (47, 48). Rim15 cytoplasm–nucleus shuttling is similarly regulated by TORC1 as that of Msn2/4, and it has been shown to be phosphorylated directly by the TORC1 substrate Sch9 (49). Thus, TORC1-Sch9-Rim15 constitutes a pathway that regulates stress-induced transcription through Msn2/4. Different from Gln3, the PP1 but not Tap42 and Sit4 are involved in the dephosphorylation of Msn2/4 (47, 50). There is still debate over whether the 14-3-3 proteins are involved in retaining Msn2/4 in the cytoplasm, and the link between Rim15 and Msn2/4 is still missing. Detailed mechanisms of TORC1 regulation of Msn2 and Msn4 may be more complex than previously thought.

4. How Does TOR Sense Environmental Nutrients?

TOR senses the quality and quantity of environmental carbon and nitrogen sources. The regulatory pathways upstream of TOR have been vigorously explored in mammalian cells. Despite the clear understanding of the pathway by which insulin is regulating mTOR, consensus on nutrient regulation of mTOR has not been reached (51). Amino acid signals may go through hVps34, a class III PI3K (Phosphoinositide 3-kinase) to affect mTOR activity. The Rag GTPases (Rag A/B/C/D) are also proposed to exert amino acid regulation of mTOR. Glucose is generally believed to regulate mTOR through AMPK (AMP-activated protein kinase), which further impinges on TSC1/2 complex. In yeast, however, little is known about how nutritional conditions are sensed and how the information is transmitted to TOR. Nutrient uptake is the first step of TOR regulation. In both mammalian and yeast cells, amino acid and glucose transporters are involved in regulation of TOR (51). In mammalian cells, for example, the L-glutamine transporter SLC1A5 is critical for mTOR activation (52). In yeast, Mep2 is required for ammonium uptake and cell growth regulation, and the amino acid permease Ssy1 and glucose sensors Snf3 are involved in cell growth regulation (53). Presumably, the extracellular nutrient availability is reflected by the intracellular nutrients transported by these nutrient transporters. The intracellular nutrients are further sensed by unknown mechanisms in yeast cells and then integrated into TOR pathway to regulate cell growth. This model is consistent with that of mammalian cells, where amino acid and glucose signals are sensed and transmitted to mTOR through Rag GTPases (Rag A/B/C/D) and AMPK, respectively (51). Moreover, the intracellular levels of ATP, phosphatidic acid, and inorganic polyphosphate are among the potential regulators of mTOR activity (4), further suggesting that intracellular nutrients are key to TOR regulation. In yeast cells, the vacuolar compartment is an intracellular nutrient source that regulates TOR activity. The vacuolar membrane-associated EGO (exit from rapamycin-induced growth arrest) protein complex, which consists of Ego1, Ego3, Gtr1, and Gtr2, could transmit critical nutrient signals to TORC1 (54). It turns out that the small GTPases Gtr1 and Gtr2 are homologous to mammalian Rag GTPases (Rag A/B/C/D), which is critical for amino acid regulation of mTOR (51).

5. Conclusion

Growth regulation remains an exciting area to explore in the near future. By taking advantage of budding yeast as an excellent genetic and cellular model, researchers have been able to examine the mechanisms underlying the mystery of growth control (4). Through a broad range of regulators and effectors, TOR senses and transmits the environmental growth stimuli to machineries that regulate multiple aspects of growth. Growth control is intimately coupled to ribosome biogenesis and TORC1 turns out to coordinate all three major RNA polymerases for this mission (5). In addition to a myriad of regulators and effectors, emerging data suggest an even greater complexity of TOR regulation, with TORC1 kinase subjecting not only to enzymatic activity control but also to multi-faceted subcellular localization control (5, 18). Understanding the detailed signaling pathway should hold promise for treatment of many human diseases in the future. Despite tremendous progress in understanding of TOR pathway and clinical application of rapamycin for various human disease conditions, challenges are also emerging as our knowledge expands (3). For example, cancer drug resistance as a result of rapamycin administration hinders the clinical use of this drug. Further dissecting the regulatory network of TOR will absolutely contribute to the outcome of cancer treatment.

References

1. Hall, M. N. (1996) The TOR signalling pathway and growth control in yeast. *Biochem. Soc. Trans.* **24**, 234–239.
2. Thomas, G., and Hall, M. N. (1997) TOR signalling and control of cell growth. *Curr. Opin. Cell Biol.* **9**, 782–787.
3. Tsang, C. K., Qi, H., Liu, L. F., and Zheng, X. F. (2007) Targeting mammalian target of rapamycin (mTOR) for health and diseases. *Drug Discov. Today* **12**, 112–124.
4. De Virgilio, C., and Loewith, R. (2006) Cell growth control: little eukaryotes make big contributions. *Oncogene* **25**, 6392–6415.
5. Tsang, C. K., and Zheng, X. F. (2007) TOR-in(g) the nucleus. *Cell Cycle* **6**, 25–29.
6. Keith, C. T., and Schreiber, S. L. (1995) PIK-related kinases: DNA repair, recombination, and cell cycle checkpoints. *Science* **270**, 50–51.
7. Wullschleger, S., Loewith, R., and Hall, M. N. (2006) TOR signaling in growth and metabolism. *Cell* **124**, 471–484.
8. Zheng, X. F., Florentino, D., Chen, J., Crabtree, G. R., and Schreiber, S. L. (1995) TOR kinase domains are required for two distinct functions, only one of which is inhibited by rapamycin. *Cell* **82**, 121–130.
9. Loewith, R., Jacinto, E., Wullschleger, S., et al. (2002) Two TOR complexes, only one of which is rapamycin sensitive, have distinct roles in cell growth control. *Mol. Cell* **10**, 457–468.
10. Barbet, N. C., Schneider, U., Helliwell, S. B., Stansfield, I., Tuite, M. F., and Hall, M. N. (1996) TOR controls translation initiation and early G1 progression in yeast. *Mol. Biol. Cell* **7**, 25–42.
11. Hay, N., and Sonenberg, N. (2004) Upstream and downstream of mTOR. *Genes Dev.* **18**, 1926–1945.
12. Urban, J., Soulard, A., Huber, A., et al. (2007) Sch9 is a major target of TORC1 in *Saccharomyces cerevisiae*. *Mol. Cell* **26**, 663–674.
13. Cosentino, G. P., Schmelzle, T., Haghighat, A., Helliwell, S. B., Hall, M. N., and Sonenberg, N. (2000) Eap1p, a novel eukaryotic translation initiation factor 4E-associated protein in *Saccharomyces cerevisiae*. *Mol. Cell. Biol.* **20**, 4604–4613.
14. Cherkasova, V. A., and Hinnebusch, A. G. (2003) Translational control by TOR and TAP42 through dephosphorylation of eIF2alpha kinase GCN2. *Genes Dev.* **17**, 859–872.
15. Hardwick, J. S., Kuruvilla, F. G., Tong, J. K., Shamji, A. F., and Schreiber, S. L. (1999) Rapamycin-modulated transcription defines the subset of nutrient-sensitive signaling pathways directly controlled by the Tor proteins. *Proc. Natl. Acad. Sci. USA* **96**, 14866–14870.
16. Powers, T., and Walter, P. (1999) Regulation of ribosome biogenesis by the rapamycin-sensitive TOR-signaling pathway in *Saccharomyces cerevisiae*. *Mol. Biol. Cell* **10**, 987–1000.
17. Wei, Y., Tsang, C. K., and Zheng, X. F. (2009) Mechanisms of regulation of RNA polymerase III-dependent transcription by TORC1. *EMBO J.* **28**, 2220–2230.
18. Li, H., Tsang, C. K., Watkins, M., Bertram, P. G., and Zheng, X. F. (2006) Nutrient regulates Tor1 nuclear localization and

association with rDNA promoter. *Nature* **442**, 1058–1061.
19. Zaragoza, D., Ghavidel, A., Heitman, J., and Schultz, M. C. (1998) Rapamycin induces the G0 program of transcriptional repression in yeast by interfering with the TOR signaling pathway. *Mol. Cell. Biol.* **18**, 4463–4470.
20. Warner, J. R. (1999) The economics of ribosome biosynthesis in yeast. *Trends Biochem. Sci.* **24**, 437–440.
21. Cardenas, M. E., Cutler, N. S., Lorenz, M. C., Di Como, C. J., and Heitman, J. (1999) The TOR signaling cascade regulates gene expression in response to nutrients. *Genes Dev.* **13**, 3271–3279.
22. Moss, T., and Stefanovsky, V. Y. (2002) At the center of eukaryotic life. *Cell* **109**, 545–548.
23. Jorgensen, P., Rupes, I., Sharom, J. R., Schneper, L., Broach, J. R., and Tyers, M. (2004) A dynamic transcriptional network communicates growth potential to ribosome synthesis and critical cell size. *Genes Dev.* **18**, 2491–2505.
24. Martin, D. E., Soulard, A., and Hall, M. N. (2004) TOR regulates ribosomal protein gene expression via PKA and the Forkhead transcription factor FHL1. *Cell* **119**, 969–979.
25. Lempiainen, H., Uotila, A., Urban, J., et al. (2009) Sfp1 interaction with TORC1 and Mrs6 reveals feedback regulation on TOR signaling. *Mol. Cell* **33**, 704–716.
26. Pascual-Ahuir, A., and Proft, M. (2007) The Sch9 kinase is a chromatin-associated transcriptional activator of osmostress-responsive genes. *EMBO J.* **26**, 3098–3108.
27. Rohde, J. R., and Cardenas, M. E. (2003) The tor pathway regulates gene expression by linking nutrient sensing to histone acetylation. *Mol. Cell. Biol.* **23**, 629–635.
28. Humphrey, E. L., Shamji, A. F., Bernstein, B. E., and Schreiber, S. L. (2004) Rpd3p relocation mediates a transcriptional response to rapamycin in yeast. *Chem. Biol.* **11**, 295–299.
29. Claypool, J. A., French, S. L., Johzuka, K., et al. (2004) Tor pathway regulates Rrn3p-dependent recruitment of yeast RNA polymerase I to the promoter but does not participate in alteration of the number of active genes. *Mol. Biol. Cell* **15**, 946–956.
30. Mayer, C., Zhao, J., Yuan, X., and Grummt, I. (2004) mTOR-dependent activation of the transcription factor TIF-IA links rRNA synthesis to nutrient availability. *Genes Dev.* **18**, 423–434.
31. Tsang, C. K., Bertram, P. G., Ai, W., Drenan, R., and Zheng, X. F. (2003) Chromatin-mediated regulation of nucleolar structure and RNA Pol I localization by TOR. *EMBO J.* **22**, 6045–6056.
32. Yorimitsu, T., and Klionsky, D. J. (2005) Autophagy: molecular machinery for self-eating. *Cell Death Differ.* **12**(Suppl 2), 1542–1552.
33. Kamada, Y., Sekito, T., and Ohsumi, Y. (2004) Autophagy in yeast: a TOR-mediated response to nutrient starvation. *Curr. Top. Microbiol. Immunol.* **279**, 73–84.
34. Kamada, Y., Funakoshi, T., Shintani, T., Nagano, K., Ohsumi, M., and Ohsumi, Y. (2000) Tor-mediated induction of autophagy via an Apg1 protein kinase complex. *J. Cell Biol.* **150**, 1507–1513.
35. Jung, C. H., Jun, C. B., Ro, S. H., et al. (2009) ULK-Atg13-FIP200 complexes mediate mTOR signaling to the autophagy machinery. *Mol. Biol. Cell* **20**, 1992–2003.
36. Chan, T. F., Bertram, P. G., Ai, W., and Zheng, X. F. (2001) Regulation of APG14 expression by the GATA-type transcription factor Gln3p. *J. Biol. Chem.* **276**, 6463–6467.
37. Magasanik, B., and Kaiser, C. A. (2002) Nitrogen regulation in *Saccharomyces cerevisiae*. *Gene* **290**, 1–18.
38. Ozcan, S., and Johnston, M. (1999) Function and regulation of yeast hexose transporters. *Microbiol. Mol. Biol. Rev.* **63**, 554–569.
39. Schmidt, A., Beck, T., Koller, A., Kunz, J., and Hall, M. N. (1998) The TOR nutrient signalling pathway phosphorylates NPR1 and inhibits turnover of the tryptophan permease. *EMBO J.* **17**, 6924–6931.
40. De Craene, J. O., Soetens, O., and Andre, B. (2001) The Npr1 kinase controls biosynthetic and endocytic sorting of the yeast Gap1 permease. *J. Biol. Chem.* **276**, 43939–43948.
41. Carvalho, J., and Zheng, X. F. (2003) Domains of Gln3p interacting with karyopherins, Ure2p, and the target of rapamycin protein. *J. Biol. Chem.* **278**, 16878–16886.
42. Bertram, P. G., Choi, J. H., Carvalho, J., et al. (2000) Tripartite regulation of Gln3p by TOR, Ure2p, and phosphatases. *J. Biol. Chem.* **275**, 35727–35733.
43. Kuruvilla, F. G., Shamji, A. F., and Schreiber, S. L. (2001) Carbon- and nitrogen-quality signaling to translation are mediated by distinct GATA-type transcription factors. *Proc. Natl. Acad. Sci. USA* **98**, 7283–7288.
44. Liu, Z., and Butow, R. A. (2006) Mitochondrial retrograde signaling. *Annu. Rev. Genet.* **40**, 159–185.
45. Beck, T., and Hall, M. N. (1999) The TOR signalling pathway controls nuclear

localization of nutrient-regulated transcription factors. *Nature* **402**, 689–692.
46. Mayordomo, I., Estruch, F., and Sanz, P. (2002) Convergence of the target of rapamycin and the Snf1 protein kinase pathways in the regulation of the subcellular localization of Msn2, a transcriptional activator of STRE (Stress Response Element)-regulated genes. *J. Biol. Chem.* **277**, 35650–35656.
47. Pedruzzi, I., Dubouloz, F., Cameroni, E., et al. (2003) TOR and PKA signaling pathways converge on the protein kinase Rim15 to control entry into G0. *Mol. Cell* **12**, 1607–1613.
48. Swinnen, E., Wanke, V., Roosen, J., et al. (2006) Rim15 and the crossroads of nutrient signalling pathways in *Saccharomyces cerevisiae. Cell Div.* **1**, 3.
49. Wanke, V., Cameroni, E., Uotila, A., et al. (2008) Caffeine extends yeast lifespan by targeting TORC1. *Mol. Microbiol.* **69**, 277–285.
50. De Wever, V., Reiter, W., Ballarini, A., Ammerer, G., and Brocard, C. (2005) A dual role for PP1 in shaping the Msn2-dependent transcriptional response to glucose starvation. *EMBO J.* **24**, 4115–4123.
51. Goberdhan, D. C., Ogmundsdottir, M. H., Kazi, S., et al. (2009) Amino acid sensing and mTOR regulation: inside or out? *Biochem. Soc. Trans.* **37**, 248–252.
52. Nicklin, P., Bergman, P., Zhang, B., et al. (2009) Bidirectional transport of amino acids regulates mTOR and autophagy. *Cell* **136**, 521–534.
53. Schneper, L., Duvel, K., and Broach, J. R. (2004) Sense and sensibility: nutritional response and signal integration in yeast. *Curr. Opin. Microbiol.* **7**, 624–630.
54. Dubouloz, F., Deloche, O., Wanke, V., Cameroni, E., and De Virgilio, C. (2005) The TOR and EGO protein complexes orchestrate microautophagy in yeast. *Mol. Cell* **19**, 15–26.

Section III

Computational Systems Biology: Computational Studies and Analyses

Chapter 19

Computational Yeast Systems Biology: A Case Study for the MAP Kinase Cascade

Edda Klipp

Abstract

Cellular networks and processes can be mathematically described and analyzed in various ways. Here, the case example of a MAP kinase (MAPK) cascade is used to detail steps in the formulation of a system of ordinary differential equations governing the temporal behavior of a signal transduction pathway after stimulation. Different analysis methods for the model are explained and demonstrated, such as stoichiometric analysis, sensitivity analysis, or studying the effect of deletions and protein overexpression. Finally, a perspective on standards concerning modeling in systems biology is given.

Key words: Computational yeast systems biology, mathematical models, signal transduction, high osmolarity glycerol (HOG) pathway, model development and analysis.

1. Introduction

The yeast *Saccharomyces cerevisiae* has been intensively used as model system to understand basic cellular processes. This also holds true for computational approaches. The basis for successful modeling of various processes relies on one hand on an active scientific community with exchange and common activities, and on the other hand on standardization efforts as well as on the availability of many experimental data and databases. *S. cerevisiae* is also easy to cultivate, harmless, and genetically well defined. It is interesting from the modeling point of view since it is an eukaryotic system with many conserved signaling pathways and essential processes in common with mammalian cells. Areas of intensive systems biology approaches comprise among others the following fields: metabolism, signaling, cell cycle, and gene expression

J.I. Castrillo, S.G. Oliver (eds.), *Yeast Systems Biology*, Methods in Molecular Biology 759,
DOI 10.1007/978-1-61779-173-4_19, © Springer Science+Business Media, LLC 2011

regulation. It turned out that mathematical modeling can be very helpful for discovering and understanding biological processes and organization principles, e.g., (i) it forces the investigator to formulate hypotheses and insights in a clear-cut and formal way; (ii) it may allow for the representation and evaluation of system compounds that are experimentally not accessible; (iii) it allows to explore many scenarios or parameter values in less time and cheaper than in experiments; (iv) it may help to extract structural dependencies or mathematical and physical relation that are hard to find by biological intuition. Still, mathematical models are no magic wand, able to produce more information than provided by the qualitative and quantitative experimental data.

In general, modeling requires abstraction and focus on important aspects. First of all, modeling requires the clear definition of the objective, i.e., a question or problem to be solved. This problem then determines which type of modeling approach might be suitable and which type of modeling result may satisfy the investigator. For example, in cell cycle modeling driving questions may be as follows: (i) which protein–protein interaction network can generate cell cycle oscillations? (ii) what is the role of cyclins and cyclin-dependent kinases? (iii) is the set of known proteins and their interactions sufficient to explain observed cycling and the effect of mutants? (iv) what may be the role of critical network features such as negative feedback? (v) what makes cell cycle progression irreversible? In the analysis of signaling pathways typical questions are as follows: (i) which proteins contribute to the pathway, and is the known wiring able to bring about the observed behavior? (ii) which parts of the pathways are responsible for modulating the amplitude or the duration of the signal? (iii) which network motifs such as negative or positive feedback are involved? (iv) is there crosstalk among various pathways and how is this realized?

2. Methods – How to Decide on the Modeling Framework

In general, every abstracted description of the system that can be challenged by specific queries could be a suitable model approach for applied biological questions. Here, we will give an overview on some established frameworks that are used given specific types of data and trying to solve specific types of questions.

Please note that the following lists do not (and cannot) aim for completeness.

- *Protein–Protein Interaction (PPI) Networks*: The information used to construct protein-protein interaction networks comprises signals for physical interaction of individual proteins, e.g., form yeast two-hybrid screens, co-immunoprecipitation, Förster resonance energy transfer (FRET), tandem affinity purification (TAP), and similar techniques commonly used. The easiest way to construct the network is to successively join next neighbors. The possible output comprises static aspects such as neighborhood relations, shortest paths from one protein to another, clusters, functional relations such as formation of large functional complexes like the ribosome or being part of a signaling pathway (1, 2). PPI can also be used as basis for dynamic models of signaling pathways as detailed below. As an example, **Fig. 19.1** shows a protein–protein interaction network relevant for the high osmolarity glycerol (HOG) signaling pathway.

- *Gene regulatory networks (GRN)* connect genes via the transcription factors that are encoded by one gene and that regulate the expression of the next gene. A prominent example is the sea urchin development GRN, which is constantly updated as soon as more information becomes available. The dynamics of GRN can be modeled in various ways, for example, with Boolean networks, with stochastic simulation, or with ordinary differential equations.
- *Boolean networks* describe cell processes with graphs consisting of nodes and edges connecting the nodes. Each node can assume one out of two states, e.g., 1 and 0, representing in a very simplified way an "on" or "off" state of genes. The states are updated in discrete time steps according to Boolean functions taking into account the value of the nodes connected by one edge. For example, the AND function turns a gene on, when both its neighbors are on at the previous time step. An example for a Boolean network applied to yeast cell cycle was presented in (3).

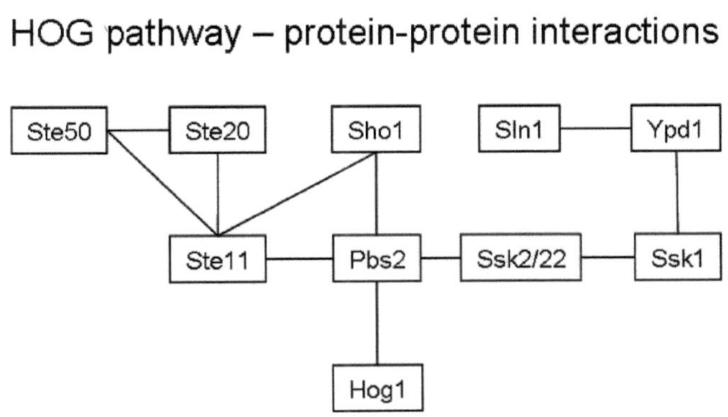

Fig. 19.1. Protein–protein interaction network involved in the high osmolarity glycerol (HOG) pathway.

- *Petri nets* also describe the states of individual nodes in a discrete fashion and these states are updated along a discrete time axis according to the rules assigned to the edges. Petri nets are bipartite graphs, distinguishing between places (e.g., states of a gene or amount of a protein) and transitions (e.g., reactions converting one compound into another). Tokens are assigned to the places and redistributed according to the rules of each transition. The pheromone pathway of yeast was modeled with a Petri net approach (4). More sophisticated approaches tend to consider more different states and update rules.
- *Ordinary differential equations*: A frequent approach (5) is the description with ordinary differential equations (ODEs), where the state space is continuous (concentrations or activities) and the time is continuous. In the following we will focus on the ODE model approach. Among others, stoichiometric analysis and steady-state analysis will be explained.
- *Stoichiometric analysis* uses as information just the stoichiometric properties, i.e., reactions and their substrates as well as the molecular characteristics of substrates in each reaction. As detailed below, methods comprise the calculation of steady-state fluxes (null space of matrix $\boldsymbol{N}$) and the calculation of conservation relation (null space of transpose of matrix $\boldsymbol{N}$).
- *Flux balance analysis (FBA)* uses information about the stoichiometric properties of a network. In addition, an objective function must be defined, e.g., the maximization of a certain flux or minimization of a byproduct. FBA is mostly applied to metabolic networks (6). FBA is based on the relations revealed for fluxes in steady state. To elucidate operation modes of the cell under different environmental conditions or to suggest such modes for biotechnological processes, it calculates from all possible steady-state fluxes that set of fluxes that maximizes or minimizes a certain function of these fluxes, e.g., by linear programming.
- *Stochastic modeling*. The dynamics on a continuous time scale can be simulated in a stochastic manner, e.g., with one of Gillespie's methods (e.g., (7)) by assuming discrete state values, e.g., molecule numbers.

Choosing the most appropriate method to model a given biological system is often difficult because each method is adapted to a particular type of data and can answer a limited and specific set of questions. Depending on the available experimental information, the purpose of modeling, the experience and preference of the modeler, signaling pathways can be described with different techniques.

3. Modeling: Mathematical Techniques and Tools

3.1. Purpose of Modeling

The development of a model serves the abstract and condensed representation of facts in order to allow for the analysis of their relations and to gain understanding about their internal organization and their communication with the environment.

Although the number of data in biological research currently explodes, such data are useless without sufficient interpretation. A computational model can on the one hand serve the data interpretation; on the other hand it can point to biological aspects that are still not sufficiently experimentally resolved. Within the field of systems biology, the view has been established that experimental research and model development should go hand in hand in an iterative manner including formulation of an initial model, hypothesis generation, experimental testing of hypotheses, model-based experimental design, model refinement upon new data.

The iterative modeling and experimentation process is hard to follow in publications, since they often only represent the final results. Model improvement with time and with accumulating experimental information is documented, e.g., for yeast cell cycle ((8) and others (9, 10)) and for signaling pathways (11–14).

3.2. Model Development

A typical situation in biological research is that an experimental observation inspires the formulation of a hypothesis (**Fig. 19.2**). As a next step one defines the questions a model is supposed to answer, i.e., the *scope* of the model. The scope determines which components and processes the model will take into account or omit and it defines the system's boundaries. Omitting certain processes from the models even though they might play a role is based on the assumption that they have only a minor influence on the event under study, that their values remain constant in the experimental setup, or that they simply cannot be described with the currently available means. For example, in the modeling of metabolic networks the effect of regulated gene expression has been neglected for a long time, although researchers were

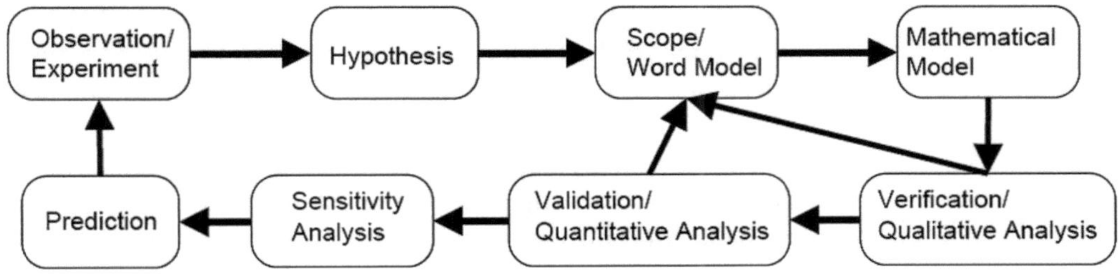

Fig. 19.2. Model development flow chart.

certainly aware of production and degradation of enzymes. However, different time scales of protein turnover and metabolic reactions frequently justify such simplifications. The initial model is usually formulated as a word model, which itself is also subjected to a process of refinement and sophistication in the course of model development. A diagram for graphical representation of the model structure is also helpful.

In a further step, the word model is translated into a mathematical model. This model must be verified, meaning to test whether the model is in principle able to answer the initial question independently of choice of specific parameter values, i.e., in a qualitative way. Verification of the model structure is an important step in the process of model development because it can save much time and effort later on. When the model is not able to fit observed data, this might be a general problem of the model structure. Having checked this in advance we can avoid validating a model in vain. Generally, it is also desirable to learn more about general properties of the model, like, e.g., existence of steady states and bifurcation points.

Model validation as a next step means assuring that the model can also reproduce observations in a quantitative manner. This is generally achieved by adjusting the model parameters such that the components of the model match observed data. Often, unique determination of parameter values from data is impossible. Therefore, it is advisable to further support for the model by testing whether it is also able to reproduce independent data without changing the fitted parameters. Independent in this sense means that the data were used neither to fit the parameters nor to develop our model. This can be achieved by dividing available data into a training data set and test data set or by producing completely new data. The test data generally describe the same phenomena but under slightly different conditions. It is a prerequisite for a sound model validation that the model is able to reproduce observed data under different conditions but with the same parameters that were used to reproduce the training data set. This is supposed to reflect the fact that our model accurately describes the intrinsic structure of the studied system and, like nature, is able to adequately adjust its reaction to a changing environment/input without changing internal structure and interactions.

Every model is only an abstracted representation of the processes under study. Hence, it is important to know the limits of applicability of a model. They determine to what extent possible predictions and conclusion hold. Moreover, it is important to know to which parameters the model output is sensitive, i.e., which parameter changes have a substantial impact on the systems behavior, and thus have to be determined most accurately. This is studied by sensitivity analysis. Local sensitivity analysis considers changing one parameter value at a time and looking at the resulting change of a specific output variable. A classical measure of sensitivity is the relative sensitivity S that is defined as

$$S = \frac{\Delta O}{O} \cdot \frac{p}{\Delta p} \qquad [1]$$

where $\Delta O/O$ is the relative change of some output O of interest and $\Delta p/p$ is the relative parameter change, compared to the initial state of parameter, respectively. S is easy to interpret, as $S = 1$ means that a certain percentage change of a parameter yields the same percentage change of the considered output. Classical sensitivity analysis studies the reaction of one or more output variable to the change of one parameter at a time. Generally, it cannot be assumed that parameters have an independent influence on the considered output. In most cases the sensitivity to one parameter depends on the state of one or more other parameters, which can be studied by global sensitivity analysis.

Sensitivity to parameters gives us important information about the system. It shows where small measurements errors can have drastic consequences for the system behavior and where additional research or measurements might be adequate. It also indicates interesting targets for drug developers as it makes sense to manipulate a system where it is most sensitive. The other way around, sensitivity analysis tells us something about the robustness and resilience of the system.

After formulation, verification, and validation of the model, it can be used to explore more systematically the regions of the state space that are of particular interest, i.e., make predictions. The model ideally should be able to predict future experiments. Correct predictions of the experiments support the confidence in the model and also in the original hypothesis. Moreover, the model can be used to design future experiments. The results of sensitivity analysis indicate where additional measurements give us the most information about the system.

The case that the model – although carefully designed – does not correctly predict the experiments is also very interesting. Now it has to be checked whether the experiments still comply with the original hypothesis. If so, the model must be modified, otherwise the hypothesis must be modified. Both ways, we close the cycle.

4. Model Development Step by Step

We will describe here step by step how to develop a dynamic model for a cellular network. As a running example, a very simplified version of a *M*itogen-*A*ctivated *P*rotein *K*inase (MAPK) cascade is used.

The following steps are considered:

- Define network structure.
- Analyze the network structure.
- Describe dynamics with a mathematical model, e.g., set of ODEs.
- Estimate parameters.
- Test model against data (does the output describes data input satisfactorily?).
- Test model against new experimental scenarios.
- Make predictions for hitherto untested (but in principle testable) scenarios.

4.1. Definition of Network Structure

For describing, for example, the dynamics of a MAP kinase (K) cascade, we first collect information about the reactants and the connecting reactions, such as phosphorylation of MAP kinase kinase kinase (MAPKKK, also named S_1 hereafter) occurs after activation of a receptor via a number of proteins interactions (indirectly) or by an upstream kinase (directly) to yield the phosphorylated form, MÀPKKKP (S_2). The next steps are phosphorylation of MAPKK (S_3) by MAPKKK yielding MAPKKP (S_4), and phosphorylation of MAPK (S_5) by MAPKK resulting in MAPKP (S_6). We observe as well dephosphorylation of all forms of phosphorylated kinases by phosphatases. For simplicity we neglect all further interactions of the proteins with cell processes. In addition, we assume that the receptor (S_0) directly induces phosphorylation of MAPKKK.

The network for the MAPK dynamics looks

$$\begin{aligned} &\text{MAPKKK} \xrightarrow{\text{Receptor}} \text{MAPKKKP} \\ &\text{MAPKKKP} \xrightarrow{\text{Phosphatase}} \text{MAPKKK} \\ &\text{MAPKK} \xrightarrow{\text{MAPKKKP}} \text{MAPKKP} \\ &\text{MAPKKP} \xrightarrow{\text{Phosphatase}} \text{MAPKK} \\ &\text{MAPK} \xrightarrow{\text{MAPKKP}} \text{MAPKP} \\ &\text{MAPKP} \xrightarrow{\text{Phosphatase}} \text{MAPK} \end{aligned} \qquad [2]$$

We can make various assumptions about the dynamics of the receptor. After an external stimulus, it may remain activated or become degraded. In the later case, we can consider an additional component of the network:

$$\text{Receptor} \rightarrow \qquad [3]$$

Determining the network topology defines the limits of the systems under study and enables the integration of stoichiometric information and experimental parameters.

4.2. Stoichiometric Analysis

The network presented in equation [2] can be characterized by its stoichiometric matrix $\boldsymbol{N}$. To this end, we consider all the catalytic processes as independent reactions. Matrix $\boldsymbol{N}$ contains the stoichiometric coefficients of all compounds S_1 through S_6 (represented by the rows) in all reactions v_1 to v_6 (represented by the columns).

$$\boldsymbol{N} = \begin{pmatrix} -1 & 1 & 0 & 0 & 0 & 0 \\ 1 & -1 & 0 & 0 & 0 & 0 \\ 0 & 0 & -1 & 1 & 0 & 0 \\ 0 & 0 & 1 & -1 & 0 & 0 \\ 0 & 0 & 0 & 0 & -1 & 1 \\ 0 & 0 & 0 & 0 & 1 & -1 \end{pmatrix} \qquad [4]$$

This way, matrix $\boldsymbol{N}$ is purely based on the stoichiometry of the network, i.e., on its wiring. The analysis of matrix $\boldsymbol{N}$ employs linear algebra. The linear dependence of rows of the stoichiometric matrix points to moiety conservation in the system, i.e., it reveals which compounds or moieties are neither produced nor degraded by the network in total, such as the sum of differently modified forms of a protein. In mathematical terms, one has to find a regular matrix $\boldsymbol{G}$ such that $\boldsymbol{G} \cdot \boldsymbol{N} = 0$. Then the expression $\boldsymbol{G} \cdot \boldsymbol{S} = \text{const.}$ states the conservation relations. The equation $\boldsymbol{G} \cdot \boldsymbol{N} = 0$ is solved by the matrix

$$\boldsymbol{G} = \begin{pmatrix} 1\,1\,0\,0\,0\,0 \\ 0\,0\,1\,1\,0\,0 \\ 0\,0\,0\,0\,1\,1 \end{pmatrix} \qquad [5]$$

or each row vector presenting linear combinations of the row vectors forming $\boldsymbol{G}$. Here, every column corresponds to a compound; every row corresponds to a conservation relation. Inspection of matrix $\boldsymbol{G}$ reveals that the sum of both forms of each kinase is constant over time, even in non-steady state, such as

$$\text{MAPKP}\,(t) + \text{MAPK}\,(t) = \text{const.} \qquad [6]$$

The linear dependence of columns of $\boldsymbol{N}$ ($\boldsymbol{N} \cdot \boldsymbol{K} = 0$ with regular matrix $\boldsymbol{K}$) reveals the dependence of fluxes in steady state, i.e., steady-state fluxes are linear combinations of the columns of matrix $\boldsymbol{K}$. For example, in an unbranched pathway, all fluxes must be the same in case of steady state. The kernel matrix of the stoichiometric matrix $\boldsymbol{N}$ for the MAPK cascade is

$$K = \begin{pmatrix} 1 & 0 & 0 \\ 1 & 0 & 0 \\ 0 & 1 & 0 \\ 0 & 1 & 0 \\ 0 & 0 & 1 \\ 0 & 0 & 1 \end{pmatrix} \quad [7]$$

Again, each linear combination of the columns of $\boldsymbol{K}$ is also a solution. Here, columns refer to different possible steady-state flux combinations, rows refer to compound. Solution [7] can be interpreted as follows: in steady state the rates of phosphorylation and dephosphorylation of each kinase is the same, preventing the accumulation of one or the other form. This holds only in steady state, but will be violated in dynamic regimes.

4.3. Mathematical Description of Dynamic Processes

In the frame of modeling with ordinary differential equations, the dynamics of the biochemical reaction network is expressed by the balance equations

$$\frac{d\boldsymbol{S}(t)}{dt} = \boldsymbol{N}\boldsymbol{v}\left(\boldsymbol{S}(t), \boldsymbol{p}\right) \quad [8]$$

where $\boldsymbol{S}$, $\boldsymbol{v}$, and $\boldsymbol{p}$ denote the vectors of concentrations, reaction rates, and parameters of the system, respectively, and t is the time. $\boldsymbol{N}$ is the stoichiometric matrix introduced above.

The set of equations describing the dynamics of the MAPK cascade reads

$$\begin{aligned} \frac{d\text{MAPKKKP}}{dt} &= -\frac{d\text{MAPKKK}}{dt} = v_1 - v_2 \\ \frac{d\text{MAPKKP}}{dt} &= -\frac{d\text{MAPKK}}{dt} = v_3 - v_4 \\ \frac{d\text{MAPKP}}{dt} &= -\frac{d\text{MAPK}}{dt} = v_5 - v_6 \end{aligned} \quad [9]$$

Typical expressions for the reaction rates are (i) mass action rate law

$$v_i\left(S_j\right) = k_i \cdot S_j, \quad [10]$$

with k_i being the rate constant or by (ii) Michaelis–Menten kinetics

$$v_i\left(S_j\right) = \frac{V_{\max} S_j}{K_{\text{M}} + S_j} \quad [11]$$

with $V_{\max}$ and K_M denoting the maximal velocity and the half-saturation constant or by (iii) Hill kinetics

$$v_i(S_j) = \frac{V_{\max} S_j^n}{K_{0.5}^n + S_j^n} \quad [12]$$

with n denoting the so-called Hill coefficient.

The mass action law implies a linear dependence of rate on substrate concentration, while hyperbolic Michaelis–Menten kinetics and sigmoid Hill kinetics show saturation. Note that more elaborated kinetic mechanisms have been reported, especially for the case of multiple substrates and for reversible reactions (15).

For the MAPK cascade, we will assume for simplicity that all reaction follow mass action kinetics. In this special case, one must take into account that each reaction is catalyzed either by the upstream kinase or by a phosphatase, leading to the following kinetics:

$$\begin{aligned} v_1 &= k_1 \times \text{MAPKKK} \times \text{Receptor} \\ v_2 &= k_2 \times \text{MAPKKKP} \times \text{Phosphatase} \\ v_3 &= k_3 \times \text{MAPKK} \times \text{MAPKKKP} \\ v_4 &= k_4 \times \text{MAPKKP} \times \text{Phosphatase} \\ v_5 &= k_5 \times \text{MAPK} \times \text{MAPKKP} \\ v_6 &= k_6 \times \text{MAPKP} \times \text{Phosphatase} \end{aligned} \quad [13]$$

In the following, we will also assume that the receptor, once activated, gets degraded with rate:

$$v_0 = k_0 \times \text{Receptor} \quad [14]$$

Assigning kinetics to the individual reactions allows simulating the systems dynamics. To this end, one must determine the parameter values from experimental data.

4.4. Parameter Estimation

Parameter estimation is the attempt to determine the values of the kinetic constants or other parameters such as K_M values of a specific model from experimental data. This is a crucial step in model construction, at the same time it is a step which has to face uncertainty on different levels: uncertainty about the network structure of our model, uncertainty in the experimental data, and uncertainty in the conditions under which the experimental data have been measured. Essentially, we can employ both quantitative and qualitative data to restrict the range of possible parameter values. In most cases, it is not possible to determine parameter values uniquely. Often, we find many sets of parameter values that can equally well explain the data. To discriminate among those sets, one may employ additional investigations of the model under various conditions, as outlined below. What in theory is done is to maximize the likelihood of the model explaining the data

$$L(y, p) = P(y|p) \quad [15]$$

given as the probability P to observe the data set $\boldsymbol{y}$ from the model with parameter vector $\boldsymbol{p}$. In practice, we intend to minimize the sum of squared distances of experimental data $y_{i,j}$ and simulation results x_i given as

$$Z = \sum_{i,j,k} \left(x_i(t_k) - y_{i,j}(t_k)\right)^2 \quad [16]$$

where index i runs over all compounds, index j over all experiments, and index k over all measurement time points.

A number of computational tools are currently available to enable parameter estimation for users with different mathematical background. Among them is the frequently used tool Copasi (16), which offers different methods for both simulation and parameter estimation. For the data presented in **Table 19.1**, the steepest descent method with iteration limit 500 yields the following (rounded) parameter values:

$$k_0 = 1.2,\ k_1 = 0.96,\ k_2 = 1.18,\ k_3 = 1,\ k_4 = 1,\ k_5 = 1,\ k_6 = 1 \quad [17]$$

The time course plot of the model with the estimated parameters provided by Copasi looks as shown in **Fig. 19.3**. For the following types of analysis, we will assume that all parameter

Table 19.1
Fictive experimental time course data for the phosphorylated MAPKs (arbitrary units)

Time	MAPKKKP	MAPKKP	MAPKP
0	0	0	0
1	0.291	0.138	0.047
2	0.216	0.183	0.117
3	0.126	0.152	0.135
4	0.065	0.105	0.118
6	0.014	0.035	0.06
8	0.003	0.009	0.021
10	0	0.002	0.006

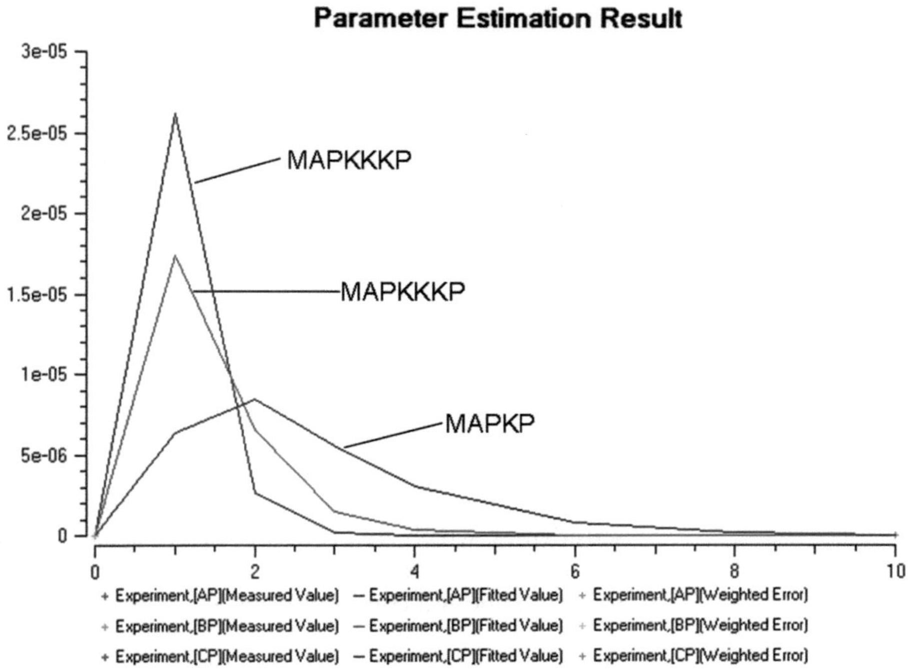

Fig. 19.3. Copasi-provided time course representation of the MAPK model with parameters estimated from data given in **Table 19.1** (arbitrary units).

values are equal to 1, i.e., $k_i = 1$, for $i = 0, \ldots, 6$ (compare **Fig. 19.7a** or **Fig. 19.8a**).

4.5. Analysis of Models

The model can be analyzed in various ways, first to test whether its behavior really reflects the aspects that we wanted to represent, second to deduce predictions based on a presumably appropriate description.

(1) ***Sensitivity analysis for steady states***: The aim is to check the dependence of the steady-state behavior on parameter values. For small parameter perturbations, one can use metabolic control analysis and calculate response coefficients, flux control coefficients, or concentration control coefficient. The response coefficients have the form given in equation [1], where O denotes a steady-state variable and p a parameter value. For control coefficients, O represents either a steady flux or a steady-state concentration and p stands for the rate of an individual reaction.

Metabolic control analysis (MCA) seeks to quantify the impact of individual rates or parameters on the steady-state values of variables by calculating the respective derivative (17, 18). We can consider so-called flux and concentration control coefficients, respectively:

$$C_{v_k}^{J_j} = \frac{v_k}{J_j} \cdot \frac{\partial J_j}{\partial v_k} \qquad [18]$$

and

$$C_{v_k}^{S_j} = \frac{v_k}{S_i} \cdot \frac{\partial S_i}{\partial v_k} \quad [19]$$

as well as response coefficients of fluxes and concentration with respect to parameter values, respectively:

$$R_{p_k}^{J_j} = \frac{p_k}{J_j} \cdot \frac{\partial J_j}{\partial p_k} \quad [20]$$

and

$$R_{v_k}^{S_j} = \frac{p_k}{S_i} \cdot \frac{\partial S_i}{\partial p_k} \quad [21]$$

The MCA theorems (19) establish a relation between these sensitivities, which are properties of the whole system, and the local sensitivities of the individual rates v_i with respect to the compound concentrations S_j

$$\varepsilon_{S_j}^{v_k} = \frac{S_i}{v_k} \cdot \frac{\partial v_k}{\partial S_i} \quad [22]$$

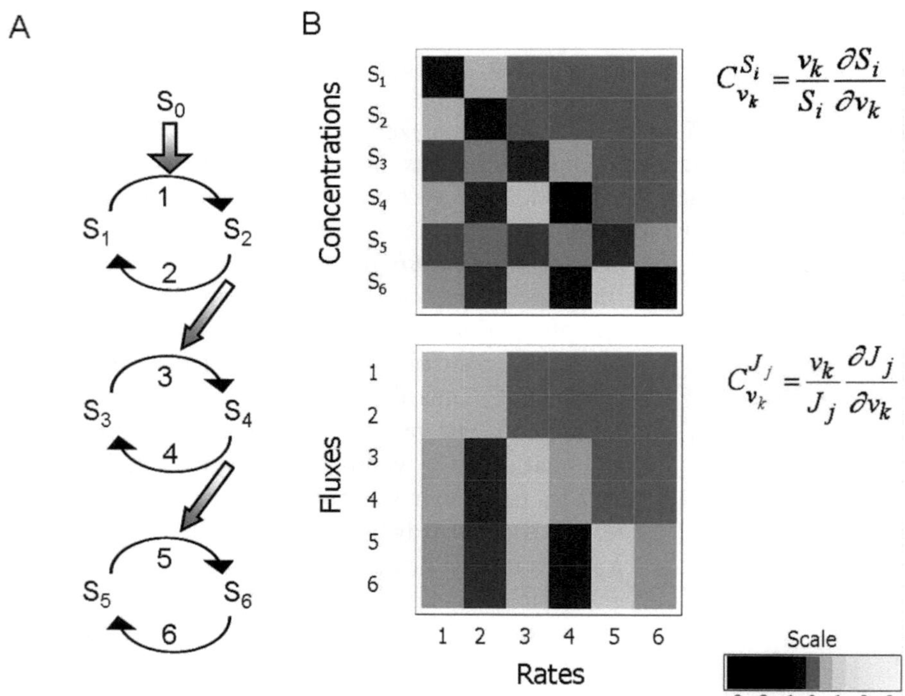

Fig. 19.4. (**a**) Structure of the MAPK cascade shown in [13]. (**b**) Flux control coefficients (*lower panel*) and concentration control coefficients (*upper panel*) as defined in equations [18] and [19] in color-code representation. Rates 1, 3, and 5 have positive control over subsequent reactions and subsequent phosphorylated forms (S_j), but negative control over non-phosphorylated compounds ($S_{j,0}$). Rates have no control over upstream reactions (*yellow* areas). Parameters: $k_i = 1,\ i = 0,..,6,\ S_0(0) = S_1(0) = S_3(0) = S_5(0) = 1,\ S_2(0) = S_4(0) = S_6(0) = 0.$

and the network stoichiometry N. For further details, refer to textbooks (20, 21). The flux and concentration control coefficients for the MAPK in [13] are shown in **Fig. 19.4**.

To assess the effect of wiring in a MAPK we can compare the structure presented in [13] and shown in **Fig. 19.4** with the case that we consider the binding of the upstream kinase to the downstream kinase explicitly (and not only as catalyst of the phosphorylation reaction). The structure and the effect on the control pattern are shown in **Fig. 19.5**. Most prominently, in this case all reactions have control over all rates and compound concentrations.

(2) ***Check the effect of large parameter changes.*** In case, check whether the number or stability of steady states change upon parameter variation. For example [9, 13, 14], there is only one stable steady state,

$$S_2^{ss} = \frac{k_1 S_0}{k_2 + k_1 S_0}, \qquad S_4^{ss} = \frac{k_1 k_3 S_0}{k_2 k_4 + k_1 (k_3 + k_4) S_0},$$
$$S_6^{ss} = \frac{k_1 k_3 S_0}{k_2 k_4 k_6 + k_1 (k_3 k_5 + k_3 k_6 + k_4 k_6) S_0} \qquad [23]$$

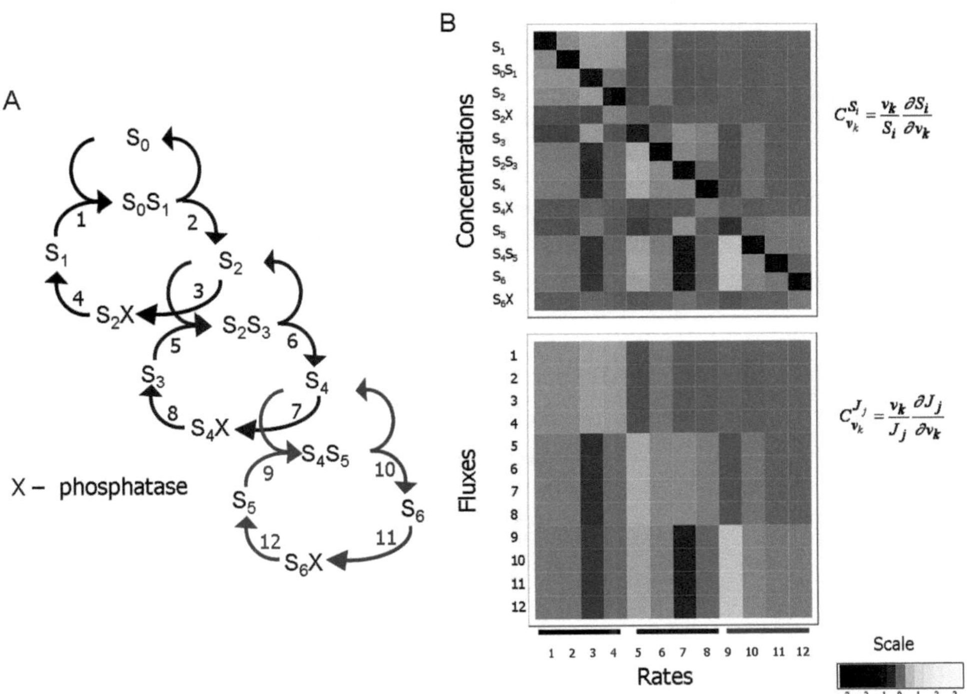

Fig. 19.5. (**a**) Structure of the MAPK cascade with explicit consideration of complex formation of upstream kinases with downstream kinases. (**b**) Flux and concentration control coefficients (equations [18] and [19]) in color-code representation. Parameters: $k_i = 1,\ i = 0,..,6$, $S_0(0) = S_1(0) = S_3(0) = S_5(0) = 1$, $S_2(0) = S_4(0) = S_6(0) = 0$.

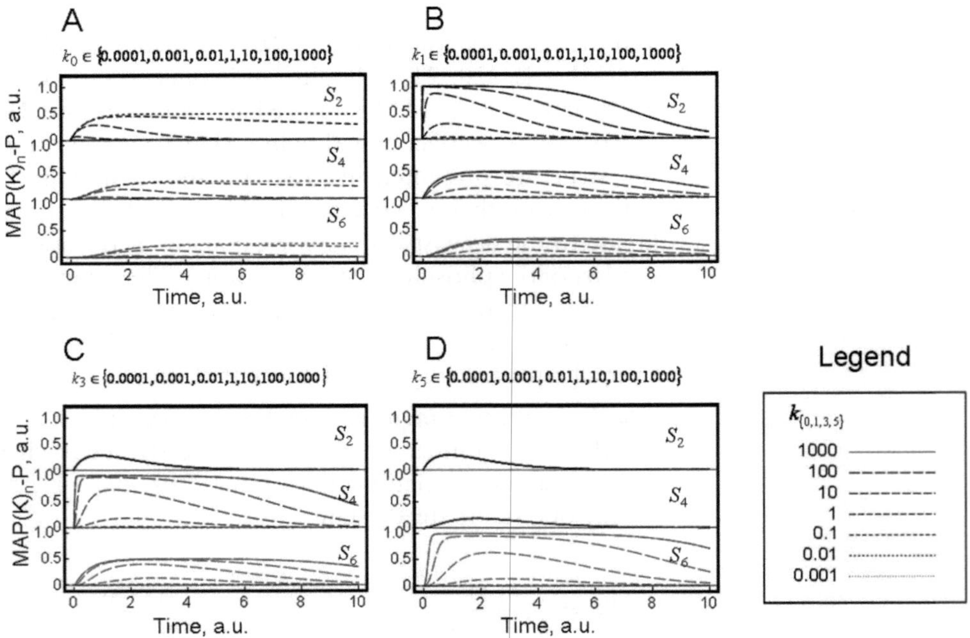

Fig. 19.6. Effect of parameter variation on the time courses of the MAP kinases. S_2, S_4, and S_6 are presented in stacked panels to prevent overlap. Parameter values influence time course concentrations depending on whether they affect producing or degrading reactions and on the distance of the affected reaction to the considered compound. For example, increasing k_1 to 1,000 has only minor effect on the amplitude of S_6 and induces some prolongation of its activation, while increasing k_5 has strong impact on the amplitude of S_6 and a comparable effect on the duration of activation (a.u., arbitrary units). Parameters: $k_i = 1$, $i = 0, .., 6$ (except of varied values), $S_0(0) = S_1(0) = S_3(0) = S_5(0) = 1$, $S_2(0) = S_4(0) = S_6(0) = 0$.

i.e., if the receptor is inactive, then all phosphorylated kinases vanish, if the receptor assumes a basal level, then the phosphorylated kinases also assume a basal level determined by the rate constants. The effect of parameter variation is demonstrated in **Fig. 19.6**. We see that varying the kinase strength has different impact on MAPK depending on the level of the kinase. As predicted by MCA, the closer the variation to the MAPK the stronger the effect.

(3) ***Sensitivity analysis for time-dependent states***: Check the effect of parameter changes on the temporal behavior. Especially interesting for signaling pathways is the analysis of time-dependent response coefficients

$$R_{p_k}^{S_j(t)} = \frac{p_k}{S_i(t)} \cdot \frac{\partial S_i(t)}{\partial p_k} \qquad [24]$$

which show the impact of a parameter value on the dynamics of a compound, and not only on its steady-state value (22). The effect of different parameter values and initial

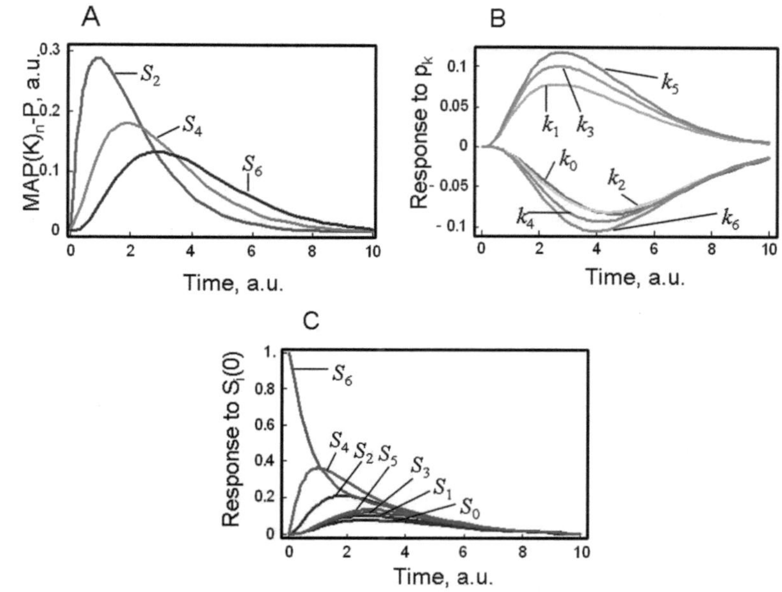

Fig. 19.7. Time-dependent response: (**a**) Time course of the MAPK cascade as shown in **Fig. 19.4** (**b** and **c**) time-dependent response coefficients for the lowest phosphorylated kinase (S_6) for parameters (*middle*) and initial concentrations (*right*). The kinase parameters show temporarily positive response, which declines toward the end of simulation time; the phosphatase parameter shows negative response. All initial concentrations show positive response, but with some time shift from the lowest to the uppermost compounds (a.u., arbitrary units). Parameters: $k_i = 1,\ i = 0,..,6$, $S_0(0) = S_1(0) = S_3(0) = S_5(0) = 1$, $S_2(0) = S_4(0) = S_6(0) = 0$.

concentrations on the time course of the phosphorylated MAPK is shown in **Fig. 19.7**.

4.6. Making Predictions

A model shall describe known facts in a precise and formalized way. An important additional value of a model is that it can be used to make new predictions. The way to challenge a model in order to predict hitherto unexpected is in the hands and the imagination of the modeler. But there are some ways that are nowadays becoming standards.

(4) ***Deletion analysis.*** Check steady states or temporal behavior for the effect of deleting an element, e.g., the deletion of a gene.

(5) ***Overexpression analysis.*** Check steady states or time course patterns for the effect of overexpression of specific genes.

Some examples for deletion and overexpression are shown in **Fig. 19.8**.

4.7. More Advanced Model Analysis Measures

(6) ***Rewiring***: Check steady states or time course behavior for changes in wiring, e.g., representing the exchange of promoters or transcription factor binding sites (TFBS).

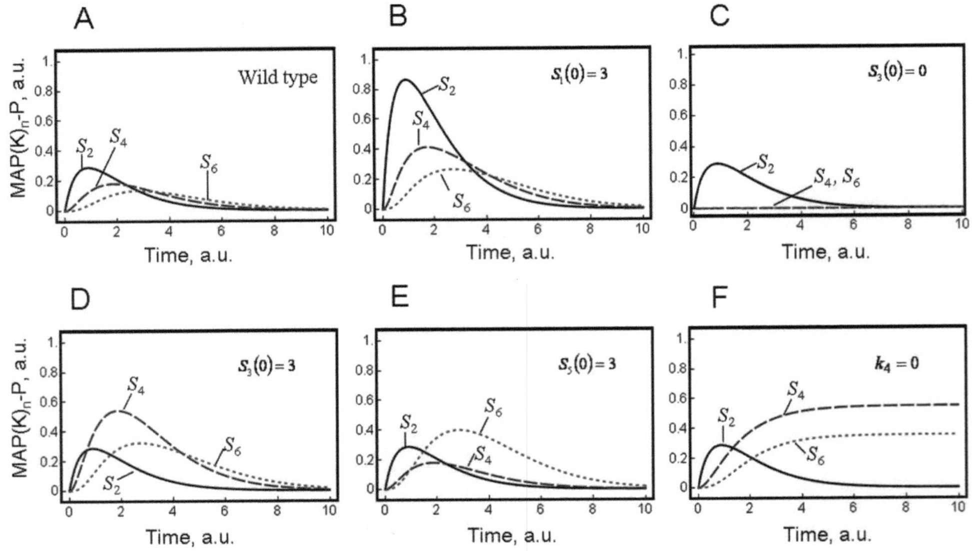

Fig. 19.8. Time course of the MAPK cascade for different scenarios. (**a**) "Wild type," (**b**), (**d**), (**e**) overexpression of the individual kinases S_1, S_3, and S_5, respectively. (**c**) Deletion of kinase MAPKK (S_4). (**f**) Deletion of the phosphatase reaction 4. Parameters: $k_i = 1,\ i = 0, .., 6,\ S_0(0) = S_1(0) = S_3(0) = S_5(0) = 1,\ S_2(0) = S_4(0) = S_6(0) = 0$, (except of varied values).

(7) ***Effector addition***: Check steady states or time course behavior for effect of added compounds (inhibitors, activators).

(8) ***Effect of small molecule numbers***: Use stochastic simulations to study the effect of small molecule numbers, which may occur in signaling or regulatory networks.

5. Discussion

For the example of a simplified MAPK cascade, we discussed general steps in formulating, simulating, and analyzing a model. Both model construction and model analysis concern the network structure and the parameter values and kinetics of individual reactions.

There are a number of reasons why it might be impractical to follow the above recipe in a forward manner. Among them are the most important ones that the network structure is not sufficiently well known or that there is no sufficient data to estimate all involved parameters. In the first, one might want to test different versions of the model with the same complexity, in the second case one might want to find an easier model that still covers main features of the process, but can be explained by the data.

There are a number of examples for models of different complexity describing essentially the same process. In order to study the question how cell cycle oscillations may be explained based on known protein–protein interactions, Goldbeter and coworkers presented in 1991 a model with five components, a cyclin as well as a cyclin-dependent kinase and a protease, each one in two different forms (23). Employing the principles of delay and negative feedback, this small reaction system shows limit cycle-type oscillations and is able to explain some observations, such as the staggered activation of the cyclin, the kinase, and the protease. Later models of cell cycle became much more comprehensive. For example, a more complex model comprises about 50 components and reactions and can now explain the behavior of many mutants (8).

In the analysis of the response of yeast cells to osmostress, a first model (Klipp et al. (24)) was very comprehensive. This includes about 40 variables and explains the role of the glycerol transporter as well as the different timing of regulation – short-term regulation through reduction of glycerol export, long-term regulation through gene expression regulation, and increased glycerol production. New insights are the role of the aquaglyceroporin Fps1 for glycerol regulation and short-term response, the role of turgor pressure, and the effect of repeated stimulation. Drawbacks of big models are many parameters that cannot be determined from experimental data. Mettetal and colleagues reduced the description of the same biological process as a control system and presented a model with only two variables (25). Among the advantages of such a reduced model one of them is the usability of the established generalized system response to the design of the control system. Furthermore, it has only few parameters, which can be reliably estimated from experimental data. However, this approach is only applicable to small perturbations and has some limitations for the prediction of phenotypes for specific mutations, knockouts, or overexpression experiments.

An example for model testing different architectures of the same network has been presented by Hao et al. (26). They analyzed three wiring versions of the Sho1 branch of the HOG pathway asking which one could explain experimental data representing successive osmotic stresses. By a combined modeling and experimental approach based on experimental data, they could favor a feedback phosphorylation of Sho1 by Hog1.

5.1. Design Principles and Features

Mathematical models can do more than just explain data and predict new experiments. They can also elucidate so-called design principles relating cellular function to its evolution. Among those principles is feedback regulation, positive or negative, short ranging or long ranging. For signaling pathways, one may ask for signal encoding: duration, strength, timing, amplification – what is relevant for the cell to respond? Does timing matter? These questions shall be left open for further analysis. Such analysis needs on one hand, reliable quantitative data, together with models that can be further refined by incorporation of new amounts of data at different biological levels. To this end, both areas of systems biology need to rely on standards.

5.2. Model Standards

Models are created and simulated in order to understand biological observations. Their analysis supports the design of future experiments. Standardization plays a crucial role in enabling the exchange and interpretation of scientific research results, and in particular in computational modeling (5). A number of computer languages and formats have been designed by different communities to encode the structure of mathematical models, such as the Systems Biology Markup Language (SBML) (27). To allow reuse and exchange of models, standards for representation of a model are formulated in MIRIAM (28). Further standards for the annotation of compounds and biological process and for describing experiments are under development. An example is the annotation of compounds in models. For example, if we identify the MAPK (S_5) in [13] with the protein Hog1p, and the MAPKP (S_6) with phosphorylated Hog1p, then the assignments given in **Table 19.2** would relate these components and their parts to entries of the databases SGD (http://www.yeastgenome.org) and ChEBI (http://www.ebi.ac.uk/chebi), where each compound has a unique identifier (*see* **Table 19.2**). Assigning such annotation can be done manually or using computational tools such as semanticSBML (http://www.semanticsbml.org/).

Table 19.2
Examples for compound assignments

Compound	Qualifier	ID	From database
Hog1p	is	S000004103	SGD
Hog1pPP	isVersionOf	S000004103	SGD
	hasPart	35780	ChEBI

Acknowledgments

I thank Dr. Matteo Barberis and Dr. Wolfram Liebermeister for helpful discussions. This work has been supported by the European Community through the projects Quasi, Cellcomput, and UNICELLSYS (http://www.unicellsys.eu/).

References

1. Han, J. D., Bertin, N., Hao, T., et al. (2004) Evidence for dynamically organized modularity in the yeast protein-protein interaction network. *Nature* **430**, 88–93.
2. Uetz, P., and Pankratz, M. J. (2004) Protein interaction maps on the fly. *Nat. Biotechnol.* **22**, 43–44.
3. Bulashevska, S., and Eils, R. (2005) Inferring genetic regulatory logic from expression data. *Bioinformatics* **21**, 2706–2713.
4. Sackmann, A., Heiner, M., and Koch, I. (2006) Application of Petri net based analysis techniques to signal transduction pathways. *BMC Bioinformatics* 7, 482.
5. Klipp, E., Liebermeister, W., Helbig, A., Kowald, A., and Schaber, J. (2007) Systems biology standards – the community speaks. *Nat. Biotechnol.* **25**, 390–391.
6. Varma, A., Boesch, B. W., and Palsson, B. O. (1993) Stoichiometric interpretation of Escherichia coli glucose catabolism under various oxygenation rates. *Appl. Environ. Microbiol.* **59**, 2465–2473.
7. Gillespie, D. T. (1977) Exact Stochastic Simulation of coupled chemical-reactions. *J. Phys. Chem.* **81**, 2340–2361.
8. Chen, K. C., Calzone, L., Csikasz-Nagy, A., Cross, F. R., Novak, B., and Tyson, J. J. (2004) Integrative analysis of cell cycle control in budding yeast. *Mol. Biol. Cell* **15**, 3841–3862.
9. Chen, K. C., Csikasz-Nagy, A., Gyorffy, B., Val, J., Novak, B., and Tyson, J. J. (2000) Kinetic analysis of a molecular model of the budding yeast cell cycle. *Mol. Biol. Cell* **11**, 369–391.
10. Novak, B., Toth, A., Csikasz-Nagy, A., Gyorffy, B., Tyson, J. J., and Nasmyth, K. (1999) Finishing the cell cycle. *J. Theor. Biol.* **199**, 223–233.
11. Bhalla, U. S. (2002) Biochemical signaling networks decode temporal patterns of synaptic input. *J. Comput. Neurosci.* **13**, 49–62.
12. Bhalla, U. S. (2004) Models of cell signaling pathways. *Curr. Opin. Genet. Dev.* **14**, 375–381.
13. Bhalla, U. S., and Iyengar, R. (1999) Emergent properties of networks of biological signaling pathways. *Science* **283**, 381–387.
14. Bhalla, U. S., and Iyengar, R. (2001) Robustness of the bistable behavior of a biological signaling feedback loop. *Chaos* **11**, 221–226.
15. Cornish-Bowden, A. (2004) *Fundamentals of Enzyme Kinetics*. London: Portland Press.
16. Hoops, S., Sahle, S., Gauges, R., et al. (2006) COPASI – a COmplex PAthway SImulator. *Bioinformatics* **22**, 3067–3074.
17. Heinrich, R., and Rapoport, T. A. (1974) A linear steady-state treatment of enzymatic chains. General properties, control and effector strength. *Eur. J. Biochem.* **42**, 89–95.
18. Kacser, H., and Burns, J. A. (1973) The control of flux. *Symp. Soc. Exp. Biol.* **27**, 65–104.
19. Reder, C. (1988) Metabolic control theory: a structural approach. *J. Theor. Biol.* **135**, 175–201.
20. Heinrich, R., and Schuster, S. (1996) *The Regulation of Cellular Systems*. New York, NY: Chapman & Hall.
21. Klipp, E., Liebermeister, W., Wierling, C., Kowald, A., Lehrach, H., and Herwig, R. (2009) *Systems Biology. A Textbook*. Weinheim: Wiley-VCH.
22. Ingalls, B. P, and Sauro, H. M. (2003) Sensitivity analysis of stoichiometric networks: an extension of metabolic control analysis to non-steady state trajectories. *J. Theor. Biol.* **222**, 23–36.
23. Goldbeter, A. (1991) A minimal cascade model for the mitotic oscillator involving cyclin and cdc2 kinase. *Proc. Natl. Acad. Sci. USA* **88**, 9107–9111.
24. Klipp, E., Nordlander, B., Krüger, R., Gennemark, P., and Hohmann, S. (2005). Integrative model of the response of yeast to osmotic shock. *Nat. Biotechnol.* **23**, 975–982.
25. Mettetal, J. T, Muzzey, D., Gomez-Uribe, C., and van Oudenaarden, A. (2008) The frequency dependence of osmo-adaptation in *Saccharomyces cerevisiae*. *Science* **319**, 482–484.
26. Hao, N., Behar, M., Parnell, S. C., et al. (2007) A systems-biology analysis of feedback inhibition in the Sho1 osmotic-stress-response pathway. *Curr. Biol.* **17**, 659–667.
27. Hucka, M., Finney, A., Sauro, H. M., et al. (2003) The systems biology markup language (SBML): a medium for representation and exchange of biochemical network models. *Bioinformatics* **19**, 524–531.
28. Le Novere, N., Finney, A., Hucka, M., et al. (2005) Minimum information requested in the annotation of biochemical models (MIRIAM). *Nat. Biotechnol.* **23**, 1509–1515.

Chapter 20

Standards, Tools, and Databases for the Analysis of Yeast 'Omics Data

Axel Kowald and Christoph Wierling

Abstract

One of the major objectives of systems biology is the development of mathematical models for the quantitative description of complex biological systems, such as living cells. Biological data and software tools for the design, analysis, and simulation of models are two basic ingredients for the new field of systems biology. In this chapter we give an overview of databases and repositories that provide valuable information for the integrative analysis and modeling of data generated by the different omics techniques. We also provide a review of the most popular software tools currently used in computational systems biology studies. Standards for the annotation of biological data and for the analysis and exchange of models are fundamental for the success of systems biology and provide the glue that connects experimental data with mathematical models. We also discuss some broad trends regarding where systems biology is heading to.

Key words: Systems biology, databases, standards, data integration, modeling, simulation, software tools.

1. Introduction

Comprehensive high-throughput experiments and techniques are allowing the generation of large quantities of data for different components of a biological system (e.g., cell, tissue, and organism). The data analyzed by the specific techniques provide information on different classes of cellular entities at different levels, such as genomics, transcriptomics, proteomics, and metabolomics. These levels of information are usually subsumed under the term 'omics. From here, curated databases are needed to store and manage the information, software tools in order to analyze the data and develop models, and standards to ensure

J.I. Castrillo, S.G. Oliver (eds.), *Yeast Systems Biology*, Methods in Molecular Biology 759,
DOI 10.1007/978-1-61779-173-4_20, © Springer Science+Business Media, LLC 2011

data compatibility and facilitate the exchange of omics data and computer models.

Computer models can show whether the mathematical description, assumptions, and parameters of a specific model can reproduce the observed behavior of the biological system correctly and help to increase the knowledge of its dynamic behavior. Thus, a computer or in silico model can be used to predict the outcome of different perturbations (e.g., change of an extrinsic parameter, gene knockout, mutations, or inhibition). A mathematical model of a cellular system is based on the known reaction network of the particular system. To perform realistic simulations additional information about the kinetics of the reactions is needed. Later on, comprehensive experiments can be performed in order to obtain reliable values for the kinetic parameters (i.e., parameter estimation). Eventually, the model can be used for the generation of quantitative predictions which are subject to experimental validation and subsequent model refinement.

A critical issue when working with omics data is data integration. Basically, from reliable curated databases, data integration deals with the integration of heterogeneous data into databases with the aim to query for specific information or to parse data from these databases for data mining or mathematical modeling. Here, a (sometimes overlooked) prerequisite is that the data have to be compatible (i.e., obtained from the same biological system under equivalent environmental conditions). Data integration requires, technically, the definition of data exchange standards and protocols and the implementation of parsers that ensure the correct data exchange between databases and tools. A large number of public databases for storage of experimental data and pathways have been implemented, covering a broad spectrum of data resources. These include among others databases of genomics, transcriptomics, proteomics and metabolomics data, as well as pathway information from textbooks and primary sources.

In this chapter we give an overview of omics-related databases, introduce standards that are frequently used in this context, and discuss software tools that are commonly used in systems biology for data integration and the development, simulation, and analysis of mathematical models.

2. Databases and Data Repositories for Systems Biology

The use of databases in biological studies has not ceased to increase since the advent of genomic sequence data. The National Center for Biotechnology Information (NCBI) (http://www.ncbi.nlm.nih.gov/) and the European Bioinformatics Institute

(EMBL-EBI) (http://www.ebi.ac.uk/) provide access to multiple sequence databases such as the Genetics Sequence database (GenBank), the Reference Sequence database (RefSeq), and UniGene (these ones at the NCBI). Among the most relevant are the EMBL-EBI, the EMBL Nucleotide database, and the Ensembl automatic genome annotation database (1). Besides this, the NCBI and the EMBL-EBI provide access to databases specialized on other biological information, such as protein sequences and their annotation (e.g., Swiss-Prot, TrEMBL (2), UniProt (3)), protein families and domains (e.g., InterPro (4, 5)), and protein structures (e.g., Protein Data Bank (PDB) (6)). Many databases usually cover data from different organisms or species. In contrast to this, there are also organism-specific databases, such as the Saccharomyces Genome Database (SGD; http://www.yeastgenome.org/), a comprehensive, periodically curated, resource for access to *Saccharomyces cerevisiae* (budding yeast) functional genomics data (7). **Table 20.1** shows relevant omics- and systems biology-related databases. A more comprehensive list of databases and

Table 20.1
Selected databases for systems biology

Data resource	URL
Sequences	
GenBank	http://www.ncbi.nlm.nih.gov/Genbank/
EMBL Nucleotide Database	http://www.ebi.ac.uk/embl/
UniGene	http://www.ncbi.nlm.nih.gov/unigene
Ensembl	http://www.ensembl.org/index.html
Pathway and interaction databases	
BioCyc	http://www.biocyc.org
KEGG	http://www.genome.jp/kegg
Reactome	http://www.reactome.org
IntAct	http://www.ebi.ac.uk/intact
DIP	http://dip.doe-mbi.ucla.edu
MINT	http://mint.bio.uniroma2.it/mint
MPact	http://mips.gsf.de/genre/proj/mpact
ConsensusPathDB	http://cpdb.molgen.mpg.de
Kinetics databases	
BRENDA	http://www.brenda-enzymes.org
SABIO-RK	http://sabio.villa-bosch.de/SABIORK
Expression data resources	
Gene Expression Omnibus (GEO)	http://www.ncbi.nlm.nih.gov/projects/geo
ArrayExpress	http://www.ebi.ac.uk/arrayexpress/index.html
Systems biology model repositories	
BioModels	http://www.biomodels.org
JWS	http://jjj.biochem.sun.ac.za

resources can be found in (8). Furthermore, the January issue of *Nucleic Acids Research* (database issue) provides each year an updated overview of relevant biological databases.

In addition to sequence data, gene expression data (e.g., RNA expression data from microarrays or deep sequencing techniques) constitute a valuable resource for the analysis of gene function and expression patterns at the transcriptome level. Gene Expression Omnibus (GEO) (9) at NCBI and ArrayExpress at EMBL-EBI (10) are two central repositories which freely distribute and share access to comprehensive gene expression data sets.

Another group of databases that provide a rich resource, especially for the functional interpretation of experimental data and the development of models for systems biology, are pathway databases. Pathway databases comprise different aspects of cellular processes, like metabolic reactions, signal transduction pathways, or gene regulatory events. Pathguide (http://www.pathguide.org/) provides a comprehensive list of many pathway-related databases.

One of the first databases on metabolic reactions is EcoCyc. EcoCyc integrates genomic and gene products information into the metabolic and regulatory network of *Escherichia coli* (11). Similarly, databases with cellular processes collected in other organism-specific pathway databases have been implemented. Thus, BioCyc (http://pathway.yeastgenome.org/biocyc/) is a rich resource of *S. cerevisiae*-related metabolic pathways. Relevant pathway databases covering pathway information from *S. cerevisiae* and other organisms are The Kyoto Encyclopedia of genes and genomes (KEGG) and the Reactome database (**Table 20.1**).

KEGG provides a broad spectrum of genomic data from multiple species and organisms (12). It provides information on pathways of different organisms, presented as reference pathway maps. The reference maps cover metabolic pathways and signal transduction pathways and constitute a valuable resource of the reaction network of a specific organism (e.g., yeast). Another pathway database focusing on human cellular and biochemical processes is Reactome (13). Reactome is a curated database using a detailed ontology. It provides a comprehensive description of molecular and cellular processes, such as metabolic and signal transduction pathways. An overview of the different pathways covered by Reactome is presented at the database homepage (http://www.reactome.org). Although the focus of this database is mainly on human biological processes, it also provides information on organisms with orthologous genes and conserved pathways such as yeast.

Besides curated data of individual reactions, interaction data such as protein/protein interactions (PPIs) are also stored in databases. For instance, the MPact database is a large collection of curated experimental data of PPIs in yeast (14). Other PPI databases that cover not only yeast but also other species are IntAct (15), DIP (16), and MINT (17).

Interaction and pathway databases often target different networks (e.g., the metabolic network, the PPIs, or the gene regulatory network). In many cases a comprehensive analysis of all these functional data becomes necessary. This is when an integration of interaction and pathway databases becomes relevant. One approach to data integration is provided by the ConsensusPathDB database (18). Currently, this database integrates interaction data from 24 publicly accessible databases. It provides functionalities for the graphical representation of the interaction data and for the analysis of omics-based data (e.g., overrepresentation analysis).

Other relevant resources for the setup of systems biology models are repositories with information on reaction kinetics of biochemical reactions. BRENDA is the largest freely available database providing biochemical and molecular information on all classified enzymes for a large set of different organisms (19). The database covers manually curated data from more than 79,000 primary literature references. The BRENDA web site (http://www.brenda-enzymes.org/) supports the user by searching detailed information on a particular enzyme and its catalyzed reaction(s). In addition to details as substrates, products, stoichiometries, and others, BRENDA also provides functional details on reaction kinetics, such as kinetic parameters for different conditions (e.g., pH and temperature). This kind of data is needed for the development of quantitative mathematical models. Another database containing manually extracted kinetic data is SABIO-RK (20, 21). SABIO-RK tries to meet the requirements of the systems biology community that are needed for setting up in silico models of biochemical reaction networks.

In addition to this, databases collecting mathematical models of biological systems have been implemented. For example, Java Web Simulation (JWS) provides a collection of different kinetic models (22). Furthermore, the JWS web site also provides basic functionalities for the simulation and analysis of models. Another repository of mathematical models is BioModels (23). BioModels database already provides a large set of models that can be downloaded in different formats.

3. Standards for Systems Biology

The accumulated data in functional genomics studies is very heterogeneous and often documented and archived in very diverse formats. This makes difficult the reutilization of such data and the conversion can be error prone. To overcome these difficulties, standards are necessary. Standard formats for generation and storage of experimental data from high-throughput experiments have been implemented, with the extensible markup language (XML) becoming a flexible tool for the definition of standard formats (http://www.w3.org/XML).

The development of a standard proceeds via four different steps: (i) the informal design of a conceptual model, (ii) the formalization, (iii) the development of a data exchange format, and (iv) the implementation of supporting tools (24). A concept for the microarray domain is defined by the minimum information about a microarray experiment (MIAME) (25). For the standardized collection, integration, storage, and dissemination of proteomics data guidelines are defined by the minimum information about a proteomics experiment (MIAPE) (26). Similar to specifications for experimental data, concepts for the mathematical description of biological systems are also defined by the minimum information requested in the annotation of biochemical models (MIRIAM) (27).

Different XML-based data formats have been implemented for the exchange of data. As an example, microarray and gene expression markup language (MAGE-ML) is designed for the description of data from microarray experiments. Standards for the description of pathway data and mathematical models are, for instance, SBML (28, 29), CellML (30), BioPAX (31), and PSI-MI (32). While the focus of PSI-MI and BioPAX is more on the detailed description of PPIs and cellular reaction networks, respectively, SBML and CellML are basically designed for the description of mathematical models. The systems biology markup language (SBML) is already used by a large set of software applications (see http://sbml.org/SBML_Software_Guide/SBML_Software_Summary) and its benefits are discussed below in the description of software tools. Furthermore, a standard for the graphical representation of biochemical reaction systems has been defined. It is called the systems biology graphical notation (http://www.sbgn.org) and it defines nodes, edges, and further graphical elements for the graphical representation of cellular reaction networks. This standard is already used by some software applications (e.g., CellDesigner, see below).

Besides standards for the storage and exchange of data between different tools, standards for the communication between software applications are also necessary. For this purpose, application programmer interfaces (APIs) are used. An API defines a set of methods along with their parameters that are provided by a certain software tool. These methods can be called by other applications. In recent years more and more software applications make use of WebServices. A WebService is a well-defined XML-based API that follows the W3C recommendations (http://www.w3.org/2002/ws/). Several databases and web-based applications already provide WebServices (e.g., KEGG, SABIO-RK, and BRENDA). A tool for the integration of WebServices-based functionalities of different applications into a workflow is Taverna (33).

4. Software Tools

Different high-throughput techniques generate different types of data, such as sequence data, interaction data, concentrations, or information on the intracellular localization of biomolecules. Although the majority of systems biology models aim at a deeper understanding of the dynamical behavior of the biological system, a considerable number of programs have been developed to deal with the different data types. A comprehensive review of all the software tools would be beyond the scope of this chapter, and we will highlight some of the most commonly used tools instead. For a more in-depth analysis the reader is referred to (34, 35). **Figure 20.1** presents an overview of the most popular tools currently in use. The data are derived from a questionnaire (34) and are explained in the following sections.

4.1. General-Purpose Tools

Programming languages like C++, Java, or Python belong to the general-purpose tools, together with software packages like Mathematica and MATLAB, which not only provide a complete programming language but are also capable of symbolic manipulation of equations and the visualization of the results.

Mathematica is produced by Wolfram Research (http://www.wolfram.com) and currently exists as version 7 for all the major operating systems. Mathematica consists of two components, the kernel that runs in the background and the graphical user interface (GUI) that communicates with the kernel. The GUI has the form of a so-called notebook that contains all the input, output, and graphics. Apart from its numerical calculation and graphics capabilities Mathematica allows to perform advanced symbolic calculations. For many specialized topics Mathematica packages are available that provide additional functionality. Mathematica can also communicate with Java, .NET, or C/C++ code. This means that Mathematica can access external code written in one of these languages and that the Mathematica kernel can actually be called from other applications. Besides an excellent help utility, there are also many sites on the Internet that provide additional help and resources. The site http://mathworld.wolfram.com contains a large repository of contributions from Mathematica users all over the world. Before implementing

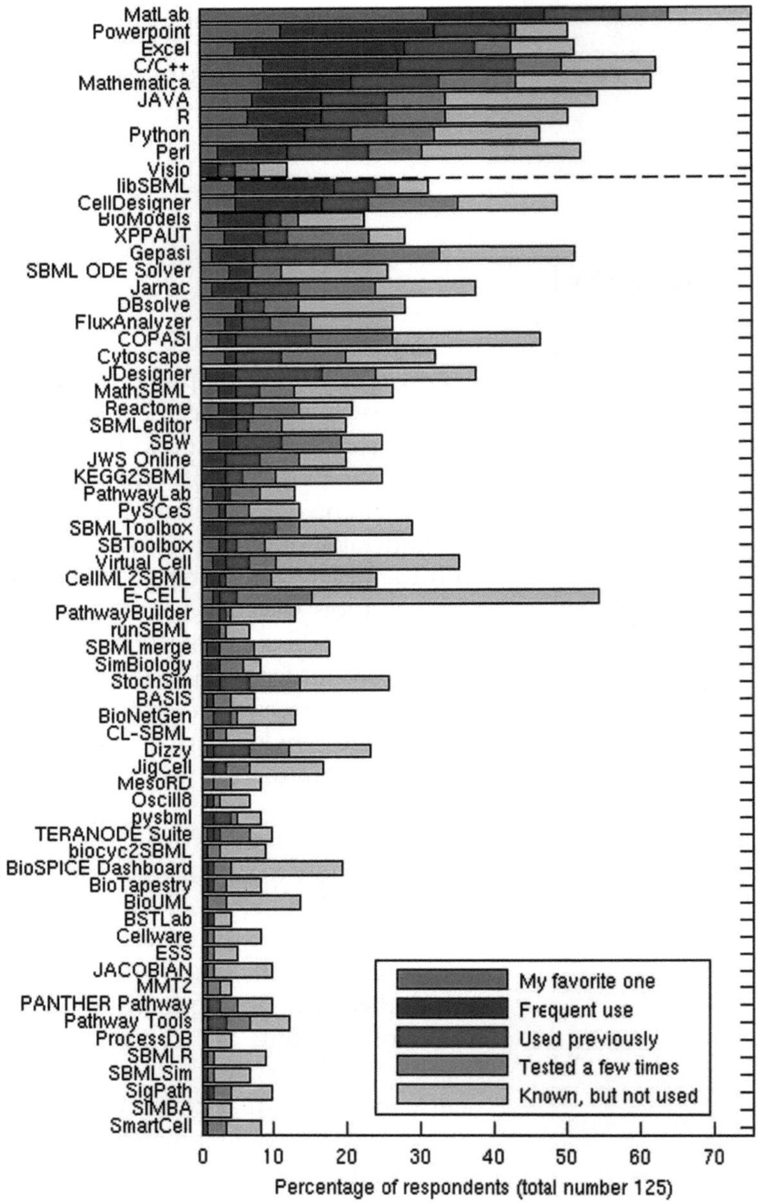

Fig. 20.1. Usage of software tools and databases for systems biology. Results of a questionnaire among 125 researchers regarding their use of software tools. The *upper part* shows general-purpose tools and the *lower part* specialized tools. Tools are sorted according to the sum of the frequencies for "My favorite one" and "Frequent use." Reproduced from (34) with permission from Nature Publishing Group (NPG).

a new function or algorithm it is worthwhile to check this site. If questions and problems arise during the use of Mathematica a useful source is the newsgroup news://comp.soft-sys.math.mathematica.

Another relevant tool, and major rival of Mathematica, is MATLAB R2008b, produced by MathWorks (http://www.mathworks.com). Both products are very similar in many aspects and the choice rests with the user. MATLAB is available for the same platforms as Mathematica has strong numerical and graphical capabilities. It also has its own programming language and programs are stored in so-called M-files. Toolboxes (special M-files) add additional functionality to the core MATLAB distribution, and like Mathematica, MATLAB can be called by external programs to perform high-level computations. A repository exists for user-contributed files (http://www.mathworks.com/matlabcentral/fileexchange and http://www.mathtools.net/MATLAB/toolboxes.html) as well as a newsgroup (news://comp.soft-sys.matlab). Albeit those similarities there are also differences. **Table 20.2** gives a summary of the main characteristics and differences of each tool.

4.2. Specialized Tools

The general-purpose tools are powerful packages for the numerical, symbolical, and visual analysis of arbitrary mathematical problems. They have some limitations, however, since they require considerable time and effort to get started. To overcome

Table 20.2
Main characteristics of computational software platforms for programming, modeling, and visualization of results in systems biology studies

Topic	Mathematica	MATLAB
Debugging	Since version 6 basic debugging capabilities are provided	Dedicated debugger allows to single step through M-files using breakpoints
Add-ons	Many standard packages ship with Mathematica and are included in the price	Many important toolboxes have to be bought separately
Deployment	User needs Mathematica to run the calculations specified in notebooks	Separately available compiler allows to produce stand-alone applications
Symbolic computation	Excellent built-in capabilities	Possible with commercial toolbox
Storage	All input, output, and graphics are stored in a single notebook.	Functions are stored in individual M-files. A large project can have hundreds of M-files
Graphics	Graphics is embedded in notebook	Graphic appears in a separate window
ODE model building	Differential equations are specified explicitly	Dynamical processes can be graphically constructed using Simulink, a companion product of MATLAB

this, many specialized tools have been developed that are very restricted in their application range, but much easier to use. Typical applications are the construction of biochemical reaction networks, the analysis of reaction networks (stability, flux analysis, and metabolic control theory), parameter fitting, or the simulation of stochastic reactions.

4.2.1. CellDesigner

Figure 20.1 shows that CellDesigner (36) is the most popular specialized stand-alone tool for systems biology (libSBML is not an application itself, but a programming library for handling SBML files). It is a freely available, easy to use, application (current version 4.0.1) for the creation and simulation of all sorts of biochemical reaction networks. It is written in Java and hence runs on most operating systems. The native model format is SBML, with layout information being stored in "annotation" tags. The GUI consists of different sub-windows that can be hidden or moved around (**Fig. 20.2**).

In CellDesigner a model is constructed by selecting icons from a large number of predefined shapes for different types of molecules such as proteins, ion channels, metabolites (**a**), and dragging these onto the model pane (**c**). These icons can be modified to indicate post-translational modifications like phosphorylation or methylation. SBML models are built using species (i.e., biomolecules), compartments, and reactions. This structure can be viewed in panel (**b**). Selecting a component in this panel highlights the corresponding element in the model pane (**c**) and in panel (**e**), which shows further details of the selected component. Although several automatic layout algorithms are available it is usually better to manually arrange the components in the model pane (**c**). This is also the place to add a commentary text or arrows, which are stored in an additional picture layer (as shown in panel **d**) and that can be switched on and off. Additional text notes can be attached to each model component (panel **f**) and are stored in the SBML file. This is very helpful, for example, to store from which source (e.g., publication and database) the kinetic data of a component (enzyme concentration and reaction rates) have been obtained (see also the CellDesigner "Worked Example"). As a final step, kinetic data and an explicit kinetic law are required to turn the graphical model into a dynamical model. CellDesigner allows to enter any mathematical expression as a kinetic law. Since version 4 it has also a collection of predefined kinetic laws (e.g., mass action and Michaelis–Menten types). Once this information is entered, simulations can be performed to obtain quantitative data on how the concentration of the model species changes over time. The simulation can be performed using the built-in solver or the Copasi solver (see next section). The

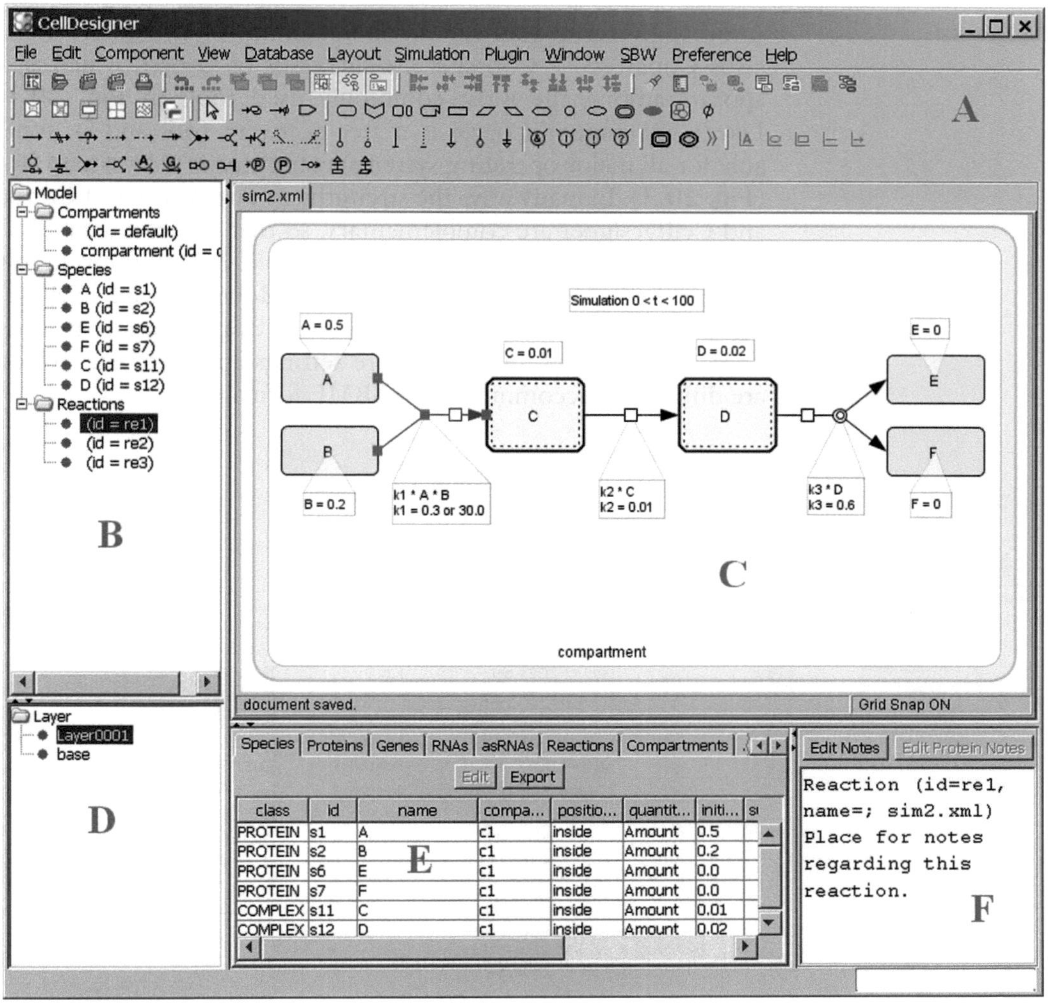

Fig. 20.2. Screenshot of CellDesigner 4.0.1 (available from http://www.celldesigner.org) a popular tool for the creation and simulation of dynamical models in systems biology. The graphical model is automatically converted into a set of ordinary differential equations that can then be integrated over time. The content of the different sub-windows (**a–f**) is explained in the text.

Copasi solver enables CellDesigner to perform stochastic simulations in addition to deterministic ones.

Apart from the core functionality just described, CellDesigner has further features like a free plugin API, the possibility to download models and species information from different databases, and the option to connect to the systems biology workbench (*see* CellDesigner "Other Tools"). Video tutorials with an introduction to the basic capabilities of CellDesigner are also available (http://www2.hu-berlin.de/biologie/theorybp/video_tutorials.php).

4.2.2. Copasi

Another powerful tool for the investigation of dynamical models is Copasi (37). Copasi is the successor of Gepasi, which also appears among the most commonly used tools in **Fig. 20.1**. Like CellDesigner Copasi is a free software tool and is available for all major operating systems from http://www.copasi.org (**Fig. 20.3**). In many ways the strengths and weaknesses of Copasi and CellDesigner are complementary, so that it is often useful to use both programs together. This is possible since both programs can import and export models in SBML. However, while CellDesigner uses SBML as native model format, Copasi has also its own XML-based model format to store settings and metadata, which are difficult to accommodate in SBML format.

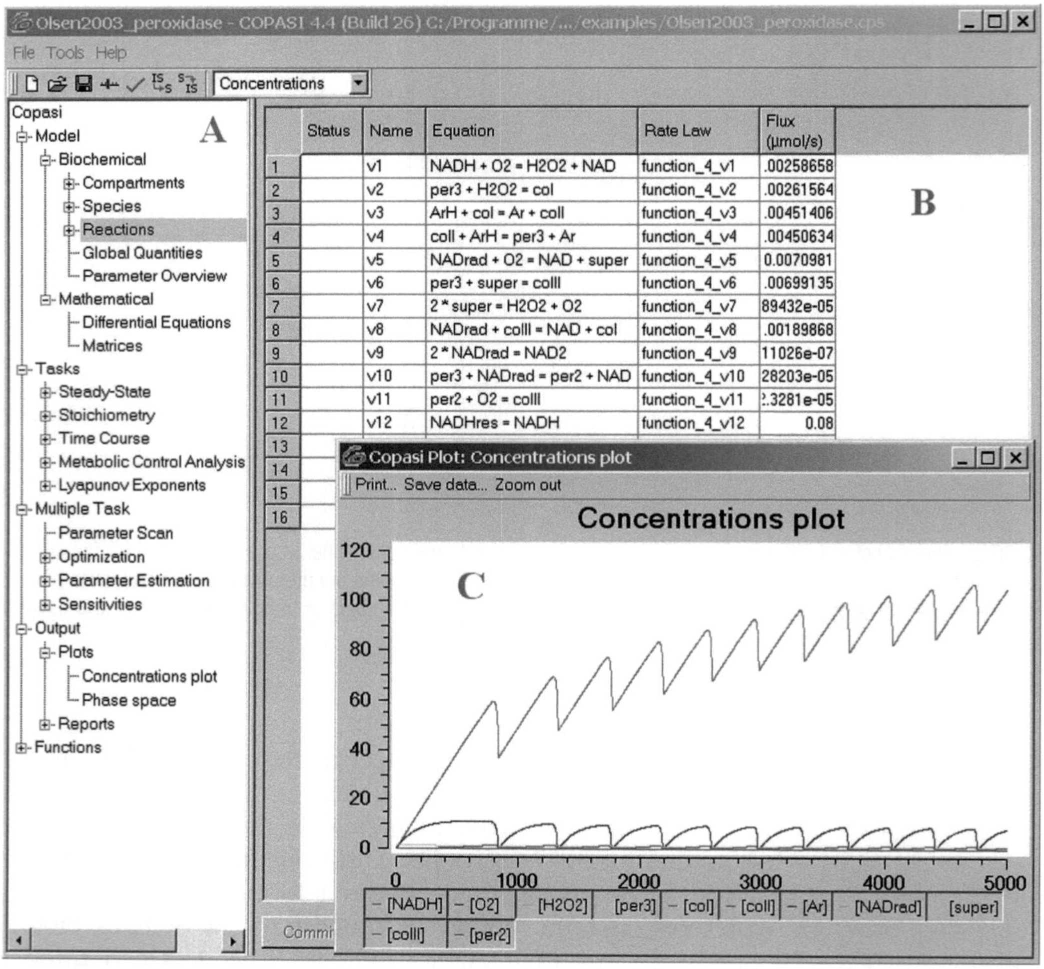

Fig. 20.3. Screenshot of Copasi 4B26 (available from http://www.copasi.org), a powerful tool for the analysis and simulation of dynamical models. Like most tools for system biology Copasi can read and write models in SBML format, but additionally stores model also in a proprietary format. The content of the different sub-windows (**a–c**) is explained in the text.

While CellDesigner is particularly good at constructing graphs and displaying biochemical models, Copasi is rather austere and simply provides a list of the reactions (**Fig. 20.3**, panel **b**). The strengths of Copasi, however, lie in its analysis tools, fitting and simulation capabilities which can be accessed via panel **a**. Copasi not only allows browsing of the SBML model structure (compartments, species, reactions) but it also allows to display the underlying set of differential equations. As with CellDesigner, it is possible to integrate the system over time and create plots of concentration vs. time (**Fig. 20.3**, panel **c**). Copasi is also capable of stochastic model simulations, an option not included in CellDesigner. To enable CellDesigner to perform stochastic simulations, using the Java language bindings (available for the latest stable release 4B30), CellDesigner can use Copasi's time course solvers. To make it work (under Windows) download the language bindings from http://www.copasi.org/tiki-index.php?page_ref_id=108 and copy CopasiJava.dll into the CellDesigner root folder. After starting CellDesigner the "Simulation" menu has now a second entry called "COPASI GUI."

Copasi also excels at other complex analysis tools that are not available in CellDesigner. For instance, it is possible to compute how the steady-state concentrations of all variables change depending on a specific parameter (sensitivity analysis) and perform simulations to estimate which parameter values would in principle better reproduce the experimental results (to be confirmed by independent experiments). Furthermore, Copasi is capable of metabolic control analysis (MCA), which involves calculating elasticities as well as flux and concentration control coefficients. Other types of analysis not covered in this chapter, including an introduction to the basic capabilities of Copasi, are available from http://www2.hu-berlin.de/biologie/theorybp/video_tutorials.php. Finally, there is also a forum available to ask questions regarding the use of Copasi (http://www.copasi.org/tiki-view_forum.php?forumId=1).

Besides stand-alone tools, like CellDesigner and Copasi, web-based applications for the development, simulation, and analysis of cellular reaction networks are also available. Web-based tools operate through the web browser and are easily accessible on different platforms (38). It is not necessary to install a local copy of the software or subsequent upgrades although they may be limited in speed. A relevant web-based tool is PyBioS (http://pybios.molgen.mpg.de (39)). PyBioS provides functionalities for the development, simulation, visualization, analysis, and storage of large computer models. For model development, PyBioS has interfaces to external pathway databases, such as KEGG, Reactome, and ConsensusPathDB. It also can visualize plots of time course simulations within automatically generated graphs of the reaction network that makes it easy to track fluxes.

4.3. New Tools and Trends

The programs described above are only a small selection of tools used in computational systems biology studies (**Fig. 20.1**). More comprehensive lists can be found at http://sbml.org/SBML_Software_Guide/SBML_Software_Summary and in (40). In this dynamic field, new data, computing power, and experimental techniques are promoting the appearance of new tools. Although it is difficult to predict the future, main trends can be summarized here.

4.3.1. Stochastic Modeling

The main goal in systems biology is to increase our knowledge of the dynamic behavior of biological systems. Deterministic approaches are the most commonly used but progressively more software tools are being developed to account for stochastic effects. Deterministic simulations using ordinary differential equations (ODEs) assume that the number of molecules in the system is so large that it can be treated as a continuous variable. A second assumption is that the system is completely deterministic. Random fluctuations are not considered. However although most molecules in a cell exist in large numbers, some others are very rare and the smaller the number of molecules, the more unrealistic those assumptions become. For example, transcription factors exist in low numbers in cells (e.g., approximately 10 molecules of Lac repressor in 1 *E. coli* cell (41)). Proteins involved in signal transduction pathways are also very rare, as are defective mitochondria, which are relevant for the aging process. Under those circumstances it becomes important that 4 or 5 molecules might be in a cell, but not 4.325 (as is possible with ODEs). Of special importance can be the difference between 0 and 1 if this item is a self-reproducing object. For example, if modeled with ODEs, it is practically impossible to obtain zero defective mitochondria, since a small amount (well below 1) will always appear in the ODE model. Because of their self-reproducing property a population of defective mitochondria could always re-grow from this artefact. If modeled stochastically, however, all defective organelles will disappear (zero concentration) once the last one is destroyed and thus they cannot re-grow.

If the simulated system has more than one possible steady state there can also be qualitative differences between a deterministic and a stochastic simulation. In an ODE model the system will settle into one of the possible steady states and remain there. However, if modeled stochastically the system can jump from one steady state to the other if they are closely enough.

Several stochastic simulation algorithms have been developed to calculate the change of the number of molecules during the time course of a chemical reaction (42–44). Depending on the complexity of the system the calculations are very time consuming and to obtain statistically meaningful results often many random trajectories have to be computed. However, the progressive increase in computing power is making stochastic simulators more accessible. As described above, Copasi is capable of stochastic time course simulations. It can also be run from the command line (allowing batch processing) and can be instructed to calculate many random trajectories.

The stochastic simulation tool Dizzy has been developed by Stephen Ramsey from the Institute for Systems Biology in Seattle (45) and is freely available from http://magnet.systemsbiology.net/software/Dizzy. The software is written in Java and is available for all platforms with a Java virtual machine. Models can be imported either in SBML format or in a proprietary scripting language. Once loaded, the model can be simulated by a variety of stochastic and deterministic algorithms. Dizzy can be controlled using a graphical user interface, but it can also be started from the command line, since GUI and the number processing machinery are separated. For a complex model describing the accumulation of defective mitochondria (46) we performed 1,000 repetitions of the model simulation on a Linux cluster, using the command line version of Dizzy. Although SBML is the de facto standard for models in systems biology, the model definition language of Dizzy has powerful features for handling arrays of variables and defining large number of reactions that make it interesting for specialized problems. Using these special language constructs we were able to define 4,950 reactions with only 5 lines of code for the above-mentioned mitochondrial model.

The Systems Biology Workbench (SBW) at http://www.sys-bio.org (47) is a software infrastructure that enables different tools to communicate with each other (see also next section). A powerful SBW module that provides stochastic simulation and analysis functions has been developed by the Sauro group and can be downloaded from http://public.kgi.edu/~rrao.

The software packages discussed so far are only suitable for spatially homogeneous models, which implies a constant and immediate mixing of all participating species. For stochastic simulations of 3D systems, including compartments and reaction diffusion systems MesoRD (48) is suitable. It is a free software tool, written in C++ which implements the next subvolume method to simulate the Markov process corresponding to the reaction diffusion equation. It can be obtained from http://mesord.sourceforge.net.

4.3.2. Integration and Connectivity of Models and Databases

Before the advent of SBML as model storage format it was very difficult to test models developed using different software packages. Re-implementation was often the only option. By adhering to the SBML format, models can now be easily exchanged, tested, and expanded even if different tools are used. CellDesigner and Copasi are a good example on how this exchangeability increases the value of both tools. Thus, in many cases it is more convenient to design and build a model using CellDesigner and then to do sophisticated simulations and analyses in Copasi. SBML is also having an enormous relevance in the area of dynamic modeling, and the development of SBML is not finished yet. Currently efforts are underway to standardize the graphical representation of biochemical networks using Systems Biology Graphical Notation (http://sbgn.org) and its integration into the SBML format will be a major step forward. Similarly, it will be important to accommodate models of spatially heterogeneous systems into SBML. This will make it possible to model other molecular entities and barriers, such as membranes, large protein complexes, or the effects of high viscosity using partial differential equations or stochastic simulations.

We have already discussed the importance of databases in systems biology. The connectivity of the modeling tools with these databases will constitute a relevant topic in the near future. Currently scientists have to search the literature or the databases manually in order to extract the necessary 'omics and kinetic data. Although the search for the latest 'omics data in literature databases and scientific sources will remain necessary, it would save time and reduce errors if the high amount of curated data annotated in databases (e.g., kinetic data) could be accessed directly from within the simulation tool. There is, for example, PyBioS that provides functionalities for the direct import of reactions from public pathway databases, such as KEGG, Reactome, or ConsensusPathDB. Also CellDesigner shows how useful such a connection can be. It allows to directly download models from http://biomodels.net, which is a model repository web site. Furthermore, after selecting a species or reaction, it is possible to connect to different databases (e.g., SGD, DBGET, iHOP, and PubMed) and look up information regarding this item. Although this feature needs further improvement, it clearly shows the trend. It would be, for instance, very useful to have a direct connection to a well-curated database for kinetic data, since a major problem in model building is to obtain reliable well-documented values for kinetic constants, reaction rates, and species concentrations. Although initial steps in this direction exist (e.g., http://www.brenda-enzymes.info and http://sabio.villa-bosch.de), more work is required, which will involve standardizing the naming scheme for genes, RNAs, proteins, and metabolites.

When developing a program that performs an analysis, a common difficulty is that there is an additional task involved with programming parts external to the core function (for example, to adapt the format to the core program). That means, although a program might be specialized in analyzing the topology of a reaction network, it also has to provide means for the input and output of the reaction details. An answer to this difficulty could be the Systems Biology Workbench (SBW) (47, 49), which is a software system that enables different tools to communicate with each other (http://sbw.sourceforge.net/). Thus, SBW-enabled tools can use services provided by other modules and in turn advertise their own specialized services. At the center of the system is the SBW broker that receives messages from one module and relays them to other modules. The list of SBW-enabled programs contains programs specialized in the graphical creation of reaction networks (JDesigner and CellDesigner), simulation tools (Jarnac and TauLeapService), analysis and optimization tools (Metatool, Bifurcation and Optimization), and utilities like the inspector module, which provides information about other modules.

5. Guidelines for the Construction of Models

The following guidelines are intended to provide a basic guide for the construction of dynamic models for non-experts in the field, using the CellDesigner tool. For more detailed information on modeling and systems biology models, see specific chapters in this volume.

1. Use CellDesigner to construct the model layout via drag and drop.
2. Do not use one of the automatic layout wizards. It is usually better to arrange the species manually with "Grid Snap" switched on.
3. In SBML, molecular species have a name that you choose when constructing the model and an ID that is automatically provided by CellDesigner (*see* **Fig. 20.2**, panel **e**). When entering the kinetic law that ID has to be used. To avoid confusion it is recommendable that name and ID are identical. After the layout of the model is finished, click on "Edit/Replace Species ID..." to make all IDs identical to the corresponding names.
4. Use "Edit/Add Layer" to add an additional layer for comments (text) and graphical elements (arrows, lines) that are not part of the actual model, but provide additional information for the user. If desired the elements in the layer can be made invisible by right clicking the layer (**Fig. 20.2**, panel **d**).

5. After the basic reaction scheme has been constructed, each reaction needs a kinetic law (a mathematical expression which properly describes the conversion of substrates into products). Unless other information is available from the literature, it is possible to choose between a mass action (mass balance) law or an irreversible Michaelis–Menten kinetics. The advantage of the mass action law is that the model will always reach a steady state and it will need less parameters than the Michaelis–Menten kinetics. The disadvantage is that it is less realistic since it does not cover all possibilities (e.g., enzyme saturation effects).
6. Once a kinetic law is chosen, numerical values for the parameters (reaction rates, half-lives, and saturation constants) are required. This is one of the most important parts of model building (together with choosing a kinetic law). The best approach would be to design and perform controlled experiments to determine the kinetic parameters of the specific system under the conditions tested. Alternatively, data are already available for enzymes of many organisms using different experimental conditions (e.g., through http://www.brenda-enzymes.info). If a specific value has been obtained it is advisable to save information about the original source (**Fig. 20.2**, panel **f**). This helps other colleagues to understand the numerical basis used for the model.
7. For stochastic simulations the species (molecules) need to have the correct units. After the species is placed on the model pane, do a right click and select "Edit Species...." If one enters "item" as substance unit in the emerging dialog box, the number under "initial..." is now interpreted as molecules and not as concentration. If the Copasi plugin is loaded a stochastic simulation can now be performed from within CellDesigner, otherwise the SBML file has to be opened with Copasi.

6. Concluding Remarks

The increasing amounts of data being generated by well-designed high-throughput experiments and techniques will allow the development of dynamic models and simulations with a high degree of complexity. However, to make efficient use of these possibilities, the data have to be structured and formatted in a way that they can be easily accessible for modeling. Not only have the data to be curated and deposited in a database but the data will need to be presented in a machine-readable form. When programs like CellDesigner can automatically retrieve the relevant data it will eventually much easier to construct and compare large quantitative models. This also requires that there is a unique and standardized naming scheme for genes, enzymes and reactions. From a modelers point of view kinetic constants like reaction rates and Michaelis Menten constants are essential for a realistic simulation of a biochemical model, but unfortunately there are currently no high throughput methods for the measurement of enzyme kinetics. The available data are often old, measured with different techniques under different conditions in different organisms and hidden in the scientific literature.

At the modeling level relevant progress has been made with the development of the systems biology markup language (SBML). This standard is a major step forward since it breaks down the barriers between the different software tools and enables the free exchange of models. By exchanging models it is then possible to use different tools for different tasks (e.g., layout, analysis, or simulation of the model).

An important issue to take into account at the early stages of the construction of the biological model is, what is the objective of the model? Will it be used just to try to reproduce the already reported data? Will it be a dynamic model able to predict the output of new experiments, help in the design of new tests, confirm or discard hypotheses before the need to be tested in the laboratory? Biochemical systems are inherently dynamic, with concentration and location of metabolites and enzymes constantly changing depending on the environmental conditions. Will the model be able to increase our knowledge of the dynamic behavior of the system?

Quantitative dynamical models calculate how the system changes over time and can simulate system behavior. A well-formulated quantitative model can be used, for example, to study and answer questions that are experimentally inaccessible or too expensive or simulate changes in concentrations and see what consequences can be expected. It is also possible to look at time periods that are too short or too long to be resolved experimentally. For humans it is notoriously difficult to make predictions about the behavior of nonlinear dynamical systems. Models are therefore a way to test hypotheses by making hidden assumptions explicit and simulating the often opposing effects of the system under investigation. The insights gained from such simulations are then valuable hints for the design of new experiments which will continue to increase our knowledge of the biological system, one of the main goals of systems biology.

Acknowledgments

This work was supported by the German Federal Ministry of Education and Research (BMBF) with the project Mutanom (01GS08105).

References

1. Hubbard, T., Barker, D., Birney, E., et al. (2002) The Ensembl genome database project. *Nucleic Acids Res.* **30**, 38–41.
2. Boeckmann, B., Bairoch, A., Apweiler, R., et al. (2003) The SWISS-PROT protein knowledgebase and its supplement TrEMBL in 2003. *Nucleic Acids Res.* **31**, 365–370.
3. Apweiler, R., Bairoch, A., Wu, C. H., et al. (2004) UniProt: the universal protein knowledgebase. *Nucleic Acids Res.* **32**, D115–119.
4. Biswas, M., O'Rourke, J. F., Camon, E., et al. (2002) Applications of InterPro in protein annotation and genome analysis. *Brief Bioinform.* **3**, 285–295.
5. Mulder, N. J., Apweiler, R., Attwood, T. K., et al. (2003) The InterPro database 2003 brings increased coverage and new features. *Nucleic Acids Res.* **31**, 315–318.
6. Berman, H. M., Westbrook, J., Feng, Z., et al. (2003) The protein data bank. *Nucleic Acids Res.* **28**, 235–242.
7. Cherry, J. M., Ball, C., Weng, S., et al. (1997) Genetic and physical maps of *Saccharomyces cerevisiae*. *Nature* **387**(6632 Suppl), 67–73.
8. Galperin, M. Y., and Cochrane, G. R. (2009) Nucleic acids research annual database issue and the NAR online molecular biology database collection in 2009. *Nucleic Acids Res.* **37**, D1–4.
9. Barrett, T., Troup, D. B., Wilhite, S. E., et al. (2007) NCBI GEO: mining tens of millions of expression profiles – database and tools update. *Nucleic Acids Res.* **35**, D760–765.
10. Brazma, A., Parkinson, H., Sarkans, U., et al. (2003) ArrayExpress – a public repository for microarray gene expression data at the EBI. *Nucleic Acids Res.* **31**, 68–71.
11. Karp, P. D., Ouzounis, C. A., Moore-Kochlacs, C., et al. (2005) Expansion of the BioCyc collection of pathway/genome databases to 160 genomes. *Nucleic Acids Res.* **19**, 6083–6089.
12. Kanehisa, M., Araki, M., Goto, S., et al. (2008) KEGG for linking genomes to life and the environment. *Nucleic Acids Res.* **36**, D480–484.
13. Matthews, L., Gopinath, G., Gillespie, M., et al. (2009) Reactome knowledgebase of human biological pathways and processes. *Nucleic Acids Res.* **37**, D619–622.
14. Güldener, U., Münsterkötter, M., Oesterheld, M., et al. (2006) MPact: the MIPS protein interaction resource on yeast. *Nucleic Acids Res.* **34**, D436–441.
15. Kerrien, S., Alam-Faruque, Y., Aranda, B., et al. (2007) IntAct – open source resource for molecular interaction data. *Nucleic Acids Res.* **35**, D561–565.
16. Salwinski, L., Miller, C. S., Smith, A. J. et al. (2004) The database of interacting proteins: 2004 update. *Nucleic Acids Res.* **32**, D449–451.
17. Chatr-aryamontri, A., Ceol, A., Palazzi, L. M., et al. (2007) MINT: the Molecular INTeraction database. *Nucleic Acids Res.* **35**, D572–574.
18. Kamburov, A., Wierling, C., Lehrach, H., and Herwig, R. (2009) ConsensusPathDB – a database for integrating human functional interaction networks. *Nucleic Acids Res.* **37**, D623–628.
19. Chang, A., Scheer, M., Grote, A., Schomburg, I., and Schomburg, D. (2009) BRENDA, AMENDA and FRENDA the enzyme information system: new content and tools in 2009. *Nucleic Acids Res.* **37**, D588–592.
20. Wittig, U., Golebiewski, M., Kania, R., et al. (2006) SABIO-RK: integration and curation of reaction kinetics data. In: Leser, U., Naumann, F., and Eckman, B. (eds.), *Proceedings of the 3rd International workshop on Data Integration in the Life Sciences 2006 (DILS'06).* Hinxton, UK. Lecture Notes in Bioinformatics Vol. 4075, pp. 94–103. Berlin, Heidelberg: Springer.
21. Rojas, I., Golebiewski, M., Kania, R., et al. (2007) SABIO-RK: a database for biochemical reactions and their kinetics. *BMC Syst. Biol.* **1**(Suppl 1), S6.
22. Oliver, B. G., and Snoep, J. L. (2004) Web-based kinetic modelling using JWS Online. *Bioinformatics* **20**, 2143–2144.

23. Le Novere, N., Bornstein, B., Broicher, A., et al. (2006) BioModels database: a free, centralized database of curated, published, quantitative kinetic models of biochemical and cellular systems. *Nucleic Acids Res.* **34**, D689–691.
24. Brazma, A., Krestyaninova, M., and Sarkans, U. (2006) Standards for systems biology. *Nat. Rev. Genet.* 7, 593–605.
25. Brazma, A., Hingamp, P., Quarckenbush, J., et al. (2001) Minimum information about a microarray experiment (MIAME) – towards standards for microarray data. *Nat. Genet.* **29**, 365–371.
26. Taylor, C. F., Paton, N. W., Lilley, K. S., et al. (2007) The minimum information about a proteomics experiment (MIAPE). *Nat. Biotechnol.* **25**, 887–893.
27. Le Novere, N., Finney, A., Hucka, M., et al. (2005) Minimum information requested in the annotation of biochemical models (MIRIAM). *Nat. Biotechnol.* **23**, 1509–1515.
28. Hucka, M., Finney, A., Sauro, H. M., et al. (2003) The systems biology markup language (SBML): a medium for representation and exchange of biochemical network models. *Bioinformatics* **19**, 524–531.
29. Hucka, M., Finney, A., Bornstein, B. J., et al. (2004) Evolving a lingua franca and associated software infrastructure for computational systems biology: the systems biology markup language (SBML) project. *Syst Biol.* **1**, 41–53.
30. Lloyd, C. M., Halstead, M. D. B., and Nielsen, P. F. (2004) CellML: its future, present and past. *Prog. Biophys. Mol. Biol.* **85**, 433–450.
31. Luciano, J. S. (2005) PAX of mind for pathway researchers. *Drug Discov. Today* **10**, 937–942.
32. Hermjakob, H., Montecchi-Palazzi, L., Bader, G., et al. (2004) The HUPO PSI's molecular interaction format – a community standard for the representation of protein interaction data. *Nat. Biotechnol.* **22**, 177–183.
33. Hull, D., Wolstencroft, K., Stevens, R., et al. (2006) Taverna: a tool for building and running workflows of services. *Nucleic Acids Res.* **34** (Web Server issue), 729–732.
34. Klipp, E., Liebermeister, W., Helbig, A., Kowald, A., and Schaber, J. (2007) Systems biology standards – the community speaks. *Nat. Biotechnol.* **25**, 390–391.
35. Klipp, E., Herwig, R., Kowald, A., Wierling, C., and Lehrach, H. (2005) *Systems Biology in Practice*. Weinheim: Wiley-VCH.
36. Funahashi, A., Matsuoka, Y., Jouraku, A., Morohashi, M., Kikuchi, N., and Kitano, H. (2008) CellDesigner 3.5: a versatile modeling tool for biochemical networks. *ProcIEEE* **96**, 1254–1265.
37. Hoops, S., Sahle, S., Gauges, R., et al. (2006) COPASI – a COmplex PAthway SImulator. *Bioinformatics* **22**, 3067–3074.
38. Lee, D., Saha, R., Yusufi, F. N. K., Park, W., and Karimi, I. A. (2008) Web-based applications for building, managing and analysing kinetic models of biological systems. *Brief Bioinform.* **10**, 65–74.
39. Wierling, C., Herwig, R., and Lehrach, H. (2007) Resources, standards and tools for systems biology. *Brief Funct. Genomic Proteomic* **6**, 240–251.
40. Klipp, E., Liebermeister, W., Wierling, C., Kowald, A., Lehrach, H., and Herwig, R. (2009) *Systems Biology. A Textbook*. Weinheim: Wiley-VCH.
41. Levin, B. (1999) *Genes VII*. Oxford: Oxford University Press.
42. Gillespie, D. T. (2001) Approximate accelerated stochastic simulation of chemically reacting systems. *J. Chem. Phys.* **115**, 1716–1733.
43. Gillespie, D. T. (1977) Exact stochastic simulation of coupled chemical reactions. *J. Phys. Chem.* **81**, 2340–2361.
44. Gibson, M. A., and Bruck, J. (2000) Efficient exact stochastic simulation of chemical systems with many species and many channels. *J. Phys. Chem.* **104**, 1876–1889.
45. Ramsey, S., and Orrell, D. (2005) Dizzy: stochastic simulations of large-scale genetic regulatory networks. *J. Bioinform. Comput. Biol.* **3**, 1–21.
46. Kowald, A., Jendrach, M., Pohl, S., Bereiter-Hahn, J., and Hammerstein, P. (2005) On the relevance of mitochondrial fusions for the accumulation of mitochondrial deletion mutants: a modelling study. *Aging Cell* **4**, 273–283.
47. Hucka, M., Finney, A., Sauro, H. M., Bolouri, H., Doyle, J., and Kitano, H. (2002) The ERATO systems biology workbench: enabling interaction and exchange between software tools for computational biology. *Pac. Symp. Biocomput.* 450–461.
48. Hattne, J., Fange, D., and Elf, J. (2005) Stochastic reaction-diffusion simulation with MesoRD. *Bioinformatics* **21**, 2923–2924.
49. Sauro, H. M., Hucka, M., Finney, A., et al. (2003) Next generation simulation tools: the systems biology workbench and BioSPICE integration. *OMICS* 7, 355–372.

Chapter 21

A Computational Method to Search for DNA Structural Motifs in Functional Genomic Elements

Stephen C.J. Parker, Aaron Harlap, and Thomas D. Tullius

Abstract

The rapidly increasing availability of DNA sequence data from modern high-throughput experimental techniques has created the need for computational algorithms to aid in motif discovery in genomic DNA. Such algorithms are typically used to find a statistical representation of the nucleotide sequence of the target site of a DNA-binding protein within a collection of DNA sequences that are thought to contain segments to which the protein is bound. A major assumption of these algorithms is that the protein recognizes the primary order of nucleotides in the sequence. However, proteins can also recognize the three-dimensional shape and structure of DNA. To account for this, we developed a computational method to predict the local structural profiles of any set of DNA sequences and then to search within these profiles for common DNA structural motifs. Here we describe the details of this method and use it to find a DNA structural motif in the *Saccharomyces cerevisiae* yeast genome that is associated with binding of the transcription factor RLM1, a component of the protein kinase C-mediated MAP kinase pathway.

Key words: Motif discovery, hydroxyl radical, DNA structure, Gibbs sampling, transcription factor, RLM1.

1. Introduction

Recent technological advances have led to the ability to experimentally determine where proteins bind to DNA throughout most of the yeast genome (1). These methods are based on immunoprecipitating proteins bound to chromatin, followed by hybridizing the recovered DNA fragments to a microarray (ChIP-chip), or by directly sequencing the bound DNA fragments using next-generation sequencing methods (ChIP-seq). One limitation

J.I. Castrillo, S.G. Oliver (eds.), *Yeast Systems Biology*, Methods in Molecular Biology 759,
DOI 10.1007/978-1-61779-173-4_21,

of these methods is the low resolution at which the protein-binding site can be determined. DNA-binding proteins typically recognize 10–20 bp of DNA (2), while a typical ChIP-chip catalog (1) used microarray probes with an average size of 480 bp. To find the actual binding site of the protein, computational algorithms are used to search for smaller motifs that are statistically enriched within the relatively large immunoprecipitated DNA fragments.

Conventional approaches to discovering sequence motifs are implicitly based on the assumption that proteins interact directly with the DNA bases, a phenomenon referred to as direct readout. However, protein–DNA interactions are actually much more complex. Besides making direct hydrogen bonds with the DNA bases, a protein also can make contacts with the deoxyribose sugars and the phosphate backbone of DNA, thereby sensing the local variation in the shape of the DNA duplex (3). This alternate mode of recognition is called indirect readout (4–6). We suggest that the direct readout perspective that is central to most computational motif discovery algorithms may lead to a biased view when searching for biologically functional motifs in genomic DNA.

In earlier work, we introduced the use of the hydroxyl radical (·OH) as a chemical probe of the shape of DNA in solution (7, 8). We showed that the quantitative extent of hydroxyl radical cleavage at each nucleotide of duplex DNA is proportional to the solvent accessible surface area of the deoxyribose residue that is part of that nucleotide (9). Therefore, the hydroxyl radical cleavage pattern represents the shape of the surface of a DNA molecule, at single-nucleotide resolution (10).

Using a database of experimentally determined hydroxyl radical cleavage patterns for a variety of DNA sequences, we developed an algorithm to accurately predict the cleavage pattern for any DNA sequence (11). We have shown previously that these cleavage patterns can be used to detect DNA structural motifs in biologically functional regions of the human genome (12). We call these motifs common OH radical cleavage signatures (CORCS) and use a computer program called CORCS Screening Utility version 2 (CORCSScrU2) to discover them. CORCSScrU2 uses Gibbs sampling, a statistical method that allows for the rapid exploration of large search spaces (13), to find common structural motifs in sets of DNA sequences.

Here, we illustrate the application of this method by using it to find DNA structural motifs associated with the binding of the yeast transcription factor RLM1, a component of the protein kinase C-mediated MAP kinase pathway, using ChIP-chip data from a recent study (1). No sequence-based motifs were found for this protein in the cited study or in a subsequent study that reanalyzed the original ChIP-chip data using newer computational motif discovery algorithms (14). In contrast, the CORCSScrU2 program discovered statistically significant DNA structural motifs, demonstrating the utility of this approach.

2. Materials

2.1. Computing Environment

1. Our computational method requires a UNIX-like operating system. Mac OSX and most Linux distributions will work. Windows distributions may also work, but have not been tested by us.
2. The computer must have the Perl programming language installed. To verify this, open a terminal window and type the command:

   ```
   perl -v
   ```

   This command will report which version of Perl is installed, if any.
3. The CORCSScrU2 program depends on the Bio::SeqIO module from the Bioperl Toolkit (15) (*see* **Note 1**).
4. Download the CORCSScrU2 set of programs from here http://dna.bu.edu/corcsscru2/CORCSScrU2_programs.tar.gz
5. Uncompress the downloaded file by issuing the following command in the terminal:

   ```
   tar -zxvf CORCSScrU2_programs.tar.gz
   ```

6. Open the newly created CORCSScrU2_programs directory, where you will find a host of scripts that will facilitate the use of the method.
7. In a terminal window, navigate to the above directory and enter the following command:

   ```
   perl CORCSScrU2.pl
   ```

8. If you have correctly installed Bioperl, this command will result in a help message from the CORCSScrU2.pl Perl script.
9. Your computing environment is now ready to use.

2.2. DNA Sequence Data

To search for DNA structural motifs, the first step is to collect a set of DNA sequences. As an example of the application of the method, we used sequences found to bind to RLM1 from ChIP-chip experiments performed by Harbison et al. (1).

We decided to use binding data for the RLM1 protein because, as described above, no motifs were discovered when sequence-based computational methods were employed. In addition, a recent study that used protein-binding microarrays to analyze the DNA-binding specificities of a large number of yeast transcription factors was unable to obtain reliable specificity data for RLM1 (16).

The co-crystal structure of an RLM1 homolog bound to DNA indicates that the protein makes many contacts with the DNA backbone (17), suggesting that the local shape of the DNA-binding site may be important for recognition irrespective of the underlying sequence. Recent work showed that by taking the structure of the protein–DNA complex into account, a low-information RLM1 sequence motif could be statistically discovered (18). However, such a method requires prior knowledge of the structural details of the protein–DNA complex, which may not always be available. By contrast, our method uses only DNA structural patterns, which can be accurately predicted for any DNA sequence (11) without regard for the structure of the protein–DNA complex.

Many other types of DNA sequence data can be analyzed with our method. For example, we were successful in detecting DNA structural motifs in DNaseI hypersensitive sites in the human genome (12). We also note that some genes with similar expression profiles have been shown to contain common DNA sequence motifs (19). We suggest that regulation of other genes may depend on the presence of a common structural motif in genomic DNA, which could be found by using our method to search for structural motifs in promoter regions of genes with similar expression patterns.

The DNA sequences used in our method must be fasta formatted – that is, each sequence should have a header line starting with ">" followed by subsequent lines that contain the DNA sequence.

1. Download or create a fasta file that contains the set of DNA sequences of interest.
2. Save this file in the directory created in **Section 2.1**.
3. To allow the user to test the method, we have provided the fasta file used for our example analysis of RLM1. In the CORCSScrU2_programs directory, find the file named RLM1_YPD.fa. This file contains 57 DNA sequences, with an average length of 899 bp, that represent regions in the yeast genome that were identified by ChIP-chip experiments to be bound by RLM1 (1).

3. Methods

3.1. Running CORCSScrU2

1. Open a terminal window and navigate to the CORCSS-crU2_programs directory that was created in **Section 2.1**.
2. The fasta file described in **Section 2.2** that contains the RLM1-bound DNA sequences (RLM1_YPD.fa) should also be in this directory.
3. Execute the command:

```
Perl CORCSScrU2.pl -w 8 -f RLM1_YPD.fa >
     RLM1_YPD.fa.real.1.out
```

4. This command will run the CORCSScrU2 program on the supplied fasta file and search for a common 8 bp DNA structural motif. The length of the motif is specified by the -w option (in this case, -w 8). The results of the run will be stored in the file RLM1_YPD.fa.real.1.out. We note that the above command runs CORCSScrU2 with default options. Sometimes these options are not optimal. *See* **Note 2** for further details. Each output file contains the nucleotide sequence alignment and the hydroxyl radical cleavage pattern alignment that were found by Gibbs sampling in that run of CORCSScrU2. The position of each motif within each individual sequence is indicated. The final score for the alignment (Score) and the average percent information (API) for the motif are listed at the end of the output file.
5. Total compute time for one run of CORCSScrU2 on the example RLM1 data set should be tens of minutes. The time will vary depending on the number and length of the sequences being analyzed (*see* **Note 3**), and on the computer on which the program is run.
6. Other motif lengths can be searched for, by changing the -w option.
7. Run CORCSScrU2 100 times as described in Step 3 above, each time changing the output file to a different name (*see* **Note 4**).
8. Repeat Step 7, only this time using the -r option, and alter the output file name(s) accordingly. This option randomizes the input sequences, as a control.

3.2. Analyzing CORCSScrU2 Results

1. Since Gibbs sampling is a stochastic process, the exact same motif may not be discovered twice in different CORCSSS-crU2 runs using the same data set. However, if a strong motif is present, the distribution of all the CORCSScrU2

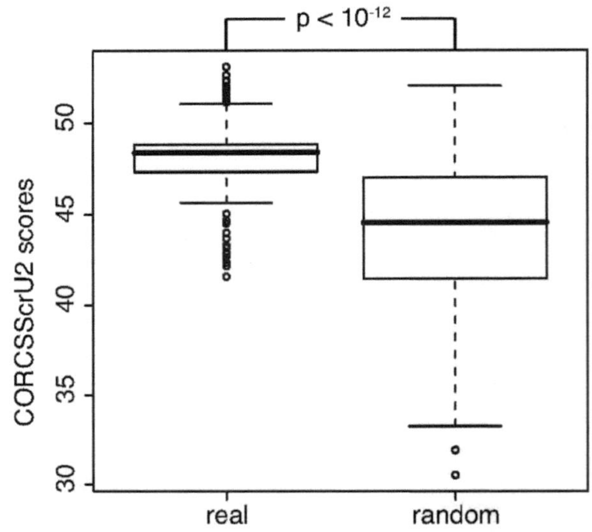

Fig. 21.1. Box plots comparing the distributions of CORCSScrU2 scores for real protein-binding sequences and random DNA sequences. Each distribution represents 100 CORCSScrU2 runs using either real RLM1-binding site data or randomized data. The statistical difference between the two distributions was calculated using a Kolmogorov–Smirnov test (*see* text).

scores for the real compared to the randomized DNA sequences should be shifted toward higher values.

2. Gather the scores for the 100 CORCSScrU2 runs on real sequences and compare them to the 100 runs on random sequences. The score for each CORCSScrU2 run is on the next to last line of the output file (Score).
3. We plot the results of this comparison in **Fig. 21.1**, for RLM1. Note that the distribution of scores for the real sequences is significantly shifted to higher scores compared to the distribution of scores for the random sequences (the control) ($p < 10^{-12}$; Kolmogorov–Smirnov test; http://www.physics.csbsju.edu/stats/KS-test.html).
4. To visualize a DNA structural motif, we recommend creating a heatmap, as depicted in **Fig. 21.2a**.
5. To do this, first use the included accessory script to convert the output from CORCSScrU2 to a matrix representation, by running the following command on one of the CORCSScrU2 output files. It is best to do this on an output file with a high CORCSScrU2 score (*see* **Note 5**):

```
perl  transform_CORCSScrU2_output_to_matrix.pl -f
      RLM1_YPD.fa.real.37.out                       >
      RLM1_YPD.fa.real.37.out.mtx
```

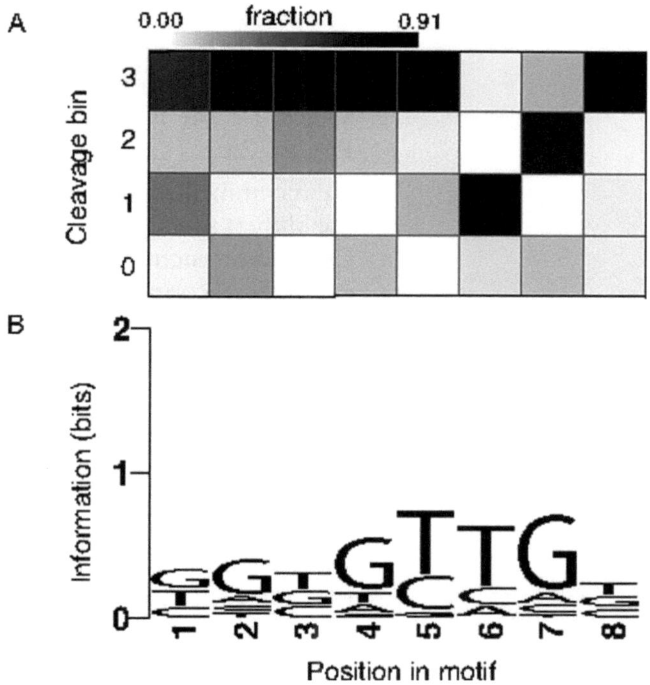

Fig. 21.2. Heatmap and sequence logo of a structural motif. (**a**) Heatmap representation of the highest scoring structural motif, where the *x*-axis is the position in the motif and the *y*-axis represents the hydroxyl radical cleavage intensity bin (0 = lowest; 3 = highest). The level of shading indicates the fraction of patterns with a given cleavage bin at each position in the motif. (**b**) Sequence logo for the motif depicted in (**a**). Note that this sequence logo shows that different DNA sequences can result in the structural pattern shown in **a**.

6. This command will convert the output from CORCSScrU2 into a format amenable to creating a heatmap using the matrix2png utility (20) (*see* **Note 6**).
7. The resulting heatmap image can be viewed using any graphics program that is compatible with the png file format.
8. To compare the heatmap that represents the DNA structural motif to the underlying nucleotide sequence, we recommend creating a sequence logo, which is a representation of the amount of nucleotide sequence information that a motif contains at each position (21, 22).
9. To do this, use the included accessory script to convert the output from CORCSScrU2 into a fasta file:

```
perl  transform_CORCSScrU2_output_to_fasta.pl  -f
    RLM1_YPD.fa.real.37.out                      >
    RLM1_YPD.fa.real.37.out.fa
```

10. The resulting fasta file can then be converted into a weblogo using an online utility (*see* **Note 7**). We show in **Fig. 21.2b** the sequence logo that is produced by CORCSScru2 for the RLM1 structural motif that is depicted in **Fig. 21.2a**.
11. We recommend repeating Steps 5–10 for the top 5–10% of the highest scoring motifs from all the CORCSScrU2 runs. Common structural motifs are likely to be found among the highest scoring results. One should be careful to dismiss motifs resulting from poly A/T tracts, which is a common occurrence for yeast (*see* **Note 8**).
12. The structural motif that is presented in **Fig. 21.2** was found to occur frequently in the top scoring CORCSScrU2 runs using RLM1 DNA-binding data. To compare the amount of information embodied in the DNA structural motif to the information embodied in the nucleotide sequence motif, we measured the amount of information for each type of data at each position in the motif (**Fig. 21.3**). The information content at each position represents how well defined is the motif. Motifs with a low degree of degeneracy have high information content. Information content calculations were performed as previously described (12). The results presented in **Fig. 21.3** show that there is more DNA structural information than nucleotide sequence information at each position in the RLM1 motif. Previous studies found either no nucleotide-based motifs for RLM1-bound regions (1, 14) or, after incorporating protein structure properties, a motif with

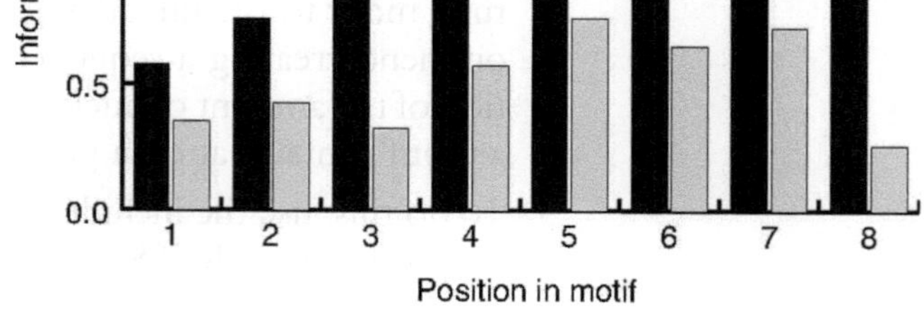

Fig. 21.3. Motif information content. The amount of information, in bits, was calculated for the motif depicted in **Fig. 21.2**. Note that every nucleotide position has more information in the hydroxyl radical cleavage pattern than in the underlying nucleotide sequence.

low information content (18). Our results suggest that the local DNA structure may be important for RLM1-binding specificity.

3.3. Scanning a Structural Motif Across a DNA Sequence

1. Once a structural motif has been discovered, that motif can be scanned across a set of DNA sequences to determine where else it occurs in the genome. For example, the structural motif could be scanned across the entire yeast genome, which is available for download (in fasta format) from

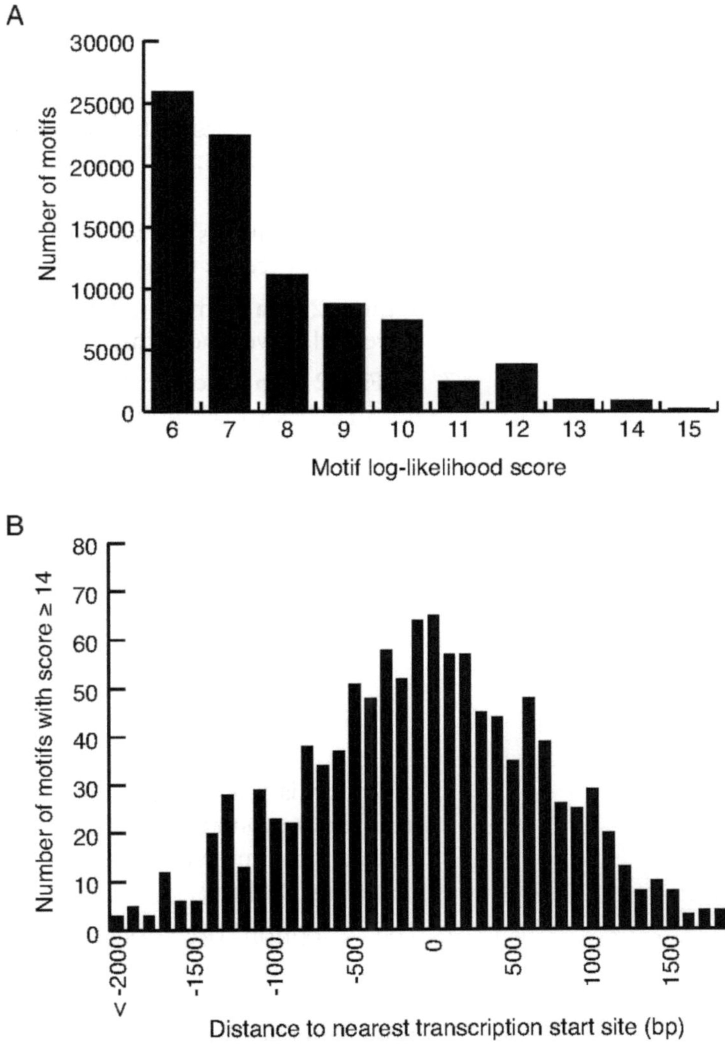

Fig. 21.4. (**a**) The number of instances of the RLM1 motif that is found in the yeast genome at different score thresholds. (**b**) High scoring RLM1 structural motifs cluster near transcription start sites (TSS). Negative numbers indicate the distance in base pairs upstream relative to the nearest TSS, while positive numbers indicate the distance downstream of the TSS.

the UCSC Genome Browser (http://genome.ucsc.edu/) (23, 24).

2. We provide a script, CORCSScrU2_matrix_scanner.pl, that will scan a CORCSScrU2-discovered structural motif across a set of input DNA sequences and report the positional hits based on a score threshold.
3. We use a log-likelihood ratio to determine high-scoring instances of the motif. This ratio is computed by comparing the structural profile of each window in the input sequence to the provided structural motif versus random expectation and calculating the $\log_2$ ratio of these probabilities.
4. Run the CORCSScrU2_matrix_scanner.pl script as follows (*see* **Note 9**):

```
perl CORCSScrU2_matrix_scanner.pl -m
    RLM1_YPD.fa.real.37.out.mtx -f yeast_genome.fa
```

5. This script outputs the location of matches to the structural motif that exceeds the specified score threshold.
6. **Figure 21.4a** demonstrates that as the score threshold is increased, fewer motifs are discovered in the yeast genome.
7. In **Fig. 21.4b** we compare the positions of the highest scoring matches to the positions of all yeast transcription start sites. This analysis reveals that the RLM1 structural motif occurs most often nearby transcription start sites, suggesting that it may be involved in transcriptional regulation.

4. Notes

1. If Bioperl is not installed on your system, general information can be found at http://www.bioperl.org. Installation instructions can be found at http://www.bioperl.org/wiki/Installing_BioPerl.
2. The -i option (iterations) is set to a default value of 15,000. We recommend setting this parameter to at least the number of input sequences multiplied by 100. So, for 25 input sequences, use -i 2,500. The -v option (convergence criterion) is set to a default value of 1,500. This option specifies the number of iterations past the -i threshold in which the score does not increase, in order to consider a motif to be fully converged. We recommend setting this option to at least the number of input sequences multiplied by 10. Other

CORCSScrU2 options can easily be explored by rerunning the program with different parameters.

3. For the example set of RLM1-bound sequences, we ran CORCSScrU2 with the -i option set to 30,000, which took about 1 h to run on one node in our computer cluster. The cluster we used is an IBM eServer xSeries with 1 GHz Intel Pentium III processors. More modern processors should operate much faster.
4. CORCSScrU2 can be run fewer than the suggested 100 times, but this will result in decreased ability to detect a statistical difference between real and random motifs. We suggest changing the output file names by incrementing a counter encoded in the file name itself. We also suggest batching CORCSScrU2 runs on a computer cluster, so that many runs can be performed simultaneously. Consult your system administrator to find out how to run batches of jobs on your system.
5. Here we assume that the output file with the name "RLM1_YPD.fa.real.37.out" has the highest score (next to last line of output) for that set of CORCSScrU2 runs. This will likely be different for another set of CORCSScrU2 runs.
6. The matrix2png utility (20) can either be installed locally on your computer, or used online at the following URL: http://chibi.ubc.ca/matrix2png/bin/matrix2png.cgi. We include the matrix file RLM1_YPD.fa.real.37.out.mtx, so that the user can test the procedure to create a heatmap using the matrix2png utility.
7. The weblogo utility (22) can be used online to create sequence logos. For further instructions, see http://weblogo.berkeley.edu/logo.cgi. We include the fasta file RLM1_YPD.fa.real.37.out.fa, so that the user can test the procedure to create a sequence logo using the weblogo utility.
8. Poly A/T tracts are known to be enriched in many yeast promoter regions and are thought to be involved in the regulation of nucleosome phasing (25). Since poly A/T tracts will have common DNA structural properties, these regions may crop up from time to time in CORCSScrU2 results for yeast sequences. The occurrence of this phenomenon will depend on the input sequences used.
9. The fasta file yeast_genome.fa can be downloaded from the UCSC Genome Browser at the following URL: http://genome.ucsc.edu/. The default score threshold for the script CORCSScrU2_matrix_scanner.pl is set to 12, but can be changed by using the -s option. For example, to change the score threshold to 14, add "-s 14" to the command line.

Acknowledgments

We thank Eric Bishop for providing the Perl module that is used to predict hydroxyl radical cleavage patterns for any DNA sequence. SCJP was the recipient of a National Academies Ford Foundation Dissertation Fellowship. This work was supported by an ENCODE Technology Development Grant from the National Human Genome Research Institute of the National Institutes of Health to TDT (HG003541).

References

1. Harbison, C. T., Gordon, D. B., Lee, T. I., et al. (2004) Transcriptional regulatory code of a eukaryotic genome. *Nature* **431**, 99–104.
2. Stormo, G. D. (2000) DNA binding sites: representation and discovery. *Bioinformatics* **16**, 16–23.
3. Sathyapriya, R., Vijayabaskar, M. S., and Vishveshwara, S. (2008) Insights into protein–DNA interactions through structure network analysis. *PLoS Comput. Biol.* **4**, e1000170.
4. Otwinowski, Z., Schevitz, R. W., Zhang, R., et al. (1988) Crystal structure of trp repressor/operator complex at atomic resolution. *Nature* **335**, 321–329.
5. Brennan, R. G., and Matthews, B. W. (1989) Structural basis of DNA-protein recognition. *Trends Biochem. Sci.* **14**, 286–290.
6. Gartenberg, M. R., and Crothers, D. M. (1988) DNA sequence determinants of CAP-induced bending and protein binding affinity. *Nature* **333**, 824–829.
7. Price, M. A., and Tullius, T. D. (1992) Using hydroxyl radical to probe DNA structure. *Methods Enzymol.* **212**, 194–219.
8. Price, M. A., and Tullius, T. D. (1993) How the structure of an adenine tract depends on sequence context: a new model for the structure of T_nA_n DNA sequences. *Biochemistry* **32**, 127–136.
9. Balasubramanian, B., Pogozelski, W. K., and Tullius, T. D. (1998) DNA strand breaking by the hydroxyl radical is governed by the accessible surface areas of the hydrogen atoms of the DNA backbone. *Proc. Natl. Acad. Sci. USA* **95**, 9738–9743.
10. Jain, S. S., and Tullius, T. D. (2008) Footprinting protein-DNA complexes using the hydroxyl radical. *Nat. Protoc.* **3**, 1092–1100.
11. Greenbaum, J. A., Pang, B., and Tullius, T. D. (2007) Construction of a genome-scale structural map at single-nucleotide resolution. *Genome Res.*, **17**, 947–953.
12. Greenbaum, J. A., Parker, S. C. J., and Tullius, T. D. (2007) Detection of DNA structural motifs in functional genomic elements. *Genome Res.* **17**, 940–946.
13. Lawrence, C., Altschul, S., Boguski, M., Liu, J., Neuwald, A., and Wootton, J. (1993) Detecting subtle sequence signals: a Gibbs sampling strategy for multiple alignment. *Science* **262**, 208–214.
14. MacIsaac, K. D., Wang, T., Gordon, D. B., Gifford, D. K., Stormo, G. D., and Fraenkel, E. (2006) An improved map of conserved regulatory sites for Saccharomyces cerevisiae. *BMC Bioinformatics* **7**, 113.
15. Stajich, J. E., Block, D., Boulez, K., et al. (2002) The Bioperl toolkit: Perl modules for the life sciences. *Genome Res.* **12**, 1611–1618.
16. Zhu, C., Byers, K., McCord, R., et al. (2009) High-resolution DNA binding specificity analysis of yeast transcription factors. *Genome Res.* **19**, 556–566.
17. Santelli, E., and Richmond, T. J. (2000) Crystal structure of MEF2A core bound to DNA at 1.5 Å resolution. *J. Mol. Biol.* **297**, 437–449.
18. Morozov, A. V., and Siggia, E. D. (2007) Connecting protein structure with predictions of regulatory sites. *Proc. Natl. Acad. Sci. USA* **104**, 7068–7073.
19. Spellman, P. T., Sherlock, G., Zhang, M. Q., et al. (1998) Comprehensive identification of cell cycle–regulated genes of the yeast Saccharomyces cerevisiae by microarray hybridization. *Mol. Biol. Cell* **9**, 3273–3297.
20. Pavlidis, P., and Noble, W. S. (2003) Matrix2png: a utility for visualizing matrix data. *Bioinformatics* **19**, 295–296.
21. Schneider, T. D., and Stephens, R. M. (1990) Sequence logos: a new way to display

consensus sequences. *Nucleic Acids Res.* **18**, 6097–6100.

22. Crooks, G. E., Hon, G., Chandonia, J., and Brenner, S. E. (2004) WebLogo: a sequence logo generator. *Genome Res.* **14**, 1188–1190.
23. Kent, W. J., Sugnet, C. W., Furey, T. S., et al. (2002) The human genome browser at UCSC. *Genome Res.* **12**, 996–1006.
24. Karolchik, D., Kuhn, R. M., Baertsch, R., et al. (2008) The UCSC genome browser database: 2008 update. *Nucleic Acids Res.* **36**, D773–779.
25. Segal, E., and Widom, J. (2009) Poly(dA:dT) tracts: major determinants of nucleosome organization. *Curr. Opin. Struct. Biol.* **19**, 65–71.

Chapter 22

High-Throughput Analyses and Curation of Protein Interactions in Yeast

Shoshana J. Wodak, Jim Vlasblom, and Shuye Pu

Abstract

The yeast *Saccharomyces cerevisiae* is the model organism in which protein interactions have been most extensively analyzed. The vast majority of these interactions have been characterized by a variety of sophisticated high-throughput techniques probing different aspects of protein association. This chapter summarizes the major techniques, highlights their complementary nature, discusses the data they produce, and highlights some of the biases from which they suffer. A main focus is the key role played by computational methods for processing, analyzing, and validating the large body of noisy data produced by the experimental procedures. It also describes how computational methods are used to extend the coverage and reliability of protein interaction data by integrating information from heterogeneous sources and reviews the current status of literature-curated data on yeast protein interactions stored in specialized databases.

Key words: Protein–protein interaction (PPI) networks, protein complexes, yeast interactome.

1. Introduction

Most of the recent technological advances in large-scale identification of protein interactions have been championed in the yeast *Saccharomyces cerevisiae*. Thanks to these advances we now have increasingly accurate, although not yet comprehensive, system-level views of the interaction proteome – or interactome – of this model organism. Depending on the type of technique used for probing interactions different views are obtained, prompting further investigations into the origins of these differences and the principles that govern protein interaction affinity and specificity in the living cell. The fact that yeast is one of the most

J.I. Castrillo, S.G. Oliver (eds.), *Yeast Systems Biology*, Methods in Molecular Biology 759,
DOI 10.1007/978-1-61779-173-4_22, © Springer Science+Business Media, LLC 2011

extensively studied model organisms has been particularly helpful. In addition to its fully sequenced and well-annotated genome and proteome, various genome-scale studies produced invaluable data on gene expression (1–3), transcriptional regulation (4–6), cellular localization of proteins (7, 8), phenotypic interactions (9–11), protein half-life (12), and protein abundance (13). In addition, a large number of hypothesis-driven small-scale studies have provided invaluable information on the function and regulation of specific proteins and key cellular processes. All these data and accumulated knowledge have been most instrumental as supporting evidence for identified protein interactions and for interpreting the interaction landscape in terms of cellular processes and pathways.

This chapter summarizes the major experimental techniques that have been used for large-scale identification of protein–protein interactions (PPI) and protein complexes in yeast. It highlights their complementary nature and discusses the data they produce and some of the biases from which these data may suffer. An important focus of this chapter is on the key role played by computational methods for processing, analyzing, and validating data on tens of thousands of proteins and their interaction partners produced by the different high-throughput techniques. It furthermore describes how computational methods can be used to extend our knowledge of the interactome by integrating data from heterogeneous sources and discusses the current status of literature-curated data on yeast protein interactions stored in specialized databases.

2. Methods for Analysis of Protein Interactions

2.1. High-Throughput Techniques for Characterizing Protein Interactions

Several techniques have been instrumental in enabling large-scale characterization of protein interactions in yeast. These techniques fall into two major categories: methods such as tandem affinity purification combined with mass spectrometry (TAP/MS), which detect multi-protein complexes, and techniques that systematically identify binary interactions. Methods of the latter category include the yeast two-hybrid (Y2H) (14) and split-ubiquitin (15) screens and various protein complementation assays (PCA) (16).

2.1.1. Tandem Affinity Purification and Mass Spectrometry (TAP/MS)

TAP/MS is perhaps one of the most powerful methods for detecting protein complexes. It involves tagging a given protein with a C-terminal TAP tag and expressing it at physiological concentration under the control of its endogenous promoter. The tagged protein (bait) and its associated interacting partners (preys) are subsequently co-purified using a two-step affinity capture procedure (17), and the identity of purified proteins is resolved using mass spectrometry. This procedure identifies mainly stably associated subunits but sometimes also more weakly interacting proteins. Data from multiple baits and multiple purifications of the same bait are processed computationally to produce a network of binary associations between proteins, from which information on protein complexes is derived (18–20). Only a fraction of these binary associations actually correspond to physical interactions between the polypeptides, whereas the remainder represent indirect interactions mediated by other proteins or by nucleic acids (RNA or DNA), which are not eliminated by the current high-throughput TAP protocols. Non-native spurious interactions may also occur due to the presence of contaminants, which consist of polypeptides that tend to either bind to the column matrices used for affinity purification or form non-specific interactions with many different proteins. The fact that cells are lysed prior to purification also brings in contact proteins that would not normally interact in vivo. Lastly, the TAP/MS protocols applied so far provide only a component list for each complex, but no information on stoichiometry. **Table 22.1** summarizes the various descriptions of the yeast interactome generated using the TAP/MS procedures. The most comprehensive of these were generated in 2006 by two independent studies (18, 19). Computational procedures were subsequently used to combine the raw data produced by both studies and derive significantly larger and more accurate sets of binary associations and complexes (18, 19, 21, 22).

2.1.2. Yeast Two-Hybrid (Y2H) Screens

In contrast to TAP/MS, the yeast two-hybrid (Y2H) and other binary screens detect only binary interaction partners. Although it is generally assumed that these partners are engaged in direct physical interactions, associations mediated by other protein components cannot be ruled out. In Y2H screens (23–26) the interaction partners are fused with a DNA-binding domain and a transcription-activating domain, respectively, and expressed in yeast. The interaction between the proteins of interest, when it takes place, reconstitutes a transcription factor that activates the expression of reporter genes. Earlier large-scale Y2H studies (23–25) were marred by a high rate of false-positive interactions, often due to bait-mediated auto-activation, and their coverage was rather limited, leading to a very small overlap between the interactions detected in different studies (24). More recent versions of the technique are better at detecting false-positive interactions (26) but there remain other outstanding issues. In particular, interactions identified by Y2H are all formed in the nucleus, which is the natural environment of nuclear proteins, but not of proteins usually found in other cellular compartments, although these proteins are also commonly assayed using this technique.

Table 22.1
Yeast high-throughput PPI data sets

Authors	Year	Methods	# Proteins	# Interactions	# Complexes
Krogan et al. (19)	2006	TAP-MS	2708	7123	547
Gavin et al. (18)	2006	TAP-MS	1430	6532	491
Collins et al. (21)	2007	TAP-MS/PE	1622	9074	400[a]
Yu et al. (26)	2008	Y2H (union)	1278 (2108)	1809 (2930)	NA
Uetz et al. (25)	2000	Y2H (screen)	1004 (817)	957 (692)	NA
Ito et al. (24)	2001	Y2H (core)	3278 (797)	4549 (841)	NA
Tarassov et al. (27)	2008	PCA	1124	2770	NA
Miller et al. (28)	2005	Split ubiquitin	536	1985	NA

For Uetz et al. (25) and Ito et al. (24), the figures in parentheses indicate number of proteins and interactions in the high-quality portions of the data sets. For Yu et al. (26), the same figures indicate the union of their data with the high-quality portions of Uetz et al. (25) and Ito et al. (24).

[a]The set of 400 complexes was generated by Pu et al. (22) from a PPI network with 12,035 interactions among 1,921 proteins, obtained by thresholding the data set of Collins et al. (21) with a slightly lower cut-off of the protein enrichment (PE) score, associated with each link, than in Ref. (21).

TAP/MS tandem affinity purification/mass spectrometry; Y2H yeast two-hybrid screens, *PCA* protein complementation assays, *Split ubiquitin* split-ubiquitin membrane yeast two-hybrid system.

Screening interactions in a context that differs from that in which they occur in the cell may affect both their affinity and specificity, with the latter depending crucially on competing cellular components (including other proteins) present as the interaction takes place. It is not too surprising therefore that the interactions identified by the Y2H technique tend to display little overlap with the pairwise associations produced by the TAP/MS studies, as recently highlighted (26) and illustrated in **Fig. 22.1**. It has been argued that in comparison to TAP/MS PPIs, those identified by Y2H are enriched in transient and condition-specific interactions between complexes and pathways (26). The various interaction data sets produced by the Y2H screens are summarized in **Table 22.1**. Noteworthy is the high confidence binary interaction data set recently derived by combining the reliable portions of two earlier proteome-wide Y2H screens (24, 25) with additional screens (26).

2.1.3. Protein Complementation Assays (PCA)

Protein complementation assays (PCA) are analogous to Y2H, except that they probe interactions that occur in the cytoplasm. In PCA, the bait and prey proteins are, respectively, fused with fragments of a reporter protein. When the fragments are brought into proximity by the interacting proteins, they associate and fold into the native structure of the reporter, thereby reconstituting its activity, which is either essential to cell growth or can be monitored spectroscopically (16). A recent proteome-wide PCA

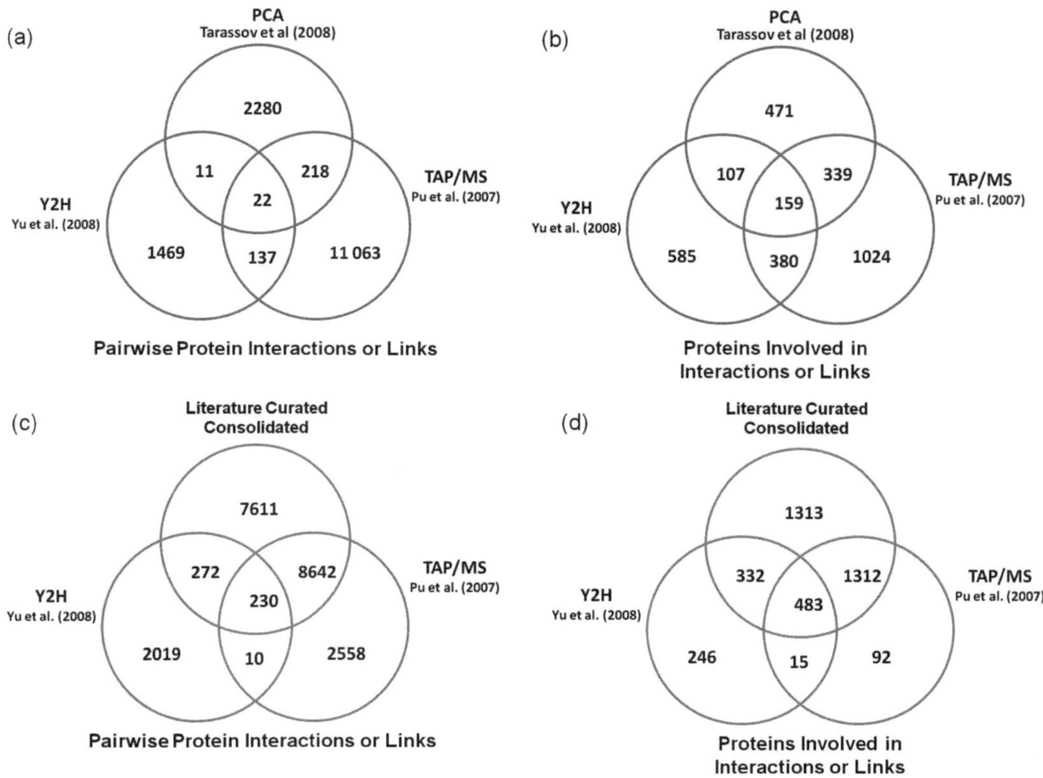

Fig. 22.1. Overlap between protein interaction data sets derived from high-throughput studies and literature curation for the yeast *Saccharomyces cerevisiae*. The following Venn diagrams illustrating the overlap of protein interactions and proteins are shown: (**a**) Overlap between the high confidence portion the protein interaction data from Yu et al. (26), derived using yeast two-hybrid screens (Y2H), the protein interactions from Tarassov et al. (27) identified using protein complementation assays (PCA), and the consolidated high confidence protein interaction data set from Pu et al. (22) obtained using the TAP/MS technique (*see* legend of **Table 22.1** for details). Interactions derived from the latter technique represent links between co-purified proteins, whereas those derived from the former two techniques represent binary interactions. (**b**) Overlap between the proteins involved in the interactions of the same data sets as in (**a**). (**c**) Overlap of literature-curated data on protein interactions consolidated from nine different databases using the iRefIndex procedure (73) and made accessible by the iRefWeb interface (our unpublished results, see text), with protein interaction data from Yu et al. (26) and Pu et al. (22), derived using Y2H screens and the TAP/MS technique, respectively. The literature-curated interactions include both binary interactions and co-complex links. (**d**) Overlap between the proteins involved in the same interaction data sets as in (**c**).

analysis in yeast used the essential enzyme dihydrofolate reductase as the reporter and identified 2,770 interactions among 1,124 proteins (27). About 80% of these interactions are novel. Only 10% of PCA interactions fall within complexes defined by Pu et al. (22). While 15% of PCA PPIs involve proteins in functionally related complexes, 36% and 38% are protein pairs where one protein or both are not in the complex data set, respectively (27). The origin of these discrepancies is presently unclear, although it should be noted that the PCA study as designed can detect protein pairs within a distance of ~80 Å from one another, which implies that such pairs may not necessarily be in physical contact.

2.1.4. Split-Ubiquitin Screens

A screening technique related to both Y2H and PCA is the split-ubiquitin yeast two-hybrid assay also denoted as 'membrane yeast two-hybrid' (MYTH) (15). This technique does not require the interacting proteins to localize to the nucleus and has proven to be particularly suitable for the identification of binary interactions involving integral membrane proteins, which are poorly covered by the other techniques. In this assay the bait and prey proteins are fused to the N- and C-terminal moieties of ubiquitin, respectively. The C-terminal moiety is also fused to a transcription factor. Upon the association of the bait and prey proteins, the reconstituted ubiquitin molecule is the target of a ubiquitin-specific protease, which cleaves off the transcription factor allowing it to activate a reporter gene (15). Using this technique 1,985 putative interactions, involving over 536 integral membrane proteins in yeast, have been identified (28) (*see* **Table 22.1**). Processing these data further using machine learning techniques (*see* (29) and **Section 2.6**) indicated that a majority of these putative interactions are of low or medium confidence (1,085 interactions in the first case and 677 in the second), leaving only 131 high confidence interactions in the entire data set.

The overlap between the most recent comprehensive PPI data sets of *S. cerevisiae*, derived using the TAP/MS, Y2H, and the PCA techniques, respectively, is illustrated in **Fig. 22.1**. This overlap is relatively limited, underscoring the fact that these techniques probe distinct and complementary features of the yeast interactome.

2.2. Role of Computational Procedures in Deriving Interactome Descriptions

Of the various high-throughput techniques described above, TAP/MS is the most computationally intensive, an aspect not generally appreciated. The sets of protein components identified by this technique in thousands of purification runs are not the final product, but the raw material that must be processed further in order to derive reliable information on protein complexes that form in the cell.

This data processing task is a key operation that involves two main steps, as recently reviewed (20). First, the lists of components derived from different purifications are converted into a network of binary protein–protein links, with each link quantified by a score reflecting the confidence with which the link has been detected in the experiments. This network of weighted binary links is thresholded to exclude low scoring links (*see* **Section 4.1**). Second, the resulting high confidence (HC) network is partitioned into densely connected regions (30), yielding the final complexes – each described by a list of components. This two-step protocol involves complex computational procedures, and the use of different computational methods can have a profound effect on the resulting interactome descriptions. This was illustrated by a detailed analysis of the raw data and the information on complexes derived from these data in the two most recent comprehensive yeast TAP/MS studies (18, 19). It was shown (22) that although there was substantial overlap between the sets of tagged proteins produced by both studies and between their co-purified partners, the final sets of complexes differ significantly (**Fig. 22.2a**). In particular, the complexes of the Heidelberg study

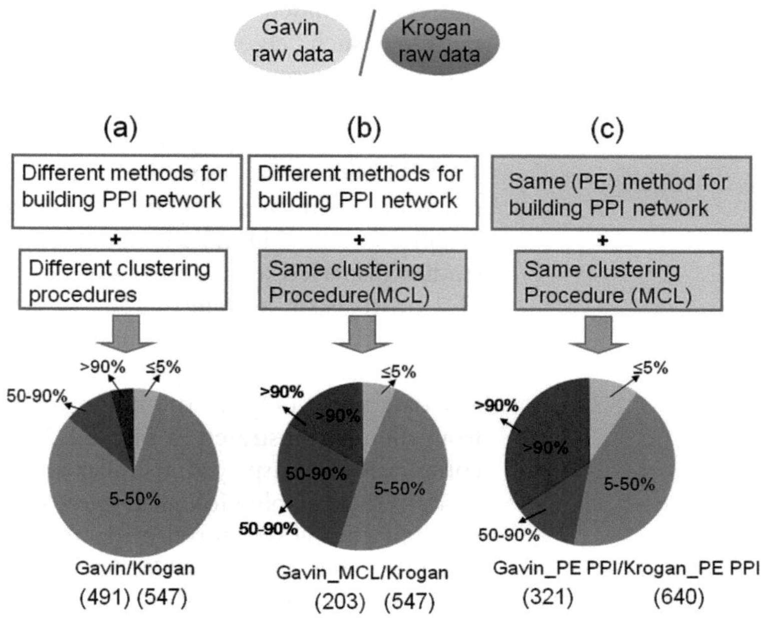

Fig. 22.2. Convergent descriptions of the yeast interactome derived from independent high-throughput TAP/MS studies, thanks to uniform computational procedures. Pie charts (*bottom*) illustrating the overlap between the components of complexes derived from different reprocessed versions of the raw data from the two recent TAP/MS studies by Gavin et al. (18) and Krogan et al. (19). (**a**) Overlap between the components of the complexes as published in the two original studies. In this case both the methods for computing the protein networks and for deriving complexes from these networks differ. (**b**) Overlap between the components of the complexes obtained by applying the Markov clustering algorithm (MCL) (see text) to the protein interaction network derived by Gavin et al. (18). In this case the procedures for computing the networks differ, but the method for computing complexes from the two networks is the same. (**c**) Overlap between the components of complexes obtained by using the same procedure to derive the PPI network from the raw data of each study and applying the MCL procedure to derive complexes from the computed networks. In this case, both PPI networks were computed using the protein enrichment (PE) score of Collins et al. (21), applied as described in Pu et al. (22). Each set of derived complexes was subdivided into four categories according to the fraction *f* of overlapping components: $f > 90\%$; *f* between 50 and 90%; *f* between 5 and 50%; and $f < 5\%$ (represented by wedges of decreasing gray scale) in complexes of the data set derived from the study of Krogan et al. (19) and the maximum matching complex from the study of Gavin et al. (18). Computing the reciprocal match yields essentially the same results (22). The number of complexes computed from the two TAP/MS data sets in each of the three scenarios are indicated in parentheses below the corresponding pie charts.

(18) shared a large fraction of their subunits with one another, whereas those of the Toronto group (19) displayed no overlap. It was established that these differences were mainly due to the use of different clustering procedures to partition the network of binary links constructed from the raw data, into multiprotein complexes (22). Clustering is an optimization problem commonly solved using heuristic algorithms. There is a wealth of algorithms to choose from, but their ability to recover meaningful clusters very much depends on the problem at hand (30).

One such algorithm, the Markov clustering procedure (MCL), was recently shown to be particularly effective in partitioning noisy protein interaction networks (30, 31). Applying this algorithm followed by a post-processing step to the two published TAP/MS networks, respectively (22) (*see* **Section 4.2**), yielded a much more similar set of complexes than those published in the original studies (**Fig. 22.2b**). Furthermore, complexes from both studies displayed a modest degree of overlap with one another (1.2 proteins on average) (22). An even better correspondence between the two sets of complexes was achieved when exactly the same computational procedures were used to reprocess the raw data sets, compute the networks, and derive the complexes from them as illustrated in **Fig. 22.2c**. The resulting new sets of complexes also displayed a similar and improved level of consistency with available biological knowledge, including information on yeast complexes stored in databases, indicating that they are of comparable accuracy (22). Applying exactly the same computational procedures to the raw data produced by the two independent TAP/MS studies thus yields essentially convergent descriptions of the yeast interactome.

2.3. Error Rates of Interactions and Complexes

2.3.1. Error Rates of Interactions

A key question to ask when considering interactome descriptions produced by high-throughput techniques is to what extent these descriptions reflect biological reality. A way of addressing this question is to estimate their error rate. Unfortunately, there presently are no standard measures for estimating the error rate of the identified binary links and complexes, and the development of such measures is an active area of research (26). Since complementary laboratory experiments can be carried only on selected subsets of the data, preference is given to quantitative criteria that measure consistency with prior knowledge.

A particularly important role is played by comparisons against a set of highly reliable, or 'Gold Standard,' literature-curated interactions stored in databases. Using such Gold Standards, error rates for the interaction data can be estimated. These error rates are often defined as the fraction of binary links deemed to be spurious (false positives) or alternatively as the ratio of true-positive/false-positive (TP/FP) interactions. But these estimates may vary widely, depending on how the Gold Standard

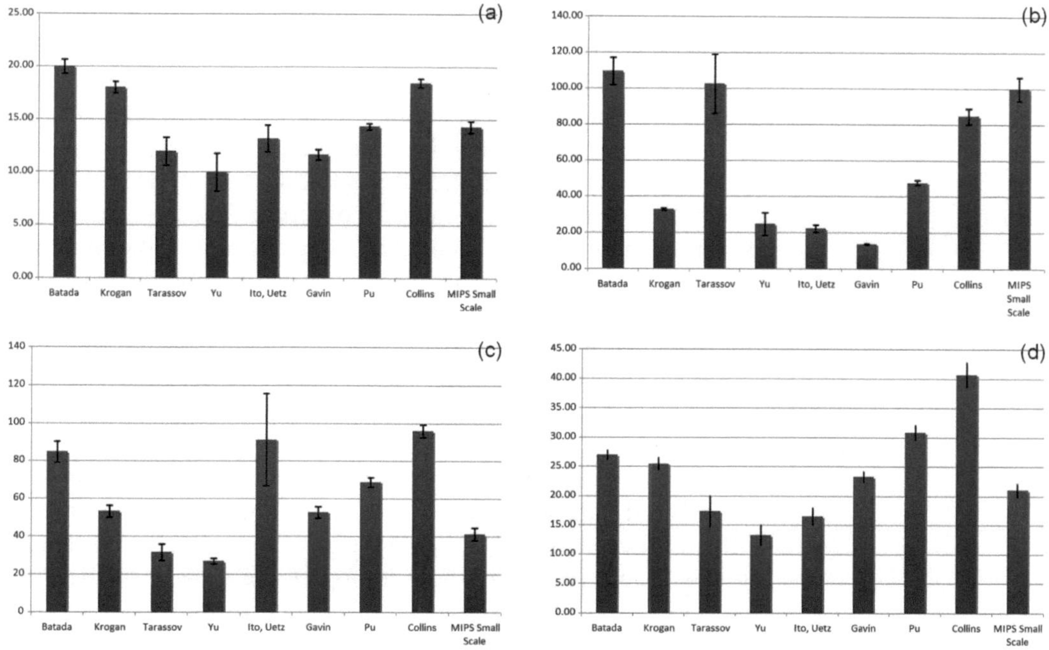

Fig. 22.3. Error rate estimates for major protein interaction data sets derived from various high-throughput studies and literature curation in yeast: the role of the Gold Standard (GS) data set. The panels display vertical bars representing the estimated error rates (*vertical axis*) defined as the ratio true-positive (TP)/false-positive (FP) protein interactions computed on the basis of different Gold Standard (GS) data sets. Error rates were estimated for nine different PPI data sets listed on the *horizontal axis*: Batada (91), Krogan (19), Tarassov (27), Yu (26), Ito/Uetz (23–25), Gavin (18, 19), Pu (22), Collins (21), and MIPS small scale (34). Only the high confidence portions of the different PPI data sets were evaluated, whenever available. For the various GS data sets, the TP examples consisted of protein pairs where both members of the pair are part of known multi-protein complexes. Two sets of known complexes were considered: the older set from MIPS (34) and the newer CYC2008 set (35). The TN (true negative) examples were protein pairs drawn randomly from all the proteins in the positive set minus any pairs that were in that set. In all the calculations the number of TN examples was 10 times larger than that of the TP examples. For each of the evaluated PPI data set, 10 different random draws were carried out to select the TN examples and the results were averaged. When a draw yielded a pair whose members do not overlap with proteins in the evaluated data set, the draw was repeated. The average error rate (*thick bars*) and standard error (*thin bars*) are indicated for each data set. (**a**) Error rates (average TP/TN ratios) for the different PPI data sets estimated using a Gold Standard derived from the MIPS complexes, as described above. (**b**) Error rates for the different PPI data sets estimated using a Gold Standard derived from the MIPS complexes, as in (**a**) but adding the condition that the drawn TN examples not be co-localized to the same cellular compartment. Co-localization data were obtained from (7). (**c**) Error rates estimated using a Gold Standard derived from the MIPS complexes, as in (**a**), but omitting ribosomal proteins from the analysis. (**d**) Error rates computed as in (**c**), but using the CYC2008 complexes to define the Gold Standard data set.

itself is defined (**Fig. 22.3** and (22, 26)). While the TP interactions are defined directly from the Gold Standard as the reliable set of known interactions, a particularly contentious point is the definition of 'true-negative' (TN) interactions, or interactions that do not, or are unlikely to occur. Of the various ways in which such true negatives are defined, the random selection of protein pairs that are never reported to interact or to belong to the same reported complex is deemed the least biased (32). Although there is no guarantee that such randomly selected pairs are not 'contaminated' by true, yet undiscovered interactions, the amount of contamination is likely to be small (32), provided that the pairs are selected from a large enough pool of proteins. These methods also need to compensate for the fact that the set of true-negative interactions defined in this manner tends to be significantly larger than the set of proteins known to interact.

Thus, comparing error rates for different data sets is meaningless unless these rates are estimated using the same Gold Standard and calculation method, and even in the most favorable case, only the relative magnitudes may be informative as discussed in (22) and illustrated in **Fig. 22.3**.

This figure also illustrates the biases introduced by particular definitions of the Gold Standard. For example, requiring that the TN examples not be co-localized to the same sub-cellular compartment (**Fig. 22.3b**) significantly boosts the TP/FP ratios of all the analyzed data sets, relative to those computed in the absence of this requirement (**Fig. 22.3a**). More importantly still, applying this requirement strikingly improves the quality of the PCA data set (27), relative to the PPIs obtained using other methods (*see* **Section 4.3**). Likewise, including (**Fig. 22.3a**) or excluding (**Fig. 22.3c**) ribosomal proteins from the Gold Standard alters significantly the relative magnitudes of the estimated TP/FP for several of the PPI data sets, and so does the choice of the specific Gold Standard itself (**Fig. 22.3c**, **d**). Thus, some of the high error rates estimated for the yeast interaction networks derived from the latest TAP-MS studies (33) should be critically reviewed in this light.

Another problem is that Gold Standard data sets for yeast (usually derived from complexes of the MIPS catalogue (34)) have been outdated until the recent report of an updated list of 408 complexes (CYC2008) (35).

It has also been argued that there is no unique organism-specific Gold Standard. Indeed, a recent systematic validation (26) of binary protein interactions in *S. cerevisiae* using yeast two-hybrid screens, protein complementation assays (36), and the mammalian protein–protein interaction trap technique (37) provides compelling evidence that interactions of this type tend to differ from those identified using purification methods, as suggested by the poor overlap between the interactions identified using the different methods (**Fig. 22.1**). Lower error rates of a given type of method can therefore be derived by comparison to Gold Standards that are appropriate for the same type of methods (26).

2.3.2. Validation of Protein Complexes

Protein complexes identified from various PPI networks are also commonly validated against known complexes annotated in databases taken as the Gold Standard. Typically the components of the newly identified complexes are systematically compared to those of the annotated set using various measures quantifying the overlap. Several different measures have been proposed, with some being more stringent (19, 30) than others (38). Other independent reliability measures are often used in addition. These involve evaluating consistency with available information on cellular localization (7), the functional annotations of individual protein components stored in databases (using the GO hierarchy (39)), and often also on the expression profiles of the corresponding genes (40), as illustrated in **Fig. 22.4**. Reliability criteria based on inferred domain–domain interactions (41) or on annotated interactions between paralogous proteins (40) have also been proposed. But these criteria involve assumptions and approximations and their usefulness is further limited by the quality and particular biases in the available information (e.g., subsets of proteins and interactions covered).

2.4. Estimating Protein–Protein Interaction (PPI) Coverage

The main goal of high-throughput methods is to provide interactome descriptions that are as unbiased and comprehensive as possible. But this goal may be an elusive one, considering that most current methods are not very good at capturing transient interactions and those involving very low abundant proteins. Even for the more stable and abundant complexes in yeast, achieving comprehensive coverage remains a challenge. A case in point is the still limited coverage of membrane proteins (18, 19). Furthermore, there are ample reasons to believe that changes in cellular conditions may significantly influence the composition of some

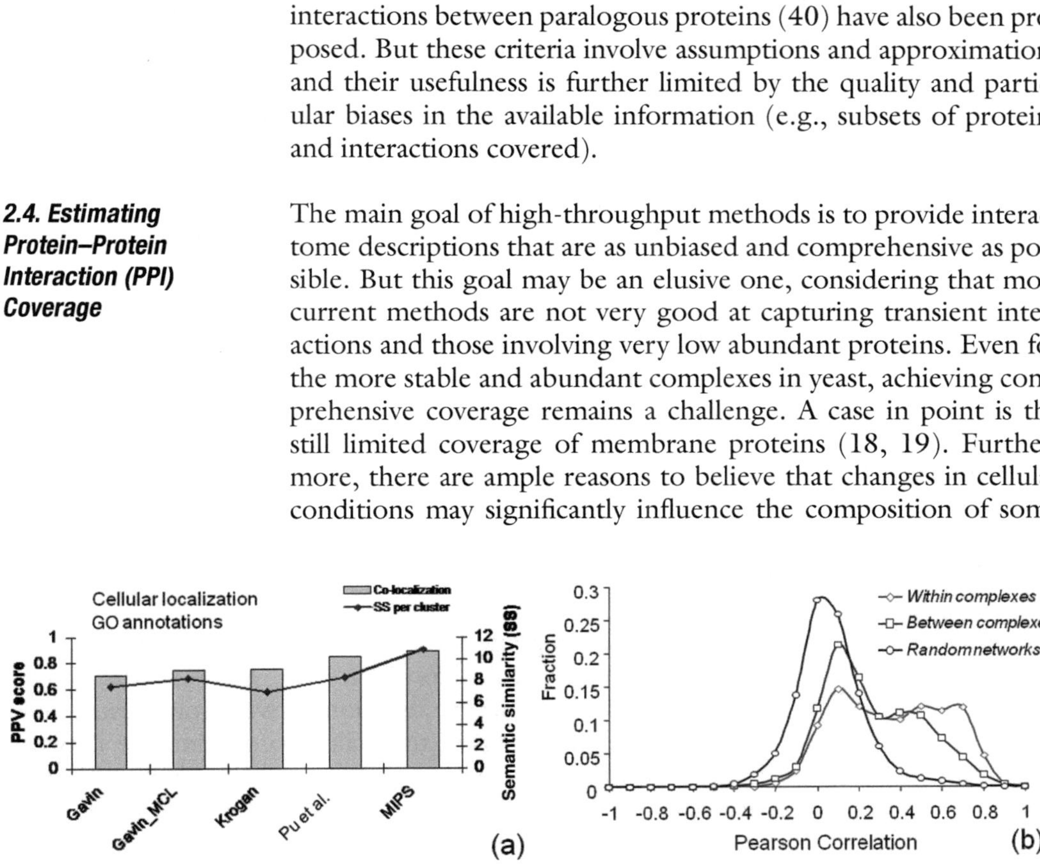

Fig. 22.4. Validation of yeast protein complexes derived from high-throughput studies and literature curation. (**a**) Similarity of the functional annotations and extent of sub-cellular co-localization for proteins within complexes. The similarity in functional annotations is estimated from the gene ontology (GO) annotations using the semantic similarity (SS) measure (92) (*vertical axis, right-hand side of the panel*). The extent to which components within complexes have been assigned to the same sub-cellular compartment is evaluated using the PPV (positive predictive value) score (*vertical axis, left-hand side of the panel*), defined as the fraction of protein pairs within complexes that are assigned to the same cellular compartment. Sub-cellular localization data for *Saccharomyces cerevisiae* are those of (7, 8). The validated data sets of yeast protein complexes are the published complexes of Gavin et al. (18, 19), the complexes derived by applying the MCL procedure to the published PPI network of Gavin et al. (as described in (22) and in **Fig. 22.2b**), the published complexes of Krogan et al. (19), those derived by Pu et al. (22) from the consolidated data sets of the two most recent Tap/MS studies, and the MIPS complexes (34). (**b**) Distributions (ordinate) of the pairwise Pearson correlation coefficient (abscissa) of the mRNA expression profiles of *S. cerevisiae* proteins of different categories (see legend). The complexes are from the MIPS catalog. The mRNA expression data are those from (1).

complexes. However, few if any of the high-throughput techniques have so far explored any significant range of conditions encountered by living cells.

Estimating the current level of coverage of the yeast interactome is therefore extremely difficult without at least a good guess of what the size of the complete interaction set might be. The way interactions are defined (binary interactions or co-complex links) will also influence the projections. Recent estimates, which suggest that 50% of all possible interactions in yeast have been identified (by all types of experimental methods combined) (33), are overly optimistic. One problem with these estimates is that they ignore the fact that the set of known interactions is not a random sample of the entire network (42, 43). More conservative estimates based on a careful statistical analysis suggest that only about 15–20% of 'all' binary interactions in yeast have so far been mapped (44). Similar figures were reported on the basis, among other things, of the observed rates at which known interactions remain undetected by yeast two-hybrid methods (26). The same report projects the total number of binary interactions in *S. cerevisiae* to be between 18,000 and 30,000. In comparison the number of binary interactions in human was recently estimated at ~600,000, with a current coverage of less than 1% (45).

2.5. Increasing PPI Coverage and Accuracy Through Data Integration

While estimating the current rate of PPI coverage remains an issue, approaches to maximizing it by combining evidence for interactions from different sources, and using this information to build consolidated interactome descriptions, have been very useful.

Such data integration can be achieved using a variety of machine learning (ML) algorithms (29). Many of these algorithms optimize an objective function derived from various lines of evidence, by maximizing the ability to discriminate between the true-positive and true-negative interactions in a Gold Standard data set of exactly the same type and which suffers from the same caveats, as those used for error rate estimations (*see* **Section 4.4**).

One of the more successful approaches, loosely based on a naïve Bayes classifier, was used to derive a consolidated protein interaction network for *S. cerevisiae* from the raw data sets of the two recent TAP/MS studies (21). This consolidated network comprises 1,622 proteins and 9,074 protein associations, significantly more than each of the two studies taken individually (**Table 22.1**). The estimated average error rate of the high confidence (HC) portion of this network is also lower than for several recently derived high confidence networks and similar to that of a data set of binary interactions identified by small-scale experiments and annotated by the Munich Information Center for Protein Sequences (MIPS) (34). Very similar high-accuracy consolidated yeast protein interaction networks were derived using related computational approaches (46, 47).

A very useful feature of ML algorithms is that they are flexible enough to accommodate many different types of evidence for PPIs even when those are individually not very reliable. It is hence common to consider data on co-localization, co-expression, co-regulation, and GO annotations, in addition to data from different high-throughput experiments (TAP/MS and Y2H) and literature curation (48). However, since supporting evidence tends to be available for the same set of interactions (stable associations or those involving abundant proteins), the gains in coverage, particularly in terms of the number of new proteins involved, often remain modest.

The inference of protein–protein interactions can be further strengthened by considering properties computed from an underlying PPI or genetic interaction network. Examples include the Pearson correlation between the interaction profiles of two proteins (38) and metrics of shared connectivity (49) in the PPI network. More complex graph analysis approaches, which consider both the local and global topologies of the underlying networks to rank the likelihood that two proteins interact, were shown to yield further improvements (47, 50). Combining evidence from heterogeneous sources, with inferences based on network properties, has recently produced a set of 19,258 protein–protein associations for *S. cerevisiae*, whose estimated false-positive rate is ~10% (47).

2.6. Mapping Protein Complexes into Pathways

2.6.1. Epistatic Interactions

Genetic interactions represent phenotype alterations produced by the deletion or mutation of one gene in the background of a mutation (or deletion) of another gene (51). These interactions are considered to be aggravating when the fitness of the double mutant (usually measured as its growth rate) is worse than expected, where the expectation is usually defined as the combined effects of the double deletion. Alleviating interactions are those where the double mutant is fitter than expected. Earlier technologies such as synthetic genetic array (SGA) and diploid-based synthetic lethality analysis by microarray (dSLAM) relied solely on synthetic lethality (52, 53). A recent variant of SGA, the epistatic miniarray profile (E-MAP) (9), and an improved SGA method (54) comprehensively measure both aggravating and alleviating interactions on a continuous scale. Data from these studies provide a complementary view to the PPI network and are particularly useful for organizing functional modules (such as protein complexes) into pathways.

2.6.2. Deriving Clues on Protein Interactions, Complexes, and Gene Function from Epistatic Interactions

It has been shown that alleviating interactions frequently occur between the subunits of a complex, whereas aggravating interactions tend to occur between proteins belonging to functionally parallel pathways (9). Proteins belonging to the same complex or the same pathway tend to exhibit highly correlated genetic interaction profiles (9, 55, 56). Furthermore, it is not uncommon to find that the genetic interaction profile of a gene shows either alleviating or aggravating interactions with genes involved in diverse cellular processes, a strong indication that multi-functional genes are prevalent in the cell. Until recently, however, computational methods were incapable of teasing out information on the multi-functional roles of genes from the epistatic miniarray profile (E-MAP) data. These data were commonly analyzed using hierarchical clustering and related methods, which partition genes into disjoint groups. Such methods were instrumental in mapping genes into pathways (55, 56), dividing large protein complexes into functional sub-modules and identifying epistasis groups that are otherwise not involved in physical interactions (9). Elaborating on an earlier study (57), E-MAP data were recently combined with protein–protein interaction data to improve identification of protein complexes and to delineate functional relationships among them (58, 59).

2.6.3. Local Coherence of Epistatic Profiles Reveals Functional Modules and Pleiotropic Gene Function

To investigate the pleiotropic function of genes from these data sets involves a different approach, which makes use of a biclustering algorithm (60). This algorithm termed local coherence detection (LCD) (61) exploits the fact that a group of multi-functional genes may display a tight coherence over a fraction of their E-MAP interaction profiles if they operate in a common process under specific conditions, but behave distinctly otherwise. Unlike hierarchical clustering methods, which rely on correlating global interaction profiles, LCD is capable of utilizing both global and local profile similarities (*see* **Section 4.5**).

The application of LCD to several E-MAP data sets (61) for the yeast *S. cerevisiae* grouped the corresponding genes into many known functional modules and protein complexes reported previously. In addition it uncovered a number of recently documented and novel multi-functional relationships between genes and gene groups. For example, LCD was able to recapitulate previously reported multiple links between the Lge1/Bre1 ubiquitination complex and complexes involved in various cellular processes, including transcription, chromatid cohesion, and DNA damage repair (61).

By exploiting partial profile similarity, LCD was also able to group together multiple complexes or genes into a single cluster, indicating that these complexes or genes might operate in compensatory processes activated in response to deletion of a common subset of genes. In a number of cases, complexes performing quite diverse roles are found in a cluster, suggesting new roles for known complexes. For instance, the Compass complex, whish is involved in transcription initiation, co-clusters with gene involved in sister chromatid cohesion (Ctf4/Ctf8/Ctf18) and DNA repair

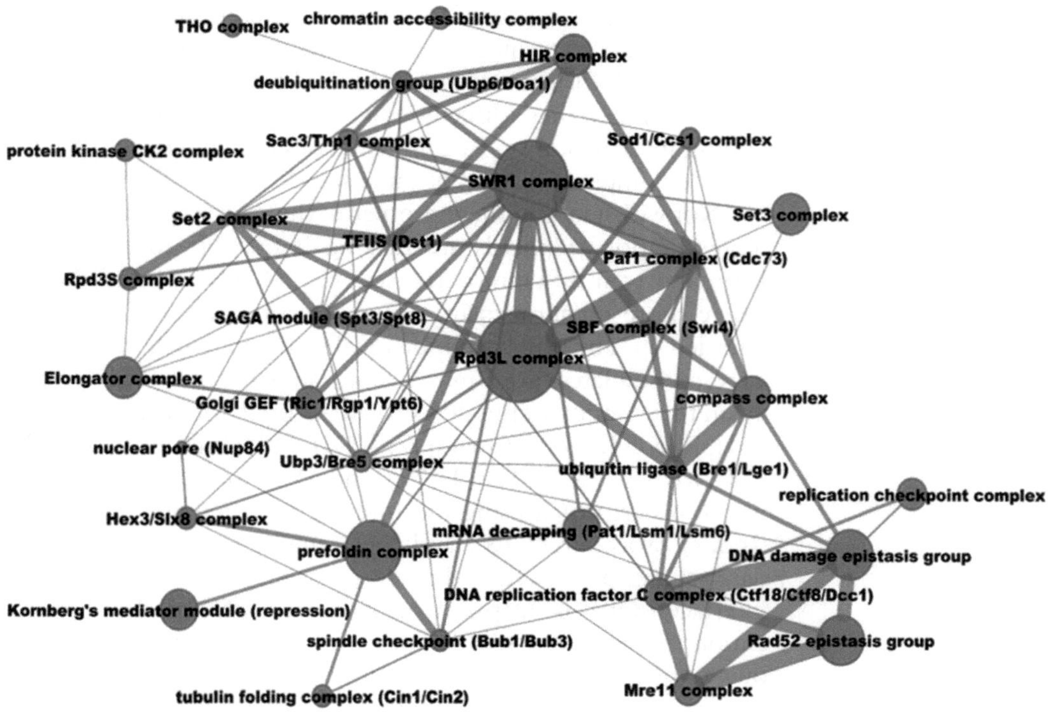

Fig. 22.5. Functional modules involved in yeast chromosome biology featuring coherent epistatic interaction profiles. The network of functional modules identified by the local coherence detection (LCD) procedure (61) from the epistatic miniarray profile (E-MAP) data of (9) is depicted. Nodes represent complexes or known epistasis groups identified as modules displaying a coherent epistatic interaction profile. The node size is proportional to the number of genes and the edge thickness is proportional to the number of time the linked nodes are identified in the same module by the LCD algorithm (61). Individual genes can be part of several modules.

(*see* **Fig. 22.5**) suggesting a novel role of this complex in the later two processes, which is currently being verified experimentally.

2.7. Literature-Curated Data on Yeast Protein Interactions

Hypothesis-driven small-scale studies involving specific systems or focusing on specific cellular processes are another valuable source of information on protein interactions and complexes in yeast and other model organisms. This information is typically reported in the scientific literature, usually in text form. With the growing interest in protein interactions, efforts have been undertaken by various groups worldwide to curate this information and make it available in databases specialized in protein–protein interactions (62, 63) or focusing on specific model organisms. Software tools to interactively 'mine' the scientific literature for protein interactions in an automatic fashion have also been developed (64, 65).

It has been assumed until recently that literature-curated (LC) protein interactions data are of higher accuracy than those produced by high-throughput studies, because they are derived from carefully crafted focused investigations. But recent analyses of the LC PPI data are suggesting that this is no longer the case, due in part not only to improvements in high-throughput methods but also to inherent difficulties in extracting and archiving relevant information from published text (66, 67). Efforts to address some of these difficulties by adopting standards for representing information on PPIs in databases are underway (68), but the actual implementation of these standards is far from uniform.

Not withstanding these problems, the currently available data representation standards are making it possible to consolidate PPI data from different databases (62, 64, 69–73) and thereby derive more comprehensive views of various interactomes than afforded by any database individually. In one such effort, data were consolidated from nine major databases that focus primarily on the curation of physical protein–protein interactions: BIND (74), BioGRID (62), Corum (75), DIP (63), IntAct (76), HPRD (77), MINT (78), MPACT (79), and MPPI (80). The consolidation was performed using the Interaction Reference Index (iRefIndex) procedures (73), and the resulting non-redundant PPI data set can be viewed and analyzed using the web-based query interface iRefWeb (http://wodaklab.org/irefweb). The consolidation methodology stands out from similar efforts by its rigorous, publicly documented procedure for resolving redundancies. In particular, it examines amino acid sequences to establish the identity of interacting proteins and groups of different isoforms of the same protein. This allows the iRefIndex to reliably combine records from different databases that use different types of protein identifiers to support the same PPI or complex.

The latest release of the iRefWeb/iRefIndex resource lists a total of 72,410 non-redundant PPI interactions, made by 6,188 proteins for the yeast *S. cerevisiae*. These interactions are collectively supported by 6,656 publications cited by the different databases. These numbers of PPIs and corresponding proteins clearly do not reflect biological reality. Indeed, they indicate that the PPIs have been identified for as many or more proteins than the yeast genome codes for, and that the total number of curated yeast interactions is about—five to six times larger than those identified by all high-throughput studies combined (our unpublished results). These excessive numbers are mainly due to the fact that some databases include unprocessed low confidence data made available by authors of various high-throughput studies.

Ideally, to obtain a more realistic description of the consolidated yeast interactome, data integration procedures of the type described above would need to be applied. Their application in turn requires at the very least some type of ranking of the literature-curated PPIs in terms of their reliability. Such ranking is systematically provided by the confidence scores associated with protein links derived from TAP/MS studies or machine learning-based data consolidation procedures (21). With a few exceptions (such as MINT) (78), it is however not generally available from the source PPI databases.

It has been pointed out that PPI identified by several different publications or experiments are more likely to be biologically relevant. Requiring that PPI be supported by several publications has been a common approach for scoring PPI data in public databases, as it is easy to interpret and the bias toward well-studied interactions is immediately evident. As examples, IRefIndex stores several attributes that indicate how many publications support a given interaction (73), and BioGRID hosts sets of interactions supported by multiple publications – or by certain select publications that the annotators deem reliable (62).

When requiring any of the consolidated *S. cerevisiae* PPI to be supported by at least two publications, and filtering out PPI data deemed unreliable stored by some of the source databases (e.g., raw data sets from certain high-throughput studies) the number of consolidated interactions and proteins in the iRefWeb/iRefIndex system is reduced significantly to 16,755 and 3,440, respectively, which is much more realistic and better reflects our current knowledge of the yeast interactome. Interestingly, 85–92% of the interactions in this filtered data set are reported by high-throughput studies (corresponding to publications reporting at least 10 and 20 interactions, respectively). However, it should be realized that PPI data filtered in this manner may still include low confidence interactions, as pruning those would require more sophisticated analyses.

3. Conclusions

Despite many outstanding issues, high-throughput methods have achieved real progress in providing increasingly accurate descriptions of the *S. cerevisiae* interactome. Obtaining such accurate descriptions involves the use of increasingly sophisticated computational procedures and methods for integrating the results of the high-throughput experiments with the vast body of heterogeneous data and knowledge accumulated on yeast over the years. Although the system-level views afforded by the current interactome descriptions yield useful insight into aspects of molecular and cellular function and may provide clues on evolutionary processes, they have very serious limitations. The network model of protein links is far too abstract, as it ignores key attributes of the underlying phenomena, such as the stoichiometry of the interacting components, or the temporal and spatial constraints that govern their formation. Furthermore, our current view of the interactome is clearly biased toward interactions that are stable enough to be detected and those that can presently be probed by the methods at hand.

The major challenge now facing the field is to translate this abstract description into more detailed models describing real physical interactions at the near atomic or atomic scale. Such models can help reveal the physical contacts that can actually be made between proteins and suggest how these contacts might have evolved (81). They can furthermore provide atomic descriptions of the interacting interfaces that are useful for various mechanistic investigations as well as for drug design (82).

With the development of new methods for deriving quantitative information on complex stoichiometry in TAP/MS studies (83) and for gaining information on the internal organization of complexes (84), accurate models may in the future be built with the help of the increasing body of data on known protein structures (PDB) and computational procedures to optimally dock the components to each other. These so-called docking procedures are becoming increasingly powerful and accessible to non-specialists (85) as recently reviewed (86, 87).

4. Notes

4.1. Defining the High Confidence Portion of a PPI Network

Gold standard data sets such as those described in **Section 2.3** are routinely used in defining the high confidence portion of the PPI networks derived from the TAP/MS data. This usually involves the following steps. First, the raw mass spectrometry data from different purifications of the same complex and different complexes are combined into a single confidence score for each detected PPI. Examples are the socio-affinity (18) or the purification enrichment (PE) (21) scores. Next an ROC (receiver operator curve) (29) or equivalent is plotted. This curve indicates how the rate of FP (false-positive) interactions varies with the confidence score level at which the network is thresholded. In some approaches, the confidence scores themselves are derived using machine learning procedures of the type discussed in **Section 2.6**. In this case the ML algorithm (*see* **Section 4.4**) is 'trained' on a portion of the PPI data set, to maximize their ability of discriminating between true-positive (TP) and true-negative (TN) interactions, as defined by a Gold Standard data set. The trained model is then applied to the full data set to produce the confidence scores that are used to plot a receiver operator curve (ROC). In both scenarios, the ROC is used to settle on an acceptable FP rate, which defines the confidence score threshold above which interactions should be considered for further analysis. Typically the high confidence (HC) portion of PPI data sets derived from TAP/MS experiments represents only a small fraction of the full data set (<5%). The HC portion of the network is less densely connected and therefore more readily processed by clustering algorithms such as Markov clustering (MCL).

4.2. Post-processing Step to Define Protein Complexes with Shared Membership

The Markov clustering algorithm (MCL) partitions the PPI network into non-overlapping clusters. When these clusters are taken to represent multi-protein complexes as in (19), complex membership is exclusive: a given protein can be part of one complex only. However, analysis of the PPI network and available information on known complexes clearly indicates that such exclusive membership poorly reflects biological reality. To correct for that, Pu et al. (22) mapped the complexes/clusters computed by MCL back into the PPI network and subjected them to a post-processing step, in which components of a given complex observed to also form dense interaction patterns with proteins in other complexes were assigned to multiple complexes simultaneously. Applying this post-processing step to complexes derived from the PPI network described in (22) yielded a total of 78 (19.5%) complexes, sharing at least 1 component with at least 1 other complex, with the remaining majority (322) containing only uniquely assigned components. The overlapping complexes were found to share on average 2.5 genes with 0.48 other complexes. A similar level of overlap occurs in the hand-curated complexes in the MIPS and SGD archives, where overlapping complexes share 2.3 genes with 0.86 complexes on average. But the fraction of complexes displaying these overlaps is lower than in the curated complexes, reflecting some inherent limitations of the PPI network approach as discussed in (22).

4.3. Specific Biases in Evaluating PPI Error Rates

The way in which the 'true-negative' TN interactions are defined in the Gold Standard data set can introduce important biases in evaluating the error rate in PPI data sets (*see* discussion in **Section 2.3**). Requiring that the TN interactions in the Gold Standard not be localized to the same sub-cellular compartment significantly improves the quality (measured as the TP/FP ratio) of several of the data sets, while markedly enhancing the differences between them (**Fig. 22.3b** vs. **Fig 22.3a**). In particular this ratio is increased nearly 10-fold for the PCA data set in (27), relative to the same ratio computed when the TN interactions are simply randomly selected (**Fig. 22.3a**). We believe that this is due to the fact that by design, the PCA method detects exclusively co-localized partners, whereas TAP/MS or Y2H screens may detect interactions between proteins localized to different cellular compartments, for example, due to non-specific binding during purification (TAP/MS) or to expressing the two proteins in the nucleus (Y2H). Thus, the constrained definition of TN may not be appropriate for evaluating the real error rate of methods

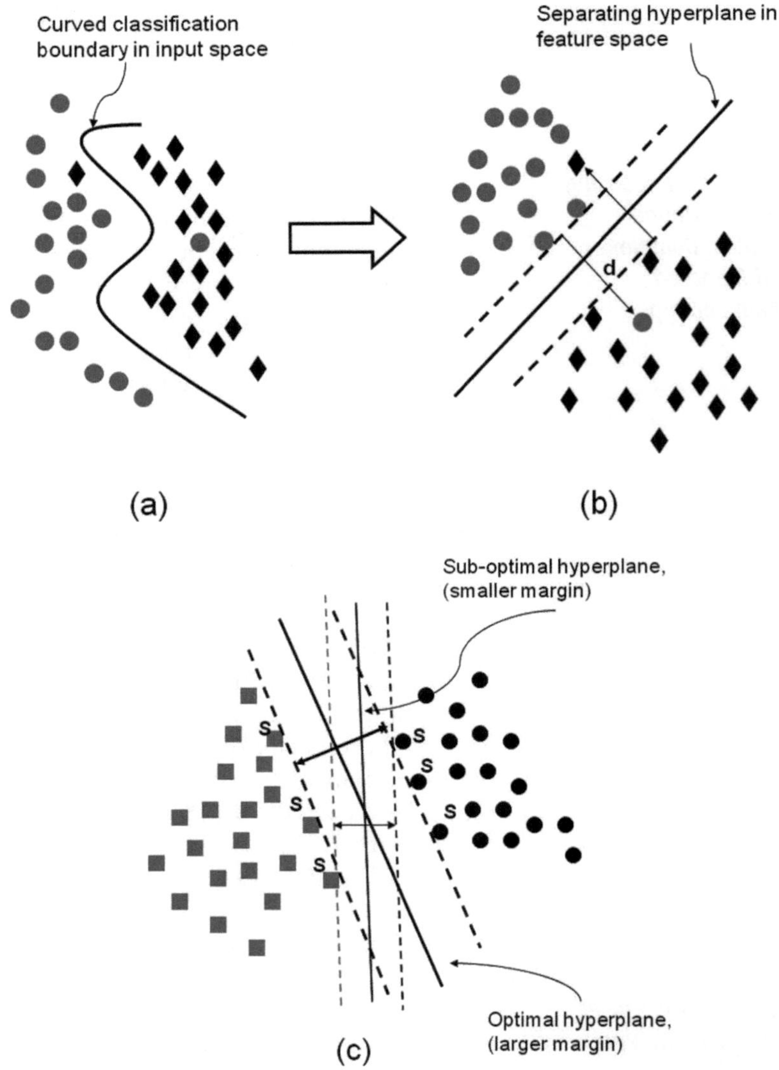

Fig. 22.6. Graphical illustration of the principles underlying support vector machine (SVM) procedures. (**a**) An idealized two-dimensional representation of the space of input examples. These examples belong to two categories: *diamonds* and *circles*, respectively. Examples of the two categories are clearly separated by a curved boundary, except for one example from each category, which is misclassified. (**b**) An idealized two-dimensional representation of the vectors representing the examples from (**a**) in the *feature* space, after the mapping performed by the SVM. Also shown is the *separating line* (hyperplane) identified by the SVM, which optimally segregates the vectors from the two categories, save for those corresponding to the misclassified examples in (**a**). The hyperplane translates into the curved boundary in the space of input examples shown in (**a**). Any points that fall within the margin or that are misclassified are accompanied by a penalty term commonly defined as the error cost *C* multiplied by the distance *d* of the point to the margin corresponding to their correct class. (**c**) An idealized two-dimensional representation illustrating two solutions produced by the SVM procedure. A sub-optimal solution showing a hyperplane with narrow margins bounded by only one

such as PCA – where one expects a relatively high fraction of the total number of FP interactions to be concentrated among co-localized proteins.

4.4. Support Vector Machine (SVM) as One of the Most Versatile Machine Learning Procedures for PPI Prediction

In PPI prediction, the goal of supervised classifiers like the support vector machine (SVM) is to distinguish between true and spurious PPIs on the basis of various lines of biological evidence. This involves optimizing an objective function that depends on the considered evidence (see below), by training on known true-positive (TP) and true-negative (TN) examples of PPIs. These examples are typically derived from a Gold Standard data set (*see* **Section 2.3**). To perform this optimization (training) the SVM transforms the lines of evidence for each input training example into a vector in a space of potentially very high dimension (*feature* space) and finds a hyperplane that optimally separates the TP examples from the TN examples within this space (**Fig. 22.6a, b**). It has been applied to a wide array of problems in biology including PPI prediction (28, 47, 50), as recently reviewed (88, 89), and has been shown to be robust to noise (32) in this context. Importantly, the SVM does not explicitly transform each example into the feature space. Instead, it uses a so-called *kernel* function that maps two input examples to the dot product of their corresponding vector representations in feature space. Since the SVM objective function is formulated as a linear combination of dot products, it is not necessary to explicitly compute the vector representation of each input example. Not only does this prevent the intractable problem of materializing a vector in high or infinite dimensions, it also allows to employ evidence with no obvious vectorial representation – such as protein sequence similarity or PPI network connectivity. These kernels have an intuitive interpretation as a measure of similarity between two examples. Depending on the kernel used, finding a decision hyperplane in the feature space can correspond to arbitrarily complex non-linear decision surfaces in the space of the input examples (input space) (**Fig. 22.6a**).

A large number of kernel mapping functions can be used, but the choice of the appropriate kernels can greatly benefit from prior knowledge about a particular problem. The radial basis function (RBF) kernels often give acceptable performance for many applications including PPI prediction and may also be useful if little

Fig. 22.6. (continued) vector from each category and the optimal solution showing a hyperplane with wider margins bounded by three vectors from each category. The vectors bounding the margins are called *support vectors*, denoted as *S* in the figure. More generally, support vectors are all vectors that crucially influence the solution found by the SVM procedure. In the case depicted in (**b**) they will also include the misclassified vectors.

is known about a problem. Recently, these more generic kernels have been combined with the so-called *diffusion* kernels that measure the connectivity of proteins in an underlying physical or genetic interaction networks to improve the prediction of PPIs (47, 50) or genetic interactions (50). Heterogeneous lines of evidences can be combined (*see* **Section 2.6**) with an appropriate kernel or by combining kernels. Although SVMs are generally resistant to noise, eliminating redundant or highly correlated evidence can improve performance (90). To achieve perfect accuracy on the training examples, the hyperplane produced following the kernel function mapping should completely separate the feature vectors into two non-overlapping groups. Depending on the kernel, such a result may not be possible or it may be overly complex and not generalize well to data not included in training. To combat this *overfitting* problem, the SVM maximizes the distance (margin) between the decision surface and the closest positive and negative examples (**Fig. 22.6c**) – called *support vectors.* A large margin in feature space corresponds to a smoother decision surface in input space, and hence the SVM can permit some misclassification errors in order to increase the width of this margin (*see* **Fig. 22.6b**). The cost parameter, *C*, controls the relative trade-off between error rate and model complexity. High values of *C* increase the misclassification cost, leading to a model that will do comparatively well on classifying the training data, but that may suffer from over-fitting.

Lastly, many kernels also have additional parameters. Finding the optimal values for these and the *C* parameter often involves some type of search algorithm guided by the success rate in classifying examples not used in the training phase (cross validation).

For a more rigorous introduction to the SVM in the context of computational biology, *see* (88). Links to SVM software are available on the web site http://www.kernel-machines.org.

4.5. The Local Coherence Detection (LCD) Algorithm

The local coherence detection algorithm is a biclustering procedure that groups genes based on their genetic (epistatic) interaction profiles (*see* (61) and http://wodaklab.org/biclustering/). The main difference between LCD and hierarchical clustering is that it considers the similarity of both the profiles as a whole and portions of it, whereas hierarchical clustering considers only the similarity of global profiles. LCD is also unique in handling all numeric data types, encompassing positive, negative, and missing values, and thus is able to take full advantage of the rich information afforded by the continuous-scale, real-valued, high-density E-MAP data such as that of (9). Exploiting the similarities between portions of the epistatic profiles involves using stringent fitness scores derived on the basis of solid statistical criteria. Those are established using minimally randomized background models, where only edge weights of the genetic interaction (GI) network are swapped while the edge weight distribution, node degree distribution, node labels, and network topology were preserved (61). These models afford the detection of gene groups with coherent GI patterns with a negligible rate of false positives. Such gene groups are defined as functional modules.

Acknowledgments

The authors thank Brian Turner and Andrei Turinsky for their work on the consolidation of protein interaction data and for help with iRefWeb. Funding is acknowledged from the Canadian Institutes of Health Research (CIHR MOP#82940), the Ontario Research Fund, and the SickKids Foundation. SJW is Canada Research Chair, Tier 1.

References

1. Gasch, A. P., Spellman, P. T., Kao, C. M., et al. (2000) Genomic expression programs in the response of yeast cells to environmental changes. *Mol. Biol. Cell* **11**, 4241–4257.
2. Hughes, J. D., Estep, P. W., Tavazoie, S., and Church, G. M. (2000) Computational identification of cis-regulatory elements associated with groups of functionally related genes in *Saccharomyces cerevisiae*. *J. Mol. Biol.* **296**, 1205–1214.
3. Spellman, P. T., Sherlock, G., Zhang, M. Q., et al. (1998) Comprehensive identification of cell cycle-regulated genes of the yeast *Saccharomyces cerevisiae* by microarray hybridization. *Mol. Biol. Cell* **9**, 3273–3297.
4. Harbison, C. T., Gordon, D. B., Lee, T. I., et al. (2004) Transcriptional regulatory code of a eukaryotic genome. *Nature* **431**, 99–104.
5. Lee, T. I., Rinaldi, N. J., Robert, F., et al. (2002) Transcriptional regulatory networks in *Saccharomyces cerevisiae*. *Science* **298**, 799–804.
6. Wingender, E., Chen, X., Hehl, R., et al. (2000) TRANSFAC: an integrated system for gene expression regulation. *Nucleic Acids Res.* **28**, 316–319.
7. Huh, W. K., Falvo, J. V., Gerke, L. C., et al. (2003) Global analysis of protein localization in budding yeast. *Nature* **425**, 686–691.
8. Kumar, A., Agarwal, S., Heyman, J. A., et al. (2002) Subcellular localization of the yeast proteome. *Genes Dev.* **16**, 707–719.
9. Collins, S. R., Miller, K. M., Maas, N. L., et al. (2007) Functional dissection of protein complexes involved in yeast chromosome biology using a genetic interaction map. *Nature* **446**, 806–810.
10. Hillenmeyer, M. E., Fung, E., Wildenhain, J., et al. (2008) The chemical genomic portrait of yeast: uncovering a phenotype for all genes. *Science* **320**, 362–365.
11. Tong, A. H., Lesage, G., Bader, G. D., et al. (2004) Global mapping of the yeast genetic interaction network. *Science* **303**, 808–813.
12. Belle, A., Tanay, A., Bitincka, L., Shamir, R., and O'Shea, E. K. (2006) Quantification of protein half-lives in the budding yeast proteome. *Proc. Natl. Acad. Sci. USA* **103**, 13004–13009.
13. Ghaemmaghami, S., Huh, W. K., Bower, K., et al. (2003) Global analysis of protein expression in yeast. *Nature* **425**, 737–741.
14. Fields, S., and Song, O. (1989) A novel genetic system to detect protein-protein interactions. *Nature* **340**, 245–246.
15. Stagljar, I., Korostensky, C., Johnsson, N., and te Heesen, S. (1998) A genetic system based on split-ubiquitin for the analysis of interactions between membrane proteins in vivo. *Proc. Natl. Acad. Sci. USA* **95**, 5187–5192.
16. Morell, M., Ventura, S., and Aviles, F. X. (2009) Protein complementation assays: approaches for the in vivo analysis of protein interactions. *FEBS Lett.* **583**, 1684–1691.

17. Puig, O., Caspary, F., Rigaut, G., et al. (2001) The tandem affinity purification (TAP) method: a general procedure of protein complex purification. *Methods* **24**, 218–229.
18. Gavin, A. C., Aloy, P., Grandi, P., et al. (2006) Proteome survey reveals modularity of the yeast cell machinery. *Nature* **440**, 631–636.
19. Krogan, N. J., Cagney, G., Yu, H., et al. (2006) Global landscape of protein complexes in the yeast *Saccharomyces cerevisiae. Nature* **440**, 637–643.
20. Wodak, S. J., Pu, S., Vlasblom, J., and Seraphin, B. (2009) Challenges and rewards of interaction proteomics. *Mol. Cell. Proteomics* **8**, 3–18.
21. Collins, S. R., Kemmeren, P., Zhao, X. C., et al. (2007) Toward a comprehensive atlas of the physical interactome of *Saccharomyces cerevisiae. Mol. Cell. Proteomics* **6**, 439–450.
22. Pu, S., Vlasblom, J., Emili, A., Greenblatt, J., and Wodak, S. J. (2007) Identifying functional modules in the physical interactome of *Saccharomyces cerevisiae. Proteomics* **7**, 944–960.
23. Ito, T., Tashiro, K., Muta, S., et al. (2000) Toward a protein-protein interaction map of the budding yeast: A comprehensive system to examine two-hybrid interactions in all possible combinations between the yeast proteins. *Proc. Natl. Acad. Sci. USA* **97**, 1143–1147.
24. Ito, T., Chiba, T., Ozawa, R., Yoshida, M., Hattori, M., and Sakaki, Y. (2001) A comprehensive two-hybrid analysis to explore the yeast protein interactome. *Proc. Natl. Acad. Sci. USA* **98**, 4569–4574.
25. Uetz, P., Giot, L., Cagney, G., et al. (2000) A comprehensive analysis of protein-protein interactions in *Saccharomyces cerevisiae. Nature* **403**, 623–627.
26. Yu, H., Braun, P., Yildirim, M. A., et al. (2008) High-quality binary protein interaction map of the yeast interactome network. *Science* **322**, 104–110.
27. Tarassov, K., Messier, V., Landry, C. R., et al. (2008) An in vivo map of the yeast protein interactome. *Science* **320**, 1465–1470.
28. Miller, J. P., Lo, R. S., Ben-Hur, A., et al. (2005) Large-scale identification of yeast integral membrane protein interactions. *Proc. Natl. Acad. Sci. USA* **102**, 12123–12128.
29. Hastie, T., Tibshirani, R., and Friedman, J. (2001) *The Elements of Statistical Learning: Data Mining, Inference, and Prediction*. New York, NY: Springer.
30. Brohee, S., and van Helden, J. (2006) Evaluation of clustering algorithms for protein-protein interaction networks. *BMC Bioinformatics* 7, 488.
31. Vlasblom, J., and Wodak, S. J. (2009) Markov clustering versus affinity propagation for the partitioning of protein interaction graphs. *BMC Bioinformatics* **10**, 99.
32. Ben-Hur, A., and Noble, W. S. (2006) Choosing negative examples for the prediction of protein-protein interactions. *BMC Bioinformatics* 7(Suppl 1), S2.
33. Hart, G. T., Ramani, A. K., and Marcotte, E. M. (2006) How complete are current yeast and human protein-interaction networks? *Genome Biol.* 7, 120.
34. Mewes, H. W., Frishman, D., Guldener, U., et al. (2002) MIPS: a database for genomes and protein sequences. *Nucleic Acids Res.* **30**, 31–34.
35. Pu, S., Wong, J., Turner, B., Cho, E., and Wodak, S. J. (2009) Up-to-date catalogues of yeast protein complexes. *Nucleic Acids Res.* **37**, 825–831.
36. Remy, I., and Michnick, S. W. (2004) Mapping biochemical networks with protein-fragment complementation assays. *Methods Mol. Biol.* **261**, 411–426.
37. Eyckerman, S., Verhee, A., der Heyden, J. V., et al. (2001) Design and application of a cytokine-receptor-based interaction trap. *Nat. Cell Biol.* **3**, 1114–1119.
38. Wang, H., Kakaradov, B., Collins, S. R., et al. (2009) A complex-based reconstruction of the *Saccharomyces cerevisiae* interactome. *Mol. Cell. Proteomics* **8**, 1361–1381.
39. Ashburner, M., Ball, C. A., Blake, J. A., et al. (2000) Gene ontology: tool for the unification of biology. The gene ontology consortium. *Nat. Genet.* **25**, 25–29.
40. Deane, C. M., Salwinski, L., Xenarios, I., and Eisenberg, D. (2002) Protein interactions: two methods for assessment of the reliability of high throughput observations. *Mol. Cell. Proteomics* **1**, 349–356.
41. Deng, M., Mehta, S., Sun, F., and Chen, T. (2002) Inferring domain-domain interactions from protein-protein interactions. *Genome Res.* **12**, 1540–1548.
42. Gentleman, R., and Huber, W. (2007) Making the most of high-throughput protein-interaction data. *Genome Biol.* **8**, 112.
43. Hakes, L., Pinney, J. W., Robertson, D. L., and Lovell, S. C. (2008) Protein-protein interaction networks and biology-what's the connection? *Nat. Biotechnol.* **26**, 69–72.
44. Huang, H., Jedynak, B. M., and Bader, J. S. (2007) Where have all the interactions gone?

Estimating the coverage of two-hybrid protein interaction maps. *PLoS Comput. Biol.* **3**, e214.
45. Stumpf, M. P., Thorne, T., de Silva, E., et al. (2008) Estimating the size of the human interactome. *Proc. Natl. Acad. Sci. USA* **105**, 6959–6964.
46. Hart, G. T., Lee, I., and Marcotte, E. R. (2007) A high-accuracy consensus map of yeast protein complexes reveals modular nature of gene essentiality. *BMC Bioinformatics* **8**, 236.
47. Qiu, J., and Noble, W. S. (2008) Predicting co-complexed protein pairs from heterogeneous data. *PLoS Comput. Biol.* **4**, e1000054.
48. Reguly, T., Breitkreutz, A., Boucher, L., et al. (2006) Comprehensive curation and analysis of global interaction networks in *Saccharomyces cerevisiae*. *J. Biol.* **5**, 11.
49. Goldberg, D. S., and Roth, F. P. (2003) Assessing experimentally derived interactions in a small world. *Proc. Natl. Acad. Sci. USA* **100**, 4372–4376.
50. Qi, Y., Suhail, Y., Lin, Y. Y., Boeke, J. D., and Bader, J. S. (2008) Finding friends and enemies in an enemies-only network: a graph diffusion kernel for predicting novel genetic interactions and co-complex membership from yeast genetic interactions. *Genome Res.* **18**, 1991–2004.
51. Boone, C., Bussey, H., and Andrews, B. J. (2007) Exploring genetic interactions and networks with yeast. *Nat. Rev. Genet.* **8**, 437–449.
52. Pan, X., Yuan, D. S., Ooi, S., et al. (2007) dSLAM analysis of genome-wide genetic interactions in *Saccharomyces cerevisiae*. *Methods* **41**, 206–221.
53. Tong, A. H. Y., and Boone, C. (2006) Synthetic genetic array analysis in *Saccharomyces cerevisiae*. *Methods Mol. Biol.* **313**, 171–192.
54. Costanzo, M., and Boone, C. (2009) SGAM: an array-based approach for high-resolution genetic mapping in *Saccharomyces cerevisiae*. *Methods Mol. Biol.* **548**, 37–53.
55. Schuldiner, M., Collins, S. R., Thompson, N. J., et al. (2005) Exploration of the function and organization of the yeast early secretory pathway through an epistatic miniarray profile. *Cell* **123**, 507–519.
56. Segre, D., Deluna, A., Church, G. M., and Kishony, R. (2005) Modular epistasis in yeast metabolism. *Nat. Genet.* **37**, 77–83.
57. Kelley, R., and Ideker, T. (2005) Systematic interpretation of genetic interactions using protein networks. *Nat. Biotechnol.* **23**, 561–566.
58. Bandyopadhyay, S., Kelley, R., Krogan, N. J., and Ideker, T. (2008) Functional maps of protein complexes from quantitative genetic interaction data. *PLoS Comput. Biol.* **4**, e1000065.
59. Ulitsky, I., Shlomi, T., Kupiec, M., and Shamir, R. (2008) From E-MAPs to module maps: dissecting quantitative genetic interactions using physical interactions. *Mol. Syst. Biol.* **4**, 209.
60. Prelic, A., Bleuler, S., Zimmermann, P., et al. (2006) A systematic comparison and evaluation of biclustering methods for gene expression data. *Bioinformatics* **22**, 1122–1129.
61. Pu, S., Ronen, K., Vlasblom, J., Greenblatt, J., and Wodak, S. J. (2008) Local coherence in genetic interaction patterns reveals prevalent functional versatility. *Bioinformatics* **24**, 2376–2383.
62. Stark, C., Breitkreutz, B. J., Reguly, T., Boucher, L., Breitkreutz, A., and Tyers, M. (2006) BioGRID: a general repository for interaction datasets. *Nucleic Acids Res.* **34**, D535–D539.
63. Xenarios, I., Salwinski, L., Duan, X. J., Higney, P., Kim, S. M., and Eisenberg, D. (2002) DIP, the database of interacting proteins: a research tool for studying cellular networks of protein interactions. *Nucleic Acids Res.* **30**, 303–305.
64. von Mering, C., Jensen, L. J., Snel, B., et al. (2005) STRING: known and predicted protein-protein associations, integrated and transferred across organisms. *Nucleic Acids Res.* **33**, D433–D437.
65. Hoffmann, R., Krallinger, M., Andres, E., Tamames, J., Blaschke, C., and Valencia, A. (2005) Text mining for metabolic pathways, signaling cascades, and protein networks. *Sci. STKE*, **283**, pe21.
66. Cusick, M. E., Yu, H., Smolyar, A., et al. (2009) Literature-curated protein interaction datasets. *Nat. Methods* **6**, 39–46.
67. Salwinski, L., Licata, L., Winter, A., et al. (2009) Recurated protein interaction datasets. *Nat. Methods* **6**, 860–861.
68. Orchard, S., Jones, P., Taylor, C., et al. (2007) Proteomic data exchange and storage: the need for common standards and public repositories. *Methods Mol. Biol.* **367**, 261–270.
69. Chaurasia, G., Malhotra, S., Russ, J., et al. (2009) UniHI 4: new tools for query, analysis and visualization of the human protein-protein interactome. *Nucleic Acids Res.* **37**, D657–D660.
70. Jayapandian, M., Chapman, A., Tarcea, V. G., et al. (2007) Michigan Molecular Interactions (MiMI): putting the jigsaw puzzle together. *Nucleic Acids Res.* **35**, D566–D571.

71. Kamburov, A., Wierling, C., Lehrach, H., and Herwig, R. (2009) ConsensusPathDB – a database for integrating human functional interaction networks. *Nucleic Acids Res.* **37**, D623–D628.
72. Prieto, C., and De Las Rivas, J. (2006) APID: Agile Protein Interaction DataAnalyzer. *Nucleic Acids Res.* **34**, W298–W302.
73. Razick, S., Magklaras, G., and Donaldson, I. M. (2008) iRefIndex: a consolidated protein interaction database with provenance. *BMC Bioinformatics*, **9**, 405.
74. Bader, G. D., Betel, D., and Hogue, C. W. (2003) BIND: the biomolecular interaction network database. *Nucleic Acids Res.* **31**, 248–250.
75. Ruepp, A., Brauner, B., Dunger-Kaltenbach, I., et al. (2008) CORUM: the comprehensive resource of mammalian protein complexes. *Nucleic Acids Res.* **36**, D646–D650.
76. Kerrien, S., Alam-Faruque, Y., Aranda, B., et al. (2007) IntAct – open source resource for molecular interaction data. *Nucleic Acids Res.* **35**, D561–D565.
77. Prasad, T. S., Kandasamy, K., and Pandey, A. (2009) Human protein reference database and human proteinpedia as discovery tools for systems biology. *Methods Mol. Biol.* **577**, 67–79.
78. Chatr-aryamontri, A., Ceol, A., Palazzi, L. M., et al. (2007) MINT: the Molecular INTeraction database. *Nucleic Acids Res.* **35**, D572–D574.
79. Guldener, U., Munsterkotter, M., Oesterheld, M., et al. (2006) MPact: the MIPS protein interaction resource on yeast. *Nucleic Acids Res.* **34**, D436–D441.
80. Pagel, P., Kovac, S., Oesterheld, M., et al. (2005) The MIPS mammalian protein-protein interaction database. *Bioinformatics* **21**, 832–834.
81. Kim, P. M., Lu, L. J., Xia, Y., and Gerstein, M. B. (2006) Relating three-dimensional structures to protein networks provides evolutionary insights. *Science*, **314**, 1938–1941.
82. Cochran, A. G. (2001) Protein-protein interfaces: mimics and inhibitors. *Curr. Opin. Chem. Biol.* **5**, 654–659.
83. Wepf, A., Glatter, T., Schmidt, A., Aebersold, R., and Gstaiger, M. (2009) Quantitative interaction proteomics using mass spectrometry. *Nat. Methods* **6**, 203–205.
84. Hernandez, H., Dziembowski, A., Taverner, T., Seraphin, B., and Robinson, C. V. (2006) Subunit architecture of multimeric complexes isolated directly from cells. *EMBO Rep.* 7, 605–610.
85. Kamal, J. K., and Chance, M. R. (2008) Modeling of protein binary complexes using structural mass spectrometry data. *Protein Sci.* **17**, 79–94.
86. Janin, J., and Wodak, S. (2007) The third CAPRI assessment meeting Toronto, Canada, April 20–21, 2007. *Structure* **15**, 755–759.
87. Lensink, M.F., Mendez, R., and Wodak, S. J. (2007) Docking and scoring protein complexes: CAPRI 3rd Edition. *Proteins* **69**, 704–718.
88. Ben-Hur, A., Ong, C. S., Sonnenburg, S., Scholkopf, B., and Ratsch, G. (2008) Support vector machines and kernels for computational biology. *PLoS Comput. Biol.* **4**, e1000173.
89. Noble, W. S. (2006) What is a support vector machine? *Nat. Biotechnol.* **24**, 1565–1567.
90. Guyon, I., Weston, J., Barnhill, S., and Vapnik, V. (2002) Gene selection for cancer classification using support vector machines. *Mach. Learn.* **46**, 389–422.
91. Batada, N. N., Reguly, T., Breitkreutz, A., et al. (2007) Still stratus not altocumulus: further evidence against the date/party hub distinction. *PLoS Biol.* **5**, e154.
92. Lord, P. W., Stevens, R. D., Brass, A., and Goble, C. A. (2003) Investigating semantic similarity measures across the gene ontology: the relationship between sequence and annotation. *Bioinformatics* **19**, 1275–1283.

Chapter 23

Noise in Biological Systems: Pros, Cons, and Mechanisms of Control

Yitzhak Pilpel

Abstract

Genetic regulatory circuits are often regarded as precise machines that accurately determine the level of expression of each protein. Most experimental technologies used to measure gene expression levels are incapable of testing and challenging this notion, as they often measure levels averaged over entire populations of cells. Yet, when expression levels are measured at the single cell level of even genetically identical cells, substantial cell-to-cell variation (or "noise") may be observed. Sometimes different genes in a given genome may display different levels of noise; even the same gene, expressed under different environmental conditions, may display greater cell-to-cell variability in specific conditions and more tight control in other situations. While at first glance noise may seem to be an undesired property of biological networks, it might be beneficial in some cases. For instance, noise will increase functional heterogeneity in a population of microorganisms facing variable, often unpredictable, environmental changes, increasing the probability that some cells may survive the stress. In that respect, we can speculate that the population is implementing a risk distribution strategy, long before genetic heterogeneity could be acquired. Organisms may have evolved to regulate not only the averaged gene expression levels but also the extent of allowed deviations from such an average, setting it at the desired level for every gene under each specific condition. Here we review the evolving understanding of noise, its molecular underpinnings, and its effect on phenotype and fitness – when it can be detrimental, beneficial, or neutral and which regulatory tools eukaryotic cells may use to optimally control it.

Key words: Noise in gene expression, noise control mechanisms, regulatory networks, network biology.

1. Relevance of Noise in Biological Systems

Since the earliest discoveries of the basic mechanisms that control gene expression, biologists have been intensively engaged in measuring mRNA and protein levels. Such studies, driven by a

J.I. Castrillo, S.G. Oliver (eds.), *Yeast Systems Biology*, Methods in Molecular Biology 759,
DOI 10.1007/978-1-61779-173-4_23, © Springer Science+Business Media, LLC 2011

broad range of technologies, have established links between the programs that regulate gene expression and the corresponding molecular phenotype – the level of expression of each individual gene and its response to external signals and, ultimately, to the phenotype of the organism. The advent of genomics revolutionized the study of gene expression: technologies such as DNA and protein microarrays, and most recently, RNA sequencing enables the measurement of gene expression for every gene, providing rich information about transcription (1), mRNA degradation (2), translation (3), RNA editing (4), and more. Examination of control regions of genes will enable researchers to decipher the regulatory programs that underlie the observed behavior.

However, common to all such technologies is the fact that measurements represent averaged RNA and protein levels over large populations of cells. For instance, a typical microarray experiment requires more than 100,000 cells. Such measurements thus provide very reliable estimates of the *average* level of expression of a given gene over an entire population of what are typically genetically identical cells. If all cells in a population expressed a given gene at the same or even very similar levels, then the average alone would, indeed, capture the reality well. However, if different cells in the population expressed a given gene at different levels, then information about cell-to-cell variation would be lost. If cells take control of the extent to which they allow or restrict such diversity, deciphering the mechanisms that exert such control would require alternative models and technologies.

From a theoretical perspective, what is the potential for cell-to-cell variation in gene expression levels, among genetically identical cells? Stochasticity and randomness govern the microscopic world inside cells, the world of molecular recognition that is driven by interactions between molecules in a crowded environment. The effects of such random events may be particularly dramatic when it comes to molecules that are represented in just a few copies per cell, as is the case with many regulators of gene expression. For instance, if a particular regulator is present, on average, in two copies per cell, we should not be surprised to find cells that have four copies and others with one or even zero copies of this molecule. It is also easy to imagine that the targets of such a regulator would also display corresponding, and perhaps even greater, fluctuations as a result. Cell-to-cell variation at the level of gene expression may constitute a real possibility.

What would be the interest of studying this possible variation? We might think that if the population averages out such fluctuations then they will have no functional consequence. Here we will argue for the converse. Consider, for instance, a population of genetically identical *Escherichia coli* cells that are attacked by an antibiotic drug. While the majority of the population may die, a portion may survive the attack. Note that we do not consider here the case of resistant cells that are genetically different from the majority of the population; rather, only those *persisters* (5) that are genetically identical to the rest of the cells. Further investigation revealed a stochastic split of the population into two subpopulations, one that is sensitive to the drug and another that is not, and a purely random event that allows cells to switch from one fate to the other (5). Another example, for instance, sporulation in yeast, a potential defensive response to stress, manifests itself in only a portion of genetically identical cells in a population. It turns out that a single protein stochastically expressed at varying levels in different cells determines the different cellular fates (6). To take examples from multi-cellular organisms, a combined theoretical–experimental approach involving immune T-cells found that stochastic variation in the expression of key signaling proteins generates substantial cell-to-cell variability in the antigen responsiveness needed among cells in a clonal population (7). Another example relates to populations of cells in tumors. After chemotherapy treatment the majority of the cells may die, but a smaller sub-population may survive. Recently, an individual protein was found that, due to stochastic effects, may be temporarily present at varying levels in particular cancer cells (8). As a consequence, some of the cells are rendered resistant to a drug, whereas others are unable to survive the same therapy. In all these systems, a seemingly random event at the molecular level – often the choice of expression level for key genes in particular cells – gave rise to dramatically different phenotypes at the cellular and organism levels. Without measuring the levels of relevant proteins at the population level and the average over all the cells in the population, the underpinnings of the sophisticated environmental response can be altogether missed.

The paradigm that emerges is that certain critical biological phenomena, such as drug persistence, stress response, immune response, and cancer cell proliferation, are rooted in stochastic molecular events, which ultimately lead to phenotypic variation among genetically identical cells. Yet these cases might represent the exception rather than the rule. The perception that cells tightly control the expression of their genes, to such an extent that cell-to-cell variation would be limited, may indeed be correct in the case of many genes that must be expressed at precisely fixed levels. The dozens of proteins that make up large macromolecular complexes such as the ribosome, for example, should conceivably be kept under tight control, not least to eliminate wasteful production. It is thus conceivable that cells may have evolved mechanisms to determine the extent to which they can safely permit variations in the expression of some genes in their genomes and restrict variation in others.

Stochastic variation in gene expression levels among genetically identical cells grown under the same conditions is often dubbed "noise" (9–12). Quantification of noise in gene expression often requires measurements of mRNA or protein levels in single cells. In recent years, experimental techniques to measure gene expression in single cells at high throughput and with great accuracy have matured (9–13, 18), along with theoretical models to rationalize the results (14, 15). The aims of those studying noise in gene expression are thus to measure noise in various biological setups, to decipher the regulatory means by which cells control noise, and to study the biological consequences of such non-genetic variation. Another substantial bonus from the study of noise comes from the fact that the statistical properties of noise can reveal basic principles of the molecular processes that govern gene expression. For instance, by analysis of noise spectra it was possible to deduce that gene expression occurs in bursts, whose two key parameters are the burst size and frequency (15). This information is essential to understanding noise behavior but its implications extend to the basics of gene expression mechanisms.

2. Noise in Gene Expression

Consider the expression level of protein X in a unicellular organism such as yeast in the following thought experiment: Let us measure its expression in two genetically identical cells. Since the two cells are genetically identical, we might expect that the expression level of our protein would be identical in both. Suppose that we have measured the actual copy number of protein X in each cell and found it to be twice as high in one of the cells, compared to the other. Our first suspicion might be that a measurement error occurred or, to put it more quantitatively, that the measurement error is larger than the true variation between our two cells. Let us then assume that the difference is reproducible, even after many repetitions; moreover, it is not seen with respect to another protein, Y, which on average shares the same expression level as X.

So, why the level of expression of protein X appears to differ so significantly between the two cells? First, even if our two cells are identical at the DNA level, errors do occur at the level of transcription and translation: on average, 1 in every 10,000 transcribed nucleotides, and 1 in every 1,000 translated amino acids, is expected to be wrong (16, 17). Such errors may affect the expression level of a protein, e.g., by affecting the stability of the mRNA and the protein or by affecting protein X's regulators. A dramatic phenotypic consequence of such errors was recently demonstrated in a study in which transcription error rate was increased by a positive feedback (17).

Let us assume that in the two cells, all copies of X were transcribed and translated without a single error. What other reason may be responsible for the different levels of expression of protein X? The cells might have been subjected to different "micro-environments" (for instance, one might have been closer to the edge of a colony or they might have been at different stages of the cell cycle). Let us further assume that the two cells had been exposed to the same micro-environment, they were synchronized relative to the cell cycle, and they were also equal in size and mass. Remarkably, the two cells might still express protein X in varying amounts, because they could differ at the microscopic level (for instance, the existence of two copies of a transcription factor controlling the expression of the X gene in one cell, while three in the other). Let us assume that the regulators are found at the same level in all cells too. Moreover, the two cells are identical not only with respect to DNA, mRNA, and protein sequence but also with respect to the concentration, location, and molecular dynamics of every molecule within them, including every transcription factor and ligands. From this perfectly identical starting point, let each of the cells live out its natural lifespan. After a short while, would our cells show identical or different levels of protein X? Even under these "identical" conditions, basic physical chemistry shows that differences might still be possible. All processes involving the propagation of genetic information, including the unwinding of the DNA for transcription, transcription itself, processing of RNA, degradation of transcripts, translation, protein modification, and protein degradation, are based on interactions between molecules and inevitably include a stochastic component (i.e., while in one cell the transcription factor may initiate, e.g., two transcription events at a given time interval, in the other it may occur only once. The binding constant between that factor and the promoter of the gene encoding X may be the same in the two cells, but this is merely a macroscopic constant that relates to a ratio of probabilities). Of particular interest is the fact that small stochastic differences may be amplified or canceled out due to further random events (for instance, in one cell each transcript is translated 10 times before it is degraded, resulting in 20 copies of protein per cell and only 10 in the other). An example in which the initial difference between the two cells may be diminished may be that RNA degradation, often mediated by the binding of an RNA binding protein, may occur faster in the first cell with higher transcript levels.

In summary, there are many mechanisms by which two genetically identical cells may express a specific protein (protein X) at different levels. Can we measure the effect of stochasticity if all other factors such as genotype and cell size are kept constant or due to environmental changes (for instance, by applying environmental stress)? If so, what patterns would be expected for the levels of protein X and other proteins (for example, protein Y, expressed on average at the same level as protein X across all cells but with different functions and involved in a different regulatory network, or protein Z, expressed at higher levels)? More importantly, can noise play a part in central biological processes and be responsible for cell fates and specific phenotypes that could be investigated? For example, do cells set different noise levels for each gene, according to both its function and environmental conditions? Can cells and organisms control the extent of variation and stochasticity so as to minimize it when harmful or amplify it and benefit from it when possible?

3. Measuring Noise in Living Cells

A landmark work that transformed the study of cellular stochasticity into a quantitative biophysical science was that of Elowitz et al. (18). While finding two cells with the same concentration, location, and dynamics of all molecules is still impossible, they found a clever practical solution. By using two fluorescent proteins (cyan and yellow) under the control of two identical promoters, placed in two similar locations in the *E. coli* genome, they generated a cellular environment that was practically identical for the two genes – thus, not two identical cells with the same gene as in our thought experiment (see previous section), but rather the same cell that serves the expression of two distinguishable genes. Since the two proteins fluoresce in separate colors, comparing the intensity from the two channels in the same cell was sufficient to identify differences that must be attributed, for the most part, to stochastic events that happened inside that cell. Elowitz and coworkers then used fluorescent-activated cell sorting (FACS) to measure fluorescence in the two channels that correspond to the two genes in individual cells. The difference in expression between a pair of two such proteins expressed in the same cells was termed the "intrinsic noise" of the system – only stochastic processes within each cell could give rise to differences in the expression level of these two probes in each cell. With the same experimental method, focusing on one wavelength at a time, the researchers could then compare the different cells and examine the variations among them at the level of protein expression. This variation was called the "extrinsic noise," as it captured external sources of variation, such as varying concentrations of a relevant transcription factor and number of ribosomes in each cell. It was then possible to estimate not only each noise source but its relative contribution to the final level of variation. The authors concluded that extrinsic noise was a major factor, but that intrinsic noise had a significant contribution too to the overall cell-to-cell variation. Applying a similar system to *Saccharomyces cerevisiae*, Raser and O'Shea were able to measure the relative contribution of intrinsic and extrinsic sources of noise in this model eukaryote and found that the intrinsic noise is gene-specific and may constitute an evolvable trait that can be optimized to balance fidelity and diversity in eukaryotic gene expression (19).

4. Expanding the Scope: Measuring Noise for Many Genes and Conditions

The studies described thus far have focused on exogenous genes and measured noise levels under a limited set of environmental conditions. Yet to penetrate deeper into the biological significance of noise, three additional steps were necessary: first, to measure noise of endogenous genes, insofar as was possible, with minimal interruption of their native controls; second, to scale up measurements so as to probe as many genes as possible; and third, it was essential that the noise of a given native gene would be measured under varying growth conditions. Bar-Even et al. (20) and Newman et al. (21) accomplished these aims in a complementary fashion. For this purpose, they used a library of *S. cerevisiae* strains, each expressing one of the endogenous genes of this species (22) fused to a green florescent protein (GFP). Examining the fluorescence of cells from each such strain by FACS, it was possible to measure the cell-to-cell variations in expression of each gene in the genome. Here, since a single type of fluorescent protein (GFP) was used, intrinsic and extrinsic noise were no longer separated and the integrated contribution from the two sources of variation was measured as a single number [yet when a sample of the genes was also measured in two colors, a predominant contribution from the intrinsic noise was actually found (20)]. In all, Bar-Even and colleagues studied 43 genes; yet they examined noise under a diversity of conditions. These authors selected their sample genes so as to represent several genetic modules that were originally defined, based on classical microarray experiments: these included stress-related genes, genes encoding structural constituents of the proteasome, genes involved in ergosterol metabolism, and genes responsible for the processing of ribosomal RNA (rRNA). They then exposed the cells to 11 different conditions, the majority of which were stressful, yet a few actually involved recovery from stress conditions. Finally, they measured, for each gene, the distribution of expression values under each condition at the single cell level. Newman and colleagues did not explore as many conditions; instead, they measured the expression of a most significant portion of the entire yeast genome (21). The two papers focused on two parameters that characterized each gene: the mean of the distribution and the "noise coefficient" – namely, the variance divided by the square of the mean. Plotting the noise coefficient vs the mean expression, the two groups found the same intriguing, "scaling" relationship: the noise declined as the reciprocal of the mean of the distribution. That is, not only were highly expressed genes less noisy (as might be expected, given the mean of the distribution, with the value obtained from a standard microarray experiment) it was possible to rather accurately predict the amount of noise for most genes under most conditions. Theoretical analysis in both works suggested that such scaling might result from variations in mRNA copy numbers, caused by the stochasticity of "birth and death" of mRNA molecules or from fluctuations in promoter activity (20, 21).

If noise can be inferred from the mean, could we avoid measuring noise in future experiments and settle for the more conventional measurements? The fact is that, although most genes under most conditions obeyed the scaling law, interesting deviations were found. The "noise residual" was thus defined as the difference between the amount of noise that a gene actually displayed and the amount of noise that could be predicted for the gene, given its mean expression and the general scaling between noise and mean that holds true for most genes in most conditions (**Fig. 23.1**). For instance, stress-related genes were consistently above the scale (i.e., for these genes, noise was typically higher than the expected value, given their own mean and the general scaling).

What could be the rationale behind this enhanced noise in stress genes? One intriguing possibility is that cells implement a "risk distribution strategy" with these genes – that different cells in the isogenic population provide stochastically different "responses" (i.e., expression levels of these genes). According to this hypothesis, the cells that happened to express these genes at the optimal level would be more likely to survive. In fluctuating environments such approach might, under some circumstances, constitute the most feasible strategy (23). Note, however, that since the cells are genetically identical, such changes would not be inherited (see below on the "memory" of such fluctuations and on the combination of genetic and non-genetic diversity). The fact that in the experiments where stress was alleviated (20), the stress genes typically showed reduced noise (**Fig. 23.1**) supports that control of the noise in these genes may constitute a cellular response to changes in environmental conditions from non-stressful to stressful conditions and vice versa.

Examination of the response of other genes revealed an opposite trend: negative noise residuals (**Fig. 23.1**). Take, for instance, the genes encoding constituents of the proteasome, a multi-subunit cellular complex. Under stress conditions, these

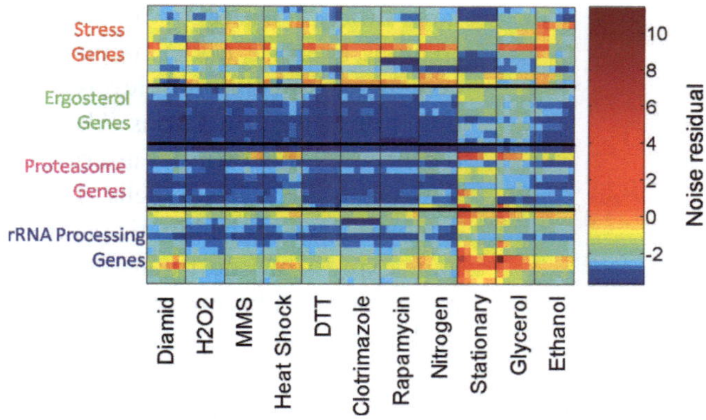

Fig. 23.1. Variations in levels of noise of specific yeast genes under different conditions. Levels of noise residuals of 43 *S. cerevisiae* genes in 11 different environmental conditions. Here, each *horizontal line* corresponds to a gene. Each condition is represented as a set of consecutive columns with each column corresponding to a time point. The noise was measured at six time points following the environmental change (20), and noise propagation can be traced for each gene at each condition. The color code depicts the "noise residual" of each gene in each condition, namely the difference between the actual noise level of a gene and the noise level expected for that gene given its mean expression level and the general scaling between noise and mean expression (20). The 43 genes were sampled from the entire yeast genome and they represent four modules: stress genes, ergosterol metabolism, constituents of the proteasome, and genes involved in the processing of ribosomal RNA (rRNA). The environmental conditions represent different perturbations and stress relaxing conditions (1st through the 7th, and the 11th sets of columns, and 8th–10th column, respectively) (20). The stress genes show higher noise levels throughout the conditions, especially under stress, while genes involved in ergosterol metabolism and proteasome and rRNA biosynthesis show noise being kept at controlled, low levels, particularly under stress conditions.

genes featured very tight distribution, with negative noise residuals. The example of genes encoding structural constituents of the proteasome shows that in some cases high levels of noise may be actually undesirable or should be kept under tight, controlled levels, for instance, when a fine coordination and stoichiometry of synthesis of specific subunits of a multi-subunit complex is necessary. In other genes in which the extent of noise implied by the mean is neither helpful nor detrimental, cells may not attempt to control the levels, and noise may be set by simple probabilistic rules (21).

5. How Cells Control Noise and Set a Desired Level for Each Gene

The aforementioned studies (18–21) show that cells appear to be able to set the noise levels and their enhanced or reduced extent compared to mean expression values. Genes that belong to the same functional category or structural complex often show similar changes in noise under a specific condition. Moreover, a given gene may show different levels of noise amplification or reduction under different conditions. This requires to be finely regulated. In recent years, new studies are revealing a rich array of strategies to regulate noise levels in cells. The means to control cellular noise can be related to the kinetic parameters of the processes governing gene expression and the topology of the regulatory networks and can sometimes be inferred from the genomic features inside and in the proximity to genes.

5.1. A Model for the Propagation of Noise

As an alternative to the works by Elowitz and colleagues (18) and Raser and O'Shea (19) which measured intrinsic and extrinsic noise, Paulsson took a different approach and developed a mathematical model that described the propagation of noise in a cellular pathway (14). Consider two molecules, A and B, such that A is either a regulator of B or that A and B are, respectively, the mRNA and protein encoded by a particular gene. In either of these cases, noisy fluctuations in A might be further propagated into B. Focusing on the downstream component B, Paulsson suggested an alternative to the dichotomy between intrinsic and extrinsic noise, realizing that the noise in B would arise from a combination of two components: the noise generated by B itself and the noise that B "inherits" from A. While Paulsson's model was predominantly theoretical, a similar conclusion was reached by Pedraza and van Oudenaarden (24) who experimentally measured expression correlations between genes in single cells. These authors also found that noise in the expression of a gene was determined by its intrinsic fluctuations, noise transmitted from upstream genes, and global noise affecting all genes.

Understanding the contribution of noise in A to noise in B is of particular interest, since such knowledge enables the description of noise propagation along genetic chains. According to Paulsson's model, one relevant parameter that governs such propagation is the response dynamics of A and B. Intuitively, if A is a very rapidly changing molecule, then B will "inherit" the fluctuation only if B, too, is rapidly fluctuating. If, on the other hand, B has a very slow rate of turnover, it will not trace the fluctuations in A over time and will thus not inherit the noise (i.e., B will be said to have "time-averaged" fluctuations in A). What governs the response times of specific molecules? Some response times may be largely governed by the degradation kinetics. Consider an mRNA and a protein encoded by a given gene as the "A" and "B" molecule in the above formalism. If the protein had was rapidly fluctuating (e.g., due to a relatively high degradation rate) then noise at the mRNA level would be effectively propagated to the protein. In recent years, techniques to measure the stability of both mRNA (2, 25) and protein (26) at the genome level have begun to mature, and comprehensive studies to test the possibility that noise propagation can be deduced from elementary parameters and other additional factors affecting the accuracy at which fluctuations might be propagated (14) are becoming more feasible.

5.2. The Relative Efficiency of Transcription vs Translation

The extent (efficiency) of transcription and translation of a given gene and the relative contribution (ratio) between them deserve special attention. As a case example, assume that a given protein is needed at an average of 200 copies per cell. Imagine two extreme (probably not very "realistic") strategies to obtain that desired level of protein expression: The first one is when the transcription of the corresponding mRNA takes place at a very low extent: so, for instance, only two mRNA molecules are present, on average, per cell, requiring each one to be translated, on average, 100 times. The other extreme is that transcription is extensive, resulting in 200 copies of the corresponding mRNA. To get the 200 proteins, in principle, it will be enough that each mRNA will be directed to the protein biosynthesis machinery only once (less if polyribosomes are acting).

The first case, with a very low transcription rate and the need for extensive translation, is economical in terms of RNA synthesis, but what will happen to the noise at the protein level? While the mRNA is present, on average, at two copies per cell, fluctuations with one, three, or four copies are likely. Translation has the potential to further amplify such fluctuation, yielding high predicted noise. Also, the need to reuse the same mRNAs for translation may lead to a slow global response, with low rates of protein biosynthesis (in some cases higher than the mRNA and protein turnover, in which case the degraded molecules will need to be re-synthesized). Target protein levels may be difficult to achieve in some cases, with high predicted noise.

On the other hand, a higher, efficient transcription (which will provide a "pool" of ready-to-use mRNAs) coupled with limited translation (which may be controlled by different mechanisms: polyadenylation, subcellular localization, and polyribosome levels) will result in few fluctuations and a relatively noise-free protein population. This strategy may also ensure fast responses in shorter times and, once target levels are obtained, the possibility of activation of mechanisms of control (e.g., feedback regulation or others, see below). Efficient, extensive transcription coupled with balanced translation can provide quick and fine control of protein levels.

From here, a simple prediction might be that cases of extensive transcription coupled with limited translation would exhibit low noise and high level of control of expression. The possibility that low levels (less efficient) of transcription coupled with extensive translation may constitute in some cases a means to enhance noise levels should not be discarded. There are several lines of support for the argument that the extent of translation (translation levels) is directly linked to noise levels. Ozbudak and coworkers independently modified the transcription and translation of a reporter gene and found that an increase in translation mainly increased noise level (11). Black et al. also provided a convincing demonstration of the idea (12), where they changed the nucleotide coding sequence of a gene without affecting amino acid sequence and shifted it toward higher or lower translation efficiency codons. They found that noise indeed increases with translation extent (efficiency), provided that the gene was not fully induced at the transcription level. Interestingly, these results could also be predicted from a theoretical model that these authors provided that deals with the propagation of noise from the mRNA to the protein level (12). In accordance with this, Fraser et al. (27) have also found a related trend in yeast: essential genes and genes involved in cellular complexes tend to minimize their predicted noise level by employing a strategy that maximized the ratio of transcription to translation. Furthermore, an inspection of the noise residuals of the 43 *S. cerevisiae* genes measured by Bar-Even et al., against their sequence-based calculation of translation efficiency (the tRNA adaptation index) (28), shows a correlation between them (**Fig. 23.2**). Genes with high

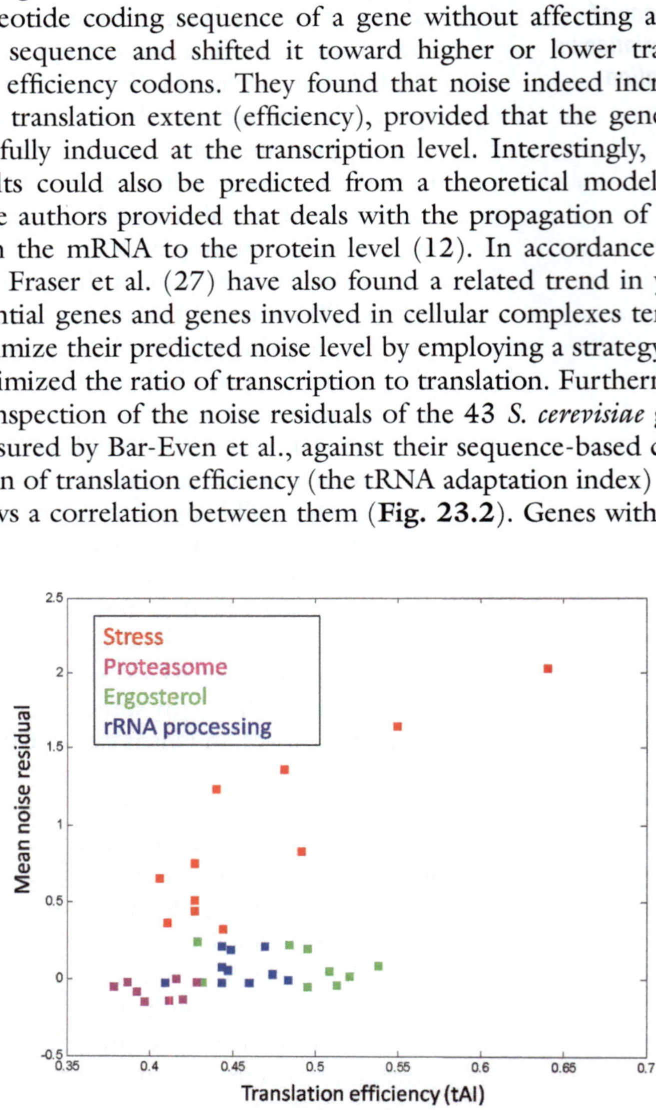

Fig. 23.2. Correlations of noise residuals and tRNA adaptation index. Noise residuals correlate with the tRNA adaptation index of yeast genes, particularly with genes involved in response to stress. Here, the same set of genes as in **Fig. 23.1** is analyzed, with colors depicting association to the four modules in **Fig. 23.1**. The noise residual of each gene is calculated as the mean of its values across all time points in all six conditions. The tRNA adaptation index (tAI) of each gene was calculated as in (28). The tAI captures the extent to which the codons in a gene are biased toward the more abundant tRNAs in the genome (high translated genes).

predicted translation efficiency (highly translated) indeed show some tendency toward higher noise residuals, whereas genes with low predicted efficiency of translation corresponded to the least noisy genes. Of particular interest are the proteasomal genes that under stress conditions display little noise, in fact even lower than expected from their means. These genes, under stressful conditions, are typically induced at the mRNA level (29). The results suggest that a means to obtain tight, noise-filtered distribution for proteasomal genes under stress is to employ the strategy of high transcription rate with limited, coupled translation. Similar conclusions have been reached by a mathematical treatment of a related system (30).

At the molecular level, the TATA box in gene promoters was found to be another key factor with which cells may tune noise levels. The TATA sequence is not present in every gene's promoter, but genes that do contain it typically show high levels of noise (19, 20, 31, 32). The prevailing explanation is that the TATA sequence amplifies fluctuations through facilitation of transcription re-initiation (33, 34).

5.3. Local Network Connectivity, Redundancy, and the Effect on Noise

An additional regulatory attribute that affects the noise level and pattern displayed by genes is their connectivity within the regulatory network. Consider two genes that are identical with respect to many of the aforementioned properties, such as the efficiency of their translation and transcription, but are nonetheless embedded in two different regulatory network motifs. In one case, the gene exerts negative feedback regulation on itself (either directly or through a mediator) and in the other, the gene exerts a positive feedback on its own level of expression. In which case would the gene manifest a higher amount of noise? Intuitively, the negative regulatory scheme would tend to counteract, or "correct," noisy fluctuations – when the amount of the autoregulated gene stochastically increases the negative regulation will counteract the fluctuation by increasing the extent of inhibition, while a fluctuation in the other direction would result in lower inhibition (10). In the positive feedback case, fluctuations in either direction are expected to intensify themselves, resulting in higher noise levels. Theoretical work, however, has recently elegantly refined these notions, predicting that while negative feedback eliminates noise, it comes at a price: reduced sensitivity to changes in environmental signals (35). When comparing circuits with the same level of sensitivity to environmental changes, it was found that a positive feedback design actually buffers noise better than negative feedback. It was further suggested that the improved capacity of a positive feedback circuit to buffer noise at a given level of environmental sensitivity comes from its time-averaging capacity (35).

Most recently, researchers faced an intriguing question pertaining to the architecture of regulatory networks: often two or more alternative circuit designs can produce the same outcome – in particular a negative regulatory effect on a gene can be obtained either by repressing the inducer of the gene or by inducing its repressor. Why is it that in reality one of the designs, and not the other, appears to operate in a particular system (36)? The authors focused on the genetic network that governs competence – the ability of microbes to uptake DNA from the environment, typically upon stress – in *Bacillus subtilis.* Like any regulated system it needs to control not only induction but also its shutoff. It turns out that the shutoff of the competence network in these bacteria is obtained by repressing the system's inducer, not by inducing a potential repressor. The authors synthesized a seemingly equivalent variation on the circuit in which shutoff happens through induction of a repressor. While the averaged properties of the native and synthetic designs were designed to be similar, the native circuit showed enhanced diversity between single cells, while the synthetic design was precise and relatively noise free. Why did evolution prefer the nosier version? The fact is that the accuracy of the synthetic design came at a price – bacteria that artificially expressed it were fit (i.e., could take up DNA from the environment) only at a narrow range of environmental parameters, compared to the cells that expressed the native system. The higher noise obtained in the native system thus appears to be adaptive as it allows higher population diversity in variable situations (36). Presumably, the negative regulation of the inducer in the native circuit is responsible for a lower expression level, possible high relative noise level, in this regulator. It was concluded that noise actually facilitates the response of the network to a variable environment. The more precise synthetic design, in which the repressor is induced, is more similar to circuits such as at the heart of the circadian clock, where accuracy and control are the main issue.

Apart from the connectivity in regulatory networks, the challenge of noise control may have constituted a driving force explaining the unexpected conservation of redundancy in biological systems. It was recently suggested that partially redundant duplicate genes may have been selected for preservation in genomes, so as to filter noise in regulatory networks (37, 38). Why are redundant genes often preserved, especially in pivotal nodes of regulatory networks, if only one member of the gene pair would suffice? Although redundant genes can often back each other up if mutations occur in one of them, it is entirely possible that the partially redundant genes will cover for each other when, due to stochasticity, one of them showed a temporary fluctuation that either increased or decreased its level, a far more likely event.

In many cases of redundancy in regulatory proteins, it was found that the regulators also negatively regulate one another (37, 38). Such a design could serve to reduce the effect of fluctuations: when the expression level of one of the two regulators goes up, it would further inhibit its partner, and when it is decreased, it would exert less of an inhibitory effect. Modeling results (37) show that the sum, or the product, of the concentrations of the two regulators may be kept relatively constant in such a regime; furthermore, if a joint target of the two regulators is only affected by the sum, or product, of the concentrations of its two upstream regulators, it may experience a low amount of noise. In other words, the negative regulatory cross-talk that is often observed between semi-redundant gene duplicates (*see* (37) for a review) may serve to transform noise fluctuations in each of them into a relatively constant sum (or product) of the two, thus minimizing further propagation of the noise. This design may also explain why, in the extreme case of a constant reduction of a gene's expression level, e.g., due to a deletion mutation in the gene, the semi-redundant duplicate may respond by increased expression, in compensation (38). Such a capacity might, in some cases, constitute a byproduct of a selectable tunable capacity to respond to random, temporal fluctuations in its counterpart's expression level (38).

5.4. A Potential Role of Non-coding RNAs

Non-coding RNAs are transcripts that are not translated into proteins. During the last decade, it has become apparent that in essentially every organism, a considerable portion of the transcribed RNAs are not translated. Many of these newly discovered non-coding RNAs function as regulators, that is, they may regulate gene expression at multiple levels. Two relevant examples of regulatory RNAs are microRNAs and antisense RNAs. Such transcripts are known to interact with their targets by means of base pair complementarity, mainly affecting the levels of stability of the RNA target (e.g., an mRNA), and the efficiency of its translation.

It is conceivable that a non-coding RNA would affect not only the average expression level of its target but also the noise that the target would display. A non-coding regulatory RNA present in excess, relative to its target, may actually serve to buffer noisy fluctuations in the target. Assume, for instance, that the regulatory RNA is present in a high average copy number, say 20 copies, in each cell in a population of isogenic cells. Now consider an mRNA target of this regulator that is gradually induced from very close to 0 copies to 50 copies per cell. The noise coefficient of that target would be high at the beginning of the induction process, due to its presence in low copy number. However, if the regulatory RNA efficiently sequesters the target when the latter is at a lower copy number, random fluctuations in the target would be dumped. Yet, as the target's concentration continues to rise, at some point it may exceed the level of the regulatory RNA, and from that point on, the buffer would be unable to effectively sequester the excess mRNAs. This would lead to a

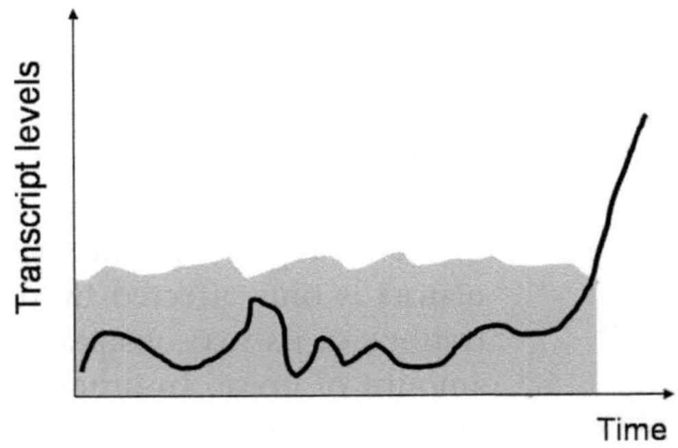

Fig. 23.3. A conceptual model of fluctuations and control of noise in RNA expression. The expression of a regulatory RNA (e.g., a microRNA or an antisense RNA) is depicted by the *gray area*, while the expression of the regulated transcript (e.g., an mRNA) is depicted with a *black line*. A regulatory RNA may filter noisy fluctuations in the levels of another regulated transcript (e.g., an mRNA). Levels of a regulatory RNA higher than the levels of the target RNA may effectively buffer noisy fluctuations (due to specific binding, sequestering, or targeting to degradation). Yet, when the levels of the regulatory RNA are lower than the target levels the difference cannot be buffered by the regulatory RNA, which will lead to fluctuations in mRNA expression.

step-like function in the concentration of the free mRNA target in the cell: it would remain close to zero, so long as its total level is below that of the regulator, but would abruptly increase, once it exceeded that level (**Fig. 23.3**). Notably, microRNAs are known to affect their targets both at the level of mRNA stability, where they destine targets to degradation, and by inhibiting their translation (c.f. (39)). It is still not clear how the choice between these two separate fates is determined for a pair of regulators and a target or why, in some cases, one fate is desired over the other. Yet the realization that the ratio of translation to transcription affects noise suggests an effect on the noise level of the target, due to a choice between the two regulatory mechanisms. A potential application arising out of such considerations lies in the emerging field of synthetic biology, in which one of the challenges is to design and build small circuits with desired properties. The synthetic use of non-coding regulatory RNAs (40) may enable researchers to "tune" the desired level of noise (either high or low) of the various components in the system.

6. Genetic vs Non-genetic Variation

It has long been known that stressful conditions increase the rate of mutations among microorganisms (41). This is rationalized as an adaptive trait of such species – leading to an increase in genetic diversity, the origin of new biological innovations, and, ultimately, survival of selected cells. Yet, depending on the selective advantage, the rate of mutations, and the effective population size, it may take tens of generations, typically many more, for specific mutations to reach a substantial portion of the population. This is a long time, compared for instance to a killing effect of the stressful condition. Thus, a faster means to increase diversity in the population may be needed.

An increase in noise levels, predominantly observed in stress-related genes under stressful conditions (20), may be envisaged as a complementary mechanism for the rapid generation of diversity. Unlike genetic diversity, which develops slowly, non-genetic diversity, or noise, is obtained instantaneously in a population (20). Furthermore, unlike mutations that are often, though not always (42), assumed to be generated at random sites, noise enhancement appears to be specifically directed toward genes that are expressed in response to changes in specific environmental parameters (43). The possibility exists that the labor involved in generating population diversity might split between non-genetic and genetic mechanisms, with the former occurring before the latter can take over. One interesting example, obtained in yeast, showed the feasibility of fitness advantage of enhanced noise under stressful conditions. In this experiment a mutation leading to increase in cell-to-cell variability in gene expression was found to be beneficial after an acute change in environmental conditions (31). In future, it would be relevant to see if spontaneous increased noise can evolve and be selected for in the lab when microorganisms adapt to stressful conditions.

If noise introduces diversity in a population, why is the need for genetic-based diversity, in the form of enhanced mutation? Noise has one obvious limitation: cells have a very short memory for noisy fluctuations; hence a stochastic increase or decrease in expression level of a gene may not be faithfully inherited to daughter cells. Since a cell that expresses a given gene at a high level is genetically identical to the rest of the cells, its descendents are expected to return to an averaged expression level in the coming generations. How long does it take the progeny to "forget" this legacy? A study in a mammalian system (44) suggested that for many genes it is one generation time.

In summary, one potential model would suggest that following an environmental change, stochasticity in gene expression may begin to diversify the population with respect to particular genes. Such diversity may provide the substrate for the selection of cells during the initial phase of coping with the stress. In parallel, as mutations begin to appear at a slower pace, they may allow sustained diversity, from which the fittest cells will become fixated in the population. Together, the two mechanisms may provide diversity and a substrate for selection at both short and long timescales.

Acknowledgments

The author thanks the European Research Council (REC) for grant support. The author also thanks Barbara Morgenstern for editorial help with the chapter.

References

1. Nagalakshmi, U., Wang, Z., Waern, K., et al. (2008) The transcriptional landscape of the yeast genome defined by RNA sequencing. *Science* **320**, 1344–1349.
2. Shalem, O., Dahan, O., Levo, M., et al. (2008) Transient transcriptional responses to stress are generated by opposing effects of mRNA production and degradation. *Mol. Syst. Biol.* **4**, 223.
3. Ingolia, N. T., Ghaemmaghami, S., Newman, J. R., and Weissman, J. S. (2009) Genome-wide analysis in vivo of translation with nucleotide resolution using ribosome profiling. *Science* **324**, 218–223.
4. Li, J. B., Levanon, E. Y., Yoon, J. K., et al. (2009) Genome-wide identification of human RNA editing sites by parallel DNA capturing and sequencing. *Science* **324**, 1210–1213.
5. Balaban, N. Q., Merrin, J., Chait, R., Kowalik, L., and Leibler, S. (2004) Bacterial persistence as a phenotypic switch. *Science* **305**, 1622–1625.
6. Nachman, I., Regev, A., and Ramanathan, S. (2007) Dissecting timing variability in yeast meiosis. *Cell* **131**, 544–556.
7. Feinerman, O., Veiga, J., Dorfman, J. R., Germain, R. N., and Altan-Bonnet, G. (2008) Variability and robustness in T cell activation from regulated heterogeneity in protein levels. *Science* **321**, 1081–1084.
8. Cohen, A. A., Geva-Zatorsky, N., Eden, E., et al. (2008) Dynamic proteomics of individual cancer cells in response to a drug. *Science* **322**, 1511–1516.
9. Raser, J. M., and O'Shea, E. K. (2005) Noise in gene expression: origins, consequences, and control. *Science* **309**, 2010–2013.
10. Thattai, M., and van Oudenaarden, A. (2001) Intrinsic noise in gene regulatory networks. *Proc. Natl. Acad. Sci. USA* **98**, 8614–8619.
11. Ozbudak, E. M., Thattai, M., Kurtser, I., Grossman, A. D., and van Oudenaarden A. (2002) Regulation of noise in the expression of a single gene. *Nat. Genet.* **31**, 69–73.
12. Blake, W. J., Kaern, M., Cantor, C. R., and Collins, J. J. (2003) Noise in eukaryotic gene expression. *Nature* **422**, 633–637.
13. Sigal, A., Milo, R., Cohen, A., et al. (2006) Dynamic proteomics in individual human cells uncovers widespread cell-cycle dependence of nuclear proteins. *Nat. Methods* **3**, 525–531.
14. Paulsson, J. (2004) Summing up the noise in gene networks. *Nature* **427**, 415–418.
15. Cai, L., Friedman, N., and Xie, X. S. (2006) Stochastic protein expression in individual cells at the single molecule level. *Nature* **440**, 358–362.
16. Rosenberger, R. F., and Hilton, J. (1983) The frequency of transcriptional and translational errors at nonsense codons in the lacZ gene of *Escherichia coli. Mol. Gen. Genet.* **191**, 207–212.
17. Gordon, A. J, Halliday, J. A., Blankschien, M. D., Burns, P. A, Yatagai, F., and Herman, C. (2009) Transcriptional infidelity promotes heritable phenotypic change in a bistable gene network. *PLoS Biol.* **24**, e44.
18. Elowitz, M. B., Levine, A. J., Siggia, E. D., and Swain, P. S. (2002) Stochastic gene expression in a single cell. *Science* **297**, 1183–1186.
19. Raser, J. M., and O'Shea, E. K. (2004) Control of stochasticity in eukaryotic gene expression. *Science* **304**, 1811–1814.
20. Bar-Even, A., Paulsson, J., Maheshri, N., et al. (2006) Noise in protein expression scales with natural protein abundance. *Nat. Genet.* **38**, 636–643.
21. Newman, J. R., Ghaemmaghami, S., Ihmels, J., et al. (2006) Single-cell proteomic analysis of *S. cerevisiae* reveals the architecture of biological noise. *Nature* **441**, 840–846.
22. Huh, W. K., Falvo, J. V., Gerke, L. C., et al. (2003) Global analysis of protein localization in budding yeast. *Nature* **425**, 686–691.
23. Kussell, E., and Leibler, S. (2005) Phenotypic diversity, population growth, and

information in fluctuating environments. *Science* **309**, 2075–2078.
24. Pedraza, J. M., and van Oudenaarden, A. (2005) Noise propagation in gene networks. *Science* **307**, 1965–1969.
25. Wang, Y., Liu, C. L., Storey, J. D., Tibshirani, R. J., Herschlag, D., and Brown, P. O. (2002) Precision and functional specificity in mRNA decay. *Proc. Natl. Acad. Sci. USA* **99**, 5860–5865.
26. Belle, A., Tanay, A., Bitincka, L., Shamir, R., and O'Shea, E. K. (2006) Quantification of protein half-lives in the budding yeast proteome. *Proc. Natl. Acad. Sci. USA* **103**, 13004–13009.
27. Fraser, H. B., Hirsh, A. E., Giaever, G., Kumm, J., and Eisen, M. B. (2004) Noise minimization in eukaryotic gene expression. *PLoS Biol.* **2**, e137.
28. dos Reis, M., Savva, R., and Wernisch, L. (2004) Solving the riddle of codon usage preferences: a test for translational selection. *Nucleic Acids Res.* **32**, 5036–5044.
29. Gasch, A. P., Spellman, P. T., Kao, C. M., et al. (2000) Genomic expression programs in the response of yeast cells to environmental changes. *Mol. Biol. Cell.* **11**, 4241–4257.
30. Rodríguez Martínez, M., Soriano, J., Tlusty, T., Pilpel, Y., and Furman, I. (2010) Messenger RNA fluctuations and regulatory RNAs shape the dynamics of a negative feedback loop. *Phys. Rev. E. Stat. Nonlin. Soft Matter Phys.* **81**, 031924.
31. Murphy, K. F., Balázsi, G., and Collins, J. J. (2007) Combinatorial promoter design for engineering noisy gene expression. *Proc. Natl. Acad. Sci. USA* **104**, 12726–12731.
32. Blake, W. J., Balázsi, G., Kohanski, M. A., et al. (2006). Phenotypic consequences of promoter-mediated transcriptional noise. *Mol. Cell* **24**, 853–865.
33. Segal, E., and Widom, J. (2009) What controls nucleosome positions? *Trends Genet.* **25**, 335–343.
34. Kim, H. D., and O'Shea, E. K. (2008) A quantitative model of transcription factor-activated gene expression. *Nat. Struct. Mol. Biol.* **15**, 1192–1198.
35. Hornung, G., and Barkai, N. (2008) Noise propagation and signaling sensitivity in biological networks: a role for positive feedback. *PLoS Comput. Biol.* **4**, e8
36. Çağatay, T., Turcotte, M., Elowitz, M. B., Garcia-Ojalvo, J., and Süel GM. (2009) Architecture-dependent noise discriminates functionally analogous differentiation circuits. *Cell* **139**, 512–522.
37. Kafri, R., Levy, M., and Pilpel, Y. (2006) The regulatory utilization of genetic redundancy through responsive backup circuits. *Proc. Natl. Acad. Sci. USA* **103**, 11653–11658.
38. Kafri, R., Springer, M., and Pilpel, Y. (2009) Genetic redundancy: new tricks for old genes. *Cell* **136**, 389–392.
39. Hendrickson, D. G., Hogan, D. J., McCullough, H. L., et al. (2009) Concordant regulation of translation and mRNA abundance for hundreds of targets of a human microRNA. *PLoS Biol.* **7**, e1000238.
40. Rinaudo, K., Bleris, L., Maddamsetti, R., Subramanian, S., Weiss, R., and Benenson, Y. (2007) A universal RNAi-based logic evaluator that operates in mammalian cells. *Nat. Biotechnol.* **25**, 795–801.
41. Sniegowski, P. D., Gerrish, P. J., and Lenski, R. E. (1997) Evolution of high mutation rates in experimental populations of *E. coli*. *Nature* **387**, 703–705.
42. Koonin, E. V., and Wolf, Y. I. (2009) Is evolution Darwinian or/and Lamarckian? *Biol. Direct.* **4**, 42.
43. Acar, M., Mettetal, J. T., and van Oudenaarden, A. (2008) Stochastic switching as a survival strategy in fluctuating environments. *Nat. Genet.* **40**, 471–475.
44. Sigal, A., Milo, R., Cohen, A., et al. (2006) Variability and memory of protein levels in human cells. *Nature* **444**, 643–646.

Chapter 24

Genome-Scale Integrative Data Analysis and Modeling of Dynamic Processes in Yeast

Jean-Marc Schwartz and Claire Gaugain

Abstract

Building a dynamic model of a complete biological cell is one of the great challenges of the 21st century. While this objective could appear unrealistic until recently, considerable improvements in high-throughput data collection techniques, computational performance, data integration, and modeling approaches now allow us to consider it within reach in the near future. In this chapter, we review recent developments that pave the way toward the construction of genome-scale dynamic models. We first describe methodologies for the integration of heterogeneous "omics" datasets, which enable the interpretation of cellular activity at the genome scale and in fluctuating conditions, providing the necessary input to models. We subsequently discuss principles of such models and describe a series of approaches that open perspectives toward the construction of genome-scale dynamic models.

Key words: Systems biology, data integration, dynamic model, modeling.

1. Introduction

To decipher the processes and mechanisms taking place in biological cells throughout a variety of conditions or disorders requires large-scale analyses at all cellular levels. This has become increasingly possible with the development of high-throughput "omics" technologies that provide rapidly expanding datasets at the (epi)genome, transcriptome, proteome, metabolome, and fluxome (study of internal fluxes) levels. Despite this wealth of data, large dynamic models of biological systems remain difficult to construct, and successful examples are still scarce. Many biological, mathematical, and computational challenges remain to be solved for the construction of genome-scale dynamic models

J.I. Castrillo, S.G. Oliver (eds.), *Yeast Systems Biology*, Methods in Molecular Biology 759,
DOI 10.1007/978-1-61779-173-4_24, © Springer Science+Business Media, LLC 2011

of biological cells. We are still missing many of the parameters needed, and, perhaps more importantly, the right conceptual framework for such models must be established.

To achieve such global understanding of cellular functions we need to associate methods from several fields, including biology, computer science, physics, mathematics, and chemistry. This is the aim of systems biology. An important precondition to the construction of large integrative models is the development of methodologies and tools for high-throughput data integration to depict the relationships between different levels of cellular processes and components. In this chapter, we review and discuss a number of recent developments paving the way toward large-scale integrative analysis and modeling of dynamic processes in biological cells. We start with a description of methodologies enabling the integration of genome-scale and heterogeneous "omics" datasets (**Section 2**), before presenting approaches leading toward the construction of genome-scale dynamic models of cellular processes (**Section 3**).

2. Data Integration and Integrative Modeling Approaches

High-throughput technologies enable the acquisition of the genome-scale data for different cellular components (RNA transcripts, proteins, metabolites, etc.). It is a difficult task to analyze information at this scale because of the large quantity and heterogeneity of the data. In order to understand cellular processes, it is necessary to combine this data. Their integration should rely on hypotheses about cellular mechanisms and serve as a preparatory step for modeling.

Data integration usually relies on simple assumptions, e.g., proteins that interact have a higher chance to be co-expressed or to participate in the same metabolic function or co-expressed genes have a higher chance to be involved in the same metabolic pathway than randomly selected genes. From a good experimental design directed to the generation of high-quality data, we should be able to substantiate these hypotheses using integrative strategies supported by statistical models. Then, data integration methodologies allow the interpretation of the results in order to elucidate biological processes and mechanisms under the conditions under investigation.

Many studies have been carried out to interpret gene expression data in a metabolic context. These studies are driven by an implicit hypothesis, which is that co-expressed genes in a specific condition should code for proteins/enzymes that are involved in specific metabolic pathways linked to the particular cellular response. If a significant correlation is observed between co-expressed genes and some specific metabolic pathways, then this information helps to understand the metabolic functions that are activated or repressed by the cell in response to the specific conditions studied (we can talk of "pathway analysis"). This type of work may be performed with any "omics" data, but the most important developments have been reported with gene expression (RNA and protein expression) data, which we describe below.

2.1. Integration of Gene Expression Data and Metabolism

There are two main difficulties to take into account in an integrative approach. First, any analysis is dependent on the quality of the datasets used (obtained from experiments under equivalent/comparable conditions) and the way they are processed before being integrated. Second, the results obtained also depend on the "metabolic model" used. In a genome-scale study, all metabolic reactions occurring in the cell should be considered as a whole. They constitute a highly complex network, which is why it is often necessary to decompose it into smaller modules (1). Many works integrating gene expression data and metabolism consider the metabolic network as a set of pathways (2–5), for example, as defined in the Kyoto Encyclopedia of Genes and Genomes (KEGG) database (6). But this representation in pathways constitutes a static and arbitrary decomposition of the metabolism of a cell. If we want to analyze data more accurately and answer biological questions, we need a representation of metabolism that is more precise, flexible, and able to reflect the dynamic fluctuations of cellular functions.

In recent years, new strategies have been proposed to exploit the metabolic network taking into account its topology and/or overcoming the problem of the decomposition in pathways. Draghici et al. (7) developed a statistical approach for pathway analysis that incorporates parameters such as the magnitude of gene expression changes, the position of these genes in the given pathways, and the topology of metabolic pathways. For example, with a dataset containing genes associated with a better survival in lung cancer, they identified the cell cycle, focal adhesion, and Wnt signaling as the most perturbed pathways. These results were already observed in experimental researches on cancer and may lead to the identification of potential drug targets. Antonov et al. (8) have developed the KEGG Spider tool that uses a global metabolic network integrating all KEGG metabolic pathways, together with a statistical treatment. Many datasets reporting variations of gene expression levels in several diseases were analyzed to extract the main metabolic pathways affected. The results gave relevant clues of metabolic changes due to diseases and thus provided insights into possible treatment. Nacher et al. (9) used a correlation-based approach to identify metabolic sub-networks associated with expression data without being constrained by pre-defined pathway boundaries, and Goffard et al. (10) proposed an "Entire Pathway" approach in PathExpress to serve a similar purpose.

We have proposed a new way to exploit the metabolic network based on the computation of elementary modes (11). Elementary modes, and the very similar concept of extreme pathways, are minimal sets of reactions that can operate in steady state in a metabolic network (12–14). The computation of elementary modes at the genome scale is seriously hampered by the problem of combinatorial explosion (functions growing very rapidly at the combinatorial level). We have developed an alternative approach that consists in (1) computing elementary modes in each metabolic map of the KEGG database and (2) assembling elementary modes that have a metabolite in common and belong to different pathways in order to overcome the constraint of pathway boundaries. We used the integration tool Blastsets (15) to map sets of up- and downregulated genes onto elementary modes in various stress conditions in the yeast *Saccharomyces cerevisiae*. We showed that a set of elementary modes that constitute a metabolic response can be linked to activated or repressed genes in specific conditions (**Fig. 24.1**). As an example, among the four elementary modes activated in response to cadmium exposure, three have cysteine as their final product. Vido et al. (16) reported that cadmium exposure increases the synthesis of cysteine and

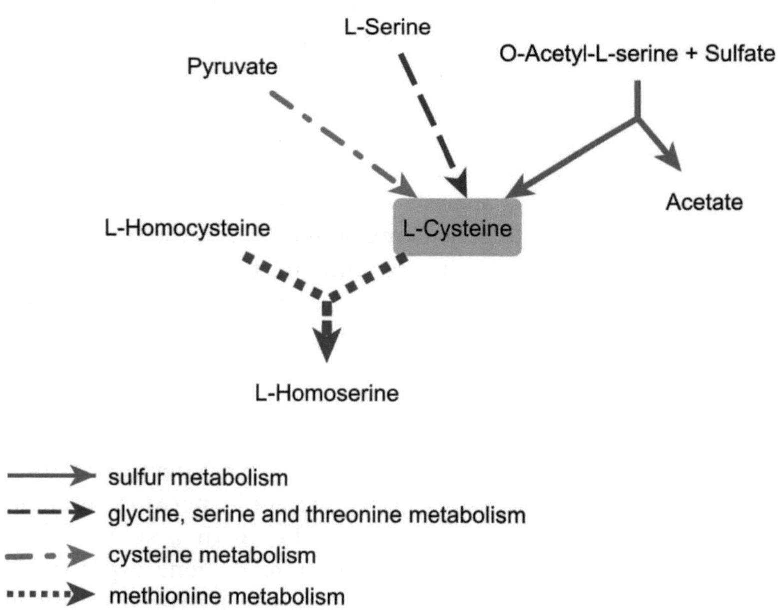

Fig. 24.1. Set of elementary modes activated in yeast cells during cadmium exposure. Each type of *arrow* represents a different metabolic pathway. The metabolites at each extremity of the *arrows* correspond to the starting and ending metabolites of an elementary mode. Cysteine is produced by three different elementary modes.

perhaps of glutathione, which are essential for cellular detoxification. The synthesis of these compounds is possible through the activation of the sulfur amino acid pathway (16). This example illustrates how this data integration technique can provide new insights into metabolic activity based on gene expression data.

Such works allow the observation of modifications at the metabolic level and the observation of dynamic processes affected by a range of conditions. They enable the identification of biologically relevant metabolic responses that may be used to define treatments for disorders or identify drug targets. It is nevertheless important to note that the approaches described so far, as well as all tools allowing the mapping of gene expression data onto pathways or ontologies, rely on a simplifying assumption that gene expression at the mRNA level may be a valid indicator of the activity or flux of metabolic reactions. In fact, there are many intrinsic mechanisms that affect the relationship between mRNA and metabolic fluxes, including mRNA stability, regulatory processes, translation efficiency, post-translational modifications, protein turnover, and others. Thus, when analyzing transcriptome data, the assumption that a high level of transcription of a specific gene leads always to a high abundance of its protein product is a simplification (see below). Comprehensive high-throughput studies are showing this and call for the development of more advanced integrative strategies.

2.2. Multi-level Integration Strategies

In order to progress toward comprehensive models of cellular functions, a combination of all available biological data (e.g., genomics, transcriptomics, proteomics, metabolomics, fluxomics, physiological data, and phenotypes) is needed. Large-scale experimental analyses were recently carried out integrating transcriptomic, proteomic, and metabolomic data for different organisms, notably *S. cerevisiae* (17) and *Escherichia coli* (18). Castrillo et al. (17) analyzed growth rate-specific patterns in yeast at several "omics" levels. They observed cell growth-associated trends (upregulation and downregulation of expression of specific groups of transcripts, proteins, and metabolite levels with increasing growth rate) occurring at all transcriptome, proteome, and metabolome levels. They investigated the congruence between the variations of the transcripts and their respective protein product(s) and showed that during balance cell growth there is only a moderate correlation due to the existence of intrinsic post-transcriptional and translational mechanisms, different for each nutrient limitation condition tested. More specifically, the control of translational efficiency is one of the main mechanisms used to finely regulate protein (e.g., enzyme) levels and final metabolic activities (17).

Post-transcriptional regulation has also been examined by Beyer et al. (19). They investigated how yeast cells use transcriptional and post-transcriptional regulation to control protein levels by determining the translational efficiency, measured by ribosome density and occupancy on the mRNAs, and protein degradation rates. They demonstrated that these mechanisms must be taken into account to better interpret protein levels in relation to mRNA levels. So far, such mechanisms are not sufficiently well characterized, to enable a systematic computation of protein abundances from mRNA levels. Such studies nevertheless are beginning to give insights into the relationships and variations of levels of cellular components. Accordingly, these works confirmed the importance of taking into account the different regulatory mechanisms existing at the mRNA and protein levels. These results can be used to determine basic principles allowing the design of more accurate large-scale cellular models. In addition to post-transcriptional control, regulatory feedback mechanisms between metabolites and transcript expression exist, that need to be characterized. Some analyses based on metabolomic and transcriptomic data have been performed to identify the relationships: first, between gene expression products and metabolite levels and second, between metabolites and transcription factors. Bradley et al. (20) have searched for a correlation between transcripts and metabolite concentrations in *S. cerevisiae*. They found a strong co-regulation that varies in nature and in strength according to the experimental conditions and the type of metabolites. These correlations cannot be quantified in general, but some trends were observed for metabolites and gene products involved in certain biological processes (e.g., the tricarboxylic acid cycle and amino acid metabolism). Moreover, they identified new interactions and relationships, either direct or indirect, between specific metabolites and transcripts.

In another study, Yeang and Vingron (21) investigated the influence of the substrates of metabolic reactions on enzyme regulation focusing on the central carbon metabolism of *E. coli*. They developed a probabilistic approach to link the metabolic network to the gene regulatory network and constructed a joint network. Perturbation datasets such as those obtained from gene knockout experiments, gene overexpression, or metabolite limitations were used to infer links that constitute paths in the joint network and enable an explanation of the observed responses (gene expression and metabolic flux data). They evaluated the explanatory power of their model by comparing the predicted response to the observed response. This integrative approach enabled the identification of regulatory feedback mechanisms between metabolites and transcription factors and leads to the construction of a model connecting the regulatory and metabolic networks of carbon metabolism.

We are still a long way from being able to infer principles of cellular activity from sets of high-throughput data. However, particular mechanisms for groups of genes, transcripts (including mRNAs and non-coding RNAs such as tRNAs, small RNAs, and microRNAs), proteins, or metabolites can be inferred and used for the construction of realistic models and simulations. As more data are being generated, properly curated and collected in data repositories, it is important to further develop integrative approaches as an essential step toward comprehensive modeling of biological systems. Data integration techniques can highlight relationships between the different cellular compounds (RNAs, proteins, metabolites, etc) and interactions between them (e.g., cooperativity, allosteric regulation, covalent modifications, and catalysis regulation). As a summary, a simple scheme is presented in **Fig. 24.2** to illustrate the links existing between molecules and the different mechanisms involved at different levels of cell organization. All these processes must be taken into account in order to better explain and understand the changes occurring in the cell in any condition. These processes are usually difficult to study at the genome scale and many more than the ones cited herein exist. Moreover, if we want to construct usable and efficient models, adequate simplifications and proper assumptions need to be made.

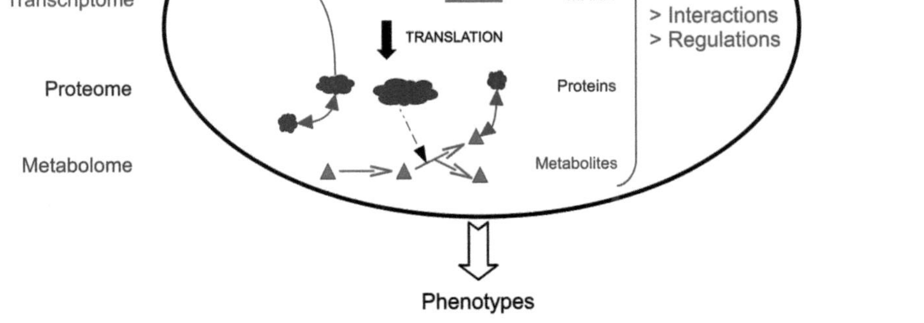

Fig. 24.2. Simplified scheme of cellular organization. Transcription enables the copy of genes into mRNAs; it is mainly regulated by proteins that bind DNA (e.g., transcription factors). The efficiency of mRNA translation into proteins is controlled by various mechanisms such as mRNA stability and ribosome density. Translation is the synthesis of proteins based on the decoding of mRNA. Enzymatic proteins catalyze metabolic reactions, which produce new molecules that can influence other cellular processes (e.g., new biochemical reactions; transcript and/or protein regulation).

3. Genome-Scale Dynamic Modeling

Even assuming that it were possible to measure the levels of all components in a biological cell with the desired accuracy and precision, we would still be far from a comprehensive understanding of how the cell functions. While data integration techniques are a necessary precondition, other techniques are needed to build informative and predictive models of cellular functions. Before looking at some of the approaches that may help us toward the construction of genome-scale dynamic cellular models, it is important to keep in mind some of the principles and point of the process of modeling itself. A model is not meant to be a totally accurate and complete representation of a real system, but an *oriented* representation that embeds our understanding of certain mechanisms which we believe are important to explain certain phenotypes. A model is a simplified representation of reality, designed to help us cope with the complexity of the real object. There is no such thing as an *exact* model; the only conceivable exact model would be a duplicate of the object itself. Therefore, there is no unique way to build a genome-scale dynamic model of a biological cell, but we should expect multiple solutions to appear in the future, focusing on different biological processes, having different levels of precision, and serving different purposes. The success of a model is not necessarily linked to its complexity and completeness; having a model that duplicates the complexity of the real system does not necessarily help to understand it. In fact, *minimal* models built from sets of principles may enable us to better comprehend the organization, dynamics, and regulation of cellular functions.

Cellular metabolism is probably the area where most progress has been made in the construction of large-scale models so far, with *S. cerevisiae* and *E. coli* being the most frequently used organisms in such works. There are many examples of static genome-scale models of biological cells, which are either purely topological (metabolic network reconstructions and stoichiometric models) or based on an assumption of steady-state fluxes (flux balance analysis, elementary modes, and flux minimization). These modeling approaches are not the object of this chapter and are properly covered in other chapters of this volume; instead, the following sections provide an overview of approaches that go beyond static representations and pave the way toward the construction of genome-scale dynamic models.

3.1. Structural and Logical Modeling

A logical model constitutes the first step from a topological to a dynamic model. These models use qualitative information about the structure of cellular networks and the nature of interactions between components to infer dynamic properties of the system. A great advantage of these approaches is that detailed kinetic parameters of reactions are not required. In a logical model, the state of a molecular compound is represented either by a Boolean value (e.g., present/absent and active/inactive) or by a discrete set of values (e.g., increasing/decreasing/constant). Qualitative rules are used to determine the change in the state of a compound as a result of its interactions with other compounds (e.g., A activates B). Logical rules may be used to represent the combined effect of different interactions (e.g., "AND" indicates that the presence of two compounds is required for a reaction to proceed). Given an initial state for all compounds, the model then enables the computation of the states reached by all compounds in the system over time, thus providing an approximation of the dynamics of the system.

While a logical model may a priori look like a very coarse approximation of a real biological system, many properties of dynamic systems are in fact strongly constrained by network topology. Logical models are therefore able to provide valuable information in a number of cases (22–24). Whelan and King (25) (*see* also **Chapter 26**, this volume) have presented what is probably the first genome-scale logical model of yeast metabolism. Their model was based on an earlier genome-scale stoichiometric network (26) augmented by data from the KEGG database (6). It was aimed at the determination of minimal medium and essential gene requirements for cellular growth. The fundamental principle used to predict the outcome of a growth experiment was based on binary associations: if a path exists from the input metabolites present in the medium to each of a set of essential metabolites for growth, then the model predicts that the yeast will grow indistinguishably from the wild type. Different medium compositions or environmental conditions were simulated by varying the list of input metabolites, and phenotypes of gene knockout mutants were simulated by removing the reaction in question from the model. The output of the model was therefore of a binary nature as well: if all essential compounds were present the model predicted continued growth, while if any essential compound was missing then arrested growth was predicted.

Two closely related formalisms have been reported to model signaling and transcriptional regulatory networks using a logical approach. Klamt et al. (27) presented a framework based on logical interaction hypergraphs, and Gianchandani et al. (28) presented a matrix-based formalism to represent logical interaction rules allowing the functional analysis of such networks. The fundamental difference between a graph and a hypergraph network representation is that the latter allows the modeling of interactions involving more than one substrate and one product, which is essential in most biological examples. For example, a metabolic

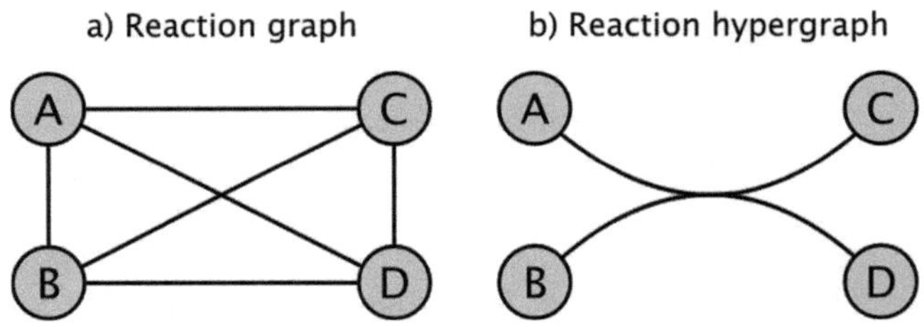

Fig. 24.3. Interaction graphs and hypergraphs. (**a**) In an interaction graph, substrates and products of a reaction A+B ↔ C+D can be connected in multiple ways and the logical dependency between compounds is not represented. (**b**) In an interaction hypergraph, a unique hyperarc connects all substrates and products, embedding the logical dependency between them.

reaction transforming two substrates A and B into two products C and D is only possible when both A and B are present, and it produces C and D in equal amounts. A graph representation is not able to capture the logical "AND" relationship (**Fig. 24.3a**), but a hypergraph embeds such a relationship by defining a unique hyperarc that links both A and B to both C and D (**Fig. 24.3b**). An "OR" relationship can be easily represented by using multiple hyperarcs connected to a common node. Such logical operators enable the modeling of complex logical interaction rules between components of metabolic or signaling networks.

An important application of logical interaction hypergraphs is the enumeration of all possible steady states of a system, that is, given a set of external input signals reaching the cell, what is the logical pattern generated by signaling and regulatory pathways and does it lead to the activation or inhibition of certain molecules? Saez-Rodriguez et al. (29) presented a logical hypergraph model of T-cell activation and were able to successfully predict unexpected signaling events after perturbation of a receptor, which were subsequently experimentally validated. Gianchandani et al. (30) used their matrix-based formalism to enumerate possible transcription states in a genome-scale model of *E. coli* and to evaluate how transcription states were affected by the availability of nutrients in the cell's environment. Both formalisms are in effect closely related, as a hypergraph can be represented by a matrix. These concepts and the methodology used for the enumeration of regulatory states were derived from elementary mode and extreme pathway analysis, which are well established in metabolic pathways (13, 14, 31).

Flux balance analysis (FBA) has been one of the most successful and widely used methods for genome-scale networks so far (32). While a detailed description of FBA is outside the scope of this chapter (as it is not a dynamic method), it is worth recalling some of its principles and mention attempts to bridge the gap between FBA and dynamic modeling. FBA relies on the topology and stoichiometry of a metabolic network, associated with the definition of an objective function, to calculate a steady-state flux distribution. The objective function reflects a hypothesis of physiological optimization of the organism, e.g., the organism is assumed to maximize the production of biomass, energy, or any other relevant product. FBA does not need kinetic parameters and does not produce any inference about metabolite concentrations, but only enables the computation of steady-state fluxes. Integrated dynamic flux balance analysis (idFBA) (33) is an extension of FBA based on the assumption that the turnover time of metabolic and signaling reactions is generally fast compared to regulatory or receptor responses. The idFBA framework proceeds by assuming quasi-steady-state conditions for fast reactions, while slow phenotypic reactions are integrated into the stoichiometric framework over time. Modeling of slow reactions is not based on detailed kinetic parameters, but only on their typical duration and delay time. Lee et al. (33) illustrated this approach using a prototypic model of *S. cerevisiae* incorporating signaling, metabolism, and transcriptional regulatory networks. The signaling network included a set of ligands binding to receptors and the subsequent internalization and phosphorylation of their complexes. The metabolic network included a set of reactions representative of glycolysis and amino acid synthesis. Regulation was modeled by a set of logical rules describing the effect of environmental factors, such as extracellular metabolites and pH values. While idFBA only provides an approximation of the dynamics of the system between different steady states, it is a fast and efficient method to model integrated systems of metabolic, signaling, and regulatory networks, which may potentially be applied to large systems.

3.2. Intermediate Modeling Approaches

Another set of methods have been presented that attempt to go beyond structural modeling and open possibilities for the construction of large-scale dynamic models. The construction of accurate kinetic models is hampered by the need to know the exact form of all rate equations and associated parameter values. Such detailed information is not generally available for all reactions and requires extensive experimental measurements. Furthermore, parameter values may vary depending on many environmental and physiological factors.

For these reasons, it is difficult to imagine building a kinetic model of an entire cell without some form of simplification and generalization of rate equations. In the structural kinetic method proposed by Steuer et al. (34), kinetic rate functions were approximated by constructing a local linear model at each point in parameter space. This local representation was embedded by a Jacobian matrix. It did not depend on the exact form of kinetic rate functions, and all terms used in the representation had a clear biochemical interpretation or were directly experimentally accessible (e.g., a flux, concentration, or degree of saturation). This approach is particularly useful to assess the stability of a physiological state in a range of parameter values or to determine bifurcation points between different types of dynamic behaviors. As an example, Steuer et al. were able to determine the required conditions for a switch between steady-state conditions and sustained oscillations in yeast glycolysis without the need to refer to the exact form of rate equations. Because it allows the identification of reaction steps and parameters that are crucial in shaping the dynamic behavior of a system, structural kinetic modeling is a useful preliminary step to a more detailed kinetic model.

A different approach was proposed by Smallbone et al. (35), who used linlog kinetics to approximate the dynamics of enzymatic reactions. In the linlog approximation, the effect of metabolite levels on flux was described by a linear sum of logarithmic terms. A generic kinetic relation was thus assumed for all enzymes and no detailed data about kinetic parameters and reaction mechanisms were needed. Nonetheless, an approximate assessment of the dynamics of fluxes and metabolite concentrations was made possible. When compared to a precise mechanistic model of yeast glycolysis (36), the linlog model was able to provide good agreement with the detailed model. Some concerns, however, are that the linlog approximation may only be reasonable over a certain range of parameter values, with the gap between linlog kinetics and real kinetics increasing rapidly outside this range. Comparisons with other detailed models and/or over wider ranges of parameter values would be needed to better assess the reliability of this approximation.

A general idea behind such intermediate approaches is to apply a generic kinetic equation to all reactions. This idea offers several advantages: (1) a simplified kinetic equation reduces the computational burden and accelerates simulations, allowing the modeling of larger systems; (2) kinetic rate equations depend on the exact enzymatic binding mechanism, which can be complex or unknown for some reactions; (3) complex reaction mechanisms result in a profusion of parameters that are not experimentally accessible. To be appropriate, a generic equation should reflect the essential mechanistic principles of the majority of reactions, but not incorporate excessive level of detail and parameters that are not experimentally tractable. It is useful in this context to think of analogies with physics and engineering, where such approaches are commonly used; the art of creating a useful model is to make the *right* approximations. A successful illustration of this principle has been reported by Ao et al. (37), who constructed a relatively large model (80 reactions and 80 metabolites) of the central metabolism of *Methylobacterium extorquens* using generic kinetic equations. The parameters used in their equations had a clear experimental interpretation, and the authors described methods to estimate biologically meaningful values of parameters when experimental data were not available. The authors showed that their generic model successfully captured the behavior of the bacterium's metabolism in steady state, as well as in response to a perturbation in formaldehyde concentration. The principles used in this work may serve as a basis toward genome-scale models based on generic kinetic equations.

3.3. Thermodynamics

The behavior of all biochemical systems obeys universal physical laws. One of them is the conservation of mass and matter, which can be translated into mass balance and flux conservation relations in the case of metabolic networks. Another universal law is the conservation of energy, which can be translated into thermodynamic laws. Incorporating thermodynamic laws into biochemical models, however, is not as straightforward as incorporating mass balance. Thermodynamic principles are therefore often ignored in biological modeling, which can result in models that are fundamentally wrong, violating fundamental natural laws. This issue is particularly crucial when a system involves cycles (as do most biological systems), because cycles may easily violate thermodynamic laws if no special care is taken.

Several authors have reported methods enabling the importation of thermodynamic principles into biochemical models. Any large-scale model should incorporate these principles to be meaningful. As explained by Ederer and Gilles (38), the second law of thermodynamics implies that an isobaric and isothermal system (e.g., cells) that is not exposed to external thermodynamic forces will reach a state of thermodynamic equilibrium. This condition introduces new relations of interdependency between the kinetic parameters of the model, restricting the range of possible parameter values. This principle was illustrated by a simple example describing the random order complexation of three compounds: in this system involving eight kinetic parameters, only seven parameters could be fixed independently of each other. When these parameters were fixed, the value of the eighth parameter was constrained by a relation of "detailed balance" arising from thermodynamic laws. If this condition was ignored, the model described a physically impossible system. Ederer and Gilles introduced a new framework for kinetic modeling that embeds detailed balance, preventing the creation of thermodynamically unfeasible models. Called "thermodynamic kinetic modeling," this framework may become valuable to construct large models that do not violate physical laws. The authors furthermore demonstrated that their formalism reduces the number of equations and parameters needed, as thermodynamic constraints are directly embedded in equations and no longer have to be separately written.

It is worth noting that thermodynamic laws must be satisfied whether the system under consideration actually reaches thermodynamic equilibrium or not. Living systems are generally not isolated from their environment but are open to exchanges of matter and energy. They rarely reach thermodynamic equilibrium and are better represented by a non-equilibrium steady state, where fluxes are stationary over time but not zero. The basic principles of non-equilibrium steady-state systems are closely related to those in equilibrium. Generalized frameworks representing the thermodynamics of biochemical systems far from equilibrium are still under development (39).

In parallel, several authors have presented methods to integrate thermodynamic constraints into FBA models, as it was found that classical FBA may result in solutions that violate thermodynamic laws. These laws impose that the direction of flux in a reaction must correspond to the direction of decreasing Gibb's free energy (40). Actual changes in free energy can be calculated from the standard free energy (which is related to the equilibrium constant of the reaction and the temperature) and metabolite concentrations. The resulting constraints on the direction of reactions are then enforced by a new set of linear constraints in the FBA framework. Hoppe et al. (40) showed that these new constraints yielded different flux distributions than classical FBA under some ranges of metabolite concentrations. Henry et al. (41) reported a systematic calculation of the Gibb's free energy of reactions in a genome-scale metabolic model of *E. coli.* They subsequently identified thermodynamically unfavorable reactions and ranges of thermodynamically feasible metabolite concentrations.

3.4. Toward Genome-Scale Dynamic Models

For a long time, the perspective of building a dynamic model of an entire biological cell was considered an unrealistic goal, not to be achieved before several decades. But recent progress in data collection techniques, computational performance, and methodologies makes this objective look more and more achievable (**Fig. 24.4**). A crucial question to keep in mind is what the purpose of the model should be. The degree of precision and the nature of processes integrated in a model must be tailored to the purpose the model is expected to serve. Depending on the level of approximation and its purpose, it is not unrealistic to expect dynamic genome-scale models to be developed within a few years.

Jamshidi and Palsson (42) described a workflow that may lead to the construction of a genome-scale dynamic model of cellular metabolism. The process starts with the construction of dynamic mass balance equations, which relate the change in metabolite concentrations to reaction fluxes by use of the stoichiometric matrix of the network. The next step is the representation of reaction kinetic functions, which describe how reaction rates depend on metabolite concentrations and kinetic parameters. These

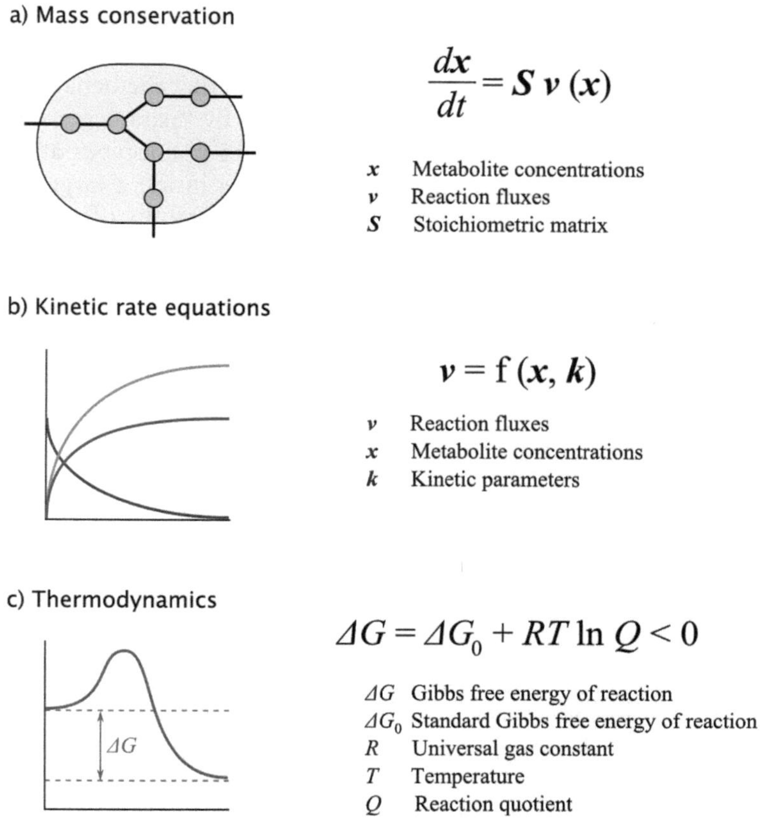

Fig. 24.4. Three types of laws must be considered in order to build a dynamic model. (**a**) Mass conservation laws link variations in metabolite concentrations to the fluxes of reactions producing and consuming them. (**b**) Kinetic rate equations describe the dependency of reaction rates on metabolite concentrations and kinetic constants (these may be dependent on the concentration of "active" protein, controlled by post-transcriptional and/or post-translational mechanisms: regulatory RNAs, covalent modifications, e.g., phosphorylation and others). (**c**) Thermodynamic laws impose that reactions proceed in the direction of decreasing Gibb's free energy.

relations are generally complex; thus to construct large-scale models a simplified representation is likely to be needed. The workflow proposed by Jamshidi and Palsson (42) relies on a linearization of reaction rates, represented by a gradient matrix that embeds kinetic and thermodynamic properties. This representation has the advantage of offering a duality between concentrations and fluxes, where either can be used as independent variables for simulation.

Important obstacles remain in the completion of such a workflow, the most often cited being the difficulty to acquire sufficient experimental data for kinetic parameter values. Nevertheless, the significance of parameter values in biological models, while important, should not be overestimated. By the analysis of elementary mode fluxes in a model of yeast glycolysis, we observed that the main features of the flux distribution were robust to large changes in parameter variations (31). The investigation of a wider range of examples confirmed that the behavior of many biochemical systems is characterized by "parameter sloppiness" (43, 44), which means that their main phenotypes are insensitive to changes in most parameter values inside a large space of variations, except for a few "stiff" combinations of parameters. Such observations suggest that biological properties may depend more fundamentally on design principles, interaction mechanisms, and scaling laws than on the fine-tuning of individual parameters. The predictive and informative value of a model may not necessarily be dependent on a high level of precision in parameter values.

References

1. Papin, J., Reed, J., and Palsson, B. Ø. (2004) Hierarchical thinking in network biology: the unbiased modularization of biochemical networks. *Trends Biochem. Sci.* **29**, 641–647.
2. Ghazalpour, A., Doss, S., Sheth, S. S., et al. (2005) Genomic analysis of metabolic pathway gene expression in mice. *Genome Biol.* **6**, R59.
3. Goffard, N., and Weiller, G. (2007) PathExpress: a web-based tool to identify relevant pathways in gene expression data. *Nucleic Acids Res.* **35**, W176–181.
4. Yang, H. H., Hu, Y., Buetow, K. H., and Lee, M. P. (2004) A computational approach to measuring coherence of gene expression in pathways. *Genomics* **84**, 211–217.
5. Ekins, S., Nikolsky, Y., Bugrim, A., Kirillov, E., and Nikolskaya, T. (2007) Pathway mapping tools for analysis of high content data. *Methods Mol. Biol.* **356**, 319–350.
6. Kanehisa, M., and Goto, S. (2000) KEGG: Kyoto encyclopedia of genes and genomes. *Nucleic Acids Res.* **28**, 27–30.
7. Draghici, S., Khatri, P., Tarca, A. L., et al. (2007) A systems biology approach for pathway level analysis. *Genome Res.* **17**, 1537–1545.
8. Antonov, A., Dietmann, S., and Mewes, H. (2008) KEGG Spider: interpretation of genomics data in the context of the global gene metabolic network. *Genome Biol.* **9**, R179.
9. Nacher, J. C., Schwartz, J. M., Kanehisa, M., and Akutsu, T. (2006) Identification of metabolic units induced by environmental signals. *Bioinformatics* **14**, e375–383.
10. Goffard, N., Frickey, T., and Weiller, G. (2009) PathExpress update: the enzyme neighbourhood method of associating gene-expression data with metabolic pathways. *Nucleic Acids Res.* **37**, 335–339.
11. Schwartz, J. M., Gaugain, C., Nacher, J. C., de Daruvar, A., and Kanehisa, M. (2007) Observing metabolic functions at the genome scale. *Genome Biol.* **8**, R123.
12. Papin, J. A., Price, N. D., Wiback, S. J., Fell, D. A., and Palsson, B. Ø. (2003) Metabolic pathways in the post-genome era. *Trends Biochem. Sci.* **28**, 250–258.
13. Schilling, C. H., Letscher, D., and Palsson, B. Ø. (2000) Theory for the systemic definition of metabolic pathways and their use in interpreting metabolic function from a pathway-oriented perspective. *J. Theor. Biol.* **203**, 229–248.
14. Schuster, S., Fell, D. A., and Dandekar, T. (2000) A general definition of metabolic pathways useful for systematic organization and analysis of complex metabolic networks. *Nat. Biotechnol.* **18**, 326–332.
15. Barriot, R., Poix, J., Groppi, A., et al. (2004) New strategy for the representation and the integration of biomolecular knowledge at a cellular scale. *Nucleic Acids Res.* **32**, 3581–3589.
16. Vido, K., Spector, D., Lagniel, G., et al. (2001) A proteome analysis of the cadmium response in *Saccharomyces cerevisiae*. *J. Biol. Chem.* **276**, 8469–8474.
17. Castrillo, J. I., Zeef, L. A., Hoyle, D. C., et al. (2007) Growth control of the eukaryote cell: a systems biology study in yeast. *J. Biol.* **6**, 4.
18. Ishii, N., Nakahigashi, K., Baba, T., et al. (2007) Multiple high-throughput analyses monitor the response of *E. coli* to perturbations. *Science* **316**, 593–597.

19. Beyer, A., Hollunder, J., Nasheuer, H. P., and Wilhelm, T. (2004) Post-transcriptional expression regulation in the yeast *Saccharomyces cerevisiae* on a genomic scale. *Mol. Cell. Proteomics* **3**, 1083–1092.
20. Bradley, P. H., Brauer, M. J., Rabinowitz, J. D., and Troyanskaya, O. G. (2009) Coordinated concentration changes of transcripts and metabolites in *Saccharomyces cerevisiae. PLoS Comput. Biol.* **5**, e1000270.
21. Yeang, C. H., and Vingron, M. (2006) A joint model of regulatory and metabolic networks. *BMC Bioinformatics* **7**, 332.
22. Sontag, E., Kiyatkin, A., and Kholodenko, B. N. (2004) Inferring dynamic architecture of cellular networks using time series of gene expression, protein and metabolite data. *Bioinformatics* **20**, 1877–1886.
23. Grimbs, S., Selbig, J., Bulik, S., Holzhütter, H. G., and Steuer, R. (2007) The stability and robustness of metabolic states: identifying stabilizing sites in metabolic networks. *Mol. Syst. Biol.* **3**, 146.
24. Conradi, C., Flockerzi, D., Raisch, J., and Stelling, J. (2007) Subnetwork analysis reveals dynamic features of complex (bio)chemical networks. *Proc. Natl. Acad. Sci. USA* **104**, 19175–19180.
25. Whelan, K. E., and King, R. D. (2008) Using a logical model to predict the growth of yeast. *BMC Bioinformatics* **9**, 97.
26. Förster, J., Famili, I., Fu, P., Palsson, B. Ø., and Nielsen, J. (2003) Genome-scale reconstruction of the *Saccharomyces cerevisiae* metabolic network. *Genome Res.* **13**, 244–253.
27. Klamt, S., Saez-Rodriguez, J., Lindquist, J. A., Simeoni, L., and Gilles, E. D. (2006) A methodology for the structural and functional analysis of signaling and regulatory networks. *BMC Bioinformatics* **7**, 56.
28. Gianchandani, E. P., Papin, J. A., Price, N. D., Joyce, A. R., and Palsson, B. Ø. (2006) Matrix formalism to describe functional states of transcriptional regulatory systems. *PLoS Comput. Biol.* **2**, e101.
29. Saez-Rodriguez, J., Simeoni, L., Lindquist, J. A., et al. (2007) A logical model provides insights into T cell receptor signaling. *PLoS Comput. Biol.* **3**, e163.
30. Gianchandani, E. P., Joyce, A. R., Palsson, B. Ø., and Papin, J. A. (2009) Functional states of the genome-scale *Escherichia coli* transcriptional regulatory system. *PLoS Comput. Biol.* **5**, e1000403.
31. Schwartz, J. M., and Kanehisa, M. (2006) Quantitative elementary mode analysis of metabolic pathways: the example of yeast glycolysis. *BMC Bioinformatics* **7**, 186.
32. Varma, A., and Palsson, B. Ø. (1994) Metabolic flux balancing: basic concepts, scientific and practical use. *Bio/Technology* **12**, 994–998.
33. Lee, J. M., Gianchandani, E. P., Eddy, J. A., and Papin, J. A. (2008) Dynamic analysis of integrated signaling, metabolic, and regulatory networks. *PLoS Comput. Biol.* **4**, e1000086.
34. Steuer, R., Gross, T., Selbig, J., and Blasius, B. (2006) Structural kinetic modeling of metabolic networks. *Proc. Natl. Acad. Sci. USA* **103**, 11868–11873.
35. Smallbone, K., Simeonidis, E., Broomhead, D. S., and Kell, D. B. (2007) Something from nothing – bridging the gap between constraint-based and kinetic modelling. *FEBS J.* **274**, 5576–5585.
36. Teusink, B., Passarge, J., Reijenga, C. A., et al. (2000) Can yeast glycolysis be understood in terms of in vitro kinetics of the constituent enzymes? Testing biochemistry. *Eur. J. Biochem.* **267**, 5313–5329.
37. Ao, P., Lee, L., Lidstrom, M., Yin, L., and Zhu, X. (2008) Towards kinetic modeling of global metabolic networks: *Methylobacterium extorquens* AM1 growth as validation. *Chin. J. Biotechnol.* **24**, 980–994.
38. Ederer, M., and Gilles, E. D. (2007) Thermodynamically feasible kinetic models of reaction networks. *Biophys. J.* **92**, 1846–1857.
39. Qian, H., and Beard, D. A. (2005) Thermodynamics of stoichiometric biochemical networks in living systems far from equilibrium. *Biophys. Chem.* **114**, 213–220.
40. Hoppe, A., Hoffmann, S., and Holzhütter, H. G. (2007) Including metabolite concentrations into flux balance analysis: thermodynamic realizability as a constraint on flux distributions in metabolic networks. *BMC Syst. Biol.* **1**, 23.
41. Henry, C. S., Broadbelt, L. J., and Hatzimanikatis, V. (2007) Thermodynamics-based metabolic flux analysis. *Biophys. J.* **92**, 1792–1805.
42. Jamshidi, N., and Palsson, B. Ø. (2008) Formulating genome-scale kinetic models in the post-genome era. *Mol. Syst. Biol.* **4**, 171.
43. Gutenkunst, R. N., Waterfall, J. J., Casey, F. P., Brown, K. S., Myers, C. R., and Sethna, J. P. (2007) Universally sloppy parameter sensitivities in systems biology models. *PLoS Comput. Biol.* **3**, 1871–1878.
44. Daniels, B. C., Chen, Y. J., Sethna, J. P., Gutenkunst, R. N., and Myers, C. R. (2008) Sloppiness, robustness, and evolvability in systems biology. *Curr. Opin. Biotechnol.* **19**, 389–395.

Chapter 25

Genome-Scale Metabolic Models of *Saccharomyces cerevisiae*

Intawat Nookaew, Roberto Olivares-Hernández, Sakarindr Bhumiratana, and Jens Nielsen

Abstract

Systematic analysis of *Saccharomyces cerevisiae* metabolic functions and pathways has been the subject of extensive studies and established in many aspects. With the reconstruction of the yeast genome-scale metabolic (GSM) network and in silico simulation of the GSM model, the nature of the underlying cellular processes can be tested and validated with the increasing metabolic knowledge. GSM models are also being exploited in fundamental research studies and industrial applications. In this chapter, the principle concepts for construction, simulation and validation of GSM models, progressive applications of the yeast GSM models, and future perspectives are described. This will support and encourage researchers who are interested in systemic analysis of yeast metabolism and systems biology.

Key words: *Saccharomyces cerevisiae*, yeast, metabolism, metabolic network reconstruction, genome-scale metabolic model (GSM), systems biology.

1. Introduction

The eukaryotic model organism *Saccharomyces cerevisiae* (budding yeast) has been used for beer, wine, and bread making since ancient times. In modern times it is also being exploited as a host for production of many compounds of industrial interest (1, 2). Yeast was proved as the main responsible for the process of fermentation by Louis Pasteur (1822–1895) and has been extensively studied at the molecular biology, biochemistry, physiology, and genetic levels. In 1996, the complete genome sequence of *S. cerevisiae* was released (first sequenced eukaryotic genome) (3, 4), and this enabled the first comprehensively study of an eukaryotic cell as a whole. Yeast shares a significant number of

J.I. Castrillo, S.G. Oliver (eds.), *Yeast Systems Biology*, Methods in Molecular Biology 759,
DOI 10.1007/978-1-61779-173-4_25, © Springer Science+Business Media, LLC 2011

cellular processes with human (e.g., more than 20% of yeast genes are human orthologs (5–7)) which makes it a good model organism for the study of human diseases and drug screening (8–10). Moreover, several high-throughput experimental platforms for generation of multilevel omics data and implementation of comprehensive databases have been established and validated in yeast (1), with methods and strategies being progressively applied to higher eukaryotes. This places *S. cerevisiae* as a reference model organism for systems biology studies of eukaryotic cells.

Metabolism constitutes a well-known, essential process of living cells. Metabolic changes play a crucial role to sustain life by adaptation to environmental changes. Together with this, changes in yeast metabolic fluxes may be exploited for the production of novel products by heterologous expression and/or metabolic engineering (e.g., polyketides (11, 12), isoprenoids (13–16), insulin (17, 18), and ethanol (19, 20)). At the clinical level, metabolic changes/imbalances are characteristic of many diseases and understanding the metabolism at a fundamental level may lead to identification of targets and drug discovery (21). The knowledge of metabolism derived from qualitative and reductionistic approaches is not sufficient to accomplish the applications mentioned above. Progressive incorporation of quantitative approaches at a holistic level is necessary. Mathematical modeling of yeast metabolism is a very good approach to quantitatively model phenotypes in various environmental conditions. Performing in silico experiments can demonstrate whether proposed molecular mechanisms are theoretically feasible and can help to guide experimental work. Once a computer model has been created and validated, it can be used to test different hypotheses and mimic experiments that are difficult (or impossible) to carry out in the laboratory. Hereby, quantitative models can help to understand the behavior of complex biological systems.

Here, we describe a conceptual framework to construct genome-scale metabolic networks, the basic mathematical formulation underlying these networks, and how they can be used for simulation of metabolism. The historic, progressive development of *S. cerevisiae* metabolic models from prototypes to genome-scale metabolic models is also presented. Finally, applications and future perspectives of yeast genome-scale metabolic models in the area of systems biology are summarized and described.

2. Genome-Scale Metabolic (GSM) Model Construction and Simulation

The biochemical information and metabolic pathways have been progressively collected from biochemical evidences in different organisms and this resulted in a complete biochemical network

2.1. Metabolic Network Reconstruction

compilation (22) that is widely used as a reference text. Nevertheless, specific metabolic networks of organisms of interest, such as *S. cerevisiae*, are needed to achieve a high-quality qualitative model that can be further used for mathematical quantitative modeling. The reconstruction of such metabolic models relies on three main levels of biological information, i.e., genomics, biochemistry, and physiology (23). Functional genomics is used mainly to identify functional annotation of open reading frames (ORFs; i.e., putative protein coding genes) encoding metabolic enzymes. Organism-specific genome databases represent valuable resources of high-quality annotations of ORFs in different genomes, for example, the *Saccharomyces* Genome Database (SGD) (24) and the Comprehensive Yeast Genome Database (CYGD) (25). Together with this, established bioinformatics approaches can be used to infer function of non-annotated ORFs from newly released genomes (26), such as Basic Local Alignment Search Tool (BLAST), Cluster of Orthologous Groups (COGs), Hidden Markov Model (HMM) of protein families and others. Metabolic databases are other useful resources for reconstruction of metabolic pathways and networks of specific organisms that have been annotated, such as the Kyoto Encyclopedia of Genes and Genomes (KEGG) (27), Reactome (28, 29), and MetaCyc (30) databases. These resources represent a good starting point for metabolic network reconstruction. However, evolution of each organism has resulted in slightly different ways to catalyze biochemical reactions or whole pathways (e.g., isoenzymes, enzyme complexes, with different substrate specificity, and/or cofactor usage). To generate a highly accurate genome-scale metabolic network, comprehensive literature surveys of evidences derived from biochemical and physiological studies, subjected to constant revision and curation, are therefore needed. For a review on comprehensive strategies for microbial metabolic network reconstruction, *see* (31).

2.2. Mathematical Formulation of a Reconstructed Metabolic Network

Once a high-quality reconstructed metabolic network has been accomplished, formulation of the network in mathematical terms is the next step. Quantitative representation of metabolism by functional integration of the reconstructed metabolic network in mathematical terms can be achieved by different strategies. Dynamic formulation based on ordinary differential equations describing the mass balances of all the metabolites and using kinetic expressions for the conversion rates has been applied to metabolism for over three decades and has led to large-scale metabolic models of red blood cells (32, 33), *Mycoplasma genitalium* (33, 34), and the hepatic lobule (35). The requirements of a large number of kinetic constants for dynamic formulation do, however, represent a major obstacle, in particular because these constants depend on environmental factors and are highly

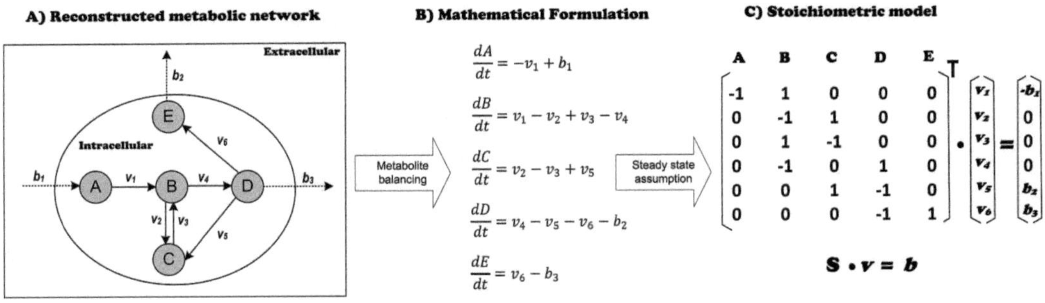

Fig. 25.1. Mathematical formulation of a simple metabolic pathway.

organism specific. To overcome this problem, static formulation of reconstructed metabolic networks has been developed. Based on metabolite balancing, the stoichiometry of reactions and thermodynamics under pseudo-steady-state assumption, a set of linear equations can be formulated. To illustrate this consider the simple metabolic network system in **Fig. 25.1a**. There are five metabolites (A, B, C, D, and E), six intracellular fluxes (v_i), and their exchange fluxes (b_i). Using metabolic balancing and pseudo-steady-state assumption, a set of linear ordinary differential equations can be expressed as illustrated in **Fig. 25.1b**. As there is a large difference in the relaxation time between enzymatic reaction and cellular growth, a pseudo-steady-state condition can be assumed for each metabolite (36). The differential equations can therefore be reduced to algebraic equations that can be represented in a matrix from $S \cdot v = b$ as illustrated in **Fig. 25.1c**.

2.3. Model Simulation

Stoichiometric models have been intensively used for quantitative understanding of metabolism. For example, metabolic flux analysis (MFA) (37), metabolic topology analysis like elementary flux mode (EFM) (38–40) or extreme pathway (EP) (38, 41–44), and flux balance analysis (FBA). Flux balance analysis, which relies on data-driven constraints and linear optimization theories, was first developed in 1994 by Vamar and Palsson (45). The approach was first applied to a GSM model of *Escherichia coli* (46) and, a few years later, to a GSM model of *S. cerevisiae* (47). The cellular constraints for FBA are classified into two major categories: adjustable and nonadjustable constraints, as summarized in **Table 25.1**. It is very important to collect a set of good constraints to bracket cellular behavior and the further consideration of metabolic adaptations during simulations. The nonadjustable constraints are inheriting properties of living cells that follow the laws of nature. On the contrary, measurable fluxes (v_m), kinetic rate constant (k), and metabolic regulations, which are obtained from experiments, are adjustable and very important for simulations because they are condition dependent. After the constraints are mathematically formulated, optimality principles (i.e., optimization of

Table 25.1
Summary of constraints used in GSM model simulations

Constraint (property)	Type[a]	Mathematical expression	Imposition of the solution space
Metabolite balancing and stoichiometry under steady-state assumption (connectivity)	HN	$S_v = 0$ (Subspace of universal space)	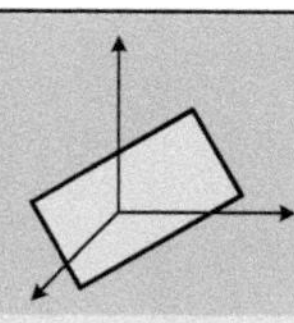
Themodynamic analysis of reversibility of biochemical fluxes (convex)	HN	$v_i \geq 0$ (convex space)	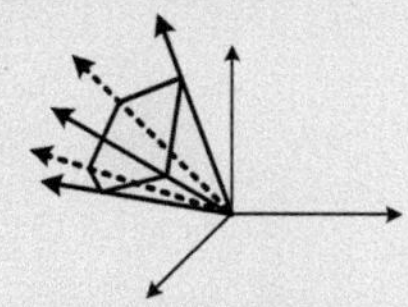
Minimum and maximum fluxes (capacity)	N/A[b]	$v_i \geq v_{i,\min}, v_i \leq v_{i,\max}$ (confined convex space)	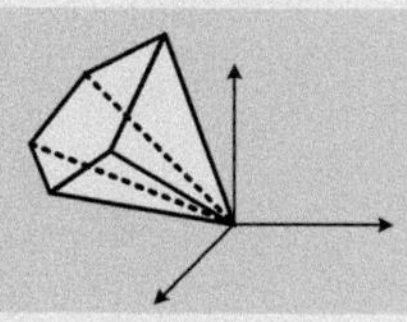
Mass action, enzyme kinetics, metabolic regulation (condition dependent)	A	$v_i = v_m$ $k_i \ll k_j$ $v_i = 0$ if $v_j \neq 0$ (operational space)	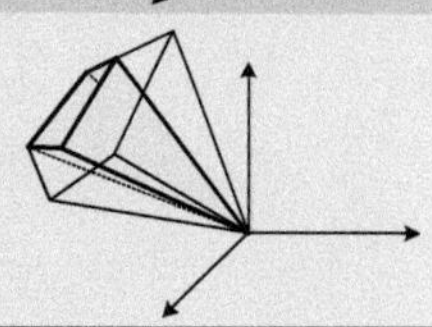

[a] *HN* = Hard nonadjustable, *N* = Nonadjustable, *A* = Adjustable. [b] Capacity property can be adjusted by cellular adaptation.

an objective function) are used, from which metabolism simulation will be performed. First, an objective function has to be defined (to determine the optimal solution from a given set of constraints). It is important to formulate a realistic objective function that can represent the physiological state of the cell in a biologically meaningful way. Different objective functions have been developed based on different biological hypotheses such as energy efficiency and production (48–51), capability to produce specific metabolic products (45), nutrient utilization (52), and the rate of biomass production, which has been reported the best objective function to describe the phenotypes of microorganisms (46, 47). Formulation of the biomass equation (i.e., how the biomass is formed from different metabolites like amino acids, fatty acids, and nucleotides) is a crucial step in order to accurately reproduce cellular phenotypes (53). To formulate the biomass equation it is necessary to take into account growth requirements such as the need for different macromolecules and free energies required for biomass synthesis as summarized in **Table 25.2**.

A few algorithms based on FBA, such as minimization of metabolic adjustment (MOMA) (54) and regulatory on/off

Table 25.2
Examples of growth requirements for formulation of the biomass equation

Growth requirements	Precursors (examples)
Macromolecules	
– Protein	Amino acids
– Carbohydrate	Trehalose, mannan, glucan, chitin
– Deoxynucleotide	dAMP, dCMP, dGMP, dTMP
– Ribonucleotide	AMP, CMP, GMP, UMP
– Lipid	Phospholipids, storage lipids, sterols, sphingolipids, fatty acids
Energy	Polymerization, maintenance, P/O ratio
Others	Intracellular metabolite pools

minimization of metabolic flux (ROOM) (55), have been implemented to improve in silico gene lethality predictions by GSM models. Comparing the prediction power of the three approaches, MOMA provides more accurate results in early state of transient growth, whereas FBA and ROOM are better to predict the outcomes after adaptation (55). Recently, incorporation of regulatory and signal transduction pathways into metabolic networks using formalisms of boolean logic together with ordinary differential equation on FBA, so-called regulatory FBA (rFBA) (56, 57) and integrated FBA (iFBA) (58, 59), has been established. These techniques have improved the capability of phenotypic prediction of FBA. Although FBA provides a good agreement between simulation results and experimental observations, it is important to realize that the fluxes calculated from FBA are based on the high degrees of freedom in the network, with not a unique solution in terms of the flux vector. For GSM simulation a MATLAB toolbox named the COBRA toolbox (60) has been developed. This toolbox is compatible with the systems biology markup language (SBML) (61, 62), and it provides a variety of useful features for GSM model simulations and sensitivity analyses.

2.4. GSM Evaluations and Examinations

Once a GSM model has been created, the first validation is to compare in silico results with various set of experimental observations, e.g., growth rates, rates of by-product secretion, substrate uptake rates, and the respiratory quotient (rate of CO_2 production divided by the rate of oxygen consumption). In *S. cerevisiae*, aerobic and anaerobic growth conditions under different nutrient limitations are good test cases because the physiological behavior is markedly different in each condition. This allows to examine the capability of the GSM model in a wide range of conditions and the necessary metabolic adaptations, with experimental data from steady-state chemostat fermentations particularly useful (47, 53) for model validation.

With the availability of the comprehensive collection of gene deletion mutants of *S. cerevisiae* (63), the comparison of in silico gene lethality predictions from GSM model against in vivo results is also a good validation test of the model (53, 64–66). Usually, in high-quality GSM models the results from in silico gene lethality predictions and in vivo results are in good agreement. However, certain amount (10–15%) of false predictions

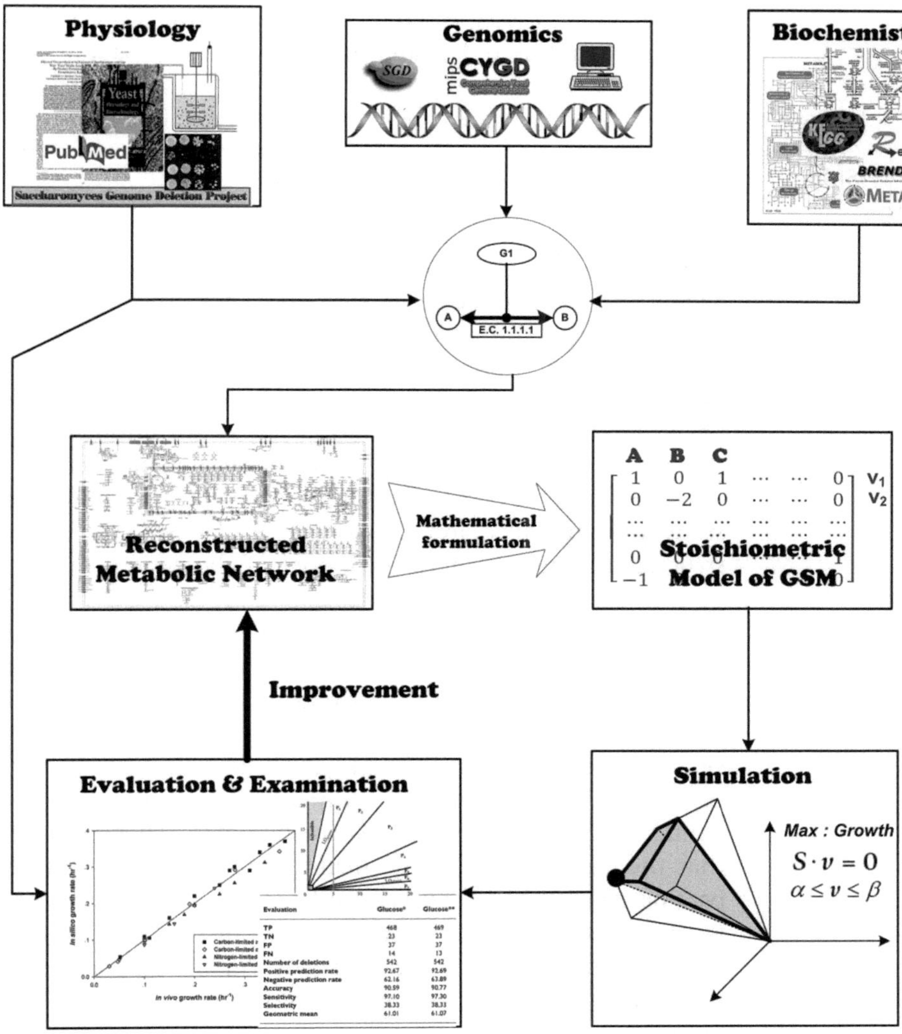

Fig. 25.2. Summary of the conceptual framework of GSM model construction and simulation. Knowledge of genomics, biochemistry, and physiology is used to collect reactions and their related information. Once a qualitative metabolic network is achieved, mathematical formulation based on metabolites balancing is applied to obtain a stoichiometric model. For simulation, constraints derived from a priori knowledge and an objective function are necessary, in order to use the FBA approach to predict phenotypes in silico. Comparisons of in vivo and in silico phenotypes lead to iterative model improvements and new hypotheses generation.

is to be expected due to missing information on, for example, regulatory mechanisms, biomass compositions, dead-end reactions, reversibility of reactions, and influence of the medium composition on simulations(64, 66). Sensitivity analyses of GSM models such as phenotypic phase plane analysis (67), robustness analysis (68), and flux viability analysis (60) are other good approaches to examine the properties and prediction capability of the reconstructed network.

An overview of GSM model construction and simulation is illustrated in **Fig. 25.2**. A GSM model is constructed by gathering and combining related data and previous knowledge, and it is generally necessary to continuously improve the GSM model using newly released data and/or adding missing data.

3. *S. cerevisiae* GSM Models

3.1. From Prototype to Genome Scale

The construction of metabolic networks based on metabolite balancing and stoichiometry to study the metabolic capabilities of *S. cerevisiae* was not a straightforward formulation. Even though metabolic models used for flux analysis of other organisms such as *E. coli* (69) and *Corynebacterium glutamicum* (70) were developed and applied in biotechnology, it was not until 1995 when the first metabolic model for *S. cerevisiae* was published.

The first reference regarding yeast metabolic models is the study of van Gulik and co-workers in 1995 (48), they constructed a metabolic network model of the central metabolism of two eukaryotic cells, *S. cerevisiae* and *Candida utilis.* Back then, using the available biochemical information, the model was used to predict the growth of the yeasts under different carbon sources. Depending on the cultivation conditions, they used different sets of reactions in the model to simulate and examine yeast physiological responses. In their formulations, one model contained 70 reactions with 86 metabolites and another model contained 81 reactions with 88 metabolites. They pointed out that the P/O ratio and the maintenance coefficients were the parameters that had the largest influence on the growth phenotypes. Considering the relevance of this first application of yeast metabolic networks, a better construction and parameters analyses were introduced by Vanrolleghenm and co-workers (71) who used many experimental datasets to calibrate the yeast metabolic network. The capabilities of the network to predict biomass yields on different substrates and the correspondence between predicted metabolic fluxes and presence or absence of the corresponding enzyme activities in cell-free extracts were examined. Their model included 78 reactions, which were adopted from the previously reported model by van Gulik et al. (48). The second metabolic model for yeast was reported by Nissen and co-workers (72), who performed a more rigorous reconstruction of the central metabolic network of *S. cerevisiae*. They applied the stoichiometric model with more extensive descriptions and introduced the application of MFA. They used their model to explore many capabilities of flux analysis in strain engineering. They proposed three major applications of metabolic models: (1) the facilitation of analysis of physiological data and the possible evaluation of intracellular fluxes under various growth conditions; (2) the impact of the presence or absence of single reactions or whole pathways in flux distributions; and (3) MFA as a mean to calculate yields of compounds that are not possible to experimentally measure. In this study the authors focused in the detailed description of the pathways and their flux distribution using anaerobic glucose limited chemostat cultures at various dilution rates (dilution rates equal to growth rates in steady-state chemostat cultures). Many other applications of the yeast prototype models were found, for instance, the calculation of the metabolic flux distribution in recombinant strains during heterologous protein production (73) and in the optimization of ethanol production (74). With the development of the analysis of carbon transitions in ^{13}C-labeled experiment by GC-MS and/or NMR, the yeast prototype models were first exploited for rigorous estimation of flux distributions to study glucose repression phenomena in *S. cerevisiae* (75). Due to the reliability and reproducibility of ^{13}C-labeled MFA, several researchers have still employed this technique together with yeast prototype network models to gain insight into the metabolism of *S. cerevisiae* (76–78). Based on a different, metabolic network topology approach, Carlson and co-workers (79) applied elementary flux mode (EFM) analysis to study a yeast strain genetically engineered to produce poly-beta-hydroxybutyrate (PHB). The authors also reconstructed a metabolic model of *S. cerevisiae* which included the pathway for PHB production. In the same year, Förster and colleagues (80) demonstrated the combination of in silico pathways analysis of the yeast prototype model and metabolome data to identify the function of orphan genes under aerobic growth conditions. This work showed that in silico analysis of metabolic pathway is an applicable tool in the field of functional genomics. Later on, the prototype model of Förster was adopted by Nookaew et al. (81) to map cellular operations under various growth conditions by extending the concept of EFM analysis.

In 2003, the first comprehensive reconstruction of the GSM network of *S. cerevisiae* was accomplished (82). The GSM model consisted of 708 ORFs accounting for 1,175 metabolic reactions and 733 metabolites including two major cellular compartments, namely the cytosol and the mitochondria. It was the first GSM network for an eukaryotic system. The in silico predictions of cellular phenotypes by simulation of the GSM model under FBA technique were in good agreement with various experimental observations and showed high accuracy in prediction of gene essentiality in large scale (47, 64).

The GSM model created by Förster et al. (82) has served as a reference for later models trying to reach more accuracy in the predictions and to include more biochemical as well as physiological evidences. For instance, the GSM model which was generated by Duarte et al. (66) included more compartments and a cell-wide proton balance. To distinguish between different GSM models, a nomenclature was established where the letter *i* to refer to the in silico model, the initial of the scientist(s) who made the reconstruction and the number of ORFs involved; for example, the model generated by Förster et al. is known as *iFF708* and the model from Duarte et al. is called *iND750*. Nearly all the genes in *iFF708* model are included in *iDN750* and the only considerable difference is the number of unique reactions which reached 1,149 compared to 842 in *iFF708*. In spite of the fact that including more compartments would capture more biological knowledge of the yeast cell, the ability to predict phenotypes, e.g., gene essentiality, was reduced compared with the *iFF708* model. The GSM model, *iND750*, was also studied for its sensitivity to display the maximum allowable phenotypes and distinct patterns of metabolic pathway utilization by phenotypic phase plane analysis (67). Furthermore, transcriptional regulatory network of nutrient controlled of 55 known transcription factors based on literature surveys (83) was combined with *iND750*. The combined model called *iMH805* showed good capability to predict growth phenotypes of regulatory gene mutants as well as gene expression profiles by rFBA simulation strategy.

After *iFF708* and *iND750* were released, two more yeast GSM models were constructed, the *iLL672* (65) and the *iIN800* (53) model. Through the integration of experimental and in silico analysis the first GSM model, *iLL672*, aimed to the elucidation of duplicated genes and their metabolic function. The GSM model included less number of reactions (1,038 reactions), but the authors claimed that the redundancy of the information, which is less relevant than dead-end reactions, was eliminated from *iFF708* leading to improved gene essentiality prediction. The second GSM model, *iIN800*, which was also derived from the original *iFF708*, had an improved biomass formation equation and extended by several reactions involved in the lipid metabolism leading to increased applicability to study lipid related processes in eukaryotes. This GSM model contains 1,446 reactions and 1,013 metabolites, and the resulting model *iIN800* was proven to have very good prediction performances and probably represents the best agreement with experimental observations. In order to overcome a problem with differences in nomenclature used in the different models a combined effort of several research groups resulted in the development of the GSM network named *iMM904* (84), which included 1,412 reactions (85). Despite the differences in size and predictive power among all these yeast GSM models, the scientific community has accepted the existence and applicability of all these models. **Table 25.3** summarizes the progressive efforts in metabolic network reconstructions and modeling of *S. cerevisiae*.

3.2. Applications of GSM Models

The applications of GSM models, not only *S. cerevisiae* models but also from other organisms, can be grouped according to three main objectives: (1) Comprehension of the topology of the metabolic network; (2) Identification of essential genes and gene deletion targets for improved by-product formation as well as prediction of growth and by-product secretion rates under various culture conditions; (3) The integration of different types of omics datasets such as transcriptome, proteome and metabolome data, thanks to the optimum development of high-throughput methods.

A first example of the use of a genome-scale model is the aid to find metabolic engineering strategies to maximize the production of desired compounds. Bro and co-workers (19) used the *iFF708* model to perform in silico flux distribution analyses to identify a number of strategies to metabolically engineer the cell to decrease glycerol production and increase ethanol production from glucose under anaerobic conditions. The authors successfully decreased the yield of glycerol by 40%, whereas they increased the ethanol yield by 3% without affecting the growth rate. In a similar way, Asadollahi et al. (86) identified new target genes for enhancing biosynthesis of sesquiterpenes by perturbation of the ammonium assimilation pathway. Yeast GSM models have also been used to analyze the redundancy and robustness of metabolic networks. Åkesson et al. (87) tried to reduce the solution space of FBA simulations by adding more constraints from transcriptome data. They showed that imposing constraints in transcriptional gene expression leads to improved metabolic predictions of *iFF708* in batch cultivations. Gene knockouts strategies for metabolic engineering based on GSM models have also been investigated. Patil et al. (88) developed an evolutionary algorithm to find the best gene deletion target(s) associated with desired optimal phenotypes. They applied the algorithm to identify potential gene targets to increase the production of succinic acid, vanillin, and glycerol. Another promising systems biology application of the GSM models in omics data integration was developed by Patil and Nielsen (89). Their algorithm called "reporter algorithm" enabled a combination of the topology of the yeast *iFF708* network and transcriptome data for uncovering metabolic hotspots in metabolic networks called reporter

Table 25.3
Genome-scale models

References	Size of the model	Experimental conditions for validations	Comments
Van Gulik et al. (48)	Five different models: (a) 70 reactions, 86 metabolites; (b) 73 reactions, 85 metabolites; (c) 74 reactions, 86 metabolites; (d) 73 reactions, 85 metabolites; (e) 81 reactions, 88 metabolites	Five different conditions: (a) anaerobic growth in glucose; (b) aerobic growth in glucose; (c) aerobic growth in ethanol; (d) aerobic growth in acetate; (e) aerobic growth in gluc.+ ethanol	First reconstruction of a metabolic network in yeast
Vanrolleghem et al. (71)	78 reactions, 98 metabolites	Aerobic growth on glucose and ethanol	The study provided the flux regime during shifting of carbon sources from glucose to ethanol
Nissen et al. (72)	37 reactions, 27 intracellular, metabolites	Anaerobic growth in glucose limited continuous culture	First consensus reconstruction of metabolic networks introducing methodically the concept of metabolic flux analysis (MFA)
Gombert et al. (75)	42 reactions and 23 lumped metabolites	Batch and chemostat growth of wild-type and glucose repression yeast mutants	^{13}C-labeling MFA of yeast central metabolic pathway
Carlson et al. (79)	64 reactions and 67 metabolites	Study of a recombinant strain for production of poly-beta-hydroxybutyrate (PHB)	The authors introduce the pathway for the production of PHB and use elementary flux modes to find metabolic engineering strategies
Förster et al. (80)	45 reactions, 42 internal metabolites, and 7 external metabolites	Aerobic growth in glucose limited culture	The model was used in combination of metabolome data to identify the function of orphan genes

(continued)

Table 25.3 (continued)

Förster et al. (64, 82) Famili et al. (47)	708 ORFs, 1,175 reactions and 733, metabolites. Two compartments: cytosol and mitochondria	Aerobic and anaerobic cultures in glucose and ethanol	First genome-scale model of an eukaryotic cell and it served for gene essentiality predictions and as a scaffold to generate new genome-scale models
Duarte et al. (66, 67)	750 ORFs, 1,149 reactions, and 646 metabolites.	Result from aerobic and anaerobic cultures in glucose and ethanol and gene essentiality predictions	The model was fully compartmentalized and proton balanced
Herrgård et al. (83)	805 ORFs, 1,149 reactions, and 646 metabolites.	Prediction of growth phenotypes of yeast mutants and their gene expression	The model was developed from *iND750* and knowledge/integration of 55 known transcription factors
Kuepfer et al. (65)	672 ORFs, 1,038 reactions, and 636 metabolites.	Growth in five different conditions: complex medium or minimal medium, glucose, galactose, glycerol, and ethanol	The predictions of the model were validated with the phenotypes from the yeast collection of knockout mutants. Used for gene essentiality predictions
Nookaew et al. (53)	800 ORFs, 1,446 reactions and 1,013 metabolites.	Different sources of data were used, including carbon and nitrogen limitations under aerobic and anaerobic conditions	More lipid pathways included compared to previous models. Improve biomass equations. Used for gene essentiality predictions.
Mo et al. (84)	904 ORFs, 1,412 reactions, and 1,013 metabolites.	Different conditions. The data was validated comparing 2,888 in silico single deletion strain growth phenotypes to experimental data	This model was the basis for the consensus jamboree network (Herrgård et al. (85))

Genome-scale model as the summit of discrete efforts in metabolic network models. The relevance of prototype metabolic models was the slow progression in the understating of the capabilities of the in silico simulation to predict flux distribution and physiological characterization of the cells metabolites. Using a similar approach, Cakir et al. (90) integrated metabolome data and the yeast GSM model to identify reporter reactions. Recently, the reporter algorithm was extended to cover other biological networks by Oliveira et al. (91). The concept has been widely applied in many applications in yeast to rule out important biological processes related to specific perturbations (92–95). In another study of the yeast metabolic network, Blank and co-workers (96) used *iLL672* combined with ^{13}C tracer experiments for genome-scale ^{13}C flux analysis. The goal of this study was to examine the robustness of various yeast mutants. Recently, *iMM904* has been applied to investigate intracellular phenotypes by random sampling of the solution space confined by given secreted metabolites. This method is a promising approach to exploit metabolome data from different physiological states using a GSM network. An overview of these different applications of yeast GSM models is summarized in **Table 25.4**.

Table 25.4
GSM models applications/studies. Two of the major applications of GSM models are the investigation of new metabolic engineering strategies and the integration of omics data in systems biology studies

References	Model Applied	Study
Patil et al. (88)	*iFF708*	The authors used genome-scale models to develop an evolutionary algorithm to identify metabolic strategies for the overproduction of targeted compounds
Patil et al. (89)	*iFF708*	Using information from the topology of the metabolic networks the authors developed an algorithm to integrate transcriptome data and indentify so-called reporter metabolites
Blank et al. (96)	*iLL672*	The model was used to identify key experiments to perform genome-scale ^{13}C flux analysis
Bro et al. (19)	*iFF708*	Improvement of ethanol production in yeast
Cakir et al. (90)	*iFF708*	Using the basis of the algorithm developed by Patil et al. (89), this new algorithm integrates metabolome data to identify reporter reactions
Asadollahi et al. (86)	*iFF708*	Together with the evolutionary algorithm developed by Patil et al. (88) the aim of this study was to increase sesquiterpenes production
Mo et al. (84)	*iMM904*	Applied random sampling of solution space to indentify significant metabolic states from genetic perturbations of the ammonium assimilation pathway

4. Current Status and Future Perspectives

As described in the previous section, five different GSM models of *S. cerevisiae* have been developed and extensively applied in many different applications. Considering the different scopes and contents as well as performances of these GSM models, the scientific community has gathered to unify the yeast reconstructed network in a consensus manner (85). Based on this jamboree work, metabolites in the consensus network were named in a standardized fashion by using international chemical (InChi) or KEGG identifiers, which allows easy comparison of all yeast reconstructed GSM networks using SMBL format. However, there is a need for developing the consensus yeast GSM network into a model that can be used for simulation, and this is the next task for the yeast scientific community. The availability of a consensus GSM model will have several advantages in terms of comparison of simulations and to be used as the common tool in yeast systems biology research.

References

1. Nielsen, J., and Jewett, M. C. (2008) Impact of systems biology on metabolic engineering of *Saccharomyces cerevisiae*. *FEMS Yeast Res.* **8**, 122–131.
2. Spier, R. E. (2000) Yeast as a cell factory. *Enzyme Microb. Technol.* **26**, 639.
3. Goffeau, A., Barrell, B. G., Bussey, H., et al. (1996) Life with 6000 genes. *Science* **274**, 546, 563–567.
4. Botstein, D., Chervitz, S. A., and Cherry, J. M. (1997) Yeast as a model organism. *Science* **277**, 1259–1260.
5. Steinmetz, L. M., Scharfe, C., Deutschbauer, A. M., et al. (2002) Systematic screen for human disease genes in yeast. *Nat. Genet.* **31**, 400–404.
6. Bassett, D. E., Jr., Boguski, M. S., and Hieter, P. (1996) Yeast genes and human disease. *Nature* **379**, 589–590.
7. Foury, F. (1997) Human genetic diseases: a cross-talk between man and yeast. *Gene* **195**, 1–10.
8. Perocchi, F., Mancera, E., and Steinmetz, L. M. (2008) Systematic screens for human disease genes, from yeast to human and back. *Mol. Biosyst.* **4**, 18–29.
9. Petranovic, D., and Nielsen, J. (2008) Can yeast systems biology contribute to the understanding of human disease? *Trends Biotechnol.* **26**, 584–590.
10. Sturgeon, C. M., Kemmer, D., Anderson, H. J, and Roberge, M. (2006) Yeast as a tool to uncover the cellular targets of drugs. *Biotechnol. J.* **1**, 289–298.
11. Mutka, S. C., Bondi, S. M., Carney, J. R., Da Silva, N. A., and Kealey J. T. (2006) Metabolic pathway engineering for complex polyketide biosynthesis in *Saccharomyces cerevisiae*. *FEMS Yeast Res.* **6**, 40–47.
12. Wattanachaisaereekul, S., Lantz, A. E., Nielsen, M. L., and Nielsen, J. (2008) Production of the polyketide 6-MSA in yeast engineered for increased malonyl-CoA supply. *Metab. Eng.* **10**, 246–254.
13. Ro, D. K., and Douglas, C. J. (2004) Reconstitution of the entry point of plant phenylpropanoid metabolism in yeast (*Saccharomyces cerevisiae*): implications for control of metabolic flux into the phenylpropanoid pathway. *J. Biol. Chem.* **279**, 2600–2607.
14. Asadollahi, M. A., Maury, J., Moller, K., et al. (2008) Production of plant sesquiterpenes in *Saccharomyces cerevisiae*: effect of ERG9 repression on sesquiterpene biosynthesis. *Biotechnol. Bioeng.* **99**, 666–677.
15. Yamano, S., Ishii, T., Nakagawa, M., Ikenaga, H., and Misawa, N. (1994) Metabolic engineering for production of beta-carotene and lycopene in *Saccharomyces cerevisiae*. *Biosci. Biotechnol. Biochem.* **58**, 1112–1114.

16. Dejong, J. M., Liu, Y., Bollon, A. P., et al. (2006) Genetic engineering of taxol biosynthetic genes in *Saccharomyces cerevisiae*. *Biotechnol. Bioeng.* **93**, 212–224.
17. Thim, L., Hansen, M. T., Norris, K., et al. (1986) Secretion and processing of insulin precursors in yeast. *Proc. Natl. Acad. Sci. USA* **83**, 6766–6770.
18. Thim, L., Hansen, M. T., and Sorensen, A. R. (1987) Secretion of human insulin by a transformed yeast cell. *FEBS Lett.* **212**, 307–312.
19. Bro, C., Regenberg, B., Förster, J., and Nielsen, J. (2006) In silico aided metabolic engineering of *Saccharomyces cerevisiae* for improved bioethanol production. *Metab. Eng.* **8**, 102–111.
20. Guo, Z. P., Zhang, L., Ding, Z. Y., Wang, Z. X., and Shi, G. Y. (2009) Interruption of glycerol pathway in industrial alcoholic yeasts to improve the ethanol production. *Appl. Microbiol. Biotechnol.* **82**, 287–292.
21. Boros, L. G., and Boros, T. F. (2007) Use of metabolic pathway flux information in anticancer drug design. *Ernst Schering Found. Symp. Proc.* **4**, 189–203.
22. Biochemical Pathways– Metabolic Pathways. (http://www.expasy.ch/cgi-bin/show_thumbnails.pl).
23. Covert, M. W., Schilling, C. H., Famili, I., et al. (2001) Metabolic modeling of microbial strains in silico. *Trends Biochem. Sci.* **26**, 179–186.
24. Weng, S., Dong, Q., Balakrishnan, R., et al. (2003) Saccharomyces Genome database (SGD) provides biochemical and structural information for budding yeast proteins. *Nucleic Acids Res.* **31**, 216–218.
25. Guldener, U., Munsterkotter, M., Kastenmuller, G., et al. (2005) CYGD: the comprehensive yeast genome database. *Nucleic Acids Res.* **33**, D364–D368.
26. Godzik, A., Jambon, M., and Friedberg, I. (2007) Computational protein function prediction: are we making progress? *Cell. Mol. Life Sci.* **64**, 2505–2511.
27. Kanehisa, M., and Goto, S. (2000) KEGG: Kyoto Encyclopedia of Genes and Genomes. *Nucleic Acids Res.* **28**, 27–30.
28. Joshi-Tope, G., Gillespie, M., Vastrik, I., et al. (2005) Reactome: a knowledgebase of biological pathways. *Nucleic Acids Res.* **33**, D428–D432.
29. Vastrik, I., D'Eustachio, P., Schmidt, E., et al. (2007) Reactome: a knowledge base of biologic pathways and processes. *Genome Biol.* **8**, R39.
30. Caspi, R., Foerster, H., Fulcher, C. A., et al. (2008) The MetaCyc Database of metabolic pathways and enzymes and the BioCyc collection of pathway/genome databases. *Nucleic Acids Res.* **36**, D623–D631.
31. Feist, A. M., Herrgård, M. J., Thiele, I., Reed, J. L., and Palsson, B. Ø. (2009) Reconstruction of biochemical networks in microorganisms. *Nat. Rev. Microbiol.* **7**, 129–143.
32. Jamshidi, N., Edwards, J. S., Fahland, T., Church, G. M., and Palsson, B. Ø. (2001) Dynamic simulation of the human red blood cell metabolic network. *Bioinformatics* **17**, 286–287.
33. Tomita, M. (2001) Whole-cell simulation: a grand challenge of the 21st century. *Trends Biotechnol.* **19**, 205–210.
34. Tomita, M., Hashimoto, K., Takahashi, K., et al. (1999) E-CELL: software environment for whole-cell simulation. *Bioinformatics* **15**, 72–84.
35. Ohno, H., Naito, Y., Nakajima, H., and Tomita, M. (2008) Construction of a biological tissue model based on a single-cell model: a computer simulation of metabolic heterogeneity in the liver lobule. *Artif. Life* **14**, 3–28.
36. Stephanopoulos, G., Aristidou, A. A., and Nielsen, J. (1998) *Metabolic Engineering: Principles and Methodologies*. Academics, San Diego, California, USA.
37. Iwatani, S., Yamada, Y., and Usuda, Y. (2008) Metabolic flux analysis in biotechnology processes. *Biotechnol. Lett.* **30**, 791–799.
38. Schilling, C. H., Schuster, S., Palsson, B. Ø., and Heinrich, R. (1999) Metabolic pathway analysis: basic concepts and scientific applications in the post-genomic era. *Biotechnol. Prog.* **15**, 296–303.
39. Schuster, S., Dandekar, T., and Fell, D. A. (1999) Detection of elementary flux modes in biochemical networks: a promising tool for pathway analysis and metabolic engineering. *Trends Biotechnol.* **17**, 53–60.
40. Schuster, S., Fell, D. A., and Dandekar, T. (2000) A general definition of metabolic pathways useful for systematic organization and analysis of complex metabolic networks. *Nat. Biotechnol.* **18**, 326–332.
41. Papin, J. A., Price, N. D., Edwards, J. S., and Palsson, B. B. Ø. (2002) The genome-scale metabolic extreme pathway structure in Haemophilus influenzae shows significant network redundancy. *J. Theor. Biol.* **215**, 67–82.
42. Papin, J. A, Price, N. D, and Palsson, B. Ø. (2002) Extreme pathway lengths and reaction participation in genome-scale metabolic networks. *Genome Res.* **12**, 1889–1900.

43. Price, N. D., Papin, J. A., and Palsson, B. Ø. (2002) Determination of redundancy and systems properties of the metabolic network of *Helicobacter pylori* using genome-scale extreme pathway analysis. *Genome Res.* **12**, 760–769.
44. Wiback, S. J., and Palsson, B. Ø. (2002) Extreme pathway analysis of human red blood cell metabolism. *Biophys. J.* **83**, 808–818.
45. Varma, A., and Palsson, B. Ø. (1994) Stoichiometric flux balance models quantitatively predict growth and metabolic by-product secretion in wild-type *Escherichia coli* W3110. *Appl. Environ. Microbiol.* **60**, 3724–3731.
46. Edwards, J. S., and Palsson, B. Ø. (2000) The Escherichia coli MG1655 in silico metabolic genotype: its definition, characteristics, and capabilities. *Proc. Natl. Acad. Sci. USA* **97**, 5528–5533.
47. Famili, I., Förster, J., Nielsen, J., and Palsson, B. Ø. (2003) *Saccharomyces cerevisiae* phenotypes can be predicted by using constraint-based analysis of a genome-scale reconstructed metabolic network. *Proc. Natl. Acad. Sci. USA* **100**, 13134–13139.
48. van Gulik, W. M., and Heijnen, J. J. (1995) A metabolic network stoichiometry analysis of microbial growth and product formation. *Biotechnol. Bioeng.* **48**, 681–698.
49. Ramakrishna, R., Edwards, J. S., McCulloch, A., and Palsson, B. Ø. (2001) Flux-balance analysis of mitochondrial energy metabolism: consequences of systemic stoichiometric constraints. *Am. J. Physiol. Regul. Integr. Comp. Physiol.* **280**, R695–R704.
50. Knorr, A. L., Jain, R., and Srivastava, R. (2007) Bayesian-based selection of metabolic objective functions. *Bioinformatics* **23**, 351–357.
51. Ebenhoh, O., and Heinrich, R. (2001) Evolutionary optimization of metabolic pathways. Theoretical reconstruction of the stoichiometry of ATP and NADH producing systems. *Bull. Math. Biol.* **63**, 21–55.
52. Oliveira, A. P., Nielsen, J., and Förster, J. (2005) Modeling *Lactococcus lactis* using a genome-scale flux model. *BMC Microbiol.* **5**, 39.
53. Nookaew, I., Jewett, M. C., Meechai, A., et al. (2008) The genome-scale metabolic model iIN800 of *Saccharomyces cerevisiae* and its validation: a scaffold to query lipid metabolism. *BMC Syst. Biol.* **2**, 71.
54. Segre, D., Vitkup, D., and Church, G. M. (2002) Analysis of optimality in natural and perturbed metabolic networks. *Proc. Natl. Acad. Sci. USA* **99**, 15112–15117.
55. Shlomi, T., Berkman, O., and Ruppin, E. (2005) Regulatory on/off minimization of metabolic flux changes after genetic perturbations. *Proc. Natl. Acad. Sci. USA* **102**, 7695–7700.
56. Covert, M. W., Schilling, C. H., and Palsson, B. (2001) Regulation of gene expression in flux balance models of metabolism. *J. Theor. Biol.* **213**, 73–88.
57. Shlomi, T., Eisenberg, Y., Sharan, R., and Ruppin, E. (2007) A genome-scale computational study of the interplay between transcriptional regulation and metabolism. *Mol. Syst. Biol.* **3**, 101.
58. Covert, M. W., Xiao, N., Chen, T. J., and Karr, J. R. (2008) Integrating metabolic, transcriptional regulatory and signal transduction models in Escherichia coli. *Bioinformatics* **24**, 2044–2050.
59. Lee, J. M., Gianchandani, E. P., Eddy, J. A., and Papin, J. A. (2008) Dynamic analysis of integrated signaling, metabolic, and regulatory networks. *PLoS Comput. Biol.* **4**, e1000086.
60. Becker, S. A., Feist, A. M., Mo, M. L., Hannum, G., Palsson, B. Ø., and Herrgård, M. J. (2007) Quantitative prediction of cellular metabolism with constraint-based models: the COBRA toolbox. *Nat. Protoc.* **2**, 727–738.
61. Hucka, M., Finney, A., Sauro, H. M., et al. (2003) The systems biology markup language (SBML): a medium for representation and exchange of biochemical network models. *Bioinformatics* **19**, 524–531.
62. Keating, S. M., Bornstein, B. J., Finney, A., and Hucka, M. (2006) SBMLToolbox: an SBML toolbox for MATLAB users. *Bioinformatics* **22**, 1275–1277.
63. Winzeler, E. A., Shoemaker, D. D., Astromoff, A., et al. (1999) Functional characterization of the *S. cerevisiae* genome by gene deletion and parallel analysis. *Science* **285**, 901–906.
64. Förster, J., Famili, I., Palsson, B. Ø., and Nielsen, J. (2003) Large-scale evaluation of in silico gene deletions in *Saccharomyces cerevisiae*. *OMICS* **7**, 193–202.
65. Kuepfer, L., Sauer, U., and Blank, L. M. (2005) Metabolic functions of duplicate genes in *Saccharomyces cerevisiae*. *Genome Res.* **15**, 1421–1430.
66. Duarte, N. C., Herrgård, M. J., and Palsson, B. Ø. (2004) Reconstruction and validation of *Saccharomyces cerevisiae* iND750, a fully compartmentalized genome-scale metabolic model. *Genome Res.* **14**, 1298–1309.
67. Duarte, N. C., Palsson, B. Ø., and Fu, P. (2004) Integrated analysis of metabolic

phenotypes in *Saccharomyces cerevisiae. BMC Genomics* **5**, 63.
68. Edwards, J. S., and Palsson, B. Ø. (2000) Robustness analysis of the *Escherichia coli* metabolic network. *Biotechnol. Progr.* **16**, 927–939.
69. Varma, A., Boesch, B. W., and Palsson, B. Ø. (1993) Biochemical production capabilities of *Escherichia coli. Biotechnol. Bioeng.* **42**, 59–73.
70. Vallino, J., and Stephanopoulos, G. (1993) Metabolic flux distributions in *Corynebacterium glutamicum* during growth and lysine overproduction. *Biotechnol. Bioeng.* **41**, 633–646.
71. Vanrolleghem, P. A., Jong-Gubbels, P., van Gulik, W. M., Pronk, J. T., van Dijken, J. P., and Heijnen, S. (1996) Validation of a metabolic network for *Saccharomyces cerevisiae* using mixed substrate studies. *Biotechnol. Progr.* **12**, 434–448.
72. Nissen, T. L., Schulze, U., Nielsen, J., and Villadsen, J. (1997) Flux distributions in anaerobic, glucose-limited continuous cultures of *Saccharomyces cerevisiae. Microbiology* **143**, 203–218.
73. Jin, S., Ye, K., and Shimizu, K. (1997) Metabolic flux distribution in recombinant *Saccharromyces cerevisiae* during foreign protein production. *J. Biotechnol.* **54**, 161–174.
74. Nissen, T. L., Kielland-Brandt, M. C., Nielsen, J., and Villadsen, J. (2000) Optimization of ethanol production in *Saccharomyces cerevisiae* by metabolic engineering of the ammonium assimilation. *Metabolic Eng.* **2**, 69–77.
75. Gombert, A. K., Moreira dos Santos, M., Christensen, B., and Nielsen, J. (2001) Network identification and flux quantification in the central metabolism of *Saccharomyces cerevisiae* under different conditions of glucose repression. *J. Bacteriol.* **183**, 1441–1451.
76. Jouhten, P., Rintala, E., Huuskonen, A., et al. (2008) Oxygen dependence of metabolic fluxes and energy generation of *Saccharomyces cerevisiae* CEN.PK113-1A. *BMC Syst. Biol.* **2**, 60.
77. van Winden, W. A., van Dam, J. C., Ras, C., et al. (2005) Metabolic-flux analysis of *Saccharomyces cerevisiae* CEN.PK113-7D based on mass isotopomer measurements of (13)C-labeled primary metabolites. *FEMS Yeast Res.* **5**, 559–568.
78. Frick, O., and Wittmann, C. (2005) Characterization of the metabolic shift between oxidative and fermentative growth in *Saccharomyces cerevisiae* by comparative ^{13}C flux analysis. *Microb. Cell Fact.* **4**, 30.
79. Carlson, R., Fell, D., and Srienc, F. (2002) Metabolic Pathway analysis of a recombinat yeast for rational strain development. *Biotechnol. Bioeng.* **79**, 121–134.
80. Förster, J., Gombert, K. A., and Nielsen, J. (2002) A functional genomics approach using metabolomics and *in silico* pathway analysis. *Biotechnol. Bioeng.* **79**, 703–712.
81. Nookaew, I., Meechai, A., Thammarongtham, C., et al. (2007) Identification of flux regulation coefficients from elementary flux modes: A systems biology tool for analysis of metabolic networks. *Biotechnol. Bioeng.* **97**, 1535–1549.
82. Förster, J., Famili, I., Fu, P., Palsson, B. Ø., and Nielsen, J. (2003) Genome-scale reconstruction of the *Saccharomyces cerevisiae* metabolic network. *Genome Res.* **13**, 244–253.
83. Herrgård, M. J., Lee, B. S., Portnoy, V., and Palsson, B. Ø. (2006) Integrated analysis of regulatory and metabolic networks reveals novel regulatory mechanisms in *Saccharomyces cerevisiae. Genome Res.* **16**, 627–635.
84. Mo, M. L., Palsson, B. Ø., and Heijnen, J. J. (2009) Connecting extracellular metabolomic measurements to intracellular flux states in yeast. *BMC Syst. Biol.* **3**, 1–17.
85. Herrgård, M. J., Swainston, N., Dobson, P., et al. (2008) A consensus yeast metabolic network reconstruction obtained from a community approach to systems biology. *Nat. Biotechnol.* **26**, 1155–1160.
86. Asadollahi, M. A., Maury, J., Patil, K. R., Schalk, M., Clark, A., and Nielsen, J. (2009) Enhancing sesquiterpene production in *Saccharomyces cerevisiae* through in silico driven metabolic engineering. *Metab. Eng.* **11**, 328–334.
87. Åkesson, M., Förster, J., and Nielsen, J. (2004) Integration of gene expression data into genome-scale metabolic models. *Metab. Eng.* **6**, 285–293.
88. Patil, K. R., Rocha, I., Förster, J., and Nielsen, J. (2005) Evolutionary programming as a platform for in silico metabolic engineering. *BMC Bioinformatics* **6**, 1–12.
89. Patil, K. R., and Nielsen, J. (2005) Uncovering transcriptional regulation of metabolism by using metabolic network topology. *Proc. Natl. Acad. Sci. USA* **102**, 2685–2689.
90. Cakır, T., Patil, K. R., Onsan, I., Ulgen, K. O., Kirdar, B., and Nielsen, J. (2006) Integration of metabolome data with metabolic networks reveals reporter reactions. *Mol. Syst. Biol.* **2**, 1–11.

91. Oliveira, A. P., Patil, K. R., and Nielsen, J. (2008) Architecture of transcriptional regulatory circuits is knitted over the topology of bio-molecular interaction networks. *BMC Syst. Biol.* **2**, 17.
92. Pizarro, F. J., Jewett M. C., Nielsen, J., and Agosin, E. (2008) Growth temperature exerts differential physiological and transcriptional responses in laboratory and wine strains of *Saccharomyces cerevisiae*. *Appl. Environ. Microbiol.* **74**, 6358–6368.
93. Vemuri, G. N., Eiteman, M. A., McEwen, J. E., Olsson, L., and Nielsen, J. (2007) Increasing NADH oxidation reduces overflow metabolism in *Saccharomyces cerevisiae*. *Proc. Natl. Acad. Sci. USA* **104**, 2402–2407.
94. Cimini. D., Patil, K. R., Schiraldi, C., and Nielsen J. (2009) Global transcriptional response of *Saccharomyces cerevisiae* to the deletion of SDH3. *BMC Syst. Biol.* **3**, 17.
95. Usaite, R., Patil, K. R., Grotkjaer, T., Nielsen, J., and Regenberg, B. (2006) Global transcriptional and physiological responses of *Saccharomyces cerevisiae* to ammonium, L-alanine, or L-glutamine limitation. *Appl. Environ. Microbiol.* **72**, 6194–6203.
96. Blank L. M., Kuepfer L., and Sauer U. (2005) Large-scale ^{13}C-flux analysis reveals mechanistic principles of metabolic network robustness to null mutations in yeast. *Genome Biol.* **6**, R49.

Chapter 26

Representation, Simulation, and Hypothesis Generation in Graph and Logical Models of Biological Networks

Ken Whelan, Oliver Ray, and Ross D. King

Abstract

This chapter presents a discussion of metabolic modeling from graph theory and logical modeling perspectives. These perspectives are closely related and focus on the coarse structure of metabolism, rather than the finer details of system behavior. The models have been used as background knowledge for hypothesis generation by Robot Scientists using yeast as a model eukaryote, where experimentation and machine learning are used to identify additional knowledge to improve the metabolic model. The logical modeling concept is being adapted to cell signaling and transduction biological networks.

Key words: Graph theory, logical models, metabolic networks, machine learning.

1. Introduction

This chapter discusses metabolic network modeling from two related perspectives: graph theory and logical modeling. It includes an informal presentation of some of the possible representations in graph theory, propositional, and first-order logic (FOL), and how these representations are related. The chapter focuses on the FOL models used as background theories for two Robot Scientists. A Robot Scientist is a combination of laboratory automation and artificial intelligence (AI) designed to automate the scientific discovery process. The discussion will highlight how the various AI methods used for hypothesis generation have constrained the available representations for models and how more recent techniques have enabled the relaxation of these constraints.

J.I. Castrillo, S.G. Oliver (eds.), *Yeast Systems Biology*, Methods in Molecular Biology 759,
DOI 10.1007/978-1-61779-173-4_26, © Springer Science+Business Media, LLC 2011

1.1. Systems Biology and the Modeling of Biochemical Networks

Systems biology (1–4) represents a shift toward a synergistic approach to whole-organism modeling, emphasizing the interactions between many interrelated components rather than the behavior of the individual components. Advances in mathematics and computer science have led to the development of diverse techniques and formalisms enabling the in silico modeling of cellular systems. All computer models represent varying degrees of abstraction from the corresponding observable phenomena, from coarse large-scale models that capture the basic interactions and components of the system, e.g., KEGG (5, 6) and EcoCyc (7), to higher fidelity representations of detailed functioning and interactions of a smaller set of components (8).

Logic and graph (LG) models are related qualitative representations for modeling metabolism, where the interactions and network structure are represented without quantitative details such as reaction kinetics/dynamics. Graph-based models are used in metabolic databases such as KEGG and EcoCyc/MetaCyc, where metabolic pathways are represented explicitly with each metabolite being a node in the graph and edges representing the chemical transformations found in the reactions comprising the pathway. Such qualitative methods differ from quantitative methods such as flux balance analysis (FBA) and ordinary differential equation (ODE) models.

Diagrams illustrating the components and interactions of graph models are easy to understand and represent a natural method for communicating the overall system structure. To take advantage of this aspect of graph models, online databases such as KEGG are almost exclusively diagram based. However, although diagrammatic representations of graph models are easy for humans to understand, it is necessary to provide a logical representation of such diagrams for computational purposes, by encoding the sets of nodes and edges implicit in the mathematical definition of the corresponding graph. Moreover, additional annotations, which play a key role in most graphical data sources, can also be represented in logic and utilized by computational inference procedures. The ability to use this and other background knowledge for learning and revision of biological networks is a key advantage of logical methods over related approaches. Logical models may use computationally efficient forms of both propositional and FOL as their representation language (9). The increased expressive power of logic can be used to extend the LG modeling paradigm beyond the mere topology of the network to allow accurate and explicit representation of the relationships between the genes, enzymes, and gene products used as annotations to the reactions, as well as various cellular compartments. This strong correspondence between graph models and logical models is described in more detail in later sections of this chapter where examples illustrate how various types of graphs can be directly coded in Prolog logic programming language (9).

Logic and graph models have been developed to study many components of systems biology including metabolism, signaling pathways, and transcription networks, for example:

- Lemke and coworkers (10, 11) have developed a graph-based model of the metabolic network of Escherichia coli and have used it to analyze how much damage the absence of each enzyme causes to the metabolic network. They defined metabolic damage as the number of metabolites that can no longer be produced by the organism. They showed that only 9% of enzymes catalyze the production of five or more metabolites but that more than 50% of essential enzymes are to be found in this group.
- The BIOCHAM system (12) is a dedicated biochemical reasoning engine that uses a rule-based temporal logic language to model and query all of the possible behaviors of a given biochemical model, and the MAP kinase (MAPK) signal transduction cascades have been used as an example.
- Random Boolean networks (13, 14) have also been used to model 106 genes comprising a yeast transcriptional network. Random Boolean networks are useful for modeling systems where interactions are not known beforehand. Kauffman et al. (14) used this process to identify those networks that were most stable, thereby representing rules of biological relevance to the regulation of gene transcription.

1.2. The Robot Scientist Concept and Functional Genomics

A Robot Scientist is a physically implemented robotic system that applies techniques from artificial intelligence to execute cycles of automated scientific experimentation (15, 16). A Robot Scientist can automatically execute cycles of hypothesis generation, selection of efficient experiments to discriminate between hypotheses, execution of experiments using laboratory automation equipment, and analysis of results.

The development of the Robot Scientist concept has seen two main stages: an initial proof-of-principle study comprising a rediscovery task (15) and a more recent study investigating the potential for the discovery of new biological knowledge using ADAM (16) – a Robot Scientist that is capable of automating the required laboratory (wet) experiments. Both investigations used Saccharomyces cerevisiae (Baker's or budding yeast) functional genomics as the application domain (*see* **Section 5.3**). The Robot Scientists used background models of yeast metabolism. The use of logic provides a common framework for the biological knowledge and the generation of hypotheses.

2. Representing Metabolic Networks Using Graph Theory and Logic

2.1. Graph Theory

In general, a graph (17) $G = \{V, E\}$ consists of a set of vertices (or nodes) V and a set of edges (or arcs) E. In biological systems, the nodes generally correspond to biological entities such as genes, proteins, or chemical compounds, and the edges correspond to interactions between the biological entities, e.g., protein interactions, gene expression interactions, and chemical transformations. A graph representing the most basic metabolic network will have chemical compounds as nodes and the chemical transformations found in biochemical reactions as edges (17). **Figure 26.1** shows how to represent the graph structure of a metabolic network. **Figure 26.1a** is a hypothetical metabolic network with six reactions by which the cell might synthesize compounds G and H from compounds D and A (although many biochemical reactions

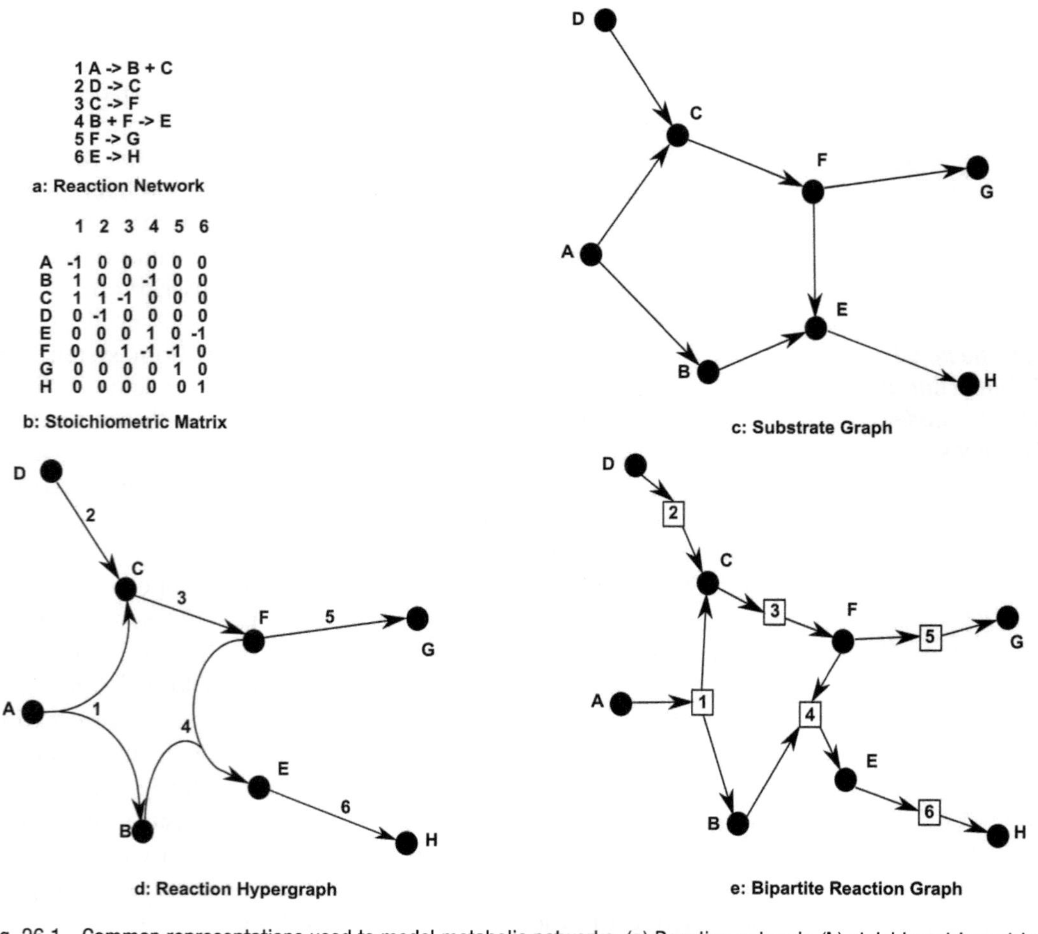

Fig. 26.1. Common representations used to model metabolic networks. (**a**) Reaction network; (**b**) stoichiometric matrix; (**c**) substrate graph; (**d**) reaction hypergraph; (**e**) bipartite reaction graph. For more information, *see* text.

are reversible, these reactions are considered non-reversible for clarity). **Figure 26.1b** shows how the reaction network can be transformed into a stoichiometric matrix, a representation central to FBA models. Each index in a stoichiometric matrix corresponds to the stoichiometry coefficient for a particular compound in a single reaction (the rows are labeled with the compounds, the columns are labeled with the reactions – all indices in a column describe the chemical transformations found in a single reaction), negative coefficients for substrates, positive coefficients for products, and a coefficient of zero indicates that the compound plays no role in a reaction. **Figure 26.1c** is a substrate graph corresponding to the chemical transformations comprising the reaction network. The substrate graph is a directed simple graph; the edges have an explicit direction and are bilateral, and each edge connects exactly two nodes. Substrate graphs are often used in metabolic network diagrams because they are clear and simple; however, there is a fundamental problem when this type of graph is used to represent a metabolic network. At first glance, the reader might think that the substrate graph effectively describes the reaction network and is equivalent to the stoichiometric matrix; however, this simple graph does not capture the interactions in the reaction network adequately. This graph suggests that there are alternative mechanisms by which molecule E can be synthesized from A, one "route" by C and another by B. In contrast, the reaction network has two reactions 1 and 4, both of which are required for synthesis of E (as well as essential reaction 3); if either reaction was missing, the cell would no longer be able to synthesize E, i.e., there is only one route to E.

If a directed hypergraph (**Fig. 26.1d**) or a bipartite graph (**Fig. 26.1e**) is used instead of the simple graph, the resulting graphs will have none of the ambiguities of the simple substrate graph. A hypergraph is a more general form of a simple graph, where each $e \in E$ is a hyperedge connecting two or more vertices V_i that correspond to a subset of V. Reaction networks require the use of directed hypergraphs where hyperarcs, denoted $a(T,H)$, connect several start nodes (the tail T) with several end nodes (the head H): T and H are subsets of V, the full set of vertices. The directed hyperarcs correspond to the direction of the reaction.

A bipartite graph is the simplest case of a k-partite graph (18). In k-partite graphs, the vertices V can be partitioned into k disjoint subsets, each subset corresponding to different type of node, i.e., a different type of biological entity. In **Fig. 26.1e**, the nodes correspond to the reaction numbers (R) and the molecules (M) – this graph can then be defined as $G = \{M, R, E\}$. Defining subsets for reaction and molecule identifiers allows this graph to again have simple or bilateral edges. An interesting and commonly found property of k-partite graphs is that the edges link nodes from different subsets; in **Fig. 26.1e**, each edge links a molecule to a reaction number or vice versa.

2.2. Propositional Logic

In propositional logic, logical formulae or rules are constructed by combining atomic propositions with logical connectives (AND ($\curlywedge$), OR ($\curlyvee$), NOT ($\neg$), and IF ($\leftarrow$)). Each proposition has a truth status, i.e., one of {true,false}. The implication connector divides each rule into a head (LHS, left-hand side) and a body (RHS, right-hand side).

If the head of the rule is limited to a single positive proposition, as is the case for rules defined using disjunctive normal form (DNF) (19), each rule can be thought of as defining how the truth value of the head variable can be determined from the truth values of the body variables. The reaction network can be written as the rule system shown in **Fig. 26.1a**, where the truth value of each proposition A,B corresponds to the presence or the absence of molecules A,B etc. It is interesting to note the similarity between these rules and the equation systems used in ODE models, where the concentration of a (head) molecule is determined by the concentrations of dependent (body) molecules and related kinetic parameters. Propositional rule sets are also equivalent to single database tables, where each column is one proposition, with as many columns as propositions in the rule set. The stoichiometric matrix in **Fig. 26.1b** is in effect a single database table.

2.3. First-Order Logic and Prolog

In brief, first-order logic (FOL) is an extension of propositional logic that, by the introduction of quantifiers and predicates, significantly extends the scope of knowledge than can be expressed using logic representations. FOL has been a core modern concept in mathematics and philosophy, and in artificial intelligence (AI). The logic programming language Prolog is a useful subset of FOL with additional support for functions such as arithmetic evaluation and default negation (9). There are now highly efficient computational theorem provers for Prolog. We will limit the discussion of modeling in FOL to Prolog.

A Prolog clause (*see* below) is similar to a rule in DNF propositional logic in that there are head and body components. Each component is termed a literal l, and literals can be atoms (a, *see* below) and other non-classical operators allowing, e.g., arithmetic calculations as well as the negation-as-failure operator (denoted not) which allows a reasoner to assume the negation of an atom in the absence of any evidence to the contrary. An atom a consists of a predicate symbol p and a set of related terms t: $p(t_1, ..., t_n)$. Terms can be atoms, lists, variables, constants, or other (nested) terms. Prolog clauses restrict the number of head literals to at most one, usually an atom, and the body literals are restricted to conjunctions (9). A clause C is an expression of the form

$a \leftarrow l_1, \ldots, l_n$, where a is the head atom and $l_1, \ldots, l_n$ are the body literals. The most relevant here is the existence of three kinds of Prolog clauses: rules, facts, and queries. Rules capture general principles and represent these as implications; facts represent specific items of knowledge and are Prolog clauses with no body literals and where no terms are variables; and queries are Prolog clauses for which a solution is requested. Theorem proving in Prolog involves finding a solution to a query by consulting a logic program and proving that the query is a logical consequence of the logic program. Logic programs consist of sets of general rules and facts (*see* example below). General concepts can be introduced in Prolog by the use of variables (9).

Predicates in Prolog are equivalent to tables in relational databases; indeed the term relation is often considered interchangeable with predicate. As a relational database can have many entries, i.e., rows in the table, a predicate (relation) can have many instances represented as facts. As an example, the reaction network in **Fig. 26.1a** can be written as the rule system presented in **Fig. 26.2** and be represented by the facts shown in **Fig. 26.3**.

The small logic program (**Fig. 26.3**) describing the reaction network is an example of a completely declarative logic program;

$B \leftarrow A$	B is present IF A is present
$C \leftarrow A$	C is present IF A is present
$C \leftarrow D$	C is present IF D is present
$E \leftarrow B \wedge F$	E is present IF B is present AND F is present
$F \leftarrow C$	F is present IF C is present
$G \leftarrow F$	G is present IF F is present
$H \leftarrow E$	H is present IF E is present

Fig. 26.2. Rules describing the six hypothetical reactions in **Fig. 26.1a**.

reaction(1,forwards).	substrate(1,'A').	product(1,'B').
reaction(2,forwards).	substrate(2,'D').	product(1,'C').
reaction(3,forwards).	substrate(3,'C').	product(2,'C').
reaction(4,forwards).	substrate(4,'B').	product(3,'F').
reaction(5,forwards).	substrate(4,'F').	product(4,'E').
reaction(6,forwards).	substrate(5,'F').	product(5,'G').
	substrate(6,'E').	product(6,'H').

Fig. 26.3. A simple Prolog program describing the six hypothetical reactions in **Fig. 26.1a** (quoted terms are text strings). For more information, *see* text.

all of the predicates are ground facts, with no use of extra-logical constructs such as lists and no nesting of functions. In most knowledge-based applications of logical models, there is a distinct separation between the "raw" knowledge, declared as sets of facts, and rules that are used to manipulate the knowledge and derive conclusions to queries. All of the logical models discussed have this separation and all are declarative to varying degrees; some of the predicates include lists and other nested constructs because of procedural considerations relating to how the model has been used.

2.4. Relating Graph and Logical Representations

As mentioned above, **Fig. 26.2** is a propositional rule set describing the reaction network in **Fig. 26.1a**. Converting the chemical reaction notation into this representation is straightforward; multiple substrate molecules can be represented as conjunctions in the body of a single rule; multiple products are translated into multiple rules, one for each product. If the reactions were reversible, new rules would be constructed with the products as conjunctions and the substrates as separate rules. This rule set is also equivalent to the hypergraph in **Fig. 26.1d**, where the substrates correspond to hyperarc heads and products to hyperarc tails. A similar process can be used to generate rules from all of the hyperarcs in the hypergraph.

These two examples illustrate how hypergraphs and propositional logic models are equivalent representations, where only one type of object or node is considered. The reaction numbers and molecule names in the bipartite graph in **Fig. 26.1e** can be most effectively represented in logic using FOL or Prolog. Extra propositional variables could be introduced to represent the reaction numbers; however, an increasing number of entity types will lead to more complex rules or more columns in the database table. The Prolog models discussed in this chapter include more entity types, allowing enzymes and open reading frames (ORFs, putative genes, e.g., YGL026C, *see* **Section 5**), and the catalyzing and coding roles of these entities to be represented in the same form as the reactions in the metabolic network. In particular, two of the models include cell compartments corresponding to the cytosol, mitochondrion, etc., often with the same reactions repeated in each compartment. In Prolog, this can be accommodated by an additional argument specifying the compartment. However, in propositional logic, a new proposition is required for each metabolite in each compartment, and an additional column is required in the stoichiometric matrix. This additional complexity reduces the ability of these propositional representations to easily communicate the knowledge contained in the network.

Figure 26.4 shows the metabolic network of a small part of the sugar metabolism of S. cerevisiae and illustrates the relationship between graph and Prolog representations. The graph is a

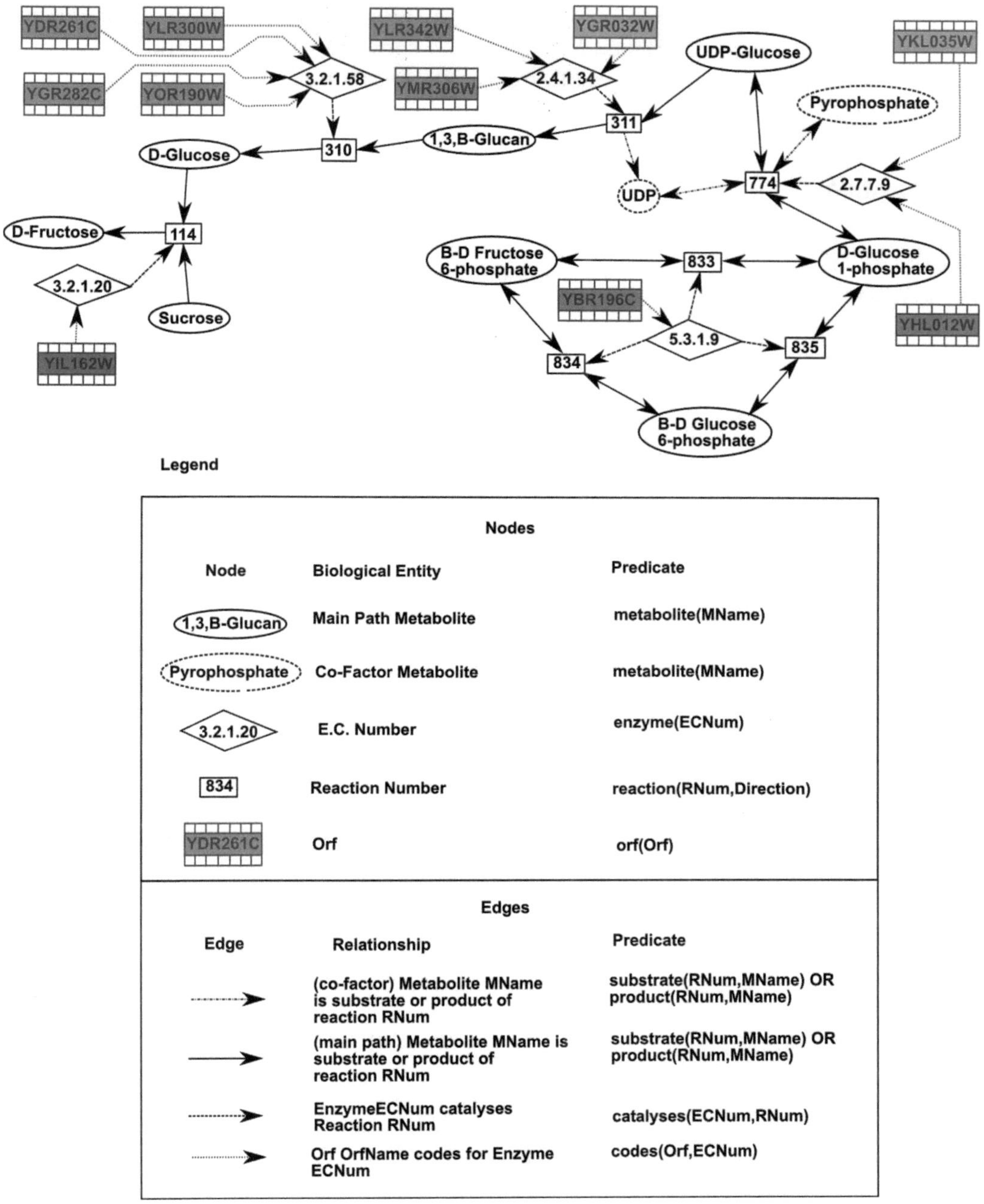

Fig. 26.4. Reactions, enzyme, and ORF information from the metabolism of sugars in S. cerevisiae with corresponding Prolog clauses (only the form of each predicate is given – the actual program would have sets of facts corresponding to each predicate). For more information, *see* text.

k-partite graph $G = \{O, EC, R, M, E\}$ corresponding to the reaction, enzyme, and ORF relationships found in the network. Each node subset corresponds to ORFs, enzymes, defined by their EC (Enzyme Commission) number, reaction numbers, and metabolites, respectively. This graph also contains only bilateral edges between members in the different node subsets. The descriptions/legends in **Fig. 26.4** show how the edges and nodes correspond to predicates in a Prolog model.

3. Simulation: Prediction of Growth Phenotypes

A major task of a logical model is to predict the change in growth phenotype for deletion mutants when one or more ORFs is/are removed from the yeast genome. This task is equivalent to finding a set of path(s) in the metabolite graph from a set of initial compounds, usually representing a (defined) growth medium, to a set of compounds deemed essential for cell growth. The task is simplified by assuming that certain common compounds such as ATP and NADP are always present in the cell. **Figure 26.5** illustrates how gene/ORF deletions can potentially block reactions in the graph. This figure assumes a direct relationship between ORFs and reactions for clarity. It also assumes that there are no cofactor metabolites present (as in **Fig. 26.4**). In practice, the existence of cofactor metabolites adds an extra complication: a path-finding algorithm relying on the simple chemical transformations as defined in a substrate graph will find shorter erroneous paths if presence of cofactors is not considered. These erroneous paths can be discarded by using reaction membership as a constraint guiding graph traversal. **Figure 26.5a** is the graph corresponding the wild type strain with no reactions missing; **Fig. 26.5b**

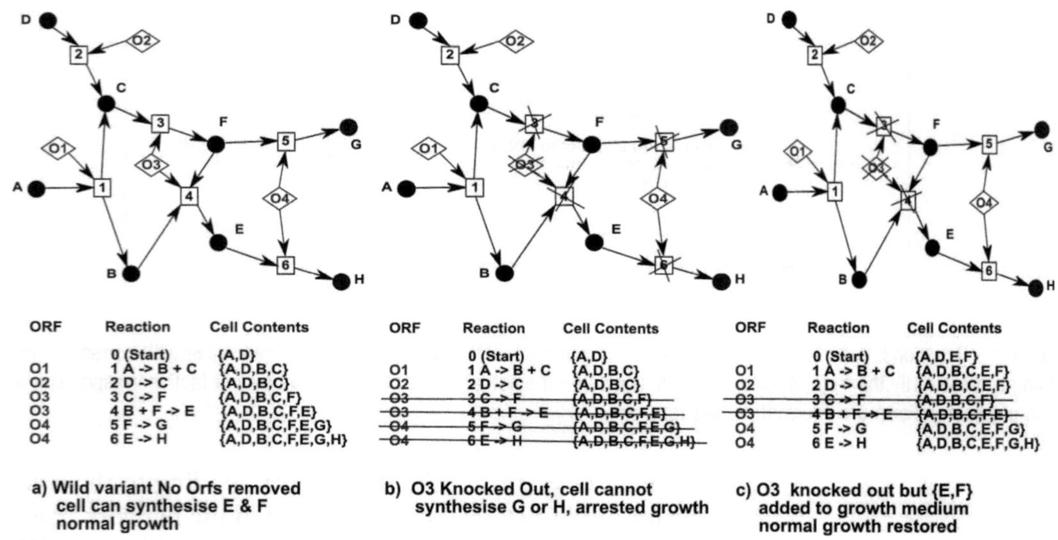

Fig. 26.5. Prediction of growth phenotypes in logical models, showing reactions (1–6), encoding ORFs (01–04), starting compounds (A and D), essential compounds (G and H), and supplemental metabolites (E and F in **c**). Crossed out reactions are those that can no longer take place (3 and 4 have no enzyme and 5 and 6 have no substrates). For more information, *see* text.

shows the reduced graph when ORF O3 is removed, resulting in arrested cell growth; and **Fig. 26.5c** has the same deletion but compounds E and F have been added to the growth medium, allowing reactions 5 and 6 to proceed again, restoring cell growth.

The *k*-partite graph representation explicitly illustrates how the relationships between ORFs, enzymes, and reactions govern cell metabolism because alternative paths to essential compounds and isoenzyme ORFs can be represented directly. An equivalent Prolog program is therefore also a direct representation. The path-following algorithm used to predict cell growth can be encoded as Prolog rules defining the conditions required for a reaction to proceed (traversal of the metabolite and reaction nodes in **Fig. 26.4**).

For a reaction to proceed, all compounds in the substrates (*or* products, if the reaction is reversible) need to be present in the cell before the reaction can occur, *and* there exists at least *one* catalyzing enzyme with at least *one* encoding ORF present in the genome (i.e., *not* removed):

```
can_proceed(RNum,Cell) :-
        all_substrates_in_cell(RNum,Cell),
        catalyses(EC,RNum),
        codes(Orf,EC),
        not(removed(Orf)).
```

This simple rule (with an equivalent rule for reaction products if the reaction is reversible – in Prolog, implication is denoted using ":-," AND is denoted by "," and the rule is terminated by ".") will typically be incorporated into a larger logic program with other rules corresponding to other semantic knowledge, such as rules for defining and proving that the essential compounds will be synthesized by the cell. (Note that this rule is a simplification of the more complex rules actually used in the models.)

The function of the removed gene can then be discovered by finding a chemical compound or a set of compounds that restore normal growth, since it can be inferred that the removed gene plays a role in the synthesis of the compounds, i.e., the removed gene codes for one or more reactions that produce the compound(s) directly or one or more reactions that produce a crucial intermediary compound.

4. Hypothesis Generation Using Logical Models

The prediction of growth phenotypes is an example of deductive inference, where the prediction is a direct consequence of the logic programs comprising the logical model. However, abductive or inductive reasoning is required to determine gene function from the experimental outcomes, i.e., forms of hypothesis generation. Robot Scientists have so far been restricted to abductive inference – the new knowledge inferred is of the form of facts or conjunctions of facts, rather than general rules as inferred by induction. Abductive inference is more relevant to the knowledge discovery task in the biological applications investigated by the Robot Scientists.

Hypothesis generation from experimental observation is a form of machine learning. There are many different machine learning paradigms, particularly in propositional logic and equivalent representations, e.g., top-down induction of decision trees (TDIDT) (20) and linear regression (21). The software package WEKA (22) includes implementations of many of these methods. Hypothesis generation in FOL or Prolog, on the other hand, requires techniques from inductive logic programming (ILP) or relational learning (23), where the hypotheses and experimental observations are expressed as logic programs. The relationship between hypotheses and experimental observations can be then expressed as $B \cup H \models E$, where H and E are the hypotheses and experimental observations, respectively; B is the background knowledge; and "$\models$" is the semantic entailment relationship. In the forward direction (deduction), this formula describes how the experimental observations can be derived from the background knowledge and the hypotheses; in the reverse direction (abduction or induction), it describes hypothesis generation. In the case of hypothesis generation from logical models, the background knowledge B is the logical model and the hypotheses are inferences by which the model can be revised so that it can correctly explain the experimental observations.

The generation of hypotheses using ILP techniques usually involves searching in a very large, possibly potentially infinite, space of potential candidates. Much ILP research (23) has been devoted to the discovery of heuristics that can direct the search to the most promising candidates, ignoring large irrelevant areas. These heuristics include an ordering of the search space, restrictions in the form of clauses and predicates that may be used to describe the hypotheses and others. Even with these heuristics, there can still be restrictions on the nature of possible hypotheses, and these restrictions have in turn governed the choice of representations for the logical models.

5. Logical Models and the Robot Scientists

There are three logical models that have been developed in conjunction with the Robot Scientists. Two of these, the original aromatic amino acid synthesis model, "Original" AAA model

(15), and the "New" AAA model (24), are fine-grained, pathway-specific models of the aromatic amino acid pathway in *S. cerevisiae* (i.e., how yeast is able to synthesize tryptophan, tyrosine, and phenylalanine); the other is "aber" (25), a whole-genome model including all of the currently known enzymatic and metabolic components and interactions in yeast metabolism. The Original AAA model was used in the proof-of-principle Robot Scientist study and was the one of the first metabolic models constructed using Prolog. The two newer Prolog models extend the concept in two important ways: (1) the "New" AAA model is a rewrite of the Original AAA model for use with a more sophisticated hypothesis generation ILP program and (2) the aber model extends the coverage of the model by several orders of magnitude. These different models have allowed our scientific discovery research to proceed in complementary ways: the pathway-specific models provide well-described, constrained test beds for the latest hypothesis generation methods, while the genome-scale model has been used for the discovery of new biological knowledge, using techniques from bioinformatics and a reliance on serendipitous aspects of the state of knowledge in the aber model, which cannot be assumed for biological models in general. This section describes these models, the hypothesis generation technique used in conjunction with each model, and how limitations of the different hypothesis generation techniques constrain the various representations used to construct the models.

5.1. Proof-of-Principle Modeling: The Original AAA Model and C-Progol

The "Original" AAA model was developed as background knowledge for the hypothesis generation steps in the proof – of-principle Robot Scientist. This was a rediscovery task that was quite limited in scope – the model was limited to a single pathway – the aromatic amino acid biosynthesis pathway in the yeast *S. cerevisiae* – and one gene was removed at a time. Hypothesis generation was undertaken by the ILP program C-Progol5 (26), an abductive variation of the C-Progol program that uses inverse entailment to induce general rules from examples specified as ground facts. C-Progol5 can abduce only one ground fact to account for any given observation, and it cannot reason abductively through negated atoms. This limitation led to a single predicate-nested list representation, effectively equivalent to the simple substrate graph in **Fig. 26.1c**:

```
enzyme(e13,['YGL026C'],['4.2.1.20','4.2.1.20','4.2.1.20'],1,1,
        [['C00065','C03506'],['C03506'],['C00065',
          'C00463']],
        [['C00001','C00078','C00661'],['C00463',
          'C00661'],['C00001','C00078']]).
```

This single fact represents a complex chemical transformation, corresponding to EC 4.2.1.20, by which tryptophan (C00078) can be synthesized either from (3-indole)-glycerol phosphate (C03506) directly or by using indole (C00463) as an intermediate compound. Although no information is lost with this representation, and it was successful for the generation of the rediscovery hypotheses, it is not the easiest form for comprehension by human scientists. However, this model also included biological concepts not included in the aber model: (1) the inhibition of enzymes by metabolites and (2) the formation of protein complexes where enzyme activity is controlled by a combination of more than one gene product. The extra numerical arguments (arguments 3 and 4) correspond to the reversibility of the reactions and the time in days for the reaction to take place. Other enzyme facts were constructed for import reactions, to account for variability in the time taken for the cell to import nutrients from the growth medium.

The C-Progol5 representation also results in the duplication of information regarding reactions catalyzed by isoenzymes, and the Prolog code for determining cell fate contains highly procedural fragments concerned with the sorting and searching of data structures which effectively limit the type of problem that can be solved. The limitations of Progol5 mentioned above further restrict the class of applicable problems to, for example, reasoning about single gene knockout experiments.

5.2. Pathway-Specific Modeling: New AAA Model and XHAIL

For the original Robot Scientist work, the laboratory automation was provided by a Beckman 2000 laboratory robot. In 2005 this was replaced by the Robot Scientist ADAM, a much more complex and fully automated system. The original rediscovery experimental work was repeated to compare ADAM with the Beckman and to determine whether the AAA model was consistent with the new results. To ensure a meaningful comparison, the rerun used the "Original" AAA model as background knowledge, with C-Progol5 for hypothesis generation. The rerun results (16) showed that ADAM was able to rediscover gene function with the same number of experimentation/hypothesis generation iterations as the Beckman robot and was more efficient in the use of laboratory resources.

To determine the consistency of the AAA model, a "New" logical representation was constructed so that XHAIL (24, 27), a recently developed ILP program, could be used for hypothesis generation. Unlike C-Progol5, XHAIL uses non-monotonic inference to reason abductively through clauses that contain negation and it can infer hypotheses with more than one clause in order to explain a single example. This non-monotonic basis gives XHAIL the ability to remove information from a model as well as add information to a model. The "New" AAA model is completely declarative, and no extra-logical terms, such as lists, were required. The resulting logic programs are therefore much closer to the *k*-partite graph in **Fig. 26.4**, with extra predicates for representing inhibition and protein complexes. Results from the ADAM rerun were mostly consistent with the Beckman results with two important exceptions: (1) growth phenotypes for the YGL026C tryptophan synthetase-deletant strain grown with indole were close to the wild type; however, the model predicts that there should be no growth because the cell can no longer synthesize tryptophan and (2) there was generally poorer growth with anthranilate than that predicted by the model. Allowing XHAIL to reason with a wider set of clauses than was possible with C-Progol5 resulted in a revised model with two new hypotheses. This revised model was consistent with the new results. The hypotheses were as follows: (1) the indole sample was contaminated by tryptophan, allowing the YGL026C mutant to grow, and (2) anthranilate was also imported into the cell slowly. Mass spectroscopy has since confirmed that the indole sample had tryptophan as an impurity.

Subsequent experiments were devised to determine whether XHAIL could infer missing enzyme complexes and reactions, more typical instances of model revision. These experiments involved creating an inconsistent model by removing crucial reactions and enzyme complexes, and creating a set of reactions and enzyme complexes that could be used by XHAIL to complete the model, and so making it again consistent with the experimental results. These sets included the removed reactions as well as reactions from other organisms, and from different pathways in yeast. In each case, XHAIL was able to select the correct reactions and enzyme complexes, and recover the consistent model. These experiments successfully demonstrate the capacity of XHAIL to discover novel reactions, encoding ORFs by using existing multi-organism online databases such as KEGG.

5.3. Whole-Genome Modeling: The Aber Model

The "aber" model (25) is a whole-metabolism model developed to extend the logical model concept from the single pathway model of the AAA model to the whole yeast metabolic network. It was constructed before the "New" AAA model and is more declarative than the Original AAA model, but reactions are still single predicates with lists corresponding to substrate and product metabolites. The model includes two separate intracellular compartments, the cytosol and the mitochondrion, as well as an external compartment representing the growth medium. Reactions can take place in either compartment or transport metabolites between compartments (external medium included). The aber model was constructed from the iFF708 model, an FBA model constructed in 2003 (28). From here, a larger set of essential compounds was used to determine growth outcome to reflect the importance of all metabolic pathways and, as a result, in

2004 additional reactions from KEGG were added to the iFF708 model.

The aber model was used as background knowledge for the hypothesis generation steps in experiments designed to extend the capabilities of ADAM to the discovery of novel biology in yeast (16). This was a relevant study, testing the ability of Robot Scientists to move beyond the proof-of-principle stage and to contribute to new scientific knowledge. Hypothesis generation from the aber model took advantage of missing knowledge in the iFF708 model. In this model, some (orphan) reactions were included based on biochemical evidence, but their encoding ORFs were not known. ADAM generated candidate ORFs/hypotheses by using sequence similarity searches. ADAM generated experiments to test these hypotheses using the model deductively. This process resulted in the generation of 20 hypotheses for genes encoding 13 enzymes for the orphan reactions, with confirmation of 12 hypotheses. Further in vitro enzyme assays confirmed the findings of the Robot Scientist and examination of the literature revealed that seven of these encoding ORFs were novel. The other six involved bioinformatics information missing from the iFF708 model. The results from ADAM were consistent with these discoveries and some of the novel hypotheses were validated using manual methods.

The series of experiments using the Robot Scientists has outlined the potential for using automated reasoning techniques to perform the hypothesis generation step in computational scientific discovery. Indeed, the original limitations of C-Progol have largely been addressed by using the more advanced reasoning capabilities of XHAIL. Although the pathway-specific models provide richly structured problems which are well-defined test beds for different ILP programs, there are still problems regarding problem size; e.g., hypothesis generation by automated reasoning using the aber model as background knowledge would result in a combinatorial increase in the number of candidate hypothesis consistent with just a single experimental observation. This is why ADAM used bioinformatics techniques to generate the hypothesis. However, such techniques are limited in scope and there is a need to make the logical approach efficient enough to work with large models.

6. Conclusions and Further Work

Graph theory and logical models focus on the structural aspects of metabolic networks, where explicit representations can capture the connections between the metabolites, as well as the enzyme and gene/ORF information. In particular, Prolog models have been central to the functional genomics investigations undertaken by the Robot Scientists. Hypothesis generation by the Robot Scientists, using the logical models as background knowledge, has allowed the discovery of new biological knowledge. These first discoveries also illustrate how revision and improvement of models can be automated, a challenging task for automated scientific discovery.

The Computational Biology group in Aberystwyth University have recently commissioned a new Robot Scientist, "Eve," designed for drug discovery. The same combination of laboratory automation and computational hypothesis generation will be used to find active compounds from a library containing thousands of chemical compounds, with potential relevance for studies on human diseases such as malaria. The logical modeling concept is also being adapted to model cell signaling networks where protein interactions, together with metabolic reactions, have a crucial role. This model will focus on three well-studied pathways in S. cerevisiae, the pheromone response pathway which controls sexual reproduction and mating, the high osmolarity glycerol (HOG) pathway which enables the yeast to adapt to hyperosmolarity in the external environment, and the filamentous/pseudohyphal growth pathway which enables adaptation to resource limitation under specific conditions.

References

1. Kitano, H. (2002) Systems biology: a brief overview. Science **295**, 1662–1664.
2. Csete, M. E., and Doyle, J. C. (2002) Reverse engineering of biological complexity. Science **295**, 1664–1669.
3. Chong, L., and Ray, L. B. (2002) Introduction to special issue. Whole-istic biology. Science **295**, 1661.
4. Davidson, E. H., Rast, J. P., and Oliveri, P., et al. (2002) A genomic regulatory network for development. Science **295**, 1669–1678.
5. Kanehisa, M., and Goto, S. (2000) KEGG: Kyoto Encyclopedia of Genes and Genomes. Nucleic Acids Res. **28**, 27–30.
6. Kanehisa, M. (1997) A database for post-genome analysis. Trends Genet. **13**, 375–376.
7. Karp, P. D., Riley, M., Paley, S. M., Pellegrini-Toole, A., and Krummenacker, M. (1996) EcoCyc: encyclopedia of Escherichia coli genes and metabolism. Nucleic Acids Res. **24**, 32–39.
8. Feng, X., and Rabitz, H. (2004) Optimal identification of biochemical reaction networks. Biophys. J. **86**, 1270–1281.
9. Bratko I. (1986) Prolog Programming for Artificial Intelligence. Reading, MA: Addison Wesley International Computer Science Series.
10. Lemke, N., Heredia, F., Barcellos, C. K., Dor Reis A. N., and Mombach, J. C. (2004) Essentiality and damage in metabolic networks. Bioinformatics **20**, 115–119.
11. Lemke, N., Heredia, F., Barcellos, C. K., and Mombach, J. C. (2003). A method to identify essential enzymes in the metabolism: application to Escherichia coli. In: Priami, C. (ed.), *Proceedings of the First International Workshop on Computational Methods in Systems Biology* (pp. 142–148). London, UK: Springer.
12. Fages, F., Soliman, S., and Chabrier-Rivier, N. (2004) Modeling and querying interaction networks in the biochemical abstract machine BIOCHAM. J. Biol. Phys. Chem. **4**, 64–73.
13. Gershenson, C.(2002) Classification of Random Boolean Networks. In: Standish, R., Abbas, H., and Bedau, M. (eds.), Artificial Life VIII. Proceedings of the Eighth

International Conference on Artificial Life (pp. 1–8). Cambridge, MA: MIT Press.
14. Kauffman, S., Peterson, C., Samuelsson, B., and Troein, C. (2003) Random Boolean network models and the yeast transcriptional network. *Proc. Natl. Acad. Sci. USA* **100**, 14796–14799.
15. King, R. D., Whelan, K. E., Jones, F. M., et al. (2004) Functional genomic hypothesis generation and experimentation by a robot scientist. Nature **427**, 247–252.
16. King, R. D., Rowland, J. J. R., Oliver, S. G., et al. (2009) The automation of science. Science **324**, 85–89.
17. Klamt, S., Haus, U. U., and Theis, F. (2005) Hypergraphs and cellular networks. PLoS Comput. Biol. **5**, 1–6.
18. Chartrand, G., and Lesniak, L. (2004) *Graphs and Digraphs*. New York, NY: Chapman and Hall/CRC.
19. Christiensen, T. S., Oliviera, A. P., and Nielsen, J. (2009) Reconstruction and logical modeling of the glucose repression signaling pathways in Saccharomyces cerevisiae. BMC Syst. Biol. **3**, 7.
20. Rokach, L., and Maimon, O. (2005) Top down induction of decision trees classifiers: a survey. IEEE *Trans. Syst. Man. Cybern. C Appl. Rev.* **v35 i4**, 476–487.
21. Montgomery, D. C., Peck, E. A., and Vining, G. G. (2001) *Introduction to Linear Regression Analysis*. New York, NY: Wiley.
22. Hall, M., Frank, E., Holmes, G., Pfahringer, B., Reutemann, P. and Witten, I. H. (2009) The WEKA data mining software: an update. SIGKDD Explorations **11**(1).
23. Muggleton, S., and DeRaedt, L. (1994) Inductive logic programming: theory and methods. J. Logic Programming **19**(20), 629–679.
24. Ray, O., Clare, A., Liakata, M., Soldatova, L., Whelan, K., and King, R. D. (2009) Towards the automation of scientific method. In: *Proceedings of IJCAI'09 Workshop on Abductive and Inductive Knowledge Development* (pp. 27–33). Pasadena, CA.
25. Whelan, K. E., and King, R. D. (2008) Using a logical model to predict the growth of yeast. BMC Bioinformatics 9, 97.
26. Muggleton, S., and Bryant, C. (2000) Theory completion using inverse entailment. In: Cussens, J. and Frisch, A. (eds.), Inductive Logic Programming, *LNCS 1866* (pp. 130–146). London, UK: Springer.
27. Ray, O. (2009) Nonmonotonic abductive inductive learning. J. Appl. Logic 7, 329–340.
28. Förster, J., Famili, I., Fu, P., Palsson B. Ø., and Nielsen, J. (2003) Genome-scale reconstruction of the Saccharomyces cerevisiae metabolic network. Genome Res. **13**, 244–253.

Chapter 27

Use of Genome-Scale Metabolic Models in Evolutionary Systems Biology

Balázs Papp, Balázs Szappanos, and Richard A. Notebaart

Abstract

One of the major aims of the nascent field of evolutionary systems biology is to test evolutionary hypotheses that are not only realistic from a population genetic point of view but also detailed in terms of molecular biology mechanisms. By providing a mapping between genotype and phenotype for hundreds of genes, genome-scale systems biology models of metabolic networks have already provided valuable insights into the evolution of metabolic gene contents and phenotypes of yeast and other microbial species. Here we review the recent use of these computational models to predict the fitness effect of mutations, genetic interactions, evolutionary outcomes, and to decipher the mechanisms of mutational robustness. While these studies have demonstrated that even simplified models of biochemical reaction networks can be highly informative for evolutionary analyses, they have also revealed the weakness of this modeling framework to quantitatively predict mutational effects, a challenge that needs to be addressed for future progress in evolutionary systems biology.

Key words: Flux balance analysis (FBA), constraint-based modeling, gene essentiality, genetic interaction, genome evolution, fitness landscape, metabolic network, *Saccharomyces cerevisiae*.

1. Introduction

Addressing many important questions in evolutionary biology relies on our understanding of the mapping between genotype and phenotype. Although evolutionary genetics analyses often use highly simplified and abstract genotype-fitness maps, recent advances in systems biology provide an unprecedented opportunity to calculate genotype–phenotype relationships using realistic mathematical models of molecular systems (1). With an increasing potential to computationally predict evolutionary relevant

J.I. Castrillo, S.G. Oliver (eds.), *Yeast Systems Biology*, Methods in Molecular Biology 759,
DOI 10.1007/978-1-61779-173-4_27,

parameters, such as the distribution of mutational effects and genetic interactions, molecular systems biology approaches begin to offer mechanistic insights into various topics from mutational robustness to genome evolution (2–6). Mathematical models that are most suited to address evolutionary questions can either directly calculate phenotypes that serve as fitness correlates (e.g., growth rate) or infer gene–gene relationships based on certain calculated phenotypes (e.g., identifying gene sets with correlated activities). These models include detailed kinetic models of metabolic pathways (7), regulatory circuits (cell cycle, circadian clock, etc.) (8), logical models of signaling networks (9), and constraint-based models of genome-scale metabolic networks (10). While most modeling approaches focus on small-scale biochemical systems (i.e., individual pathways) and characterize the mechanism of each enzymatic step, constraint-based models aim to calculate the metabolic behavior of relatively large systems (i.e., 600–1,300 genes) with relatively low data requirements. Moreover, these models are available for a number of microbial species, thereby providing a rigorous way to test evolutionary hypotheses.

The constraint-based framework uses mass balance and capacity constraints to define the space of all feasible steady-state flux distributions of the metabolic network leading from input (i.e., nutrient uptake) to output (an objective function, for instance, biomass production). Optimal network states are then identified within this space by maximizing or minimizing a certain metabolic objective function, an approach called flux balance analysis (FBA) (10, 11). However, the large size and comprehensive nature of these metabolic network models comes at a price as the framework lacks mechanistic details (e.g., kinetic rate constants and regulatory mechanisms), can only calculate steady-state patterns, and assumes that cells are fine-tuned from an evolutionary point of view. Furthermore, these models are restricted to simulate the effects of "large" mutations only (i.e., complete gene deletions or gene additions). Ultimately, the utility of genome-scale metabolic models for evolutionary analyses depends on how accurately they predict fitness correlates (e.g., growth phenotypes) and evolutionary relevant gene–gene relationships, a question that needs empirical investigation.

This chapter starts by discussing the use of constraint-based metabolic modeling in *Saccharomyces cerevisiae* to predict the effect of single and multiple mutations and hence to explore fitness landscapes (**Fig. 27.1a**). Fitness landscapes (or adaptive landscapes) visualize the relationship between genotypes and fitness and allow evolutionary biologists to investigate how mutations interact (epistasis) and which particular trajectories are taken during evolution. For example, the presence of multiple peaks on a fitness landscape indicates that some of the mutational paths to higher fitness alleles are selectively inaccessible (12) (**Fig. 27.1a**).

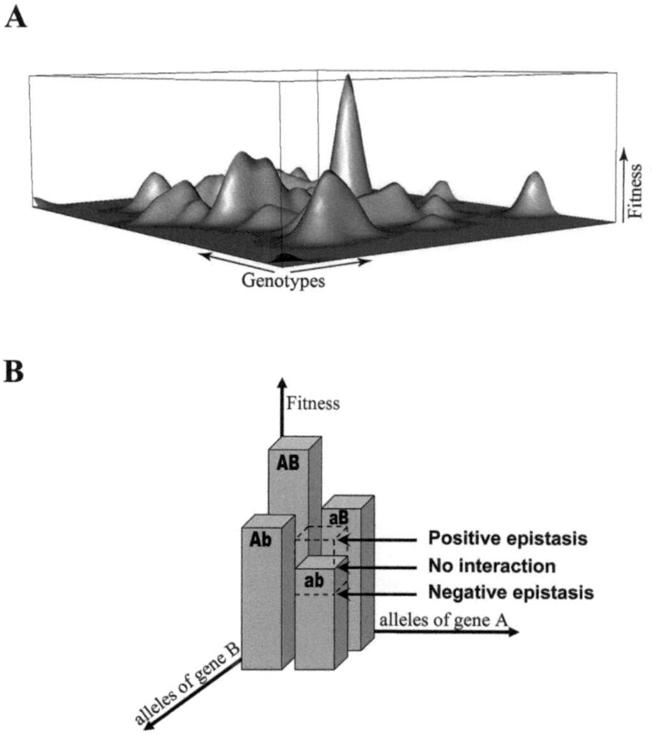

Fig. 27.1. Fitness landscape and epitasis studies. (**a**) An imaginary fitness landscape visualizing the relationship between genotype and fitness. The plane of the landscape contains all possible genotypes in such a way that similar genotypes are located close to each other on the plane and the height of the landscape reflects the fitness of the corresponding genotype. (**b**) Positive and negative epistases on a two-locus, two-allele genotype plane. Independence of gene action (no epistasis) is defined by a multiplicative model (i.e., when the fitness of the double mutant equals to the product of the fitnesses of the two single mutants).

Thus, to fully understand why particular evolutionary trajectories are realized and to what extent systems-level properties constrain the evolution of biochemical networks, we need detailed fitness landscapes of molecular systems. Next, we ask whether microbial metabolic network models have the potential to predict the outcome of evolutionary change, at least on short timescales, a question that has been addressed by laboratory evolution experiments in bacteria. Although most prior studies on evolutionary outcomes focused on *Escherichia coli* and other bacteria, we believe that similar approaches could also be adopted to analyze metabolic network evolution in baker's yeast. Finally, we analyze the shortcomings of constraint-based metabolic models to calculate weak fitness effects and genetic interactions, and discuss the utility of incorporating additional biological knowledge to increase their predictive performance. Although it is beyond the scope of the present review, we note that besides evolutionary analyses, genome-scale metabolic network reconstructions have also been employed to draw ecological inferences, for example, to infer the habitable environments of different species (13) and to investigate the ecological strategies of bacteria (14).

2. Interrogating the Fitness Landscape: Predicting Mutational Effects

A straightforward systems biology strategy to begin to explore fitness landscapes is to computationally predict the effects of single mutations and their pair-wise epistatic interactions (i.e., the non-independence between the phenotypic effects of two mutations) (*see* **Fig. 27.1b**). Besides providing an intelligible abstraction of the high-dimensional adaptive landscape, accurate prediction of single- and double-mutant phenotypes can be used, among others, to investigate the nature and evolution of genetic robustness (3, 15) and can be harnessed to identify potential novel antimicrobial drug targets (16, 17).

2.1. Computing the Growth Effect of Single-Gene Deletions

The popular approach of flux balance analysis uses optimization principles to find one particular solution (i.e., flux distribution) among all possible metabolic network states that satisfy the governing physicochemical constraints (10). The objective function of the optimization protocol may be the rate of biomass formation (growth rate) or usually the biomass yield (i.e., the rate of biomass production divided by the rate of nutrient uptake), given limiting nutrients from the environment. Thus, the phenotype of wild-type and mutant strains can be computationally characterized by their (optimal) growth rates, a phenotype that can be easily measured in laboratory experiments. This strategy formed the basis of one of the first applications of the yeast genome-scale metabolic model (18), which systematically compared in silico growth of single-gene deletant strains with in vivo growth phenotypes on a qualitative scale (i.e., lethal or viable) (19). Taking essential genes (i.e., those whose disruption leads to lethality under standard laboratory conditions) as a reference, different versions of the yeast model have been reported to predict essential and non-essential genes with 83–90% accuracy (19–21). However, this high percentage of consistent phenotypes obscures the fact that essential genes are both less frequent in vivo and more difficult to predict than non-essential ones (21–23). Plotting the true-positive and false-positive rates of essentiality predictions (**Fig. 27.2**) reveals that one model (iLL672) is clearly superior to another (iND750) in terms of its capacity to discriminate between lethal and viable knockout phenotypes across 16

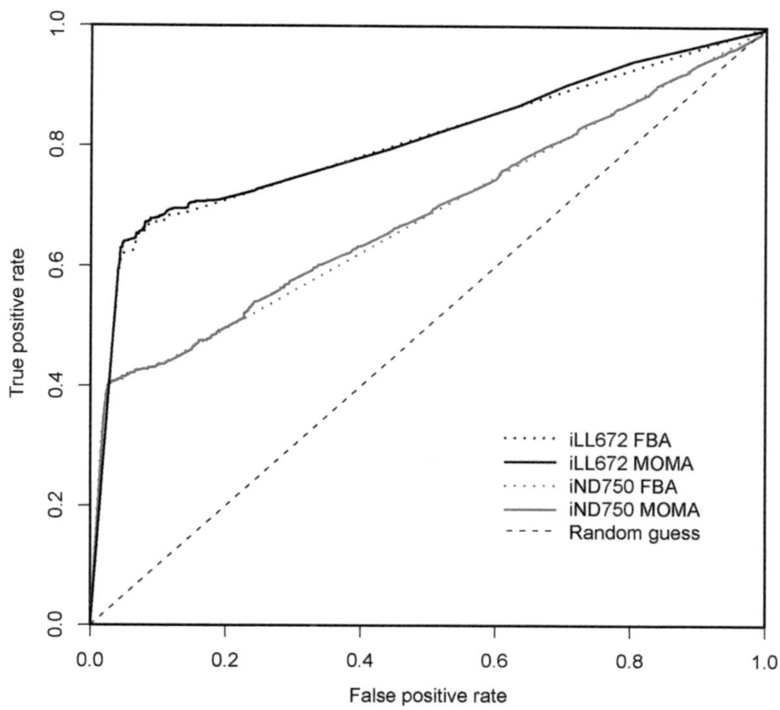

Fig. 27.2. Comparison of gene essentiality prediction performances of two different yeast metabolic network models (iND750 and iLL672) and two different optimization algorithms (FBA and MOMA) using receiver-operating characteristic (ROC) curves. ROC curves visualize classifier performance and can be employed to explore the trade-off between true-positive and false-positive rates at all possible cutoff levels (i.e., at different predicted growth rate cutoffs below which a strain is considered lethal). The closer the ROC plot is to the *upper left corner*, the higher the overall accuracy of the prediction. The *horizontal axis* represents the false-positive rate (number of true non-essential genes predicted as lethal/number of true non-essential genes), whereas the *vertical axis* represents the true-positive rate (number of correctly predicted true essential genes/number of true essential genes). We compiled gene essentiality data from (22), which measured growth phenotypes under 16 metabolically relevant conditions, and from (40), which identified genes that are essential even under nutrient-rich conditions. Computational predictions were taken from (22). Only genes present in both models were used for the comparison. We note that approximately one-third of the advantage of the iLL672 model against the iND750 model can be explained by differences in the biomass composition of the models (the area under the ROC curve of iLL672, iND750, and iND750 supplemented with the biomass composition of iLL672 is 0.8176, 0.6822, and 0.7241, respectively).

metabolically relevant growth conditions (*see* **Fig. 27.2** legend for details).

One conceptual reason why flux balance analysis might mispredict in vivo gene essentiality is that it assumes optimal network behavior even in gene knockout strains. To overcome this difficulty, other optimization criteria have been proposed that assume minimal flux reorganization in gene deletant strains with respect to the wild-type flux distribution (minimization of metabolic adjustment, MOMA, and regulatory on/off minimization, ROOM, algorithms, (24, 25)). Although this approach yields a more accurate prediction of mutant flux distributions (26), it only slightly improves gene essentiality predictions (22) (*see* continuous and dotted lines in **Fig. 27.2**). Apparently, successful prediction of essential genes more critically depends on the proper formulation of biomass compositions (i.e., model output) (21) and on the completeness of our knowledge of the metabolic processes associated with these genes (23).

While constraint-based metabolic models predict the presence or the absence of mutant growth with good accuracy, one might wonder whether these models could also capture quantitative growth differences of viable mutants. In theory, FBA or MOMA analysis of mutant strains gives growth yields as outputs, i.e., the rate of biomass production divided by the rate of limiting nutrient uptake, therefore providing quantitative predictions. The availability of large-scale competitive fitness (27) and growth curve measurement data (28) for viable yeast knockouts offers an opportunity to contrast predicted and experimentally determined growth parameters for hundreds of metabolic gene deletants. In agreement with a prior small-scale study (26), we generally find weak correlations between in vivo competitive fitness or growth rate and *in silico* biomass production (predicted biomass yield) (**Fig. 27.3a–c**). Here, it should be noted that constraint-based metabolic models compute biomass yields (29) and a comparison of growth rate data to predicted biomass yields (**Fig. 27.3c**) might not be fully descriptive. Therefore, we also plotted experimentally measured growth efficiency (a proxy for growth yield) against in silico-predicted biomass yield (**Fig. 27.3d**) which resulted in an even weaker association. This suggests that the constraint-based modeling approach fails to capture quantitative growth differences in yeast mutants, at least in batch cultures with glucose-minimal (SD) or -rich (YPD) media. It should be noted, however, that *S. cerevisiae* displays repressed respiration when grown aerobically in excess glucose (30, 31). Indeed, this regulatory effect varies across gene deletion backgrounds (32, 33) and it could strongly affect mutant growth in a way that cannot be easily captured by a stoichiometric model. It remains to be seen whether model predictions better match empirical data when mutants are grown on non-fermentable carbon sources.

2.2. Predicting Genetic Interactions

The fitness effect of mutation in one gene might be modulated by mutations in other genes, a phenomenon called genetic interaction or epistasis (**Fig. 27.1b**). Negative genetic interaction occurs when two mutations decrease fitness more than would be expected based on their individual effects (the most drastic form is referred to as synthetic lethality), and for positive interactions, the opposite is true. Epistatic relations reveal functional associations

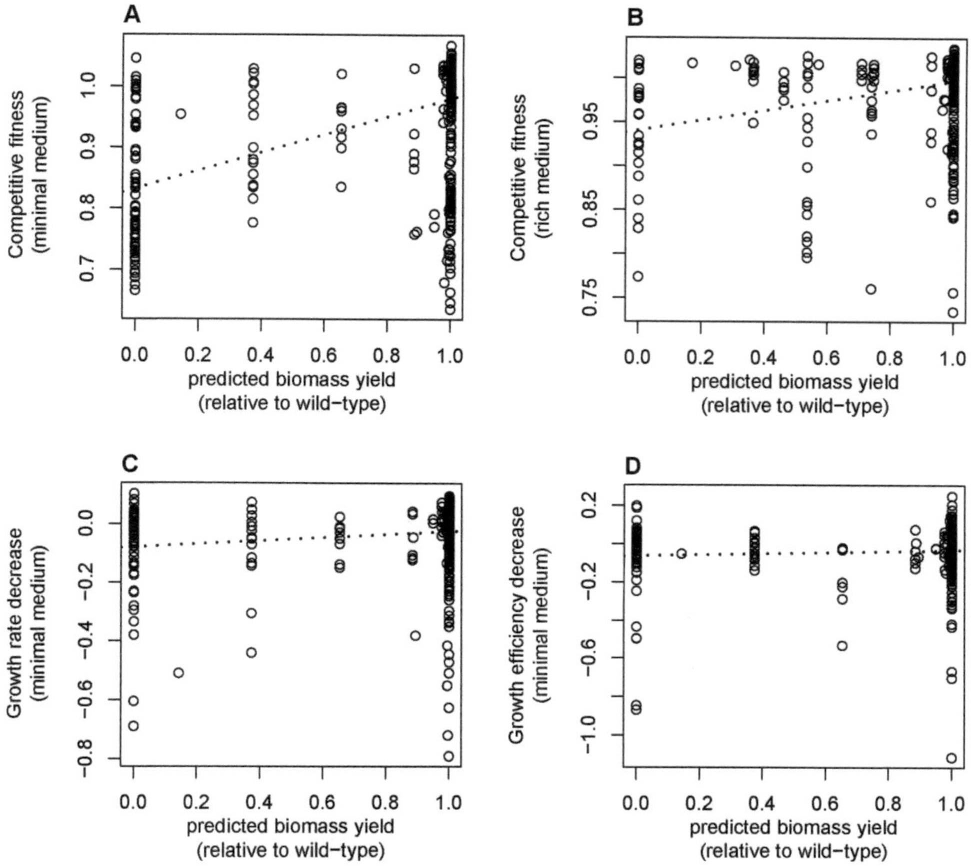

Fig. 27.3. Comparisons of in silico-predicted and experimentally determined growth parameters for viable metabolic gene deletion strains ($n = 499$). Computational predictions were taken from (22) and are based on the iLL672 model and MOMA algorithm (conclusions remain unchanged if prediction from the iND750 model or the FBA algorithm is used). (**a**) Comparison to competitive fitness data measured on glucose-minimal (SD) medium (27), Spearman's rho = 0.46, $p < 10^{-27}$. (**b**) Comparison to competitive fitness data measured on glucose-rich (YPD) medium (27), Spearman's rho = 0.26, $p < 10^{-8}$. (**c**) Comparison to growth rate data derived from growth curve measurements on SD (minimal medium) (28), Spearman's rho = 0.14, $p = 0.002$. (**d**) Comparison to growth efficiency data derived from growth curve measurements on SD (minimal medium) (28), Spearman's rho = 0.05, $p = 0.27$.

between genes (34) and influence many evolutionary processes (35), therefore it is of great importance to understand the molecular mechanisms underlying them and to develop reliable computational tools to predict them. Genome-scale metabolic models can rapidly calculate growth phenotypes for arbitrary sets of gene deletions, therefore, in principle, could be applied to systematically compute epistatic interactions between double or higher order gene knockouts. Indeed, yeast FBA models have been applied to compute both positive and negative pair-wise genetic interactions (36), and to identify synthetic lethality among multiple gene knockouts (37). However, relatively little is known about the accuracy of FBA models to capture different forms and magnitudes of in vivo genetic interactions, and one might expect that if predictions of single-gene deletion phenotypes are not perfect, then those of multiple gene deletions would be even less accurate. A small-scale experimental validation of model-predicted synthetic lethal pairs (synthetic lethals) in *S. cerevisiae* showed that almost 50% of them were correct (15), which is much higher than would be expected by chance (<1%, based on (38)). However, the FBA model missed more than 75% of published synthetic lethals in the metabolic network, suggesting that the constraint-based framework underestimates the prevalence of negative genetic interactions. Furthermore, given the apparent failure of FBA to capture quantitative growth differences of single mutants, one might speculate that positive and weak genetic interactions would be predicted with even lower success rates than synthetic lethals. Future studies using quantitative epistatic data from large-scale genetic interaction screens (39) would be needed to rigorously assess the performance and limitations of constraint-based metabolic models to predict epistasis.

2.3. Understanding Gene Dispensability and Mutational Robustness

Large-scale, single-gene deletion screens have revealed that almost 80% of protein coding genes in *S. cerevisiae* seem not to be essential for viability under standard laboratory conditions (40), an observation that tallies with results from similar analyses performed in other organisms (41). This finding raises the questions of what the mechanistic basis of gene dispensability is and whether it is the result of an evolved capacity of genetic networks to compensate for mutations. It has been suggested that the high fraction of non-essential genes might reflect mutational robustness, i.e., the capacity to compensate for mutations by using either redundant gene duplicates or alternative biochemical pathways (42). A second possibility is that seemingly dispensable genes are simply not active in the tested environmental condition(s), although they have important fitness contributions under special conditions (3). Computational systems biology models that can reliably predict the viability of single-gene deletants hold the promise to provide mechanistic explanations for gene dispensability. Indeed, a flux balance analysis model of yeast metabolism showed that a large fraction of non-essential enzymatic genes catalyze reactions that are inactive under the tested condition (i.e., carry zero flux), hence there might be no need to invoke any compensatory mechanism to explain their dispensability (3, 43). Furthermore, according to the model, genes that are active but not essential are mostly compensated by redundant gene duplicates and not by alternative pathways. It should be noted, however, that FBA models assume optimal network behavior and are therefore likely to overestimate the number of in vivo inactive reactions (i.e., suboptimal pathways are completely silenced in model solutions) (44). Nevertheless, some of the above computational predictions have been confirmed experimentally: ^{13}C-flux analysis of viable yeast knockouts showed that flux rerouting through alternative pathways explains only the minority of non-essential genes (43). Furthermore, simulating gene deletions under a number of different nutrient-limiting conditions predicted that many functionally inactive genes would become essential under other conditions (3). A large-scale chemical genomic assay in yeast provided strong support for this notion: 97% of gene deletions exhibited a measurable growth phenotype in at least one of hundreds of tested conditions compared to only 34% in rich medium (45).

Although the above computational and experimental studies suggest that gene dispensability is only apparent and is mostly explained by condition-specific gene functions, other empirical works showed that most non-essential genes display synthetic lethal interactions with some other genes (23). As synthetic lethal genetic interactions indicate compensation between two genes (i.e., mutational robustness), this raises the question as to how these seemingly contradictory findings can be reconciled. A flux balance analysis study of synthetic genetic interactions under a large number of environmental conditions demonstrated that the capacity to compensate null mutations varies substantially between different nutritional environments (15). More specifically, it has been shown computationally, and confirmed by double deletion experiments, that synthetic lethal interactions are often restricted to particular environmental conditions, partly because genes that are compensated in one condition make an essential fitness contribution in another condition. Further empirical studies on yeast gene duplicates corroborated the widespread condition dependency of mutational compensation (46, 47).

The above findings also offer indirect insights into the selective forces shaping metabolic network evolution. Instead of regarding apparent redundancies as adaptations against harmful mutations (42), the presence of distinct but functionally overlapping metabolic pathways more likely reflects the outcome of an evolutionary adaptation characterized by selection for growth in varying environments (i.e., various different nutrients). As a correlated response, some of these pathways may also increase mutational resilience under some conditions (15, 48).

3. Predicting Evolutionary Outcomes: FBA as an Evolutionary Optimization Model

How predictable is evolution? Evolutionary change is often considered to be contingent on initial conditions and chance events and therefore unique on the one hand, and replicable owing to predictable adaptive changes on the other hand (49). Systems biology modeling offers novel ways to investigate the predictability of evolutionary outcomes and organismal diversity. In particular, metabolic network models have recently been employed to test various hypotheses regarding the end state of evolutionary process. First, flux balance analysis of metabolic networks gives specific predictions on the steady-state behavior of evolutionary adapted metabolic systems. Second, by accounting for systems-level gene functions and relationships between genes, constraint-based metabolic models also have the power to predict inter-species differences in metabolic gene content, therefore to explain comparative genomics patterns. The latter includes predicting genes most likely undergoing loss and horizontal transfer events (50), asymmetric gain or loss of enzyme pairs (4), and gene contents of reduced genome endosymbiotic bacteria based on knowledge of its distant ancestors and its current lifestyle (5). Here, we restrict our attention to the use of FBA models to generate testable hypotheses on the outcome of short-term adaptive evolution.

As mentioned above, flux balance analysis uses optimization principles to find one particular network state that maximizes biomass production, that is, cellular growth. Because growth can be considered as a fitness correlate in microbes, FBA models can be seen as models about adaptation (51) in which in vivo metabolic states are sought that maximize organismal fitness. Microbial metabolism is optimized by the process of adaptive evolution, therefore FBA has, in principle, the potential to predict the outcome of evolutionary adaptation and give insight into the constraints that influence adaptation. An essential step in optimality approaches is to test the model predictions against empirical observations to reveal the particular selective forces and constraints that might have played significant roles during the evolutionary history of the organism under study.

Various experimental works have been performed to evaluate the power of FBA to predict the outcome of both natural and laboratory evolution. For example, it has been demonstrated that in vivo and *in silico* flux distributions are consistent under certain environmental conditions in *E. coli* (11, 25), suggesting that maximizing biomass production might have been an important selective force in the history of *E. coli*. However, simple FBA models fail to explain the metabolic behavior of microbes that do not metabolize nutrients most efficiently (29, 48). For instance, *S. cerevisiae* uses a mixture of respiration (high-yield route) and fermentation (low-yield route) to utilize glucose even under aerobic conditions when glucose is abundant in the medium (29). Applying alternative objective functions instead of biomass yield (52) or using game-theoretical approaches (53, 54), that is, formulating the optimization problem as frequency dependent instead of frequency independent, could help to resolve discrepancies between model predictions and experimental observations.

Suboptimal metabolic behavior of wild-type strains might also stem from incomplete adaptedness to the tested conditions and experiments have been designed to adapt strains under specific growth selection pressures in the laboratory to test whether evolved strains display *in silico*-predicted growth properties. One such adaptive evolution experiment has been performed in *E. coli*, with growth rate as the selection criterion and glycerol uptake as the main carbon source to produce biomass components (55). Cells initially grew sub-optimally on glycerol, but adapted over a period of ~60 days (1,000 generations), toward the FBA-predicted optimal behavior. One would expect that such an adaptation event should be reflected in the genotype and this has been demonstrated by whole-genome resequencing of evolved strains. In particular, mutations were identified in the glycerol kinase gene, which is clearly associated with the growth environment (56). Adaptation to utilize glycerol has also been predicted, and experimentally verified, for a lactic acid bacteria, showing extremely low initial growth rates in a glycerol environment (57). Growth of lactic acid bacteria with glycerol as the main carbon source has never been demonstrated experimentally before, even though the metabolic model predicted this output phenotype. Similarly, FBA can be used to predict the result of evolutionary adaptation in response to gene deletions, that is, the outcome of compensatory evolution. For example, adaptive evolution of *E. coli* strains carrying metabolic gene deletions resulted in increased growth rates that were similar to those predicted by FBA (in 78% of the strains tested) (58). Taken together, these studies clearly demonstrate that FBA has the potential to predict the outcome of adaptive evolution at the phenotype level.

4. Future Challenges

Constraint-based models of microbial genome-scale metabolic reconstructions present a simple computational framework to explore the metabolic capacity of wild-type and mutant strains under different environmental conditions, thereby providing a mapping between genotype and metabolic phenotype. Despite its simplicity and dependence on optimality principles, these models proved successful to compute the viability of mutant strains, to make testable predictions on gene loss and gene gain patterns on the phylogenetic tree, and, in some cases, to predict the phenotypic outcome of short-term adaptive evolution. However, the same approach appears to perform poorly in predicting weak growth effects of mutations in yeast, at least on glucose media. Furthermore, it remains to be seen how accurately in vivo genetic interactions can be captured within this framework. Given that weak fitness effects and epistatic interactions are especially important to understand the process of evolution and are more frequent than strong effects, there is a great need to develop computational approaches that can accurately describe mutations with weak phenotypic impacts (1). We now briefly discuss some possible strategies to improve the predictive power of the constraint-based framework. First, constraint-based metabolic models can be made more realistic by imposing additional relevant constraints to decrease the solution space. Some of the proposed extra constraints include thermodynamic constraints (59) (i.e., elimination of thermodynamically infeasible solutions) and regulatory constraints (60) (i.e., elimination of reactions that are repressed under a given condition). With the rapid ongoing development of high-throughput techniques that accumulate data on intracellular metabolite concentrations, mRNA, protein expression levels, and reaction fluxes (61, 62), these additional constraints could be routinely applied in future studies (*see* (63, 64)). Second, new algorithms to compute the immediate physiological effect of mutations need to be developed and tested. Clearly, the assumption of optimal growth is not tenable for mutants and some methods have been put forward to describe metabolic states after gene removal (24, 25). However, these modified optimization algorithms are largely ad hoc, and it remains to be seen whether more realistic alternative methods can be developed based on empirical data on physiological changes following gene deletions, including high-throughput data on alterations in growth properties (28), metabolic footprints (65), intracellular fluxes (43), and mRNA expression (66). Furthermore, by assuming maximal biomass production in the wild type, the FBA approach is unable to capture beneficial loss-of-function mutations (only the addition of new reactions could increase *in silico* growth in this framework). This particular shortage of FBA could be alleviated only by developing new algorithms to calculate wild-type growth behavior. Third, there are efforts underway to reconcile constraint-based and kinetic modeling approaches in order to build large-scale dynamic models of cellular metabolism without the need for extensive experimental data (67, 68). Such hybrid frameworks would allow the piecewise incorporation of both additional flux constraints and information on enzyme kinetics when they become available and would hopefully improve the predictive power of large-scale models in determining metabolic responses to perturbations. Given the tremendous efforts put into generating functional genomics and comparative data and inferring cellular networks in yeasts, we expect that *S. cerevisiae* would be at the forefront of developing new generations of genome-scale metabolic models and applying them to evolutionary questions.

Acknowledgments

We thank Csaba Pál for suggestions on the manuscript and Kiran Patil and Juan I. Castrillo for comments on the issue of quantitative fitness predictions with FBA. B.P. is supported by The International Human Frontier Science Program Organization, the Hungarian Scientific Research Fund (OTKA), the "Lendület Program," and the Bolyai Fellowship of the Hungarian Academy of Sciences. R.N. is supported by The Netherlands Genomics Initiative (NGI – Horizon grant) and The Netherlands Organisation for Scientific Research (NWO – VENI Grant).

References

1. Loewe, L. (2009) A framework for evolutionary systems biology. *BMC Syst. Biol.* **3**, 27.
2. Endy, D., You, L., Yin, J., and Molineux, I. J. (2000) Computation, prediction, and experimental tests of fitness for bacteriophage T7 mutants with permuted genomes. *Proc. Natl. Acad. Sci. USA* **97**, 5375–5380.
3. Papp, B., Pál, C., and Hurst, L. D. (2004) Metabolic network analysis of the causes and evolution of enzyme dispensability in yeast. *Nature* **429**, 661–664.
4. Notebaart, R. A., Kensche, P. R., Huynen, M. A., and Dutilh, B. E. (2009) Asymmetric relationships between proteins shape genome evolution. *Genome Biol.* **10**, R19.
5. Pál, C., Papp, B., Lercher, M. J., Csermely, P., Oliver, S. G., and Hurst, L. D. (2006) Chance and necessity in the evolution of minimal metabolic networks. *Nature* **440**, 667–670.
6. Loewe, L., and Hillston, J. (2008) The distribution of mutational effects on fitness in a simple circadian clock. *Lect. Notes Bioinf.* **5307**, 156–175.
7. Teusink, B., Walsh, M. C., van Dam, K., and Westerhoff, H. V. (1998) The danger of metabolic pathways with turbo design. *Trends Biochem. Sci.* **23**, 162–169.
8. Chen, K. C., Calzone, L., Csikasz-Nagy, A., Cross, F. R., Novak, B., and Tyson, J. J. (2004) Integrative analysis of cell cycle control in budding yeast. *Mol. Biol. Cell* **15**, 3841–3862.
9. Christensen, T. S., Oliveira, A. P., and Nielsen, J. (2009) Reconstruction and logical modeling of glucose repression signaling pathways in *Saccharomyces cerevisiae*. *BMC Syst. Biol.* **3**, 7.
10. Price, N. D., Reed, J. L., and Palsson, B. O. (2004) Genome-scale models of microbial cells: evaluating the consequences of constraints. *Nat. Rev. Microbiol.* **2**, 886–897.
11. Kauffman, K. J., Prakash, P., and Edwards, J. S. (2003) Advances in flux balance analysis. *Curr. Opin. Biotechnol.* **14**, 491–496.
12. Poelwijk, F. J., Kiviet, D. J., Weinreich, D. M., and Tans, S. J. (2007) Empirical fitness landscapes reveal accessible evolutionary paths. *Nature* **445**, 383–386.
13. Borenstein, E., Kupiec, M., Feldman, M. W., and Ruppin, E. (2008) Large-scale reconstruction and phylogenetic analysis of metabolic environments. *Proc. Natl. Acad. Sci. USA* **105**, 14482–14487.
14. Freilich, S., Kreimer, A., Borenstein, E., et al. (2009) Metabolic-network-driven analysis of bacterial ecological strategies. *Genome Biol.* **10**, R61.
15. Harrison, R., Papp, B., Pal, C., Oliver, S. G., and Delneri, D. (2007) Plasticity of genetic interactions in metabolic networks of yeast. *Proc. Natl. Acad. Sci. USA* **104**, 2307–2312.
16. Raman, K., Rajagopalan, P., and Chandra, N. (2005) Flux balance analysis of mycolic acid pathway: targets for anti-tubercular drugs. *PLoS Comput. Biol.* **1**, e46.
17. Lee, D. S., Burd, H., Liu, J., et al. (2009) Comparative genome-scale metabolic reconstruction and flux balance analysis of multiple *Staphylococcus aureus* genomes identify novel antimicrobial drug targets. *J. Bacteriol.* **191**, 4015–4024.
18. Forster, J., Famili, I., Fu, P., Palsson, B. O., and Nielsen, J. (2003) Genome-scale reconstruction of the *Saccharomyces cerevisiae* metabolic network. *Genome Res.* **13**, 244–253.

19. Forster, J., Famili, I., Palsson, B. O., and Nielsen, J. (2003) Large-scale evaluation of in silico gene deletions in *Saccharomyces cerevisiae*. *Omics* 7, 193–202.
20. Duarte, N. C., Herrgard, M. J., and Palsson, B. O. (2004) Reconstruction and validation of *Saccharomyces cerevisiae* iND750, a fully compartmentalized genome-scale metabolic model. *Genome Res.* **14**, 1298–1309.
21. Kuepfer, L., Sauer, U., and Blank, L. M. (2005) Metabolic functions of duplicate genes in *Saccharomyces cerevisiae*. *Genome Res.* **15**, 1421–1430.
22. Snitkin, E. S., Dudley, A. M., Janse, D. M., Wong, K., Church, G. M., and Segre, D. (2008) Model-driven analysis of experimentally determined growth phenotypes for 465 yeast gene deletion mutants under 16 different conditions. *Genome Biol.* **9**, R140.
23. Becker, S. A., and Palsson, B. O. (2008) Three factors underlying incorrect in silico predictions of essential metabolic genes. *BMC Syst. Biol.* **2**, 14.
24. Segrè, D., Vitkup, D., and Church, G. M. (2002) Analysis of optimality in natural and perturbed metabolic networks. *Proc. Natl. Acad. Sci. USA* **99**, 15112–15117.
25. Shlomi, T., Berkman, O., and Ruppin, E. (2005) Regulatory on/off minimization of metabolic flux changes after genetic perturbations. *Proc. Natl. Acad. Sci. USA* **102**, 7695–7700.
26. Snitkin, E. S., and Segre, D. (2008) Optimality criteria for the prediction of metabolic fluxes in yeast mutants. *Genome Inform.* **20**, 123–134.
27. Deutschbauer, A. M., Jaramillo, D. F., Proctor, M., et al. (2005) Mechanisms of haploinsufficiency revealed by genome-wide profiling in yeast. *Genetics* **169**, 1915–1925
28. Warringer, J., Ericson, E., Fernandez, L., Nerman, O., and Blomberg, A. (2003) High-resolution yeast phenomics resolves different physiological features in the saline response. *Proc. Natl. Acad. Sci. USA* **100**, 15724–15729.
29. Schuster, S., Pfeiffer, T., and Fell, D. A. (2008) Is maximization of molar yield in metabolic networks favoured by evolution? *J. Theor. Biol.* **252**, 497–504.
30. Gancedo, J. M. (1998) Yeast carbon catabolite repression. *Microbiol. Mol. Biol. Rev.* **62**, 334–361.
31. Sonnleitner, B., and Kappeli, O. (1986) Growth of *Saccharomyces cerevisiae* is controlled by its limited respiratory capacity: formulation and verification of a hypothesis. *Biotechnol. Bioeng.* **28**, 927–937.
32. Schuurmans, J. M., Boorsma, A., Lascaris, R., Hellingwerf, K. J., and Teixeira de Mattos, M. J. (2008) Physiological and transcriptional characterization of *Saccharomyces cerevisiae* strains with modified expression of catabolic regulators. *FEMS Yeast Res.* **8**, 26–34.
33. Usaite, R., Jewett, M. C., Oliveira, A. P., Yates, J. R., 3rd, Olsson, L., and Nielsen, J. (2009) Reconstruction of the yeast Snf1 kinase regulatory network reveals its role as a global energy regulator. *Mol. Syst. Biol.* **5**, 319.
34. Boone, C., Bussey, H., and Andrews, B. J. (2007) Exploring genetic interactions and networks with yeast. *Nat. Rev. Genet.* **8**, 437–449.
35. Wolf, J. B., Brodie, E. D., and Wade, M. J. (2000) *Epistasis and the Evolutionary Process*. New York, NY: Oxford University Press.
36. Segrè, D., Deluna, A., Church, G. M., and Kishony, R. (2005) Modular epistasis in yeast metabolism. *Nat. Genet.* **37**, 77–83.
37. Deutscher, D., Meilijson, I., Kupiec, M., and Ruppin, E. (2006) Multiple knockout analysis of genetic robustness in the yeast metabolic network. *Nat. Genet.* **38**, 993–998.
38. Tong, A. H., Lesage, G., Bader, G. D., et al. (2004) Global mapping of the yeast genetic interaction network. *Science* **303**, 808–813.
39. Schuldiner, M., Collins, S. R., Thompson, N. J., et al. (2005) Exploration of the function and organization of the yeast early secretory pathway through an epistatic miniarray profile. *Cell* **123**, 507–519.
40. Giaever, G., Chu, A. M., Ni, L., et al. (2002) Functional profiling of the *Saccharomyces cerevisiae* genome. *Nature* **418**, 387–391.
41. Hurst, L. D., and Pál, C. (2007) Genomic redundancy and dispensability. In: Pagel, M., and Pomiankowski, A. (eds.), *Evolutionary Genomics and Proteomics* (pp. 141–160). Sunderland, MA: Sinauer Associates Inc.
42. Wagner, A. (2000) Robustness against mutations in genetic networks of yeast. *Nat. Genet.* **24**, 355–361.
43. Blank, L. M., Kuepfer, L., and Sauer, U. (2005) Large-scale 13C-flux analysis reveals mechanistic principles of metabolic network robustness to null mutations in yeast. *Genome Biol.* **6**, R49.
44. Nishikawa, T., Gulbahce, N., and Motter, A. E. (2008) Spontaneous reaction silencing in metabolic optimization. *PLoS Comput. Biol.* **4**, e1000236.
45. Hillenmeyer, M. E., Fung, E., Wildenhain, J., et al. (2008) The chemical genomic

portrait of yeast: uncovering a phenotype for all genes. *Science* **320**, 362–365.
46. Ihmels, J., Collins, S. R., Schuldiner, M., Krogan, N. J., and Weissman, J. S. (2007) Backup without redundancy: genetic interactions reveal the cost of duplicate gene loss. *Mol. Syst. Biol.* **3**, 86.
47. Musso, G., Costanzo, M., Huangfu, M., et al. (2008) The extensive and condition-dependent nature of epistasis among whole-genome duplicates in yeast. *Genome Res.* **18**, 1092–1099.
48. Papp, B., Teusink, B., and Notebaart, R. A. (2009) A critical view of metabolic network adaptations. *HFSP J.* **3**, 24–35.
49. Travisano, M., Mongold, J. A., Bennett, A. F., and Lenski, R. E. (1995) Experimental tests of the roles of adaptation, chance, and history in evolution. *Science* **267**, 87–90.
50. Pál, C., Papp, B., and Lercher, M. J. (2005) Adaptive evolution of bacterial metabolic networks by horizontal gene transfer. *Nat. Genet.* **37**, 1372–1375.
51. Parker, G. A., and Smith, J. M. (1990) Optimality theory in evolutionary biology. *Nature* **348**, 27–33.
52. Schuetz, R., Kuepfer, L., and Sauer, U. (2007) Systematic evaluation of objective functions for predicting intracellular fluxes in *Escherichia coli*. *Mol. Syst. Biol.* **3**, 119.
53. Pfeiffer, T., and Schuster, S. (2005) Game-theoretical approaches to studying the evolution of biochemical systems. *Trends Biochem. Sci.* **30**, 20–25.
54. MacLean, R. C. (2008) The tragedy of the commons in microbial populations: insights from theoretical, comparative and experimental studies. *Heredity* **100**, 471–477.
55. Ibarra, R. U., Edwards, J. S., and Palsson, B. O. (2002) Escherichia coli K-12 undergoes adaptive evolution to achieve in silico predicted optimal growth. *Nature* **420**, 186–189.
56. Herring, C. D., Raghunathan, A., Honisch, C., et al. (2006) Comparative genome sequencing of *Escherichia coli* allows observation of bacterial evolution on a laboratory timescale. *Nat. Genet.* **38**, 1406–1412.
57. Teusink, B., Wiersma, A., Jacobs, L., Notebaart, R. A., and Smid, E. J. (2009) Understanding the adaptive growth strategy of *Lactobacillus plantarum* by in silico optimisation. *PLoS Comput. Biol.* **5**, e1000410.
58. Fong, S. S., and Palsson, B. O. (2004) Metabolic gene-deletion strains of *Escherichia coli* evolve to computationally predicted growth phenotypes. *Nat. Genet.* **36**, 1056–1058.
59. Beard, D. A., Liang, S. D., and Qian, H. (2002) Energy balance for analysis of complex metabolic networks. *Biophys. J.* **83**, 79–86.
60. Covert, M. W., Schilling, C. H., and Palsson, B. (2001) Regulation of gene expression in flux balance models of metabolism. *J. Theor. Biol.* **213**, 73–88.
61. Sauer, U. (2006) Metabolic networks in motion: 13C-based flux analysis. *Mol. Syst. Biol.* **2**, 62.
62. Ishii, N., Nakahigashi, K., Baba, T., et al. (2007) Multiple high-throughput analyses monitor the response of *E. coli* to perturbations. *Science* **316**, 593–597.
63. Akesson, M., Forster, J., and Nielsen, J. (2004) Integration of gene expression data into genome-scale metabolic models. *Metab. Eng.* **6**, 285–293.
64. Hoppe, A., Hoffmann, S., and Holzhutter, H. G. (2007) Including metabolite concentrations into flux balance analysis: thermodynamic realizability as a constraint on flux distributions in metabolic networks. *BMC Syst. Biol.* **1**, 23.
65. Allen, J., Davey, H. M., Broadhurst, D., et al. (2003) High-throughput classification of yeast mutants for functional genomics using metabolic footprinting. *Nat. Biotechnol.* **21**, 692–696.
66. Hughes, T. R., Marton, M. J., Jones, A. R., et al. (2000) Functional discovery via a compendium of expression profiles. *Cell* **102**, 109–126.
67. Smallbone, K., Simeonidis, E., Broomhead, D. S., and Kell, D. B. (2007) Something from nothing: bridging the gap between constraint-based and kinetic modelling. *FEBS J* **274**, 5576–5585.
68. Covert, M. W., Xiao, N., Chen, T. J., and Karr, J. R. (2008) Integrating metabolic, transcriptional regulatory and signal transduction models in *Escherichia coli*. *Bioinformatics* **24**, 2044–2050.

Section IV

Yeast Systems Biology in Practice: *Saccharomyces cerevisiae* as a Tool for Mammalian Studies

Chapter 28

Contributions of *Saccharomyces cerevisiae* to Understanding Mammalian Gene Function and Therapy

Nianshu Zhang and Elizabeth Bilsland

Abstract

Due to its genetic tractability and ease of manipulation, the yeast *Saccharomyces cerevisiae* has been extensively used as a model organism to understand how eukaryotic cells grow, divide, and respond to environmental changes. In this chapter, we reasoned that functional annotation of novel genes revealed by sequencing should adopt an integrative approach including both bioinformatics and experimental analysis to reveal functional conservation and divergence of complexes and pathways. The techniques and resources generated for systems biology studies in yeast have found a wide range of applications. Here we focused on using these technologies in revealing functions of genes from mammals, in identifying targets of novel and known drugs and in screening drugs targeting specific proteins and/or protein–protein interactions.

Key words: Functional complementation, conservation and divergence, interactome, chemogenomic profiling, drug targets, drug screening.

1. Introduction

The budding yeast, *Saccharomyces cerevisiae*, has long been used as a model organism to study how eukaryotic cells grow, divide, and respond to environmental cues. Conservation of these cellular activities allows functional annotation of mammalian gene products in this simple eukaryotic cell (1). The first such functional complementation experiment was performed by Kataoka et al. (2). They demonstrated that the expression of one of the human paralogous Ras proteins restores the viability of the yeast *ras1ras2* double mutants. In 1987, Lee and Nurse demonstrated that the human Cdc2 could functionally replace its counterpart in

J.I. Castrillo, S.G. Oliver (eds.), *Yeast Systems Biology*, Methods in Molecular Biology 759,
DOI 10.1007/978-1-61779-173-4_28, © Springer Science+Business Media, LLC 2011

the fission yeast *Schizosaccharomyces pombe* (3). Ever since, many human proteins have been shown to complement either the mutation or the deletion of yeast genes involved in every aspect of cellular life, including cell cycle, DNA repair, signaling, and biosynthesis of the building blocks of cells.

Since the publication of its genome sequence in 1996 (4), a wide range of techniques and resources have been generated for functional genomics and systems biology studies (*see* the review by Suter et al. (5)). As a result, more than 72% of the open reading frames (ORFs) in yeast have been characterized, as of July 2009 (SGD). For example, using the collection of the deletion strains (6) grown under multiple environmental conditions including drugs, it was estimated that more than 97% of all these ORFs have a phenotype (7–9). With the availability of these resources and technologies and, the ability of yeast cells to grow for extended periods under highly controlled conditions, *S. cerevisiae* has become a model organism for pharmacological, post-genomic (10), and systems biology studies (11).

There are now more than 600 completely sequenced genomes of all organisms (12). The availability of multiple genome sequences should allow inference about functions of proteins, based on their sequence and structure similarity (13). As a result, it was predicted that about 25% of the ORFs encoded by the yeast genome have homologues in humans ($p < e^{-30}$, euGenes, http://eugenes.org/). The percentage of yeast ORFs which have human homologues varies between different functional categories, with those in categories of transposable elements and uncharacterized proteins significantly smaller than average and those in cell cycle and DNA processing, metabolism, protein fate, protein synthesis, signal transduction and mechanisms higher than average. However, such a comparative analysis should not rely exclusively on bioinformatics, since this does not cope adequately with the multidimensional nature of the proteome and complexity of functional diversification during the course of evolution.

2. Functional Annotation Requires an Integrative Approach

Problems encountered with bioinformatics are twofold. First, not all the functional counterparts can be identified by bioinformatics analysis. For example, there is no obvious yeast homologue to F6, the bovine coupling factor 6, as predicted from primary sequence comparison of the putative polypeptides encoded by all the open reading-frames in the yeast genome. However, it has been demonstrated that expression of bovine F6 complements a null mutation in *ATP14* gene in yeast *S. cerevisiae*; Atp14p has just 14.5% amino acid sequence identity to F6 (14). Second, sequence similarity between one individual protein and other protein sequences in the public data libraries is not sufficient to determine function.

The reasons for this may be threefold. First, functionality may be assigned in biochemical terms, while giving no clear indication of the biological role of the novel protein. For instance, recognizing that an ORF encodes a protein kinase or phosphatase reveals little about the metabolic or developmental pathway in which such an enzyme may be involved. One example of this scenario is the S6 kinase 1 (S6K1) in human and the Sch9 kinase in yeast. S6K1 is a member of the AGC family of protein kinases and a direct target of mTORC1 (15) which regulates ribosome biogenesis in mammalian cells. Sch9 is also a member of the AGC family but was initially classified as the likely orthologue of mammalian PKB based on bioinformatics analysis (16). However, the finding that Sch9 is a direct target of yeast TORC1 involved in ribosome biogenesis (17) suggests that Sch9 is the yeast orthologue of mammalian S6K1. Second, the assignment of function in the organism where the gene or protein was originally discovered may have been incorrect or superficial. For instance, yeast chromosome III contains an ORF showing greater than 40% amino acid sequence identity to the NifS proteins of nitrogen-fixing bacteria (18). *S. cerevisiae* does not fix N_2, yet the *nifS* homologue is an essential gene. Similar genes have been found in a number of other bacteria, none of which fix nitrogen, and experimental and informatic analyses (19) suggest that they encode a class of transaminases that use pyridoxal phosphate as a co-factor. Third, the function assigned to genes in the lower organisms may have been lost or new function gained in the orthologues of higher organisms during the course of evolution. For example, both human and yeast Clp1 encode one of the five subunits of the conserved cleavage factor I which functions in cleavage and polyadenylation of mRNA 3′ ends (20). However, despite their high sequence and structural similarity, the expression of human Clp1, but not the yeast Clp1, is able to complement conditional and lethal mutations in the essential 5′-OH RNA kinase module of yeast or plant tRNA ligases. Moreover, human Clp1 cannot complement the growth defect of a yeast *clp1Δ* strain, indicating that yeast and human Clp1 proteins are not functional orthologues (21). These examples demonstrate that an integrated systems approach, including both "wet experiments" and bioinformatics, is necessary to uncover the functions of the novel and known genes discovered by systematic sequencing.

Nevertheless, even when functional equivalence of orthologues is reliably predicted by informatics or confirmed by "wet" experiments, yeast can still be a useful system to analyze the functions of essential amino acids or domains of proteins from higher eukaryotes, especially those involved in diseases. It is estimated that 50% of human genes implicated in inheritable diseases have yeast homologues (22). Examples of using yeast to characterize human disease genes include those implicated in congenital disorders of glycosylations (23) and in mitochondrial disorders (24).

3. Functional Conservation and Divergence

Due to its ease of manipulation and genetic tractability (25), the yeast *S. cerevisiae* has been used to analyze the functions of many proteins from mammalian cells. Cases of these studies are summarized according to their specific applications in revealing both functional conservation and divergence.

3.1. Methodology

Functional complementation experiments generally involve constructing yeast mutants which are viable conditionally and expressing the candidate gene in the mutant cells. Complementation is assumed when cell viability is observed at non-permissive conditions. There are a variety of means of generating conditional yeast mutants, including knockout mutants (6), temperature-sensitive and drug-sensitive mutants, synthetically lethal mutants (26), or mutants in which the endogenous gene transcript (27–29) and/or protein level (30, 31) is regulated.

A second class of complementation experiments is performed by transforming the candidate gene into a heterozygous yeast diploid strain followed by sporulation and tetrad dissection. Functional complementation is confirmed if expression of the candidate gene rescues the phenotype (either inviability or other conditional phenotypes) of the deletion haploid cells.

A third approach includes expression of both the yeast and the candidate genes in different plasmids in a yeast haploid strain in which the endogenous gene has been deleted from the genome. Usually, the plasmid used for expressing the yeast gene has a marker which can be counter selected, such as *URA3* in the presence of 5-FOA. A reversal of the phenotype of the deletion cells under the condition of counter-selection would demonstrate complementation due to the expression of the candidate gene. This approach has been used for characterizing the functions of both yeast (32) and human proteins (33).

Expression of mammalian genes in yeast is normally driven by strong promoters. However, using regulatable promoters, such as that of the *MET3* gene, provides additional advantage of eliminating false positives when a human expression library is screened for functional homologues to a yeast gene (27).

3.2. Essential and Minimal Functional Domains

Once the functional equivalence between yeast and mammalian proteins is established, the yeast-based system can be used to reveal essential and/or minimal functional domains. Sharma et al. (34) demonstrated that the C-terminal non-catalytic domain of WRN (encoding a RecQ helicase in humans) is sufficient to rescue the yeast *dna2* mutant and this domain physically interacts with and stimulates flap-endonuclease 1 in Okazaki fragment processing. Analysis of the members of the copper uptake transporter (CTR) revealed that the third transmembrane domain (TM3) plays an essential role in trimeric CTR assembly and function (35). Essential functional domains can also be demonstrated by using human–yeast chimeras to complement the deletion or mutation of the yeast genes (36–38).

3.3. Differentially Splicing Variants

The human proteome is significantly more complex than its genome, partly due to the differential splicing of mRNA transcribed from the same gene. A few studies have revealed functional divergences of the splice variants where only one splice variant is functional or different variants are functional at different locations. For example, Vaz et al. (39) reported that only one of several splice variants of the human TAZ gene, lacking exon 5, complemented the defective acylation phenotype and restored growth of the yeast *taz1Δ* mutant on glycerol/ethanol at 37°C. Consistently, Ma et al. (40) demonstrated that only the same variant complemented the *taz1Δ* mutant defects in energy coupling and sensitivity of mitochondria to an elevated temperature. Similarly, both splicing variants of human divalent metal transporter 1, DMT1A and DMT1B, can complement the growth defect of the fission yeast *dmt1* mutant. However, the two isoforms exhibit a differential cell type-specific expression pattern and distinct subcellular localization, indicating that they may be involved in the different iron acquisition steps from the subcellular membranes in various cell types (41).

3.4. Multiple Orthologues and Functional Divergence

Among the yeast genes which have a human homologue, around 50% of them have multiple orthologues in the human genome (euGenes, http://eugenes.org/). During the course of evolution, the function of the original yeast gene may be retained for only one orthologue, retained in all the orthologues, split among the different orthologues, or lost completely in all orthologues. For example, each of the three human adenine nucleotide carrier genes (HANC1/2/3) were able to complement the growth defect of yeast *anc2* mutant in the presence of nonfermentable carbon sources, with HANC3 being the most efficient (42). Similarly, both the human copper transporting P-type ATPases,

ATP7A and ATP7B, were shown to rescue the defect of *ccc2* yeast mutant cells grown on iron-limited medium. Conversely, bioinformatic analysis has identified two mitochondrial translation release factors in human, mtRF1 and mtRF1a (43, 44), but only mtRF1a is capable of terminating at UAA/UAG codons in yeast. mtRF1 is postulated to be a mitochondrial release factor for decoding the noncognate stop codons AGG/AGA, which are not used in either *E. coli* or yeasts (44). Chromatin assembly is mediated by histone H3–H4 chaperones, including chromatin assembly factor 1 (CAF-1) and anti-silencing function 1 (Asf1). Deletion of yeast *ASF1* renders cells sensitive to DNA-damaging reagents, unable to activate the transcription of *PHO5* and susceptible to replicational stress. Two human counterparts (hAsf1a and hAsf1b) of Asf1 were found to complement distinct functions of yeast Asf1 (45). Only hAsf1a can replace the role of Asf1 to protect cells against double-stranded DNA-damaging reagents, while only hAsf1b can substitute for the role of Asf1 in activating the *PHO5* gene and relieving cells of replicational stress. Similarly, the function of mitochondrial EF-Tu of *S. cerevisiae* in translation elongation is shared by EF-Tu and the guanine nucleotide exchange factor EF-Ts in *S. pombe* and *H. sapiens* mitochondria (46).

3.5. Protein Complexes

Despite high conservation at the sequence level, some human proteins cannot complement the functions of their counterparts in yeast. The mammalian Ssu72 shares 45% identity ($9.00e^{-42}$) with yeast Ssu72. Like its yeast counterpart, mammalian Ssu72 associates with TFIIB and the yeast cleavage/polyadenylation factor Pta1 and exhibits intrinsic phosphatase activity. However, hSsu72 was unable to rescue an *ssu72* lethal mutation in yeast (47). The loss of functional complementation can sometimes be attributed to the evolution of complex formation. The hetero-oligomeric UDP-GlcNAc transferase, composed of Alg13 and Alg14 in yeast, is essential for the synthesis of the evolutionarily conserved lipid-linked oligosaccharide (LLO) precursor for N-linked glycosylation. The human Alg13 and Alg14 orthologues fail to pair with their yeast partners, but when co-expressed in yeast can functionally complement the loss of either *ALG13* or *ALG14* (48). Similarly, when co-expressed in yeast, human Trm6p and Trm61p can rescue the temperature sensitivity of either *trm6* or *trm61* mutant in yeast, form stable complexes, and restore the activity of the tRNA 1-methyladenosine 58 methyltransferase (49). Furthermore, expression of both Mpp11 and Hsp70L of the mammalian ribosome-associated complex in yeast complements the growth defects of the *zuo1Δssz1Δ* double mutants more efficiently than expression of either of them together with a yeast partner (50). These examples demonstrated that co-evolution can also happen to domains which interact with other in a complex.

Functional conservation of complexes can also be inferred from the conserved function they have in multiple systems, although not all the subunits can be substituted. RNA polymerase III contains 17 subunits in yeasts (*S. cerevisiae* and *S. pombe*) and in human cells. Twelve of them are akin to the core RNA polymerase I or II. The other five are RNA polymerase III specific. Pair-wise complementation assays revealed that the five enzyme-specific subunits of fission yeast cannot replace their *S. cerevisiae* counterparts in vivo, contrasting with the complementation seen for 10 of 12 core enzyme subunits (51). However, homology searches in all other eukaryotic genomes revealed significant conservation to these enzyme-specific subunits, thus showing that the evolution of RNA polymerase III combines a conserved 12-subunit core with an ancient but fast evolving set of five enzyme specific subunits.

3.6. Conservation/ Divergence at the Pathway Level

Functional conservation from yeasts to mammals has been demonstrated both at the gene level and at the level of pathways involved in fundamental cellular activities, including metabolism, DNA/protein synthesis and degradation, signaling. Although it is often impossible to substitute one protein of a pathway from one species with its orthologues from a different one, functional similarity of this pathways can be demonstrated by revealing their orthologous components and functions. One good example of these is the nutrient-sensing pathway in both yeast and mammals. TORC1 is one of the two TOR complexes involved in controlling cell growth (52, 53). The TORC1 complex is made up of Tor1/2, Lst8, Kog1, and Tco89 in yeast and mTOR, mLST8, Raptor, and Deptor in mammals (54). The downstream direct targets of TORC1 consists of the Sch9 kinase in yeast (17) and S6K1 and 4EBP1 in mammals (15) which control translation and cell size. The upstream activators which respond to nutrients, such as amino acids, include the Rag GTPases formed by RagA/B and RagC/D in humans (55, 56) and by Gtr1 and Gtr2 in yeast (57). Moreover, Vam6 has been identified as the conserved guanine nucleotide exchange factor (GEF) which modulates the activity of Gtr1 in yeast (57), providing a link between amino acid availability and TORC1 activity. Similar studies are undergoing to identify the mammalian GEF(s) which regulate the activity of RagA/B. Studies on the TORC1 pathway have also revealed important differences between budding yeast and mammals. In mammalian cells, the Rag GTPases function with the TORC1 activator Rheb whose activity is negatively controlled by TSC1/TSC2 complex (58). However, in budding yeast, the Rheb orthologue may play a different role and there is no orthologous Tsc1/2 complex. These studies are likely to reveal similarities and differences of key controllers of cellular activities from lower eukaryotic cells to mammals.

3.7. Reverse Complementation

Functional equivalence has usually been demonstrated using simple model organisms as hosts due to their ease of manipulation. However, there are an increasing number of reports which demonstrate that genes from lower organisms can also functionally replace their counterparts in higher organisms. Wieland et al. (59) expressed the *S. cerevisiae* Cse4p in human cells depleted of CENP-A by RNAi and found that Cse4p is specifically incorporated into kinetochore nucleosomes and is able to complement the RNAi-induced cell cycle arrest. Expression of yeast Npc2p can efficiently revert the unesterified cholesterol and ganglioside GM1 accumulation seen in hNPC2$^{-/-}$ patient fibroblasts, demonstrating that it is a functional homologue of human NPC2 (56). Similarly, expressing the yeast adenosine triphosphate-binding cassette (ABC) transporter Ste6 in *Drosophila* mesoderm rectified the germ cell migration defect associated with mutations of *mdr49*, suggesting that Mdr49 is functionally equivalent to yeast Ste6 and acts in mesodermal cells to attract germ cells (60).

4. Applications of Yeast-Based Technologies

Yeast-based technologies developed over the past decade have not only contributed to our understanding of functions of genes of both yeast and mammals, but also found a wide-range of applications in both industry and medicine. Here we focus on recent developments on using yeast to reveal protein–protein interaction (PPI) networks of other organisms, and on using yeast to characterize drug targets, to screen for novel drugs, and to express therapeutic proteins.

4.1. The Interactome and Drug Discovery

Protein–protein interactions (PPIs) are essential in virtually every biological process. PPIs provide clues to functional relationship between interacting partners, and have emerged as drug targets (61). The yeast two-hybrid (Y2H) methodology was originally designed by Fields and Song (62), based on the reconstitution of a transcription factor upon binding of bait and prey proteins. Two technological improvements have made automation possible. One was the introduction of mating two haploid cells carrying the bait and prey vector, respectively, with survival of the diploid cells dependent on the interaction between the prey and bait proteins (63). The second was the application of the GATEWAY recombinational technology to clone large number of ORFs representing the whole genome of an organism (64). The combination of these two advances, together with development of methods for automation has made possible the interrogation of all PPIs (the interactome) within a cell.

Only about 20% overlap was observed between the two early yeast binary PPI data sets from two independent studies (65, 66), prompting debates on whether Y2H produces interaction data with a high false positive rate due to its technical limitation (67, 68). However, a second-generation Y2H study (69) confirmed that the Y2H interactions previously uncovered were of very high quality and that the binary PPIs are fundamentally different and complementary to those revealed by affinity purification followed by mass spectrometry (70, 71). It was concluded that the small overlap was due to low sensitivity (false negatives) rather than low specificity (false positives). A similar conclusion was reached when a different approach, based on protein complementary assay (72), was used to interrogate the yeast interactome. These studies suggest that a large proportion of binary PPIs (~20,000 in yeast) remained unexplored (73). Whether all interactions observed make functional sense is still debatable (74).

Nevertheless, the Y2H system has been used to generate interactomes of a few non-yeast organisms, including *Plasmodium falciparum* (75), *Caenorhabditis elegans* (76, 77), *Drosophila melanogaster* (78), and humans (79–81). Several viral–host interactomes have also been produced using the system (82). Future directions will probably include analyzing the protein interaction networks based on protein domains (83) and the interactome involved in specific pathways (84) and human diseases (85).

Revealing the protein interaction networks has emerged as a new approach to unravel the molecular basis of diseases and disease-related protein interaction networks have provided new targets toward drug discovery (86). Since most protein interactions involve large contact surfaces and the interfaces are generally flat without grooves and pockets for small molecules to bind (87), high-throughput screening does not routinely identify compounds that disrupt protein–protein interfaces (88). Nevertheless, over the last few years, there have been considerable progress in finding small molecules that disrupt protein–protein interfaces (61, 89), but many challenges remain in designing novel strategies to improve the hit rates in the future (90).

4.2. Drug Target Identification

A critical challenge in drug discovery is the identification of the mode of action of a chemical compound. As many medicinally important drugs target conserved cellular pathways, the budding yeast *S. cerevisiae* can provide clues as to their mechanism of action. This is achieved by studying the effects of chemical compounds at a genome-wide level, i.e., chemogenomic (chemical genomics) profiling. Chemogenomic profiling is the use of a genome-wide approach that allows the identification of gene products or pathways that functionally interact with bioactive molecules (91). This was made possible though the development of a series of yeast strain libraries harboring mutations, deletions, or overexpression of the majority of the yeast genome.

Chemogenomic profiling can be performed in strains with altered gene dosage, whereby the identification of direct targets of drugs are identified. In drug-induced haploinsufficiency profiling (HIP), for example, pools of heterozygous diploid strains are grown in the presence of bioactive molecules which can target gene products essential for yeast growth. Strains heterozygous for the target genes are depleted from the pool. Conversely, it is possible to overexpress genes encoding drug targets and suppress drug sensitivity. Homozygous profiling (HOP) and haploid profiling use complete loss-of-function alleles to identify pathways that confer drug sensitivity or to indentify the direct target of drugs following the principle that removing the drug target, cells will no longer be sensitive to the compound (92).

4.2.1. Chemogenomic Tools

A series of studies (6, 8, 93) from the laboratory of Ronald W. Davis described the development (and indeed applications) of a collection of yeast strains where the entire coding region of ~6,000 predicted yeast open reading frames (ORFs) were replaced by a selectable marker flanked by two unique 20 bp nucleotide sequences ("molecular barcodes"). These strains can be pooled and grown under various conditions, following which their DNA can be extracted; the barcodes amplified and hybridized against oligonucleotide arrays or sequenced through next generation sequencing, such as 454 Life Sciences or Solexa platforms. This allows the identification of mutants favorable (enriched in the pool) or deleterious (depleted in the pool) in the conditions tested. These barcoded deletion strain libraries are now commercially available as haploids of either mating type (nonessential genes), as homozygous diploids (nonessential genes), and as heterozygous diploids (essential and nonessential genes).

Recently, Yan et al. (94) generated a collection of loss-of-function barcoded strains for 87% of all essential yeast genes through decreased abundance by mRNA perturbation (DAmP), providing one more invaluable tool for yeast researchers. Ho et al. (95) present a new strategy for the identifications of drug-resistant mutations in yeast whereby they construct a molecular barcoded yeast open reading frame library (MoBY-ORF) that enables the identification of the effect of introducing individual genes into a mutant strain.

4.2.2. Examples of Chemogenomic Applications

Parsons et al. (96) have screened the single haploid deletion mutants for hypersensitivity to 82 compounds (such as benomyl, hydroxyurea, tunicamycin, and rapamycin) and natural product extracts. Two-dimensional hierarchical clustering of the data revealed overlapping patterns of compound sensitivities. With this approach, the authors were able to identify pathways targeted by the various drugs and interpret the cellular effects of novel compounds. Yu et al. (97) found that two closely related compounds, which differ by a single atom, had very different mechanisms of action in vivo, namely mitochondria disruption or nuclear DNA damaging agent. They indicate that chemogenomic profiles obtained in yeast cells can indeed be reproduced in mammalian cell cultures, suggesting similar mechanism of action of compounds in mammalian cells as well as novel structure–activity relationships.

The use of gene dosage for the identification of drug targets was first reported in the early 1980s with a study in which resistance to the drugs tunicamycin and compactin was conferred by over-expression of different genes identified from a library screen (98). Today, the use of barcoded heterozygous yeast strains has proven extremely powerful for the drug target identification in vivo (99, 100). Using either barcoded heterozygous deletion strains or barcoded DAmP strains in competition experiments, Yan et al. (94) could identify known and new targets of drugs such as sodium fluoride, fluorouracil, fluconazole, hydroxyurea, and tunicamycin. In a mammoth effort, Hillenmeyer et al. (7) performed 726 treatment experiments with heterozygous deletion strains and 418 experiments with homozygous strains using drugs approved by the World Health Organization and U.S. Food and Drug Administration, well-characterized chemical compounds as well as chemicals with unclear biological activity. With this approach, significant phenotypes were ascribed to 97% of yeast genes.

Chemogenomic profiling has enabled the mechanistic determination of a range of classical drugs such as quinine. Quinine is a major drug of choice in the treatment of malaria, a disease endemic to tropical and subtropical regions, including parts of the Americas, Asia, and Africa. Each year, there are approximately 500 million cases of malaria, killing 1–3 million people. The majority of the victims are young children in Sub-Saharan Africa. The primary mode of action of the drug is unclear and its efficacy is marred by adverse side effects among patients. To address this problem, Khozoie and coworkers (101) carried out a genome-wide quinine sensitivity screen using a yeast deletion strain collection. They found that tryptophan and tyrosine uptake via Tap2p is a major target of quinine toxicity, suggesting that dietary tryptophan supplements might help to minimize the toxic effects of quinine.

Budding yeast has also been successfully used in the identification of the mode of action of other traditional drugs, such as the antimalarial drug artemisinin (102) and the medicinal herb St. John's wort (103). St. John's wort is traditionally used as an antidepressant and for wound healing, but its mechanism of action was unknown. After conducting haploinsufficiency profiling experiments in yeast and subsequent sequence similarity comparisons, the authors could propose likely human intracellular targets.

Ericson et al. (104) performed comprehensive genome-wide screenings of *S. cerevisiae* strains susceptible to psychoactive drugs such as fluoxetine (Prozac) and cyproheptadine (Periactin). These psychoactive compounds were initially screened against wild-type yeast cells, and over 40% of them were selected for further studies due to their effects on cell growth. They were then used to screen against pools of diploid heterozygous and homozygous deletion strains. Approximately 500,000 drug-gene measurements were made, providing strong evidence for the off-target effects of these drugs on evolutionarily conserved cellular pathways, such as vesicle transport, establishment of cell polarity and chromosome biology. This study provided data supporting a system for devising derivatives of compounds with fewer side effects and for designing treatments matching individual patients' genotypes.

Genomic profiling has also been used for identifying and characterizing bacterial virulence proteins, also known as effectors (105). Siggers and Lesser (106) and Curak et al. (107) provide good overviews of the applications of *S. cerevisiae* in identifying and characterizing bacterial proteins responsible for promoting infection and causing disease. Analyses of bacterial virulence proteins has previously been limited by the lack of tractable genetic systems among higher eukaryotes, but the study of bacterial proteins in yeast provides a powerful system to determine the functions of these proteins, to probe eukaryotic cellular processes, and to model mammalian infection. Once again the vast range of tools available to yeast researchers allow for systematic screens at a genome-wide level.

4.3. Toward Therapy

A classic drug discovery process goes through a series of steps including: selection of drug targets (targets associated with a particular disease) and subsequent validation of that target; in vitro drug screens; evaluation of the specificity and safety of the new drug in terms of therapeutic use; drug testing in animal models and drug testing in pre-clinical and clinical trials. The entire process lasts approximately 15 years. In vitro drug screening is performed on purified proteins and a library of small molecules that are screen for inhibition of protein function. Even when a small molecule is found to inhibit protein function in vitro, it might not be able to enter the cell or it might be metabolized before the inhibition takes place in vivo.

Drug screening in vivo will overcome some of the disadvantages of the previous approach, as only cell-permeable inhibitors will be selected; however, the inhibition might be due to effects of different target proteins. Yeast can be used for in vivo drug screens, before any tests are performed in mammalian systems; as they are economical, easy to grow and can be genetically manipulated with greater ease. Yeast is mostly permeable to small compounds and can be further manipulated to improve this aspect (108). In the following paragraphs we will provide a flavor of the use of *S. cerevisiae* in drug design and in the production of therapeutic proteins.

4.3.1. G Protein Coupled Receptors (GPCRs)

G protein coupled receptors (GPCRs) are integral membrane proteins with seven transmembrane domains. They comprise one of the largest and most diverse protein family found in eukaryotes with hundreds of examples in humans (half of which are odorant receptors) and three in *S. cerevisiae*. GPCRs are involved in a variety of physiological processes including visual sense, sense of smell, behavioral and mood regulation, immune system activity and inflammation, autonomic nervous system, and cell density. Drugs targeting GPCRs account for about 40% of all prescription pharmaceuticals on the market, making them the single most used class of drug target. Furthermore, they are one of the most targeted protein in pharmaceutical research today; either for the development of modulators (agonists and antagonists) or as biomarkers for diagnosis and prognosis. GPCRs are potential therapeutic targets in areas including cancer, diabetes, cardiac dysfunction, neurological disorders, obesity, inflammation, and pain.

In *S. cerevisiae* the GPCRs are the mating pheromone receptors Ste2 and Ste3 (expressed in mating type *a* and *α* cells, respectively) and the nutrient responsive receptor Gpr1. Ste2 and Ste3 couple to the mating pheromone-signaling cascade via interaction with Gpa1, the G_{α} subunit of one of the two budding yeast's heterotrimeric G protein complexes. Analogously, Gpr1 relays nutritional signals to the cell via Gpa2. Upon ligand binding, the receptors undergo a conformational change, which promotes their binding of the GPCR to trimeric G proteins (comprised of G_{α}, G_{β}, and G_{γ} subunits). This allows the release of GDP and the uptake of GTP from the G_{α} subunit, which is then dissociated from the G_{β} and G_{γ} subcomplex. This dissociation activates signaling pathways that will adequately respond to the ligand stimulus. The hundreds of different human GPCRs couple to G_{α} proteins of four classes: $G_{\alpha s}$, $G_{\alpha i}$, $G_{\alpha q/11}$, and $G_{\alpha 12/13}$. Coupling to various G proteins determines the cellular changes to follow.

Several different groups have successfully managed to couple mammalian GPCRs to the yeast mating pheromone pathway. This was originally done for the classic target of drugs to treat asthma, the β2 adrenergic receptor (109), but has since been accomplished for β3 adrenergic receptor (110), D_{2S}-dopamine receptor (111), fizzled receptors (112), adenosine A2a receptor (113), CXCR4 (114), among many others. There are still a number of reasons why a heterologous GPCR might not successfully couple to the yeast pathway. These include low expression of the receptor, failure of localization of the receptor to the plasma membrane, failure of binding of the receptor to the host's trimeric G_α protein, and lack of strong or clear activation signals upon receptor binding.

Different groups resorted to a series of solutions to overcome the difficulties mentioned above. The receptors have been cloned in plasmids under very strong promoters to increase the expression levels of the GPCRs. Receptors have been expressed in strains with mutations in *ERG6*, leading to a different plasma membrane composition that facilitates the integration of the receptor. Furthermore, receptors have been coupled to fluorescent markers to verify the proper subcellular localization. One major limitation of the system is the potentially poor binding of the heterologous GPCRs to the yeast G_α subunit. This problem has been overcome by the creation of chimeric G_α proteins where domains involved in binding to GPCRs have human sequences whereas domains interacting with downstream components are derived from the host. Initial experiments were performed with long replacements of G_α fragments; however, more recently, different groups have demonstrated that the replacement of the last five amino acids of *S. pombe* or *S. cerevisiae* G_α proteins with human sequences can be sufficient for an efficient coupling of the pathways (115–117). In yeast, activation of the mating pheromone-signaling pathway can be monitored through the use of single or multiple reporter genes fused to the promoter of *FUS1*. Luciferase, *LacZ*, nutritional markers such as *HIS3,* and fluorescent markers such as GFP have all been used to this end. Furthermore, strains can be engineered as to remove negative regulators of the pathway such as the GTPase Sst2 (negatively regulator of the yeast G_α protein Gpa1) and Far1 (contributes to cell cycle arrest following *STE2* pathway activation).

Today most pharmaceutical industries have opted for the use of mammalian cultured cells in GPCR research due to the more natural context for the receptors as well as the availability of a number of commercial methods in measuring receptor activation. However, if suitably optimized, yeast cells can provide a suitable complementary screening platform in GPCR research. One of the main advantages of employing *S. cerevisiae* in GPCR research is the possibility of working with individual or specific combinations of GPCRs. The low costs of yeast media and its suitability for automation raise the possibility of developing fully automated cycles of GPCR drug discovery, both in the search for agonists and antagonists of well-defined targets and in the search for ligands for orphan GPCRs. Furthermore, heterologous GPCRs can be expressed in yeast for further purification and functional studies in vitro (118).

4.3.2. Antifungal and Other Drugs

Invasive infection by pathogenic fungi has become an increasingly common problem in critically ill and immunocompromised patients. The pathogenic yeast *Candida* sp. lives in 80% of the human population without harmful effects; however, in AIDS patients, cancer patients undergoing chemotherapy or transplant patients, it can be fatal. During the past years, in addition to *Candida*, other fungi groups such as *Aspergillus* and *Zygomycetes* have been increasingly problematic in the clinic (119). That is due to the low efficacy of current therapies against systemic fungal infections which can have a mortality rate of over 50%. As genome sequences of a increasing number of pathogenic fungi become available, *S. cerevisiae* can provide a platform for comparative genomic techniques leading to drug target identification. Furthermore, it is an ideal test bed for new experimental approaches in drug discovery (120). One major problem in the treatment of systemic fungal infections is the toxicity of the current therapies. Investigation of the mechanism of action of these therapies by using chemogenomic approaches can lead to the development of new, less toxic, and more specific antifungal drugs (121, 122).

Major projects in the pipeline will certainly further promote the use of *S. cerevisiae* as a platform for antifungal drug design. One of these approaches undertaken by the Boone laboratory (http://www.utoronto.ca/boonelab/research_projects/sigma_deletion/index.shtml) is the creation of a new library of yeast barcoded deletion collection based on Sigma 1278b, a *S. cerevisiae* strain capable of filamentous growth, a developmental process important to fungal pathogenesis.

Yeast cells have been extensively used for expression of heterologous proteins. Human proteins can generally be expressed at good levels in budding yeast from genes cloned directly from cDNA libraries. This is also true for genes derived from several human parasites such as *Trypanosoma brucei* and *Trypanosoma cruzi* (123–126), the agents causing sleeping sickness and Chagas diseases, which kill approximately 40,000 and 20,000 people each year, respectively. Proteins from *Schistosoma* sp., which cause schistosomiasis (also known as bilharzia or snail fever) affecting over 200 million people, have also been expressed in yeast providing insights into their function due to successful complementation of yeast deletion phenotypes (127, 128).

Similarly, proteins derived from *Plasmodium* sp. have also been successfully expressed in yeast where they successfully complement the yeast mutant phenotypes (129–132) and provide good protein yields for structural studies (133). Unfortunately, the expression of many *Plasmodium* proteins is not efficient due to the AT-rich nature of the parasite's genome. LaCount et al. (134) demonstrated recently that one important factor contributing to poor expression is the premature cleavage of the *Plasmodium* mRNAs through the recognition by the yeast's mRNA processing machinery as a signal for addition of polyA tails. This problem was overcome by working with strains with defects in mRNA processing. Alternatively, heterologous genes can be synthesized in vitro with codon usage optimized for yeast expression. Yeast strains expressing heterologous proteins can provide in vivo screening platforms for high-throughput drug design.

4.3.3. Therapeutic Protein Production in Yeasts

Various yeast species ranging from *S. cerevisiae*, through *Pichia pastoris* to *S. pombe* are commercially used or being developed as live factories for production of pharmaceutically relevant proteins. These proteins can be of different nature, ranging from insulin, enzymes for enzyme replacement therapy, antibodies, cytokines to viral antigens for use as vaccines, such as the hepatitis B surface antigen (135), human papillomavirus capsid protein (136, 137), and dengue virus envelope proteins (138).

The World Health Organization estimates that 2–3% of the world population suffers from some form of diabetes, a condition in which the body does not produce enough or does not respond to the hormone insulin. Diabetes is a chronic disease that is controlled through dietary changes as well as administration of insulin, which is commercially produced in *S. cerevisiae* (139).

Lysosomal storage diseases (LSDs) are a group of about 40 inherited disorders that together affect 1:5,000 to 1:10,000 of the human population. LSDs are caused by deficiencies in the production of different lysosomal enzymes and are generally treated with enzyme replacement therapy (ERT). Usual hosts for the production of enzymes for ERT are mammalian cells; however, due to their very low yields and high production costs, combined with the large amount of protein required for each therapeutic infusion (1 mg of protein/kg of body weight for fortnightly treatment of Fabry disease), treatment of LSDs are exorbitant and frequently prohibitive. Therefore, the development of methods for producing human enzymes in various yeast species has the potential of greatly improving the life expectancy and quality of life of affected patients (140–143).

Further protein therapeutics, such as cytokines and antibodies, are the largest class of drug candidates being developed by pharmaceutical companies (144). Today most of these proteins are produced in mammalian cells; however, their large-scale production could be threatened by the inadequate supply of bovine serum and by the risk of spread of viral infections. Therefore, yeast protein expression systems can once again provide a cheaper and more reliable source of therapeutic proteins. However, yeasts do not naturally produce the same type of glycoproteins as humans, but using a combination of systems biology approaches and subsequent genetic manipulations, researchers have been able to develop "humanized" yeast cells which are able to produce glycoproteins compatible with human therapies (144–152).

5. Conclusions

Decades of close collaborations among the yeast research community yielded a wide range of methods, strains, plasmids, and other resources, allowing generation of vast amount of high-throughput data to study the functions of genes in this simple eukaryotic cell. We are now in a privileged position of having these tools available to investigate and eventually manipulate cellular activities at a systems level.

Classic yeast genetics defined pathways as well as provided mechanistic insights into cellular events, most of which have been further confirmed to some extent in higher eukaryotes. With these traditional techniques combined with various systems biology technologies, yeast will continue to be a tractable model for functional studies of individual human proteins or pathways, especially those implicated in diseases. Furthermore, *S. cerevisiae* research is paving the way for genome-wide studies aiding in the understanding of modes of action of drugs, drug design, and therapeutic protein production. A systematic approach in analyzing the available data and designing new experiments will further enhance the contributions of *S. cerevisiae* toward human health.

References

1. Dolinski, K., and Botstein, D. (2007) Orthology and functional conservation in eukaryotes. *Annu. Rev. Genet.* **41**, 465–507.
2. Kataoka, T., Powers, S., Cameron, S., et al. (1985) Functional homology of mammalian and yeast RAS genes. *Cell* **40**, 19–26.
3. Lee, M. G., and Nurse, P. (1987) Complementation used to clone a human homologue of the fission yeast cell cycle control gene cdc2. *Nature* **327**, 31–35.
4. Goffeau, A., Barrell, B. G., Bussey, H., et al. (1996) Life with 6000 genes. *Science* **274**, 546, 563–567.
5. Suter, B., Auerbach, D., and Stagljar, I. (2006) Yeast-based functional genomics and proteomics technologies: the first 15 years and beyond. *Biotechniques* **40**, 625–644.
6. Winzeler, E. A., Shoemaker, D. D., Astromoff, A., et al. (1999) Functional characterization of the *S. cerevisiae* genome by gene deletion and parallel analysis. *Science* **285**, 901–906.
7. Hillenmeyer, M. E., Fung, E., Wildenhain, J., et al. (2008) The chemical genomic portrait of yeast: uncovering a phenotype for all genes. *Science* **320**, 362–365.
8. Giaever, G., Chu, A. M., Ni, L., et al. (2002) Functional profiling of the *Saccharomyces cerevisiae* genome. *Nature* **418**, 387–391.
9. Deutschbauer, A. M., Jaramillo, D. F., Proctor, M., et al. (2005) Mechanisms of haploinsufficiency revealed by genome-wide profiling in yeast. *Genetics* **169**, 1915–1925.
10. Mager, W. H., and Winderickx, J. (2005) Yeast as a model for medical and medicinal research. *Trends Pharmacol. Sci.* **26**, 265–273.
11. Oliver, S. G. (2006) From genomes to systems: the path with yeast. *Philos. Trans. R. Soc. Lond. B. Biol. Sci.* **361**, 477–482.
12. Liolios, K., Tavernarakis, N., Hugenholtz, P., and Kyrpides, N. C. (2006) The genomes on line database (GOLD) v.2: a monitor of genome projects worldwide. *Nucleic Acids Res.* **34**, D332–D334.
13. Lee, D., Redfern, O., and Orengo, C. (2007) Predicting protein function from

sequence and structure. *Nat. Rev. Mol. Cell Biol.* **8**, 995–1005.
14. Velours, J., Vaillier, J., Paumard, P., Soubannier, V., Lai-Zhang, J., and Mueller, D. M. (2001) Bovine coupling factor 6, with just 14.5% shared identity, replaces subunit h in the yeast ATP synthase. *J. Biol. Chem.* **276**, 8602–8607.
15. Dann, S. G., Selvaraj, A., and Thomas, G. (2007) mTOR Complex1-S6K1 signaling: at the crossroads of obesity, diabetes and cancer. *Trends Mol. Med.* **13**, 252–259.
16. Sobko, A. (2006) Systems biology of AGC kinases in fungi. *Sci. STKE* **2006**, re9.
17. Urban, J., Soulard, A., Huber, A., et al. (2007) Sch9 is a major target of TORC1 in *Saccharomyces cerevisiae*. *Mol. Cell* **26**, 663–674.
18. Oliver, S. G., van der Aart, Q. J., Agostoni-Carbone, M. L., et al. (1992) The complete DNA sequence of yeast chromosome III. *Nature* **357**, 38–46.
19. Ouzounis, C., and Sander, C. (1993) Homology of the NifS family of proteins to a new class of pyridoxal phosphate-dependent enzymes. *FEBS Lett.* **322**, 159–164.
20. Gross, S., and Moore, C. (2001) Five subunits are required for reconstitution of the cleavage and polyadenylation activities of *Saccharomyces cerevisiae* cleavage factor I. *Proc. Natl. Acad. Sci. USA* **98**, 6080–6085.
21. Ramirez, A., Shuman, S., and Schwer, B. (2008) Human RNA 5′-kinase (hClp1) can function as a tRNA splicing enzyme in vivo. *RNA* **14**, 1737–1745.
22. Hartwell, L. H. (2004) Yeast and cancer. *Biosci. Rep.* **24**, 523–544.
23. Grunewald, S., Matthijs, G., and Jaeken, J. (2002) Congenital disorders of glycosylation: a review. *Pediatr. Res.* **52**, 618–624.
24. Zeviani, M., and Carelli, V. (2007) Mitochondrial disorders. *Curr. Opin. Neurol.* **20**, 564–571.
25. Botstein, D., Chervitz, S. A., and Cherry, J. M. (1997) Yeast as a model organism. *Science* **277**, 1259–1260.
26. Ooi, S. L., Pan, X., Peyser, B. D., et al. (2006) Global synthetic-lethality analysis and yeast functional profiling. *Trends Genet.* **22**, 56–63.
27. Zhang, N., Osborn, M., Gitsham, P., Yen, K., Miller, J. R., and Oliver, S. G. (2003) Using yeast to place human genes in functional categories. *Gene* **303**, 121–129.
28. Wishart, J. A., Hayes, A., Wardleworth, L., Zhang, N., and Oliver, S. G. (2005) Doxycycline, the drug used to control the tet-regulatable promoter system, has no effect on global gene expression in *Saccharomyces cerevisiae*. *Yeast* **22**, 565–569.
29. Mnaimneh, S., Davierwala, A. P., Haynes, J., et al. (2004) Exploration of essential gene functions via titratable promoter alleles. *Cell* **118**, 31–44.
30. Kanemaki, M., Sanchez-Diaz, A., Gambus, A., and Labib, K. (2003) Functional proteomic identification of DNA replication proteins by induced proteolysis in vivo. *Nature* **423**, 720–724.
31. Sanchez-Diaz, A., Kanemaki, M., Marchesi, V., and Labib, K. (2004) Rapid depletion of budding yeast proteins by fusion to a heat-inducible degron. *Sci. STKE* **2004**, PL8.
32. Terziyska, N., Grumbt, B., Kozany, C., and Hell, K. (2009) Structural and functional roles of the conserved cysteine residues of the redox-regulated import receptor Mia40 in the intermembrane space of mitochondria. *J. Biol. Chem.* **284**, 1353–1363.
33. Han, G. S., Sreenivas, A., Choi, M. G., et al. (2005) Expression of Human CTP synthetase in *Saccharomyces cerevisiae* reveals phosphorylation by protein kinase A. *J. Biol. Chem.* **280**, 38328–38336.
34. Sharma, S., Sommers, J. A., and Brosh, R. M., Jr. (2004) In vivo function of the conserved non-catalytic domain of Werner syndrome helicase in DNA replication. *Hum. Mol. Genet.* **13**, 2247–2261.
35. Aller, S. G., Eng, E. T., De Feo, C. J., and Unger, V. M. (2004) Eukaryotic CTR copper uptake transporters require two faces of the third transmembrane domain for helix packing, oligomerization, and function. *J. Biol. Chem.* **279**, 53435–53441.
36. Grillari, J., Ajuh, P., Stadler, G., et al. (2005) SNEV is an evolutionarily conserved splicing factor whose oligomerization is necessary for spliceosome assembly. *Nucleic Acids Res.* **33**, 6868–6883.
37. Boocock, G. R., Marit, M. R., and Rommens, J. M. (2006) Phylogeny, sequence conservation, and functional complementation of the SBDS protein family. *Genomics* **87**, 758–771.
38. Chang, C. P., Lin, G., Chen, S. J., Chiu, W. C., Chen, W. H., and Wang, C. C. (2008) Promoting the formation of an active synthetase/tRNA complex by a nonspecific tRNA-binding domain. *J. Biol. Chem.* **283**, 30699–30706.
39. Vaz, F. M., Houtkooper, R. H., Valianpour, F., Barth, P. G., and Wanders, R. J. (2003) Only one splice variant of the

human TAZ gene encodes a functional protein with a role in cardiolipin metabolism. *J. Biol. Chem.* **278**, 43089–43094.
40. Ma, L., Vaz, F. M., Gu, Z., Wanders, R. J., and Greenberg, M. L. (2004) The human TAZ gene complements mitochondrial dysfunction in the yeast taz1Delta mutant. Implications for Barth syndrome. *J. Biol. Chem.* **279**, 44394–44399.
41. Tabuchi, M., Tanaka, N., Nishida-Kitayama, J., Ohno, H., and Kishi, F. (2002) Alternative splicing regulates the subcellular localization of divalent metal transporter 1 isoforms. *Mol. Biol. Cell* **13**, 4371–4387.
42. De Marcos Lousa, C., Trezeguet, V., Dianoux, A. C., Brandolin, G., and Lauquin, G. J. (2002) The human mitochondrial ADP/ATP carriers: kinetic properties and biogenesis of wild-type and mutant proteins in the yeast *S. cerevisiae*. *Biochemistry* **41**, 14412–14420.
43. Zhang, Y., and Spremulli, L. L. (1998) Identification and cloning of human mitochondrial translational release factor 1 and the ribosome recycling factor. *Biochim. Biophys. Acta* **1443**, 245–250.
44. Soleimanpour-Lichaei, H. R., Kuhl, I., Gaisne, M., et al. (2007) mtRF1a is a human mitochondrial translation release factor decoding the major termination codons UAA and UAG. *Mol. Cell* **27**, 745–757.
45. Tamburini, B. A., Carson, J. J., Adkins, M. W., and Tyler, J. K. (2005) Functional conservation and specialization among eukaryotic anti-silencing function 1 histone chaperones. *Eukaryot. Cell* **4**, 1583–1590.
46. Chiron, S., Suleau, A., and Bonnefoy, N. (2005) Mitochondrial translation: elongation factor tu is essential in fission yeast and depends on an exchange factor conserved in humans but not in budding yeast. *Genetics* **169**, 1891–1901.
47. St-Pierre, B., Liu, X., Kha, L. C., et al. (2005) Conserved and specific functions of mammalian ssu72. *Nucleic Acids Res.* **33**, 464–477.
48. Gao, X. D., Tachikawa, H., Sato, T., Jigami, Y., and Dean, N. (2005) Alg14 recruits Alg13 to the cytoplasmic face of the endoplasmic reticulum to form a novel bipartite UDP-N-acetylglucosamine transferase required for the second step of N-linked glycosylation. *J. Biol. Chem.* **280**, 36254–36262.
49. Ozanick, S., Krecic, A., Andersland, J., and Anderson, J. T. (2005) The bipartite structure of the tRNA m1A58 methyltransferase from *S. cerevisiae* is conserved in humans. *RNA* **11**, 1281–1290.
50. Otto, H., Conz, C., Maier, P., et al. (2005) The chaperones MPP11 and Hsp70L1 form the mammalian ribosome-associated complex. *Proc. Natl. Acad. Sci. USA* **102**, 10064–10069.
51. Proshkina, G. M., Shematorova, E. K., Proshkin, S. A., Zaros, C., Thuriaux, P., and Shpakovski, G. V. (2006) Ancient origin, functional conservation and fast evolution of DNA-dependent RNA polymerase III. *Nucleic Acids Res.* **34**, 3615–3624.
52. Loewith, R., Jacinto, E., Wullschleger, S., et al. (2002) Two TOR complexes, only one of which is rapamycin sensitive, have distinct roles in cell growth control. *Mol. Cell* **10**, 457–468.
53. Wullschleger, S., Loewith, R., and Hall, M. N. (2006) TOR signaling in growth and metabolism. *Cell* **124**, 471–484.
54. De Virgilio, C., and Loewith, R. (2006) Cell growth control: little eukaryotes make big contributions. *Oncogene* **25**, 6392–6415.
55. Kim, E., Goraksha-Hicks, P., Li, L., Neufeld, T. P., and Guan, K. L. (2008) Regulation of TORC1 by Rag GTPases in nutrient response. *Nat. Cell Biol.* **10**, 935–945.
56. Berger, A. C., Vanderford, T. H., Gernert, K. M., Nichols, J. W., Faundez, V., and Corbett, A. H. (2005) *Saccharomyces cerevisiae* Npc2p is a functionally conserved homologue of the human Niemann-Pick disease type C 2 protein, hNPC2. *Eukaryot. Cell* **4**, 1851–1862.
57. Binda, M., Peli-Gulli, M. P., Bonfils, G., et al. (2009) The Vam6 GEF controls TORC1 by activating the EGO complex. *Mol. Cell* **35**, 563–573.
58. Sancak, Y., Peterson, T. R., Shaul, Y. D., et al. (2008) The Rag GTPases bind raptor and mediate amino acid signaling to mTORC1. *Science* **320**, 1496–1501.
59. Wieland, G., Orthaus, S., Ohndorf, S., Diekmann, S., and Hemmerich, P. (2004) Functional complementation of human centromere protein A (CENP-A) by Cse4p from *Saccharomyces cerevisiae*. *Mol. Cell Biol.* **24**, 6620–6630.
60. Ricardo, S., and Lehmann, R. (2009) An ABC transporter controls export of a *Drosophila* germ cell attractant. *Science* **323**, 943–946.
61. Wells, J. A., and McClendon, C. L. (2007) Reaching for high-hanging fruit in drug discovery at protein-protein interfaces. *Nature* **450**, 1001–1009.

62. Fields, S., and Song, O. (1989) A novel genetic system to detect protein-protein interactions. *Nature* **340**, 245–246.
63. Fromont-Racine, M., Rain, J. C., and Legrain, P. (1997) Toward a functional analysis of the yeast genome through exhaustive two-hybrid screens. *Nat. Genet.* **16**, 277–282.
64. Walhout, A. J., Temple, G. F., Brasch, M. A., et al. (2000) GATEWAY recombinational cloning: application to the cloning of large numbers of open reading frames or ORFeomes. *Methods Enzymol.* **328**, 575–592.
65. Uetz, P., Giot, L., Cagney, G., et al. (2000) A comprehensive analysis of protein-protein interactions in *Saccharomyces cerevisiae*. *Nature* **403**, 623–627.
66. Ito, T., Chiba, T., Ozawa, R., Yoshida, M., Hattori, M., and Sakaki, Y. (2001) A comprehensive two-hybrid analysis to explore the yeast protein interactome. *Proc. Natl. Acad. Sci. USA* **98**, 4569–4574.
67. Mackay, J. P., Sunde, M., Lowry, J. A., Crossley, M., and Matthews, J. M. (2007) Protein interactions: is seeing believing? *Trends Biochem. Sci.* **32**, 530–531.
68. von Mering, C., Krause, R., Snel, B., et al. (2002) Comparative assessment of large-scale data sets of protein-protein interactions. *Nature* **417**, 399–403.
69. Yu, H., Braun, P., Yildirim, M. A., et al. (2008) High-quality binary protein interaction map of the yeast interactome network. *Science* **322**, 104–110.
70. Gavin, A. C., Aloy, P., Grandi, P., et al. (2006) Proteome survey reveals modularity of the yeast cell machinery. *Nature* **440**, 631–636.
71. Krogan, N. J., Cagney, G., Yu, H., et al. (2006) Global landscape of protein complexes in the yeast *Saccharomyces cerevisiae*. *Nature* **440**, 637–643.
72. Tarassov, K., Messier, V., Landry, C. R., et al. (2008) An *in vivo* map of the yeast protein interactome. *Science* **320**, 1465–1470.
73. Cusick, M. E., Klitgord, N., Vidal, M., and Hill, D. E. (2005) Interactome: gateway into systems biology. *Hum. Mol. Genet.* **14**(Spec No. 2), R171-R181.
74. Levy, E. D., Landry, C. R., and Michnick, S. W. (2009) How perfect can protein interactomes be? *Sci. Signal.* **2**, pe11.
75. LaCount, D. J., Vignali, M., Chettier, R., et al. (2005) A protein interaction network of the malaria parasite *Plasmodium falciparum*. *Nature* **438**, 103–107.
76. Li, S., Armstrong, C. M., Bertin, N., et al. (2004) A map of the interactome network of the metazoan *C. elegans*. *Science* **303**, 540–543.
77. Simonis, N., Rual, J. F., Carvunis, A. R., et al. (2009) Empirically controlled mapping of the *Caenorhabditis elegans* protein-protein interactome network. *Nat. Methods* **6**, 47–54.
78. Giot, L., Bader, J. S., Brouwer, C., et al. (2003) A protein interaction map of *Drosophila melanogaster*. *Science* **302**, 1727–1736.
79. Rual, J. F., Venkatesan, K., Hao, T., et al. (2005) Towards a proteome-scale map of the human protein-protein interaction network. *Nature* **437**, 1173–1178.
80. Stelzl, U., Worm, U., Lalowski, M., et al. (2005) A human protein-protein interaction network: a resource for annotating the proteome. *Cell* **122**, 957–968.
81. Venkatesan, K., Rual, J. F., Vazquez, A., et al. (2009) An empirical framework for binary interactome mapping. *Nat. Methods* **6**, 83–90.
82. Bailer, S. M., and Haas, J. (2009) Connecting viral with cellular interactomes. *Curr. Opin. Microbiol.* **12**, 453–459.
83. Boxem, M., Maliga, Z., Klitgord, N., et al. (2008) A protein domain-based interactome network for *C. elegans* early embryogenesis. *Cell* **134**, 534–545.
84. Markson, G., Kiel, C., Hyde, R., et al. (2009) Analysis of the human E2 ubiquitin conjugating enzyme protein interaction network. *Genome Res.* **19**, 1905–1911.
85. Lim, J., Hao, T., Shaw, C., et al. (2006) A protein-protein interaction network for human inherited ataxias and disorders of Purkinje cell degeneration. *Cell* **125**, 801–814.
86. Ideker, T., and Sharan, R. (2008) Protein networks in disease. *Genome Res.* **18**, 644–52.
87. Hopkins, A. L., and Groom, C. R. (2002) The druggable genome. *Nat. Rev. Drug Discov.* **1**, 727–730.
88. Cochran, A. G. (2000) Antagonists of protein-protein interactions. *Chem. Biol.* **7**, R85–R94.
89. Domling, A. (2008) Small molecular weight protein-protein interaction antagonists: an insurmountable challenge? *Curr. Opin. Chem. Biol.* **12**, 281–291.
90. Colas, P. (2008) High-throughput screening assays to discover small-molecule inhibitors of protein interactions. *Curr. Drug Discov. Technol.* **5**, 190–199.

91. Bharucha, N., and Kumar, A. (2007) Yeast genomics and drug target identification. *Comb. Chem. High Throughput Screen.* **10**, 618–634.
92. Hoon, S., St. Onge, R. P., Giaever, G., and Nislow, C. (2008) Yeast chemical genomics and drug discovery: an update. *Trends Pharmacol. Sci.* **29**, 499–504.
93. Giaever, G., Shoemaker, D. D., Jones, T. W., et al. (1999) Genomic profiling of drug sensitivities via induced haploinsufficiency. *Nat. Genet.* **21**, 278–283.
94. Yan, Z., Costanzo, M., Heisler, L. E., et al. (2008) Yeast Barcoders: a chemogenomic application of a universal donor-strain collection carrying bar-code identifiers. *Nat. Methods* **5**, 719–725.
95. Ho, C. H., Magtanong, L., Barker, S. L., et al. (2009) A molecular barcoded yeast ORF library enables mode-of-action analysis of bioactive compounds. *Nat. Biotechnol.* **27**, 369–377.
96. Parsons, A. B., Brost, R. L., Ding, H., et al. (2004) Integration of chemical-genetic and genetic interaction data links bioactive compounds to cellular target pathways. *Nat. Biotechnol.* **22**, 62–69.
97. Yu, L., Lopez, A., Anaflous, A., et al. (2008) Chemical-genetic profiling of imidazo [1,2-a]pyridines and -pyrimidines reveals target pathways conserved between yeast and human cells. *PLoS Genet.* **4**, e1000284.
98. Rine, J., Hansen, W., Hardeman, E., and Davis, R. W. (1983) Targeted selection of recombinant clones through gene dosage effects. *Proc. Natl. Acad. Sci. USA* **80**, 6750–6754.
99. Armour, C. D., and Lum, P. Y. (2005) From drug to protein: using yeast genetics for high-throughput target discovery. *Curr. Opin. Chem. Biol.* **9**, 20–24.
100. Giaever, G. (2003) A chemical genomics approach to understanding drug action. *Trends Pharmacol. Sci.* **24**, 444–446.
101. Khozoie, C., Pleass, R. J., and Avery, S. V. (2009) The antimalarial drug quinine disrupts Tat2p-mediated tryptophan transport and causes tryptophan starvation. *J. Biol. Chem.* **284**, 17968–17974.
102. Li, W., Mo, W., Shen, D., et al. (2005) Yeast model uncovers dual roles of mitochondria in action of artemisinin. *PLoS Genet.* **1**, e36.
103. McCue, P. P., and Phang, J. M. (2008) Identification of human intracellular targets of the medicinal Herb St. John's Wort by chemical-genetic profiling in yeast. *J. Agric. Food Chem.* **56**, 11011–11017.
104. Ericson, E., Gebbia, M., Heisler, L. E., et al. (2008) Off-target effects of psychoactive drugs revealed by genome-wide assays in yeast. *PLoS Genet.* **4**, e1000151.
105. Lesser, C. F., and Miller, S. I. (2001) Expression of microbial virulence proteins in *Saccharomyces cerevisiae* models mammalian infection. *EMBO J.* **20**, 1840–1849.
106. Siggers, K. A., and Lesser, C. F. (2008) The yeast *Saccharomyces cerevisiae*: a versatile model system for the identification and characterization of bacterial virulence proteins. *Cell Host Microbe* **4**, 8–15.
107. Curak, J., Rohde, J., and Stagljar, I. (2009) Yeast as a tool to study bacterial effectors. *Curr. Opin. Microbiol.* **12**, 18–23.
108. Zhang, M., Liang, Y., Zhang, X., Xu, Y., Dai, H., and Xiao, W. (2008) Deletion of yeast CWP genes enhances cell permeability to genotoxic agents. *Toxicol. Sci.* **103**, 68–76.
109. King, K., Dohlman, H. G., Thorner, J., Caron, M. G., and Lefkowitz, R. J. (1990) Control of yeast mating signal transduction by a mammalian beta 2-adrenergic receptor and Gs alpha subunit. *Science* **250**, 121–123.
110. Duport, C., Loeper, J., and Strosberg, A. D. (2003) Comparative expression of the human beta(2) and beta(3) adrenergic receptors in *Saccharomyces cerevisiae. Biochim. Biophys. Acta* **1629**, 34–43.
111. Sander, P., Grunewald, S., Maul, G., Reilander, H., and Michel, H. (1994) Constitutive expression of the human D2S-dopamine receptor in the unicellular yeast *Saccharomyces cerevisiae. Biochim. Biophys. Acta* **1193**, 255–262.
112. Dirnberger, D., and Seuwen, K. (2007) Signaling of human frizzled receptors to the mating pathway in yeast. *PLoS One* **2**, e954.
113. Wedekind, A., O'Malley, M. A., Niebauer, R. T., and Robinson, A. S. (2006) Optimization of the human adenosine A2a receptor yields in *Saccharomyces cerevisiae. Biotechnol. Prog.* **22**, 1249–1255.
114. Evans, B. J., Wang, Z., Broach, J. R., Oishi, S., Fujii, N., and Peiper, S. C. (2009) Expression of CXCR4, a n-coupled receptor for CXCL12 in yeast identification of new-generation inverse agonists. *Methods Enzymol.* **460**, 399–412.
115. Brown, A. J., Dyos, S. L., Whiteway, M. S., et al. (2000) Functional coupling of mammalian receptors to the yeast mating pathway using novel yeast/mammalian G protein alpha-subunit chimeras. *Yeast* **16**, 11–22.
116. Dowell, S. J., and Brown, A. J. (2009) Yeast assays for G protein-coupled receptors. *Methods Mol. Biol.* **552**, 213–229.

117. Ladds, G., Davis, K., Hillhouse, E. W., and Davey, J. (2003) Modified yeast cells to investigate the coupling of G protein-coupled receptors to specific G proteins. *Mol. Microbiol.* **47**, 781–792.
118. Niebauer, R. T., and Robinson, A. S. (2006) Exceptional total and functional yields of the human adenosine (A2a) receptor expressed in the yeast *Saccharomyces cerevisiae*. *Protein Expr. Purif.* **46**, 204–211.
119. Monk, B. C., and Cannon, R. D. (2002) Genomic pathways to antifungal discovery. *Curr. Drug Targets Infect. Disord.* **2**, 309–329.
120. Ma, D. (2001) Applications of yeast in drug discovery. *Prog. Drug Res.* **57**, 117–162.
121. Agarwal, A. K., Xu, T., Jacob, M. R., et al. (2008) Genomic and genetic approaches for the identification of antifungal drug targets. *Infect. Disord. Drug Targets* **8**, 2–15.
122. De Backer, M. D., and Van Dijck, P. (2003) Progress in functional genomics approaches to antifungal drug target discovery. *Trends Microbiol.* **11**, 470–478.
123. Balliano, G., Dehmlow, H., Oliaro-Bosso, S., et al. (2009) Oxidosqualene cyclase from *Saccharomyces cerevisiae, Trypanosoma cruzi, Pneumocystis carinii* and *Arabidopsis thaliana* expressed in yeast: a model for the development of novel antiparasitic agents. *Bioorg. Med. Chem. Lett.* **19**, 718–723.
124. Kim, Y. U., Ashida, H., Mori, K., Maeda, Y., Hong, Y., and Kinoshita, T. (2007) Both mammalian PIG-M and PIG-X are required for growth of GPI14-disrupted yeast. *J. Biochem.* **142**, 123–129.
125. Palmer, G., Louvion, J. F., Tibbetts, R. S., Engman, D. M., and Picard, D. (1995) Trypanosoma cruzi heat-shock protein 90 can functionally complement yeast. *Mol. Biochem. Parasitol.* **70**, 199–202.
126. Papageorgiou, I., De Koning, H. P., Soteriadou, K., and Diallinas, G. (2008) Kinetic and mutational analysis of the *Trypanosoma brucei* NBT1 nucleobase transporter expressed in *Saccharomyces cerevisiae* reveals structural similarities between ENT and MFS transporters. *Int. J. Parasitol.* **38**, 641–653.
127. Aguiar, P. H., Santos, D. N., Lobo, F. P., et al. (2006) Functional complementation of a yeast knockout strain by *Schistosoma mansoni* Rho1 GTPase in the presence of caffeine, an agent that affects mutants defective in the protein kinase C signal transduction pathway. *Mem. Inst. Oswaldo Cruz* **101**(Suppl 1), 323–326.
128. Santos, D. N., Aguiar, P. H., Lobo, F. P., et al. (2007) *Schistosoma mansoni*: Heterologous complementation of a yeast null mutant by SmRbx, a protein similar to a RING box protein involved in ubiquitination. *Exp. Parasitol.* **116**, 440–449.
129. Aruna, K., Chakraborty, T., Rao, P. N., Santos, C., Ballesta, J. P., and Sharma, S. (2005) Functional complementation of yeast ribosomal P0 protein with *Plasmodium falciparum* P0. *Gene* **357**, 9–17.
130. Djapa, L. Y., Basco, L. K., Zelikson, R., et al. (2007) Antifolate screening using yeast expressing *Plasmodium vivax* dihydrofolate reductase and in vitro drug susceptibility assay for *Plasmodium falciparum*. *Mol. Biochem. Parasitol.* **156**, 89–92.
131. Shams-Eldin, H., Azzouz, N., Kedees, M. H., Orlean, P., Kinoshita, T., and Schwarz, R. T. (2002) The GPI1 homologue from *Plasmodium falciparum* complements a *Saccharomyces cerevisiae* GPI1 anchoring mutant. *Mol. Biochem. Parasitol.* **120**, 73–81.
132. Wider, D., Peli-Gulli, M. P., Briand, P. A., Tatu, U., and Picard, D. (2009) The complementation of yeast with human or *Plasmodium falciparum* Hsp90 confers differential inhibitor sensitivities. *Mol. Biochem. Parasitol.* **164**, 147–152.
133. Birkholtz, L. M., Blatch, G., Coetzer, T. L., et al. (2008) Heterologous expression of plasmodial proteins for structural studies and functional annotation. *Malar. J.* **7**, 197.
134. LaCount, D. J., Schoenfeld, L. W., and Fields, S. (2009) Selection of yeast strains with enhanced expression of *Plasmodium falciparum* proteins. *Mol. Biochem. Parasitol.* **163**, 119–122.
135. Liu, R., Lin, Q., Sun, Y., et al. (2009) Expression, purification, and characterization of hepatitis B virus surface antigens (HBsAg) in yeast *Pichia Pastoris*. *Appl. Biochem. Biotechnol.* **158**, 432–444.
136. Bazan, S. B., de Alencar Muniz Chaves, A., Aires, K. A., Cianciarullo, A. M., Garcea, R. L., and Ho, P. L. (2009) Expression and characterization of HPV-16 L1 capsid protein in *Pichia pastoris*. *Arch. Virol.* **154**, 1609–1617.
137. Woo, M. K., An, J. M., Kim, J. D., Park, S. N., and Kim, H. J. (2008) Expression and purification of human papillomavirus 18 L1 virus-like particle from *Saccharomyces cerevisiae*. *Arch. Pharm. Res.* **31**, 205–209.
138. Etemad, B., Batra, G., Raut, et al. (2008) An envelope domain III-based chimeric antigen produced in *Pichia pastoris* elicits neutralizing antibodies against all four dengue virus serotypes. *Am. J. Trop. Med. Hyg.* **79**, 353–363.

139. Gerngross, T. U. (2004) Advances in the production of human therapeutic proteins in yeasts and filamentous fungi. *Nat. Biotechnol.* **22**, 1409–1414.
140. Akeboshi, H., Chiba, Y., Kasahara, Y., et al. (2007) Production of recombinant beta-hexosaminidase A, a potential enzyme for replacement therapy for Tay-Sachs and Sandhoff diseases, in the methylotrophic yeast *Ogataea minuta*. *Appl. Environ. Microbiol.* **73**, 4805–4812.
141. Chen, Y., Jin, M., Egborge, T., Coppola, G., Andre, J., and Calhoun, D. H. (2000) Expression and characterization of glycosylated and catalytically active recombinant human alpha-galactosidase A produced in *Pichia pastoris*. *Protein Expr. Purif.* **20**, 472–484.
142. Chiba, Y., Sakuraba, H., Kotani, M., et al. (2002) Production in yeast of alpha-galactosidase A, a lysosomal enzyme applicable to enzyme replacement therapy for Fabry disease. *Glycobiology* **12**, 821–828.
143. Sakuraba, H., Chiba, Y., Kotani, M., et al. (2006) Corrective effect on Fabry mice of yeast recombinant human alpha-galactosidase with N-linked sugar chains suitable for lysosomal delivery. *J. Hum. Genet.* **51**, 341–352.
144. Chiba, Y., and Akeboshi, H. (2009) Glycan engineering and production of 'humanized' glycoprotein in yeast cells. *Biol. Pharm. Bull.* **32**, 786–795.
145. Abe, H., Takaoka, Y., Chiba, Y., et al. (2009) Development of valuable yeast strains using a novel mutagenesis technique for the effective production of therapeutic glycoproteins. *Glycobiology* **19**, 428–436.
146. Graf, A., Dragosits, M., Gasser, B., and Mattanovich, D. (2009) Yeast systems biotechnology for the production of heterologous proteins. *FEMS Yeast Res.* **9**, 335–348.
147. Hamilton, S. R., and Gerngross, T. U. (2007) Glycosylation engineering in yeast: the advent of fully humanized yeast. *Curr. Opin. Biotechnol.* **18**, 387–392.
148. Rakestraw, J. A., Sazinsky, S. L., Piatesi, A., Antipov, E., and Wittrup, K. D. (2009) Directed evolution of a secretory leader for the improved expression of heterologous proteins and full-length antibodies in *Saccharomyces cerevisiae*. *Biotechnol. Bioeng.* **103**, 1192–1201.
149. Schirrmann, T., Al-Halabi, L., Dubel, S., and Hust, M. (2008) Production systems for recombinant antibodies. *Front. Biosci.* **13**, 4576–4594.
150. Takegawa, K., Tohda, H., Sasaki, M., et al. (2009) Production of heterologous proteins using the fission-yeast (*Schizosaccharomyces pombe*) expression system. *Biotechnol. Appl. Biochem.* **53**, 227–235.
151. Wildt, S., and Gerngross, T. U. (2005) The humanization of N-glycosylation pathways in yeast. *Nat. Rev. Microbiol.* **3**, 119–128.
152. Wiseman, A. (2004) Double humanized yeast makes hydrocortisone. *Trends Biotechnol.* **22**, 324–325.

Subject Index

A

B

C

J.I. Castrillo, S.G. Oliver (eds.), *Yeast Systems Biology*, Methods in Molecular Biology 759,
DOI 10.1007/978-1-61779-173-4, © Springer Science+Business Media, LLC 2011

D

E

F

G

N

O

P

Q

R

S

MIX
Papier aus verantwortungsvollen Quellen
Paper from responsible sources
FSC® C105338

If you have any concerns about our products,
you can contact us on
ProductSafety@springernature.com

In case Publisher is established outside the EU,
the EU authorized representative is:
Springer Nature Customer Service Center GmbH
Europaplatz 3, 69115 Heidelberg, Germany

Printed by Libri Plureos GmbH
in Hamburg, Germany